Dirk Steinborn

**Fundamentals of
Organometallic Catalysis**

Related Titles

Beller, M., Renken, A., van Santen, R. A. (eds.)

Catalysis

From Principles to Applications

2011

ISBN: 978-3-527-32349-4

Comba, P., Kerscher, M.

Coordination Chemistry

Concepts and Applications

2011

ISBN: 978-3-527-32300-5

Caprio, V., Williams, J.

Catalysis in Asymmetric Synthesis

2009

ISBN: 978-1-4051-9091-6

Amouri, H., Gruselle, M.

Chirality in Transition Metal Chemistry

Molecules, Supramolecular Assemblies and Materials

2008

ISBN: 978-0-470-06054-4

Crabtree, Robert H. (ed.)

Handbook of Green Chemistry - Green Catalysis

Series: Handbook of Green Chemistry (Set 1)

Series edited by Anastas, Paul T.

2009

ISBN: 978-3-527-31577-2

van Leeuwen, Piet W. N. M. (ed.)

Supramolecular Catalysis

2008

ISBN: 978-3-527-32191-9

Ribas Gispert, J.

Coordination Chemistry

2008

ISBN: 978-3-527-31802-5

Rothenberg, G.

Catalysis

Concepts and Green Applications

2008

ISBN: 978-3-527-31824-7

Niemantsverdriet, J. W.

Spectroscopy in Catalysis

An Introduction

2007

ISBN: 978-3-527-31651-9

Elschenbroich, C.

Organometallics

2006

ISBN: 978-3-527-29390-2

Hagen, J.

Industrial Catalysis

A Practical Approach

2006

ISBN: 978-3-527-31144-6

Crabtree, R. H.

The Organometallic Chemistry of the Transition Metals

2005

ISBN: 978-0-471-66256-3

Dirk Steinborn

Fundamentals of Organometallic Catalysis

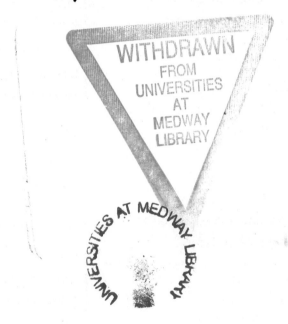

WILEY-VCH

WILEY-VCH Verlag GmbH & Co. KGaA

The Author

Prof. Dr. Dirk Steinborn
Martin-Luther-Universität Halle-Wittenberg
Institut für Chemie – Anorganische Chemie
Kurt-Mothes-Str. 2
06120 Halle
Germany

Translator:
Alan Frenkel

■ All books published by **Wiley-VCH** are carefully produced. Nevertheless, authors, editors, and publisher do not warrant the information contained in these books, including this book, to be free of errors. Readers are advised to keep in mind that statements, data, illustrations, procedural details or other items may inadvertently be inaccurate.

Library of Congress Card No.: applied for

British Library Cataloguing-in-Publication Data
A catalogue record for this book is available from the British Library.

Bibliographic information published by the Deutsche Nationalbibliothek
The Deutsche Nationalbibliothek lists this publication in the Deutsche Nationalbibliografie; detailed bibliographic data are available on the Internet at http://dnb.d-nb.de.

© 2012 WILEY-VCH Verlag & Co. KGaA, Boschstr. 12, 69469 Weinheim, Germany

All rights reserved (including those of translation into other languages). No part of this book may be reproduced in any form – by photoprinting, microfilm, or any other means – nor transmitted or translated into a machine language without written permission from the publishers. Registered names, trademarks, etc. used in this book, even when not specifically marked as such, are not to be considered unprotected by law.

Composition Thomson Digital, Noida, India
Printing and Binding betz-druck GmbH, Darmstadt
Cover Design Formgeber, Eppelheim

Printed in the Federal Republic of Germany
Printed on acid-free paper

ISBN: 978-3-527-32716-4 (Hardcover)
ISBN: 978-3-527-32717-1 (Softcover)

Contents

Preface *XIII*
Index of Frequently Used Abbreviations *XV*

1	**Introduction** *1*	
1.1	The Beginnings of Catalytic Research *1*	
1.1.1	Homogeneously Catalyzed Reactions *1*	
1.1.2	Heterogeneously Catalyzed Reactions *3*	
1.2	The Catalysis Definitions of Berzelius and Ostwald *5*	
1.2.1	Berzelius' Catalysis Concept *5*	
1.2.2	Ostwald's Definition of Catalysis *6*	
2	**Principles of Organometallic Catalysis** *9*	
2.1	Homogeneous versus Heterogeneous Catalysis *9*	
2.2	Catalytic Cycles *11*	
2.3	Activity and Productivity of Catalysts *12*	
2.3.1	Catalytic Activity *12*	
2.3.2	Catalytic Productivity *12*	
2.3.3	Conversion-Time Plots *13*	
2.4	Selectivity and Specificity of Catalysts *14*	
2.5	Determination of Catalytic Mechanisms *15*	
2.5.1	Experimental Studies *16*	
2.5.2	Theoretical Studies *17*	
2.6	Glossary for Catalysis *18*	
2.7	The Development of Organometallic Catalysis *21*	
3	**Elementary Steps in Organometallic Catalysis** *27*	
3.1	Cleavage and Coordination of Ligands *27*	
3.2	Oxidative Addition and Reductive Elimination *30*	
3.3	Oxidative Coupling and Reductive Cleavage *35*	
3.4	Olefin Insertion and β-Hydrogen Elimination *37*	

Fundamentals of Organometallic Catalysis. First Edition. Dirk Steinborn
Copyright © 2012 WILEY-VCH Verlag GmbH & Co. KGaA, Weinheim

3.5	α-Hydrogen Elimination and Carbene Insertion Reactions	40
3.6	Addition of Nucleophiles and Heterolytic Fragmentation	42
3.7	Insertion and Extrusion of CO	45
3.8	One-Electron Reduction and Oxidation	46

4 Hydrogenation of Olefins 49

4.1	Introduction	49
4.2	The Wilkinson Catalyst	50
4.2.1	Principles	50
4.2.2	Mechanism of Olefin Hydrogenation	51
4.3	Enantioselective Hydrogenation	54
4.3.1	Principles	54
4.3.2	Applications and Examples	58
4.3.2.1	Applications for Asymmetric Hydrogenation	58
4.3.2.2	Combinatorial Catalysis	59
4.3.2.3	Nonlinear Effects	61
4.3.3	Kinetically Controlled Enantioselectivity — A Closer Look	63
4.4	Dihydrogen Complexes and H_2 Activation	68
4.4.1	Dihydrogen Complexes	68
4.4.2	Activation of Dihydrogen	71
4.5	Transfer Hydrogenation	73

5 Hydroformylation of Olefins and Fischer-Tropsch Synthesis 77

5.1	Cobalt Catalysts	77
5.2	Phosphane-Modified Rhodium Catalysts	80
5.3	Enantioselective Hydroformylation	84
5.4	Significance of Hydroformylation and Outlook	88
5.4.1	Diphosphites as Ligands	89
5.4.2	Biphasic Catalysis	91
5.4.3	Synthesis of Vitamin A	93
5.4.4	Carbon Dioxide as Alternative to CO	93
5.4.5	Combinatorial and Supramolecular Catalysis	94
5.5	The Fischer-Tropsch Synthesis	95
5.5.1	Mechanism	97

6 Carbonylation of Methanol and Water-Gas Shift Reaction 101

6.1	Principles	101
6.2	The Monsanto Process	103
6.3	Synthesis of Acetic Anhydride	106
6.4	The Cativa Process	108
6.5	Water-Gas Shift Reaction and Carbon Monoxide Dehydrogenases	112
6.5.1	Water-Gas Shift Reaction	112
6.5.2	Carbon Monoxide Dehydrogenases	114

7	**Metathesis** *117*	
7.1	Metathesis of Olefins *117*	
7.1.1	Introduction *117*	
7.1.2	Mechanism *118*	
7.1.3	Catalysts *119*	
7.1.4	Mechanism – A Closer Look *123*	
7.1.5	Metathesis of Cycloalkenes *125*	
7.1.6	Metathesis of Acyclic Dienes *128*	
7.1.7	Enantioselective Metathesis *130*	
7.2	Metathesis of Alkynes *131*	
7.3	Enyne Metathesis *133*	
7.4	σ-Bond Metathesis *135*	
7.5	Metathesis of Alkanes *137*	
7.5.1	Principles *137*	
7.5.2	Mechanism *138*	
7.5.3	Alkane Metathesis Via Tandem Reactions *141*	
8	**Oligomerization of Olefins** *145*	
8.1	Ziegler Growth Reaction *145*	
8.2	Nickel Effect and Nickel-Catalyzed Dimerization of Ethene *147*	
8.3	Trimerization of Ethene *152*	
8.4	Shell Higher Olefin and α-Sablin Processes *156*	
8.4.1	The Shell Higher Olefin Process (SHOP) *156*	
8.4.2	α-Sablin Process *158*	
8.4.3	Use of Linear α-Olefins *159*	
9	**Polymerization of Olefins** *161*	
9.1	Introduction *161*	
9.2	Ethene Polymerization *162*	
9.2.1	Ziegler Catalysts *162*	
9.2.2	Mechanism – A Closer Look *165*	
9.2.3	Phillips Catalysts *167*	
9.2.4	Polymer Types and Process Specifications *169*	
9.3	Propene Polymerization *171*	
9.3.1	Regioselectivity and Stereoselectivity *171*	
9.3.2	Ziegler-Natta Catalysts *175*	
9.3.3	Polymer Types and Process Specifications *178*	
9.4	Metallocene Catalysts *179*	
9.4.1	Cocatalysts and Anion Influence *179*	
9.4.2	C_2- and C_s-Symmetric Metallocene Catalysts *182*	
9.4.2.1	Principles *182*	
9.4.2.2	Mechanism *184*	
9.4.3	Metallocene Catalysts with Diastereotopic Coordination Pockets *187*	

9.4.3.1	Principles	187
9.4.3.2	Hemitactic Polymers	190
9.4.3.3	Stereoblock Polymers	191
9.4.4	On the Significance of Metallocene Catalysts	191
9.5	Nonmetallocene Catalysts	193
9.5.1	Catalyst Systems of Early Transition Metals	194
9.5.2	Catalyst Systems of Late Transition Metals	194
9.5.3	Living Polymerization of Olefins and Block Copolymers	198
9.6	Copolymerization of Olefins and CO	200
9.6.1	Perfectly Alternating Copolymerization	200
9.6.2	Imperfectly Alternating Copolymerization	204
10	**C–C Linkage of Dienes**	**207**
10.1	Introduction	207
10.2	Allyl and Butadiene Complexes	208
10.2.1	Allyl Complexes	208
10.2.2	Butadiene Complexes	211
10.2.3	*Re/Si* and *supine/prone* Coordination of Allyl and Butadiene Ligands	213
10.3	Organometallic Elementary Steps of Allyl Ligands	214
10.3.1	Oxidative Coupling and Reductive Cleavage	214
10.3.2	Butadiene Insertion and β-Hydrogen Elimination	215
10.3.3	Allyl Insertion	215
10.3.4	Oxidative Addition and Reductive Elimination	216
10.3.5	*anti/cis* and *syn/trans* Correlations	218
10.4	Oligomerization and Telomerization of Butadiene	218
10.4.1	Cyclotrimerization of Butadiene	218
10.4.1.1	Mechanism	218
10.4.1.2	*cis/trans* Selectivity – A Closer Look	221
10.4.1.3	Industrial Synthesis of CDT	224
10.4.2	Cyclodimerization of Butadiene	224
10.4.2.1	Mechanism	224
10.4.2.2	Selectivity Control	226
10.4.3	Linear Oligomerization and Telomerization of Butadiene	230
10.5	Polymerization of Butadiene	234
10.5.1	Mechanism	234
10.5.2	Butadiene Polymerization Catalyzed by Allylnickel(II) Complexes	237
10.5.3	Synthesis and Properties of Polybutadienes and Polyisoprenes	241
11	**C–C Coupling Reactions**	**245**
11.1	Palladium-Catalyzed Cross-Coupling Reactions	245
11.1.1	Introduction	245
11.1.2	Mechanism of Cross-Coupling Reactions	246

11.1.3	Selected Types of Cross-Coupling 249
11.1.3.1	Cross-Coupling with Organolithium, Organomagnesium, and Organozinc Reagents 249
11.1.3.2	Suzuki Coupling 250
11.1.3.3	Hiyama Coupling 251
11.1.3.4	Stille Coupling 252
11.1.3.5	Sonogashira Coupling 253
11.1.3.6	Ligand Effects 254
11.1.3.7	Alkyl–Alkyl Coupling 255
11.1.3.8	Enantioselective Cross-Coupling 256
11.1.3.9	Carbonylative Cross-Coupling 258
11.2	The Heck Reaction 258
11.2.1	Mechanism of Heck Reactions 259
11.2.2	Mechanism – A Closer Look 260
11.2.3	Ligand Effects 261
11.2.4	Enantioselective Heck Reactions 263
11.3	Palladium-Catalyzed Allylic Alkylation 264
11.3.1	Principles and Mechanism 264
11.3.2	Chirality Transfer in Asymmetric Allylation 267

12 Hydrocyanation, Hydrosilylation, and Hydroamination of Olefins 271

12.1	Introduction 271
12.2	Hydrocyanation 272
12.2.1	Principles and Mechanism 272
12.2.1.1	Mechanism – A Closer Look 273
12.2.2	The DuPont Adiponitrile Process 274
12.2.3	Outlook 276
12.2.3.1	Enantioselective Hydrocyanation 276
12.2.3.2	Hydrocyanation of Alkynes 277
12.2.3.3	Hydrocyanation of Polar C=X Bonds 278
12.3	Hydrosilylation 279
12.3.1	Principles and Mechanism 279
12.3.2	Significance of Hydrosilylation and Outlook 283
12.3.2.1	Applications 283
12.3.2.2	Enantioselective Hydrosilylation 284
12.3.2.3	Hydrosilylation of Alkynes 285
12.3.2.4	σ Complexes of Silanes 286
12.4	Hydroamination 287
12.4.1	Principles 287
12.4.2	Catalyst Types 289
12.4.2.1	Alkali Metal Amides as Catalysts 289
12.4.2.2	Platinum Group Metals as Catalysts 289
12.4.2.3	Gold Complexes as Catalysts 291
12.4.2.4	Lanthanoid Complexes as Catalysts 292

13	**Oxidation of Olefins and Alkanes** 295
13.1	The Wacker Process 295
13.1.1	Introduction 295
13.1.2	Mechanism of Ethene Oxidation 297
13.1.3	Oxypalladation of Olefins 303
13.1.3.1	Types of Oxypalladation 303
13.1.3.2	Enantioselective Oxypalladation 305
13.1.3.3	Palladium Oxidase Catalysis 305
13.2	Epoxidation of Olefins 306
13.2.1	Introduction 306
13.2.2	Epoxidation of Ethene and Propene 307
13.2.2.1	O_2 and ROOH as Oxygen Transfer Agents 307
13.2.2.2	Mechanism 309
13.2.2.3	H_2O_2 as Oxygen Transfer Agent 311
13.2.3	Enantioselective Oxidation of Olefins 313
13.2.3.1	Epoxidation of Allyl Alcohols 313
13.2.3.2	Epoxidation of Nonactivated Olefins 314
13.2.4	Monooxygenases 315
13.3	C–H Functionalization of Alkanes 319
13.3.1	Introduction 319
13.3.2	C–H Activation of Alkanes 319
13.3.2.1	Cyclometallation and Orthometallation 319
13.3.2.2	Intermolecular C–H Activation of Alkanes 321
13.3.3	C–H Functionalization 323
13.3.3.1	The Shilov Catalyst System 324
13.3.3.2	The Catalytica System – Hg^{II} as Catalyst 325
13.3.3.3	The Catalytica System – Pt^{II} as Catalyst 326
13.3.3.4	Cytochrome P-450 326
14	**Nitrogen Fixation** 329
14.1	Fundamentals 329
14.2	Heterogeneously Catalyzed Nitrogen Fixation 334
14.2.1	Principles 334
14.2.2	Mechanism of Catalysis 335
14.2.3	The Industrial Catalyst 338
14.2.4	Ruthenium Catalysts 340
14.3	Enzyme-Catalyzed Nitrogen Fixation 342
14.3.1	The Fe Protein Cycle 343
14.3.2	The MoFe Protein Cycle 344
14.3.3	A Prebiotic Nitrogen-Fixing System? 347
14.4	Homogeneously Catalyzed Nitrogen Fixation 348
14.4.1	Stoichiometric Reduction of N_2 Complexes 348
14.4.2	Catalytic Reduction of Dinitrogen 352
14.4.3	Functionalization of Dinitrogen 359

Solutions to Exercises *363*

Bibliography and Sources *407*
References *408*
Further Reading *429*
Source for Structures *436*

Index *439*

Index of Backgrounds *456*

Preface

Catalysis, the basic principle for overcoming the kinetic inhibition of chemical reactions, is of fundamental importance in chemistry. This applies as much to basic and applied research as it does to industrial applications. It is estimated that today 85–90% of all products from the chemical industry are produced in catalytic processes. To understand the essence of catalysis and relationships between catalyst structure and catalytic effect on a scientific basis is not only a challenge for basic research but also an indispensable requirement for the targeted development of better and entirely novel catalysts.

Organometallic catalysis, meaning homogeneous catalysis by metal (almost always transition metal) complexes in which organometallic intermediates occur, is a relatively young area of catalysis. The discovery by Karl Ziegler in late 1953 at the Max Planck Institute for Coal Research in Mülheim/Ruhr (Germany) of low-pressure polymerization of ethene by organometallic mixed catalysts acted as an initial spark in its development. In the following decades, organometallic catalysis developed into one of the most important and innovative scientific areas in chemistry. It is an integral component of modern organic chemistry and has enabled the development of entirely novel methods of synthesis, as well as synthetic reactions of extraordinary selectivity and activity with high atom economy. Organometallic-catalyzed industrial-scale processes for the synthesis of organic industrial chemicals and polymers, as well as methods for synthesizing bioactive compounds, are cornerstones of the modern chemical, pharmaceutical, and agrochemical industries, which are oriented toward high ecological standards and economical requirements.

The inexhaustible potential of organometallic catalysis becomes clear when one looks at the great number of catalytically relevant transition metals in their multitude of oxidation states and the broad palette of coligands. Organometallic chemistry and coordination chemistry are essential scientific foundations for organometallic catalysis. The key to understanding any organometallic catalytic reaction is well-founded knowledge about the catalytic mechanism.

Accordingly, this textbook, which imparts the principles of organometallic catalysis, strives to emphasize not the details, but rather an understanding of the course of metal-catalyzed reactions. Thus, the (few) elementary organometallic steps relevant to catalysis are explained first, and from there, important organometallic-catalyzed reactions are dealt with. The focus is on mechanistic aspects. This should

Fundamentals of Organometallic Catalysis. First Edition. Dirk Steinborn
Copyright © 2012 WILEY-VCH Verlag GmbH & Co. KGaA, Weinheim

enable readers to grasp the essence of the processes and serve as a foundation for them to creatively apply what they have learned and perhaps even develop it further.

This approach is also expressed in the selection of material. Without striving to be exhaustive, it was the desire of the author that the reactions handled should reflect the full breadth of the scientific field. Emphasis is placed on industrially important processes and newer developments with interesting mechanistic aspects. Additionally, some further reactions have been included that are homogeneously catalyzed by transition metal complexes without organometallics appearing as intermediates. Among them is nitrogen fixation, which is the most instructive example for a comparative treatment of the three great areas of catalysis – the homogeneous, heterogeneous, and enzymatic one – that makes clear their similarities and differences, along with the various principles at work.

Access to further information is provided through a bibliography that focuses primarily on review articles, but also contains newer original papers. The exercises are designed not only to ask about the material already handled, but also impart deeper knowledge. Accordingly, highly detailed answers are presented at the end of the book. Knowledge from the landscape of organometallic catalysis that is important for understanding the material accompanies the main text in the form of background notes.

The present English translation is based on the second German edition, which was critically reviewed by Prof. Rudolf Taube (Halle). I owe special thanks to him and to colleagues, co-workers and students who have contributed to improvements through their advice, discussions, and suggestions. Furthermore, I would like to cordially thank Alan Frankel for the translation, Dr. Martin Graf (Wiley-VCH) for its revision and proofreading, and Dr. Steven Hawkins (Leipzig) for fruitful discussions about selected text passages. I am obliged, as well, to the publisher for the pleasant collaboration with Dr. Gudrun Walter and Dr. Waltraud Wüst.

Halle, June 2011						*Dirk Steinborn*

Index of Frequently Used Abbreviations

Ac	acetyl
Ad	adamantyl
Ar	aryl
9-BBN	9-borabicyclo[3.3.1]nonane
BD	buta-1,3-diene
BINAP	chiral bis(phosphane) ligand (p. 56)
BINAPHOS	chiral phosphane/phosphite ligand (p. 85)
bpy	2,2′-bipyridine
COD	cycloocta-1,5-diene
CDT	cyclododeca-1,5,9-triene
CODH	carbon monoxide dehydrogenase
Cp	cyclopentadienyl ligand (η^5-C_5H_5)
Cp*	pentamethylcyclopentadienyl ligand (η^5-C_5Me_5)
Cy	cyclohexyl
DACH	1,2-diaminocyclohexane
DAT	dialkyl tartrate (ROOC−CH(OH)−CH(OH)−COOR)
dba	dibenzylideneacetone (PhCH=CH−CO−CH=CHPh)
DCPD	dicyclopentadiene
DIOP	chiral bis(phosphane) ligand (p. 56)
DIPAMP	chiral bis(phosphane) ligand (p. 56)
dmpe	1,2-bis(dimethylphosphino)ethane ($Me_2P(CH_2)_2PMe_2$)
DPPF	bis(phosphane) ligand (p. 291)
dppe	1,2-bis(diphenylphosphino)ethane ($Ph_2P(CH_2)_2PPh_2$)
dppm	bis(diphenylphosphino)methane ($Ph_2PCH_2PPh_2$)
dppp	1,3-bis(diphenylphosphino)propane ($Ph_2P(CH_2)_3PPh_2$)
DuPHOS	chiral bis(phospholane) ligand (p. 56)
DVCB	1,2-divinylcyclobutane
GLUP	chiral bis(phosphinite) ligand (p. 56)
Hacac	acetylacetone
HMPA	hexamethylphosphoric acid triamide
HOTf	trifluorosulfonic acid (F_3CSO_3H)
HOTs	p-toluenesulfonic acid (p-$MeC_6H_4SO_3H$)

Fundamentals of Organometallic Catalysis. First Edition. Dirk Steinborn
Copyright © 2012 WILEY-VCH Verlag GmbH & Co. KGaA, Weinheim

H$_2$pc	phthalocyanine
H$_2$salen	N,N'-bis(salicylidene)ethylenediamine
L	ligand
LAO	linear α-olefins
Ln	rare earth metal (lanthanoid)
[M]	metal complex; see p. 27
MAO	methylaluminoxane
Mes	mesityl
MOP	chiral monophosphane ligand (p. 284)
NHC	N-heterocyclic carbene; see p. 122
Nu	nucleophile
P	polymer chain
PE	polyethylene (HDPE = high-density PE, LDPE = low-density PE, LLDPE = linear low-density PE)
PHOX	chiral 2-(phosphinophenyl)oxazoline ligand (p. 264)
PP	polypropylene (a-PP = atactic, i-PP = isotactic, s-PP = syndiotactic)
R	alkyl, aryl, H, ... (unless otherwise indicated)
s	solvent molecule (as ligand); see p. 27
TBS	tert-butyldimethylsilyl
tmeda	N,N,N',N'-tetramethylethylenediamine (Me$_2$N(CH$_2$)$_2$NMe$_2$)
Tol	tolyl
Xantphos	bis(arylphosphane) ligand (p. 291)
X	anionic ligand/substituent
VCH	4-vinylcyclohex-1-ene
ADMET	acyclic diene metathesis
ADIMET	acyclic diyne metathesis
ARCM	asymmetric RCM
AROM	asymmetric ROM
CN	coordination number
DFT	density functional theory
NLE	nonlinear effect
ON(M)	oxidation number of M
P	degree of polymerization
QM/MM	quantum-mechanics method/molecular-mechanics method
RCM	ring-closing metathesis
ROM	ring-opening metathesis
ROMP	ring-opening metathesis polymerization
TOF	turnover frequency
TON	turnover number
T_g	glass transition temperature
ve	valence electron(s)
χ(E)	electronegativity of E
□	vacant coordination site; see p. 27

Acronyms of Academic/Research Institutions and Companies

BASF	Badische Anilin- & Soda-Fabrik (Baden Aniline and Soda Factory), Ludwigshafen (Germany)
Caltech	California Institute of Technology, Pasadena (CA, USA)
CNRS	Centre National de la Recherche Scientifique (National Center for Scientific Research), France
ICI	Imperial Chemical Industries, London (UK)
IFP	Institut Français du Pétrole (French Institute of Petroleum), Rueil-Malmaison (France)
KWI	Kaiser-Wilhelm-Institut (predecessor of the MPIs), Germany
MIT	Massachusetts Institute of Technology, Cambridge (MA, USA)
MPG	Max-Planck-Gesellschaft (Max Planck Society), Germany
MPI	Max-Planck-Institut, Germany
SABIC	Saudi Basic Industries Corporation, Riyadh (Saudi Arabia)
SASOL	Suid Afrikaanse Steenkool en Olie (South African Coal and Oil), Johannesburg (South Africa)
TU	Technical University

1
Introduction

1.1
The Beginnings of Catalytic Research

In the second half of the eighteenth century, as scientists were developing an art of chemical experimentation that besides qualitative aspects, also considered quantitative issues and made use of exact measurements and mass balances for chemical reactions, an increasing number of reactions were described that marked the beginnings of catalytic research in chemistry. Table 1.1 provides an overview. At the same time, the first steps toward research into catalysis in biology and physiology were taken. From the beginning of this period, it was already clear that the phenomenon of catalysis encompassed synthesis as well as decomposition reactions, as exemplified by the catalytic formation of water from its elements (oxyhydrogen gas reaction) and the catalytic decomposition of hydrogen peroxide. The establishment of stoichiometry (J. B. Richter, 1792–1793) and the formulation of the law of constant (J. L. Proust, 1799) and multiple proportions (J. Dalton, 1803) were important foundations for recognizing two essential aspects of catalysis, namely that trace quantities of a substance ("substoichiometric quantities") can induce chemical reactions and that this substance is not consumed by the reaction.

1.1.1
Homogeneously Catalyzed Reactions

Understanding was achieved gradually. Thus, for example, it took 30 years from the observation that boiling potato starch with tartaric and acetic acids formed sugar (1781) before it was understood that the acids were not consumed. In 1801, it was discovered that other acids could also decompose starches. Experiments on the effect of the concentration of the acid led to the finding that ultimately it was water that split the starch and that "... 'boiling for a sufficient length of time' with water alone must be able to achieve the same goal!" (Döbereiner, 1808). In 1811–1812, it was finally recognized by Kirchhoff and Vogel that the sulfuric acid used in these experiments

Fundamentals of Organometallic Catalysis. First Edition. Dirk Steinborn
Copyright © 2012 WILEY-VCH Verlag GmbH & Co. KGaA, Weinheim

Table 1.1 Beginnings of catalysis research in chemistry (adapted from Mittasch [1]).

Year	Author	Discovery
1781	A. A. Parmentier	Saccharification of starches by boiling in acid
1782	C. W. Scheele	Esterification of acids (acetic acid, benzoic acid, ...) with alcohol in the presence of a mineral acid; saponification
1783	J. Priestley	Conversion of alcohol into ethene and water on heated clay (catalytic dehydration of alcohol)
1796	M. van Marum	Conversion of alcohol into aldehyde on glowing metals (catalytic dehydrogenation of alcohol)
1806	C. B. Desormes, N. Clément	Investigation of the lead chamber process; nitrogen oxides as oxygen carriers for sulfurous acid
1811	G. S. C. Kirchhoff	Extensive investigation of starch saccharification by acids and recognition that the acids are not consumed
1812	H. Davy	Recognition that lead chamber crystals (nitrosylsulfuric acid) are of substantial significance in the lead chamber process
1813	L. J. Thénard	Decomposition of ammonia on heated metals, especially iron
1815	J. L. Gay-Lussac	Cleavage of hydrogen cyanide on iron
1816	A. M. Ampère	Hypothesis of alternating formation and cleavage of metal nitrides during ammonia decomposition on metals
1817	H. Davy	Combustion of methane and alcohol on glowing platinum wire
1818	L. J. Thénard	Decomposition of hydrogen peroxide on metals, oxides, and organic substances
1821	J. W. Döbereiner	Oxidation of alcohol into acetic acid on platina mohr (platinum black) at ordinary temperature
1823	J. W. Döbereiner	Inflammation of hydrogen in the presence of platinum sponge at ordinary temperature
1824	J. S. C. Schweigger	Active sites on metal surfaces in processes defined above
1831	P. Phillips	Generation of sulfuric acid through oxidation in air of SO_2, yielding SO_3, on heated platinum
1833	E. Mitscherlich	"Contact reactions": Decomposition of H_2O_2 on Pt, Au, ... and of ClO_3^- on MnO_2; formation of ether from alcohol in the presence of acid
1835	J. J. Berzelius	Catalysis: Name assignment and definition

remained unchanged and the significance of the acid concentration for the reaction rate was emphasized, as Döbereiner had previously done (summarized and cited from [2], p. 6).

The production of sulfuric acid by burning sulfur with saltpeter, first in glass vessels and then in lead chambers, had already long been familiar. The lead chamber process was first performed in 1746 without admitting air (Roebuck) and from 1793 onwards with an air supply (Desormes and Clément). In 1806, the latter two then also described the effect of nitrous gases in the lead chamber process as an oscillating transition between two known reactions, namely the oxidation of SO_2 by NO_2 and the reoxidation of NO by O_2. In 1812, Davy described the lead chamber crystals

[(NO)HSO$_4$] as an intermediate stage. With these two findings, obtained about 200 years ago, it was first recognized that in the course of a homogeneously catalyzed reaction, the catalyst is directly involved in the reaction and opens a reaction pathway that cannot be achieved without it.

Around 1800, it was found that when diethyl ether was produced by distillation of alcohol with sulfuric acid, the acid could be used repeatedly and no component of the sulfuric acid would be contained in the diethyl ether. The opinion prevalent at the time was that sulfuric acid formed ether through its dehydrating action. The formation of ethyl sulfate, which was also observed, was ascribed to a side reaction. In 1828, Hennell characterized it as an intermediate in the formation of ether such that the alternating formation and decomposition of ethyl sulfate gave rise to the continuous formation of ether. Finally, in 1833–1834, Mitscherlich demonstrated experimentally that the dehydrating effect of sulfuric acid is not critical for formation of ether. Mitscherlich concluded that the sulfuric acid "... [derives] no 'advantage' from the result, so it behaves as an eager, selfless mediator, hence acting purely 'by contact';" he generalized that instances of "... decomposition and combination that are produced in this way occur very frequently; we will call them *decomposition and combination by contact*" (quoted from [2], p. 30).

1.1.2
Heterogeneously Catalyzed Reactions

In heterogeneous reactions, the special characteristics of catalytic reactions are more apparent. In 1783, Priestley was the first to describe the catalytic dehydration of alcohol, in a process in which alcohol vapor was directed through a heated tobacco pipe. In 1795, Deimann took this as the starting point for systematically studying the influence of the tobacco pipe – which is now known to be the catalyst – on the course of the reaction. He stated that a glass pipe by itself will not induce the reaction, but a glass pipe filled with clay fragments will. The individual components of the clay and other substances were also studied for their ability to dehydrate alcohol.

H. Davy studied the effect of platinum on combustion processes and discovered in 1816 that heated platinum wire, even below the annealing temperature, caused the (flameless!) combustion of methane (and other substances such as hydrogen, ether, alcohol) in air, and the resulting heat caused the wire to glow.[1] In 1818, Erman proved that the oxyhydrogen gas reaction only required the platinum wire to be heated to 50 °C. In 1823, Döbereiner showed that platinum sponge causes a flameless reaction of oxyhydrogen gas at ordinary temperatures, and only a few days later he reported that almost instantaneous inflammation occurs when the hydrogen

---

[1] Davy also invented (in 1815) the safety lamp for coal mines that now bears his name, in which the flame of an oil lamp is separated from the outside air by a narrow-mesh wire cylinder. Later, he incorporated a platinum spiral in his mining lamp. If the lamp was extinguished due to excessive methane content in the air, the spiral began to glow and thus allowed the miners to orient themselves in the dark.

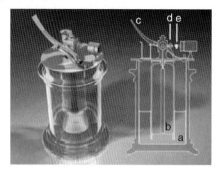

Figure 1.1 Döbereiner's lamp. It corresponds in principle to a Kipp generator filled with sulfuric acid (a), which comes into contact with the zinc rod (b) when the faucet (c) is opened. This produces hydrogen, which is released from the nozzle (d) and ignites on the platinum sponge (e).

is directed onto platinum sponge produced by thermal decomposition of "platinum salmiac" (ammonium chloroplatinate, $[NH_4]_2[PtCl_6]$) so that the hydrogen can mix with air beforehand. He stated that "... the spongy powder behaves somewhat like electrical sparks with respect to oxyhydrogen gas" (quoted from [3], p. 20). This is the basis of Döbereiner's lamp (Figure 1.1), which spread rapidly (with almost 20 000 in use by 1828). Döbereiner did not profit financially from the invention.[2]

In 1813, Thénard extensively studied the decomposition of ammonia on heated metals and found that the most pronounced effect occurred with iron, which in the process changed its appearance, becoming brittle and loose. The cause was later assumed to be the formation of a nitride as an intermediate, which was proven experimentally in 1829. Thénard obtained hydrogen peroxide by treatment of barium peroxide with hydrochloric acid/sulfuric acid in 1818. He then studied the decomposition of H_2O_2 on numerous metals and metal oxides. Ag and Ag_2O (which is reduced to silver) proved the most effective. Even organic materials such as fibrin and various kinds of tissues caused H_2O_2 to decompose. In summary, he expressed that these substances can exert their effects without themselves undergoing a change, and that behind this action is one and the same "force" for all substances, including animal and plant materials.

---

[2] Döbereiner was a professor at the University of Jena, in the Duchy of Saxe-Weimar-Eisenach, from 1810 onwards. He kept the responsible minister of state, J. W. von Goethe (author of *Faust* and other influential works, both literary and scientific), informed of his research, and served him in many ways as an advisor on chemical issues. He also presented him with one of his lighters, and Goethe wrote to Döbereiner "[I] always remember you with gratitude, since every day I have at hand the lighter you so cleverly invented." (cited from [3], p. 29).

1.2
The Catalysis Definitions of Berzelius and Ostwald

1.2.1
Berzelius' Catalysis Concept

From 1821 to 1847, Jöns Jakob Berzelius (1779–1848) compiled the new results from the physical sciences (physics, chemistry, mineralogy, geology) into annual reports. The reports were presented to the Swedish Academy of Sciences and also translated into German.[3] In the chapter on plant chemistry in his 1835 report, covering the year of 1834 (German translation dated 1836), Berzelius points to several of the phenomena mentioned previously and sums them up as follows:

"Thus, it has been demonstrated that many bodies, simple or compound, in solid or dissolved form, possess the property of exerting an influence on compound bodies quite different from conventional chemical affinity, whereby they cause in the body a conversion of the components, without themselves taking part with their components, although this can sometimes be the case.

This is a new force in both inorganic and organic nature of causing chemical activity, that may well be more widespread than previously thought, and whose nature is still hidden to us. If I call it a new force, it is in no way my belief that it should be declared a capability independent of the electrochemical relations of matter; on the contrary, I can only assume that it is a special kind of manifestation of these. As long as their interrelation may remain hidden, however, it will facilitate our research to regard it provisionally as a force of its own, and it will likewise facilitate our proceedings for us to give it a name of its own. Therefore, in order to make use of a derivation well known in chemistry, I will call it the *catalytic force* of the bodies, and will call their decomposition *catalysis*, just as we use the word analysis to indicate the separation of the components of the bodies by virtue of conventional chemical affinity. The catalytic force seems essentially to consist of the property that bodies, by their mere presence, and not by their affinity, are able to awaken affinities that are dormant at this temperature, so that ..."

In keeping with the conception of the time, a reaction between two "bodies" (modern term: "substances") presupposed their "chemical affinity." *Catalysts* were now seen as "bodies," which thanks to their *catalytic force* triggered reactions ("awaken dormant affinities"), without themselves having "affinities" to the reacting "bodies." Berzelius referred thereby not only to homogeneously *and* heterogeneously catalyzed reactions, but also included (to use the current term) enzymatically catalyzed reactions.

---

[3] J. Berzelius, *Jahres-Bericht über die Fortschritte der physischen Wissenschaften*, Tübingen (*Jber. Berz.*). From Vol. 21 (**1842**) *Jahres-Bericht über die Fortschritte der Chemie und Mineralogie*.

Berzelius contrasted the new term "catalysis" (Greek: κατάλυσις = dissolution) with "analysis." In this context, "analysis" meant a reaction induced by "conventional chemical affinity," while catalysts triggered a reaction by their mere presence.

Berzelius used the concept of catalysis to bring together a group of phenomena that could not be explained within the doctrine of reactions being caused by chemical affinity. Thus, the term was originally purely descriptive, and Berzelius intentionally refrained from attempting to explain the nature of catalysis. Berzelius did call the "catalytic force" a new force, but emphasized that he expected to be able to explain its operation within the framework of his electrochemical theory. Liebig attacked Berzelius several times for his definition of catalysis, with his criticism focusing on the "creation of a new force through a new word that does not explain the phenomenon."

1.2.2
Ostwald's Definition of Catalysis

In 1850, the physicist Ludwig Wilhelmy (1812–1864) studied the acid-catalyzed inversion of cane sugar and formulated the "law by which the action of acids on cane sugar occurs." It explicitly defines, for the first time, the chemical rate (reaction rate), which is the foundation of chemical kinetics. An exact definition of the rate of a chemical reaction is a precondition for recognizing the acceleration of a reaction by a catalyst, which is the focus of the "kinetic definition of catalysis" by Wilhelm Ostwald (1853–1932). In recognition of his work on catalysis and for his investigations into the fundamental principles governing chemical equilibria and reaction rates, he was granted the Nobel Prize in Chemistry in 1909.

With the goal of measuring the strength of acids, in the course of his experiments on the influence of acids on the hydrolysis of esters (1883), Ostwald discovered the close relationship between the strength of acids and their catalytic effect. In his experiments on the acid-catalyzed oxidation of hydrogen iodide by bromic acid (1887), he studied for the first time a system in which the reaction ran at a measurable rate even without a catalyst, so "the nature of catalysis is to be sought not in the *inducement* of a reaction, but in its *acceleration*." In the "Zeitschrift für physikalische Chemie" (**1894**, *15*, 705), in a short review of a work by F. Strohmann, Ostwald gave the definition of catalysis that has prevailed ever since:

"... If the referee were confronted with the problem of characterizing the phenomenon of catalysis in general, he would consider something like the following expression as most suitable: **Catalysis is the acceleration of a slow-running chemical reaction via the presence of a foreign substance.**

... It is thus misleading to regard the catalytic action as a force that brings about something that would not occur without the substance that acts catalytically; still less can it be assumed that the latter performs work. It will perhaps contribute to an understanding of the problem if I especially point out that time is not involved in the concept of chemical energy; thus, if the chemical energy relations are given such that a particular process must occur, then they determine only the initial and final states and the whole series of intermediate states which must be passed through, but in no way the time during which the reaction must occur. **This time depends on conditions that lie outside the two main laws of energetics. . . .**"

Ostwald thoroughly summarized the state of knowledge about catalysis in a lecture in 1901. Referring to a homogeneous system that can be transformed into products with lower free energy, he states:

". . . However, the most secure foundation for general conclusions that we know, the laws of energetics, require that the transformation actually occur. They dictate no numeric figure for the rate that must be adhered to; they only require that this rate not be strictly zero, but rather have a finite value.
   In this way, we obtain at once a definition for a catalyst for this case as well.
   A catalyst is any substance that, without appearing in the final product of a chemical reaction, changes its rate. . . ."

Already in this lecture (1901), and later in his Nobel lecture (1909), Ostwald also pointed out, regarding the theory of catalysis, that:

". . . none . . . has prospered better than Clement and Désormes' theory of *intermediate reactions*. This is based precisely on the participation of the catalyst in the reactions actually occurring, in the sum of which, however, the catalyst is not directly involved, although the partial reactions contain the catalyst as a major chemical component of the process. . . ."

Such a reaction course was initially only considered for homogeneously catalyzed reactions. At the beginning of the 1920s, Irving Langmuir (1881–1957; Nobel Prize in Chemistry, 1932) showed in his work on chemisorption that intermediate reactions are also of fundamental importance in heterogeneously catalyzed reactions.
   At the end of his Nobel lecture Ostwald asserted that ". . . until a way has been found whereby a rate of chemical reaction can generally be calculated in advance . . ., the catalysis problem cannot satisfactorily be answered." It was not until after Ostwald's death that Henry Eyring (1901–1981) provided an important theoretical foundation for an answer with transition state theory (1935).
   Ostwald's insight into the nature of catalysis, which he himself designated his "most independent and successful chemical achievement," enabled targeted research into the area of catalysis and its deliberate technical application. Ostwald studied the iron-catalyzed synthesis of ammonia starting from the known fact (see Table 1.1) that ammonia passed over weakly glowing iron decomposes into its elements almost completely, and from his recognition that a catalyst only hastens the achievement of equilibrium, and thus accelerates both the forward and reverse reaction in equal measure. In his memoirs, Ostwald wrote that his patent application, submitted in 1900, contained all the basic ideas of the Haber-Bosch process implemented in 1913. However, the lack of reproducibility of the results led Ostwald

to allow his application to lapse. In 1901, a familiar demonstration experiment, in which a glowing spiral of platinum wire continues to glow in a mixture of NH_3 and air with the formation of nitrogen dioxide, served him as the starting point for the "Ostwald process" for synthesis of nitric acid [4], which now meets practically the entire demand for HNO_3.

Wilhelm Ostwald had established catalysis as the fundamental principle for overcoming kinetic inhibition of chemical reactions, and thus created the preconditions for scientifically-based catalysis research and targeted catalyst development.

2
Principles of Organometallic Catalysis

2.1
Homogeneous versus Heterogeneous Catalysis

Homogeneous and heterogeneous catalysis are distinguished from each other according to whether or not the catalyst and reactants are present in the same phase. Biocatalysis, which indicates a biotransformation in which the substrate is initially bound noncovalently to the active site of the catalyst, is not considered here. Enzymatic catalysis is unique due to the special protein structure of the catalysts and their specific mode of action.

Homogeneous catalysis is conveniently classified according to the nature of the catalyst, which causes a specific substrate activation (Table 2.1). Brønsted acid/base catalysis, electrophilic catalysis, and nucleophilic catalysis [5] are legion in organic chemistry. Highly selective "purely" organic catalysts are also designated as organocatalysts. These include, in particular, asymmetric organocatalysts [6, 7], which behave analogously in certain ways to metal-free enzymes. The Mn^{2+}-catalyzed oxidation of oxalate with MnO_4^- and the Mo^{VI}-catalyzed formation of epoxides from olefins and hydroperoxides are introduced here as examples of redox catalysis and metal catalysis, respectively. Aside from some exceptions, the subject of the book is not those types of catalysis but rather organometallic catalysis. These are reactions catalyzed homogeneously by metal complexes – in the vast majority of cases, transition metal complexes – in which organometallic intermediates appear.

Typical in heterogeneous catalysis are gaseous reactants and solid catalysts, which are also designated as contact catalysts, especially in industry. Catalysis occurs on the (external and/or internal) surface of the catalyst, which is the reason for the designations "surface catalysis" and "contact catalysis." Diffusion processes and adsorption and desorption of reactants and products are of significance to the overall course of reaction. Typical reaction temperatures are 200–600 °C. The reactions are often carried out under high pressure.

For organometallic catalysis, it is typical for the reactants and the catalyst to be present in solution and for the reaction temperatures to be relatively low (20–150 °C). Often, homogeneous catalysts are exceptionally sensitive to oxidation and/or hydrolysis, so

Fundamentals of Organometallic Catalysis. First Edition. Dirk Steinborn
Copyright © 2012 WILEY-VCH Verlag GmbH & Co. KGaA, Weinheim

Table 2.1 Classification of homogeneously catalyzed reactions.

Catalyst	Catalyst behaves as ...[a]	Substrate activation through ...	Designation
Brønsted acid/base	PD/PA	Protonation/deprotonation	Brønsted acid/base catalysis
Lewis acid/base	EPA/EPD	Formation of a Lewis acid/base adduct	Electrophilic/nucleophilic catalysis
Metal complex	ED/EA	Electron transfer	Redox catalysis
Metal complex	EPA[b]	Coordinative interaction	Metal catalysis
Metal complex	EPA[b] (EPD)[c]	Coordinative interaction; organometallic intermediates	Organometallic catalysis[d]

a) PD/PA = proton donor/acceptor; EPD/EPA = electron pair donor/acceptor; ED/EA = electron donor/acceptor.
b) The formation of a metal–ligand bond is generally described as an EPA–EPD interaction.
c) In the formation of π and σ complexes in substrate activation, the back-donation capability of the metal is important; the metal also functions as an EPD.
d) German: metallorganische Komplexkatalyse; French: catalyse organométallique.

they must be handled under strictly anaerobic conditions. Most organometallic catalysts are transition metal compounds with defined structure and stoichiometry. For well-studied processes, the course of reaction can be (almost) completely understood on a molecular basis, which is the foundation for targeted development of "tailor-made catalysts." Typically, the reaction center is a metal atom (ion), to which are also coordinated ligands that are not directly involved in catalysis ("spectator ligands" or "control ligands"). Varying these ligands to influence the electronic and/or steric relationships at the reaction center in a targeted way ("ligand tuning"), is essential for optimizing catalysts with respect to activity, selectivity, and stability.

There are borderline cases between homogeneous and heterogeneous catalysis. One of the best known is the classical Ziegler catalyst (TiCl$_4$/AlEt$_3$), which polymerizes ethene at a high rate in an aliphatic hydrocarbon at normal pressure and room temperature. Today we know that TiCl$_3$, which is insoluble in aliphatic hydrocarbons, is formed first. The catalysis occurs on the surface of the TiCl$_3$, which has been chemically modified by the cocatalyst. However, the reaction conditions are typical for homogeneously catalyzed processes; the elementary steps can be understood, in principle, on a molecular basis, and there are analogous catalysts that work in homogeneous solution. This justifies classifying such processes as homogeneous catalysis.

The decisive advantage of homogeneously catalyzed processes is, in principle, that they are understood on a molecular level and hence support targeted catalyst development and optimization. A further essential advantage of organometallic catalysis is the breadth of variation with which catalysts can be developed, reflected in a large number of catalytically relevant metals and ligands. However, this is counteracted by the crucial disadvantage of what is sometimes an expensive and time-consuming process of catalyst separation. Developments such as the immobilization

of homogeneous catalysts on a support and biphasic catalysis accommodate in this respect the effort to combine the advantages of homogeneous catalysis with those of heterogeneous catalysis.

Homogeneous and heterogeneous catalysts are both important from an industrial perspective, and will remain so.

2.2 Catalytic Cycles

To explain the fundamental course of a homogeneously catalyzed reaction, we use as a model the reaction of two substrates S_1 and S_2 yielding a product P. Under the catalytic effect of [M], the following applies:

$$S_1 + S_2 \xrightarrow{[M]} P$$

Possible elementary steps of the catalyzed reaction are shown in Figure 2.1 as a catalytic cycle. Two disruptive side reactions are recorded, resulting in the off-loop species **5** and **6**: The reversible formation of a catalytically inactive intermediate leads to a reduction in the activity of the catalyst, since the concentration of the catalytically active complex is lowered (**3** → **5**). The irreversible decomposition of an intermediate complex leads to deactivation of the catalyst (**4** → **6**), which particularly decreases the productivity of the catalyst. The faster such reactions run, the more pronounced the decrease in productivity.

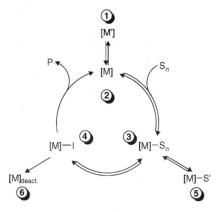

Figure 2.1 Typical elementary steps of an organometallic-catalyzed reaction. **1** → **2**: Formation of the catalyst [M] from the precatalyst [M′]. **2** → **3**: Activation of substrates S_n via formation of [M]–S_n complexes. **3** → **4**: Transformation of metal–substrate complex [M]–S_n into a metal–intermediate complex [M]–I. **4** → **2**: Cleavage of product P from [M]–I with regeneration of catalyst [M]. **3** → **5**: Reversible formation of a catalytically inactive complex [M]–S′. **4** → **6**: Irreversible decomposition of the intermediate complex [M]–I.

For every catalytic reaction, a cycle analogous to Figure 2.1 can, in principle, be formulated. Whether that is the most understandable form remains to be seen. In any case, we will not always proceed in such a way when describing the individual processes.

To understand catalysis, it is important to analyze the reversibility of the elementary steps. If one of the steps is irreversible (in the previous example, such a step is the cleavage of product from the complex, $4 \to 2$), it is guaranteed (provided the activation energies are sufficiently low) that the cycle will run through and that, in principle, a full conversion of the reactants can be achieved. If all steps are reversible, the expected turnover cannot be greater than permitted by the individual equilibria.

2.3
Activity and Productivity of Catalysts

2.3.1
Catalytic Activity

The activity of a catalyst reflects the reaction rate related to the concentration of the catalyst. One measure for the activity of a catalyst is the turnover frequency *TOF*. It is defined as the formation rate of the product P (r_P) relative to the catalyst concentration c_{cat}:

$$TOF = \frac{r_P}{c_{cat}} \quad \text{with} \quad r_P = \frac{dc_P}{dt}$$

For the sake of simplicity, the amount of product formed per amount of catalyst over a given time interval [mol product/(mol catalyst × time)] can be specified. A turnover frequency has the dimension [1/time]. Earlier, it was inappropriately designated a turnover number (*TON*), but as will be seen later, a *TON*, when correctly calculated, specifies productivity.

Alternatively, if detailed kinetic measurements exist, the rate constant k_{cat} (catalysis constant) of a catalyzed reaction can be specified. As can be seen from the Eyring equation, it depends on the Gibbs free energy of activation $\Delta G_{cat}^\ddagger$ of the catalyzed reaction (k_B = Boltzmann constant, h = Planck constant, R = gas constant, T = temperature):

$$k_{cat} = \frac{k_B \cdot T}{h} e^{-\Delta G_{cat}^\ddagger / RT}$$

2.3.2
Catalytic Productivity

The productivity of a catalyst specifies the total amount of product P that can be produced with a given quantity of catalyst (under the given reaction conditions). The figure is dimensionless (e.g., [mol product/mol catalyst]) and is called a turnover number (*TON*):

$$TON = \frac{n_P}{n_{cat}}$$

In order for information on catalytic activity and productivity to be meaningful, the reaction conditions must also be cited.

2.3.3
Conversion-Time Plots

The activity and productivity of a catalyst can be derived from a conversion-time curve (Figure 2.2). The slope at the beginning of the reaction is a measure of the (initial) activity of the catalyst. If the conversion-time plots are not hyperbolic as in curves **1** and **2** in Figure 2.2, but rather sigmoid (S-shaped) (curve **4′**), the slope at the inflection point reflects the (maximum) activity. The productivity can be read from the amount of product formed n_P at which the reaction ceases. (This assumes that there is enough substrate available during the entire course of the reaction.) Such conversion-time plots are easy to obtain experimentally, and significantly more meaningful than the statements common in patents regarding the quantity of product that has been formed at a specific point in time (e.g., 80% conversion after 5 h).

Figure 2.2a shows conversion-time plots from which statements about activity and productivity of catalysts can be made. Thus, catalysts **1** and **2** are of comparable activity, but of very different productivity. Clearly, catalyst **2** is subject to rapid deactivation. Catalysts **3** and **4** have, respectively, very low and high activity. This time window is too narrow to determine their productivity, since neither reaction has come to an end yet. However, the productivity of **4** is definitely higher than that of catalysts **1** and **2**. Figure 2.2b demonstrates that no assertions about activity and productivity can be derived from a single experimental value M stating that at timepoint t_1, amount of product n_1 has been formed. In contrast, experimentally determined conversion-time curves would make clear whether a catalyst with moderate activity but low productivity (**2**), or a catalyst with lower activity but higher productivity (**3**), is present. It would even be possible for a catalyst (**4′**) to exist with high activity and productivity, but for it to display, in contrast to **4** in Figure 2.2a, an induction period.

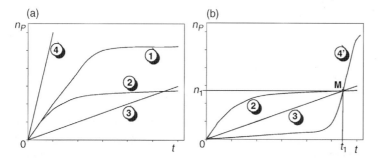

Figure 2.2 Conversion-time plots for evaluating the activity and productivity of catalysts.

2.4
Selectivity and Specificity of Catalysts

A catalyst is called selective when it preferentially or exclusively catalyzes one of several possible reactions between the reactants. If the selectivity relates to the stereochemical relationship between reactants and products, one refers to stereoselective reactions or catalysts, which can be characterized more precisely as diastereoselective or enantioselective. If a reaction permits the formation of several regioisomers but forms one preferentially or exclusively, it is a regioselective reaction. If a substrate displays various functional groups, but only one is converted within a reaction, the reaction is called chemoselective [8].

The product composition allows one to state quantitatively the extent to which a reaction occurs selectively. For enantio- and diastereoselective reactions, the enantiomeric excess (% ee = % major enantiomer − % minor enantiomer) or the diastereomeric excess (% de = % major diastereomer − % minor diastereomer) is generally quoted [9].

A catalyst that converts stereoisomeric reactants into *different* stereoisomeric products is called stereospecific. In other words, in a stereospecific reaction, distinct stereoisomers are converted into stereochemically distinct products. As with stereoselective reactions, a distinction can be drawn between enantiospecific and diastereospecific reactions. Specific reactions are necessarily selective as well, but selective reactions need not be specific.

However, occasionally – especially for enzyme catalysis – selective reactions that run strictly uniformly, and hence form one product exclusively, are designated "specific." However, we will not use the term this way and will characterize such reactions as "highly selective" [10].

Examples
- In the hydrogenation of prochiral olefins **7**, two enantiomers **8** (CIP priority: R > R′ > Me) may be formed. If a catalyst causes one of them to be created preferentially or exclusively, the reaction runs selectively (more precisely: enantioselectively).

- In the epoxidation of inner olefins **9**, two stereoisomeric epoxides **10** may be formed. If a catalyst forms the *trans* epoxide from the (E)-olefin and the *cis* epoxide from the (Z)-olefin, the reaction is stereospecific (more precisely, diastereospecific).

- The polymerization of butadiene can, via 1,2-linkage, lead to iso- (**11**) or syndiotactic 1,2-polybutadiene (**12**), or, via 1,4-linkage, to *trans*- (**13**) or *cis*-1,4-polybutadiene (**14**). The polymerization is then regioselective if either 1,2-polybutadiene (**11/12**) or 1,4-polybutadiene (**13/14**) is formed. It is, in addition, stereoselective if a single steroisomer (either **11** or **12**, or, respectively, **13** or **14**) is formed preferentially. Sometimes these polymerizations are designated stereospecific, although this is not quite correct.

- The regioselective formation of a single isomer (usually the *n*-product) is of central significance in the hydroformylation of olefins.

- If, during the hydrogenation of the enyne **15**, either **16** or **17** is formed, the reaction is chemoselective. If **17** is either the *cis,trans* (**17a**) or *trans,trans* isomer (**17b**), the reaction leading to **17** is also stereoselective.

2.5
Determination of Catalytic Mechanisms

The reaction mechanism of a homogeneously catalyzed reaction encompasses a detailed description of all partial reactions of the reaction cycle, as well as of formation, side, and decomposition reactions of the catalytically active complexes and intermediates (see Figure 2.1). That includes knowledge about reversibility, the position of equilibria, and conversion and decomposition rates for these reactions. Coordination chemistry and organometallic chemistry are indispensable foundations for understanding the reaction mechanisms in organometallic catalysis.

Once the mechanism of an organometallic-catalyzed reaction is known, it can be used to deduce relations between the structure of catalysts and intermediates and their catalytic activity and productivity, as well as their specificity and selectivity, ("structure-property relationships"). This is an essential requirement for targeted development and optimization of catalysts. Obtaining a comprehensive look at the mechanism of a metal-catalyzed reaction is a very complex task. The widest variety of experimental studies and, if at all possible, quantum-chemical calculations, must be included. Only a broad set of studies can guarantee that no conceivable mechanism is excluded inappropriately.

2.5.1
Experimental Studies

There is a wide palette of experiments that can be included to elucidate a reaction mechanism. Important experimental methods can be classified as follows:

- *Catalytic experiments* for clarification of structural and environmental influences. This includes experiments on the dependence of catalytic activity/productivity and selectivity/specificity as a function of the reaction conditions (concentration of reactants and catalyst; solvent, temperature, pressure, . . .), the substitution pattern of the reactants (for olefins, e.g., $H_2C=CHR$ vs. $H_2C=CR_2$ vs. *cis*-$RHC=CHR$ vs. *trans*-$RHC=CHR$, variation of R, . . .), and the electronic and steric properties of the coligands in the catalyst complex (for phosphanes, e.g., systematic variation of electronic and steric Tolman parameters, see Section 10.4.2.2).
- *Spectroscopic and chromatographic experiments* on catalytic solutions, which may need to be modified (e.g., with regard to concentration, in order to compensate for the detection limit of the experimental method), and studies of catalysis with isotopically labeled compounds. The goal of the experiments is to isolate or identify intermediates, side products, and decomposition products. It is essential to clarify their function in the catalytic cycle in order to, for example, distinguish the active intermediate complexes within the cycle from off-loop species (Figure 2.1, 2–4 vs. 5–6).
- *Preparative experiments* for the synthesis of postulated or identified intermediates, or of model complexes, and the study of their constitution, stability, reactivity and catalytic properties. These include synthesis of structurally defined precatalysts and studies of their structure in the solid state (single-crystal X-ray diffraction studies) and in solution (e.g., NMR spectroscopy).
- *Kinetic experiments* on the course of catalysis, with the goal of generating a quantitative reaction model. Determining the reaction partners involved in the rate-determining step is of special interest.

The starting point is, of course, a plausible hypothesis about mechanism, based on known organometallic elementary reactions. Through the complex application of the previously outlined experimental methods – and by avoiding an excessively limited selection – one can obtain an experimentally established reaction scheme. This will raise new questions that are then answered by the results of other new experiments. The result is used as feedback for the proposed mechanism and thus, an increasingly accurate and comprehensive scheme can be iteratively obtained.

2.5.2
Theoretical Studies

The development of new quantum-chemical computational methods and an enormous increase in the performance of computers have made it ever more possible to perform calculations on complex systems containing transition metals with a reasonable expenditure of time to a sufficient degree of accuracy. This has made it increasingly feasible to perform computations on all relevant intermediates *and* transition states of a catalytic cycle, and thus model the cycle entirely theoretically. In connection with profound experimental studies on mechanism, such a quantum-chemical analysis is indispensable for understanding the full complexity of most transition-metal-catalyzed reactions [M10, M13, 11].

Density functional theory (DFT) (Walter Kohn, Nobel Prize in Chemistry 1998; together with J. A. Pople) has proven to be excellent for transition-metal-catalyzed systems [12]. In many cases, a generic catalyst is taken as a starting point. For example, more complex coligands such as (o,o'-disubstituted-phenyl)phosphanes are replaced by PH_3. Then the complex cycle is modeled with good quantum-mechanical accuracy. On this foundation, computations on the actual catalyst can be performed. This is where hybrid methods (QM/MM) can be brought into play [13, 14]. In these calculations the actual reaction center is described by an accurate quantum-mechanics method (QM) and the periphery of the ligands is modeled by a molecular-mechanics method (MM), where, for example, the spatial requirements can be described adequately with low computational expenditure, even for large coligands [15, 16, M13]

The majority of computations relate to the gas phase. In general, solvent effects can be taken into account, if at all, only in the framework of relatively simple models such as Tomasi's polarized continuum model (PCM). This can mean a substantial limitation of validity for the course of reaction in solution.

In any case, quantum-chemically computed energies can only be compared when they have been calculated according to the same method. Naturally, while this is true for calculations on intermediates and transition states of one catalytic cycle by the same author, in most cases, it applies neither to calculations by different authors, nor to calculations by a single author on fundamentally different systems. For this reason, the reaction profile diagrams in this book omit the calculation method (which would be a precondition for estimating the accuracy), or any precise scaling of the energy. They are meant only to semiquantitatively represent and vividly depict the energetic course in a cycle. The interested reader is referred to the cited literature.

2.6
Glossary for Catalysis

Catalyst
A catalyst is a substance that increases the rate of a reaction without modifying the overall Gibbs free energy change in the reaction. The catalyst is both reactant and product of the reaction [17].
Catalysts are not consumed in the reaction. Thus, they do not appear in the overall reaction, but are inherently components of the reaction cycles. In general, they are very reactive and in many cases cannot be isolated; occasionally, their existence cannot even be directly demonstrated by spectroscopy.

Catalyst Complex
In organometallic catalysis, catalytically active complexes, which are the actual catalysts, are frequently called catalyst complexes.

Precatalyst
Precatalysts are compounds from which the catalysts are generated. In general, they are stable enough to isolate in a pure state.

Cocatalyst
If an additional component is necessary for generating a catalyst from the precatalyst, it is called a cocatalyst. The cocatalyst itself is not catalytically active.

Promoter (Activator)
Additives to the catalyst that increase its effectiveness (activity, productivity, selectivity, etc.) are called promoters (activators). Promoters themselves are catalytically inactive.

Initiator
A reaction can also be triggered or accelerated by initiators. In contrast to catalysts, these are not regenerated in the reaction cycle, but instead irreversibly consumed in the starting reaction.

Inhibitor
An inhibitor is a substance that reduces the rate of a catalyzed reaction (or a reaction triggered by an initiator). The process itself is occasionally called negative catalysis, even though this is not correct.

Autocatalysis
In autocatalytic reactions, a reaction product acts catalytically, so an acceleration of the reaction with increasing conversion can be observed. In complex systems, autocatalytic reaction steps can lead to oscillating reactions.
Asymmetric autocatalysis occurs when a chiral product functions as a chiral catalyst for its own synthesis [18].

Induction Period
An induction period is a very slow initial phase of a chemical reaction that, in later phases, proceeds at a higher reaction rate. The "normal" course of reaction does not begin until after the induction period. An induction period for catalytic

reactions can be caused by slow formation of catalyst and is also characteristic of autocatalytic reactions.

Reaction Profile Diagram

Diagram in which the Gibbs free energies ΔG (generally $\Delta G^\ominus$; or, as a substitute, the energies) of the reactants and products of a reaction and the intermediates and transition states are represented on the ordinate (y-axis). For clarity, the individual states are shifted horizontally; the abscissa is *not* defined. Regarding the scaling of the ordinate, see Section 2.5.2 [19].

It is clear that a comparison of ΔG values is only meaningful when they relate to the same reaction mixture, meaning the same (elementary) composition. This means that the representation of the individual reaction steps of a catalytic reaction includes, in addition to all reactants and products, the catalyst complex in molar quantity.

Rate-Determining Step

A reaction step that decisively determines the rate of the overall reaction in a reaction cycle is called rate-determining. In many cases, this is the reaction step associated with the highest-lying transition state in the reaction profile diagram. For complex reaction kinetics, there may not necessarily be a single rate-determining reaction step [20, 21].

Resting State

In many organometallic-catalyzed reactions, there is a complex that is present in significantly higher concentration than any other complex in the reaction mixture. If it is catalytically active itself, or is in equilibrium with a catalytically active complex, it is designated as the "resting state" of the catalyst.

Examples

- Homogeneous hydrogenation of olefins with the Wilkinson complex [RhCl(PPh$_3$)$_3$] (Rh^I; 16 ve; ve = valence electrons) as precatalyst, from which the catalyst [RhCl(PPh$_3$)$_2$] (Rh^I; 14 ve) is formed by cleaving the ligand (PPh$_3$) from the complex.

$$\text{C=C} + H_2 \xrightarrow{[RhCl(PPh_3)_3]} -\overset{H}{\underset{|}{C}}-\overset{H}{\underset{|}{C}}-$$

- Polymerization of ethene with metallocene catalysts such as [ZrCl$_2$Cp$_2$]/MAO (MAO = methylaluminoxane). Precatalyst is [ZrCl$_2$Cp$_2$]. The cocatalyst (MAO) acts as a methylating agent and as a Lewis acid, thus generating the catalyst complex [ZrCp$_2$Me]$^+$.

$$n\ H_2C=CH_2 \xrightarrow{[ZrCl_2Cp_2]/MAO} -(CH_2-CH_2)_n-$$

- Carbonylation of methanol, producing acetic acid, with RhCl$_3$/HI as catalytic system. Iodides such as LiI, [PR$_4$]I and [NR$_4$]I stabilize the catalyst complex and act as promoters.

$$\text{MeOH} + \text{CO} \xrightarrow[(\text{H}_2\text{O})]{\text{RhCl}_3 / \text{HI}} \text{Me-COOH}$$

- Styrene can be radically polymerized. Free radical generators such as dibenzoyl peroxide act as initiators.

$$1/2 \text{ Ph-C(O)O-O(O)C-Ph} \xrightarrow{\Delta} \text{Ph-C(O)O} \cdot \xrightarrow{\text{H}_2\text{C=CHPh}} \text{Ph-C(O)O-CH}_2\text{-}\overset{\bullet}{\text{CH}}\text{-Ph} \xrightarrow{n \text{ H}_2\text{C=CHPh}} \text{polystyrene}$$

- Radical scavengers such as aromatic amines or phenols inhibit auto-oxidation reactions of hydrocarbons.

Auto-oxidation:

$$\text{R-H} \xrightarrow[+ \text{ ini}\cdot, \text{ } - \text{iniH}]{\text{start reaction}} \text{R} \cdot \xrightarrow{+ \text{O}_2} \text{R-O-O} \cdot \xrightarrow[- \text{R} \cdot]{+ \text{R-H}} \text{R-O-OH}$$

Inhibition:

$$\text{R-O-O} \cdot + \text{HO-C}_6\text{H}_4\text{-OH} \xrightarrow{- \text{R-O-OH}} \cdot\text{O-C}_6\text{H}_4\text{-OH} \longrightarrow 1/2 \text{ O=C}_6\text{H}_4\text{=O} + 1/2 \text{ HO-C}_6\text{H}_4\text{-OH}$$

- The reaction **18** → **19** proceeds enantioselectively. It is catalyzed by **20**, formed by conversion of catalytic quantities of product **19** with Zn(i-Pr)$_2$. Thus, the reaction is asymmetrically autocatalytic, which has the advantage that no chiral compound is required other than the product itself, and there is no need to separate the product from the chiral catalyst [22].

quinoline-CHO $\xrightarrow[\text{2) H}^+]{\text{1) Zn(i-Pr)}_2}$ quinoline-CH(i-Pr)OH   quinoline-CH(i-Pr)OZn i-Pr

18   **19** (94 % ee)   **20**

- In the rhodium-catalyzed hydroformylation of olefins, the two acylrhodium(I) complexes (**21**: 16 ve; **22**: 18 ve) are in an equilibrium whose position lies to the right under catalytic conditions. Complex **22**, which itself is catalytically inactive, serves as a reservoir for the catalytically active complex **21**, and is an example of a resting state.

OC-Rh(PPh$_3$)$_2$-C(O)R $\rightleftharpoons$ (Ph$_3$P)$_2$(CO)$_2$Rh-C(O)R

21   **22**

Exercise 2.1

Suppose that a catalyst can accelerate the rate of a reaction by a factor of 10, 100, or 1000 ($T = 298$ K). Calculate the decreases in the Gibbs free energy of activation $\Delta G^\ddagger$ to which these accelerations of the reaction correspond.

2.7
The Development of Organometallic Catalysis

- 1916, *Wacker-Chemie* (Burghausen, Germany). Based on the finding by M. Kutscheroff (1881) that acetylene and water is converted to acetaldehyde in the presence of HgBr$_2$, a technical plant for the mercury-catalyzed hydration of acetylene was put into operation in 1916.

$$\equiv \quad \xrightarrow{\text{H}_2\text{O}}_{\text{Hg/H}^+} \quad \text{CHO}$$

- 1937, *Walter Reppe*, at BASF (Ludwigshafen, Germany), started to develop the synthetic potential of C–C coupling reactions of acetylene and its functionalization (starting in 1928), and founded the technological principles for safely handling acetylene under pressure ("Reppe chemistry").

$$\equiv \quad \xrightarrow{\text{CO/ROH}}_{\text{[Ni]}} \quad \diagup\!\!\diagdown\text{COOR} \qquad \equiv \quad \xrightarrow{\text{[Ni]}} \quad \text{(cyclooctatetraene)}$$

- 1938, *Otto Roelen*, at Ruhrchemie (Oberhausen, Germany), discovered the conversion of ethene with synthesis gas (CO/H$_2$) into propionaldehyde in the presence of a heterogeneous cobalt–thorium catalyst and developed the process (hydroformylation of olefins, "oxo synthesis") until it was technologically mature. At the end of the 1940s, it was shown that this is a homogeneous catalysis (precatalyst: Co$_2$(CO)$_8$).

$$\equiv \quad \xrightarrow{\text{CO/H}_2}_{\text{Co}_2(\text{CO})_8} \quad \text{CHO}$$

- 1953, *Karl Ziegler* discovered, at the MPI für Kohlenforschung (MPI for Coal Research, Mülheim/Ruhr, Germany), the low-pressure polymerization of ethene with organometallic mixed catalysts ("*metallorganische Mischkatalysatoren*"). This discovery served to ignite a rapid development of organometallic chemistry and organometallic catalysis.

$$\equiv \quad \xrightarrow{\text{TiCl}_4}_{\text{AlEt}_3} \quad (\text{polyethylene})_n$$

- 1954–1955, *Giulio Natta* (Institute of Technology, Milan, Italy) demonstrated that propene, other α-olefins, and butadiene (1955–1959) are stereoselectively polymerized with the Ziegler catalysts.

- 1954–1955, *Goodrich Gulf Chemicals* and *Phillips Petroleum (USA)*. First basic patents for the technical synthesis of highly 1,4-*cis* containing (>90%) polyisoprene and polybutadiene with Ziegler catalysts.

- 1955, *Günther Wilke* (MPI for Coal Research) developed nickel-catalyzed cyclooligomerization, linear oligomerization, and telomerization reactions of butadiene.

- 1956–1959, *Jürgen Smidt* (*Wacker process*). Based on the observation (F. C. Phillips, 1894) that $PdCl_2$ oxidizes ethene to acetaldehyde in an aqueous solution, a catalytic process for production of acetaldehyde was developed at Wacker-Chemie. This process development acquired additional significance because it occurred at a time when the chemical industries were switching from coal chemistry (acetylene-based) to petrochemistry (olefin-based) and the coal-chemistry-based Kutscheroff process ($\rightarrow$ 1916) was superseded in the modern chemical industries.

- 1963, *Karl Ziegler* and *Giulio Natta*. Nobel Prize in Chemistry for their discoveries in the field of the chemistry and technology of high polymers.
- 1965, *Geoffrey Wilkinson* (Imperial College, London) discovered a rhodium-catalyzed process for homogeneous hydrogenation of olefins, which previously had only been possible through heterogeneous catalysis by metals such as Ni (Paul Sabatier, University of Toulouse, 1912 Nobel Prize in Chemistry together with Victor Grignard).

- 1966, *Nissim Calderon* (Goodyear, Akron, Ohio) reported a homogeneously catalyzed olefin metathesis, which previously (1957) had been accomplished with heterogeneous catalysts. The mechanism was elucidated in 1971 by Y. Chauvin and J.-L. Hérisson (IFP, Rueil-Malmaison, France).

- 1966, *Hitosi Nozaki* and *Ryoji Noyori* (both Kyoto University, Japan) discovered, in the cyclopropanation of styrene, the first example of asymmetric catalysis by a structurally well-defined chiral transition-metal complex.

- 1968, *Monsanto process*. The production of acetic acid by carbonylating methanol is an important process for creating value-added compounds from inexpensive C_1 starting material. Since 1970, acetic acid has been overwhelmingly produced by the rhodium-catalyzed Monsanto process. In 1995–1996, an iridium-based process was developed (Cativa process, BP Chemicals).

- 1968, *William S. Knowles* (Monsanto, St. Louis, Missouri) and *Leopold Horner* (University of Mainz, Germany) showed independently that the Wilkinson complex, modified by a chiral phosphane, can enantioselectively hydrogenate prochiral olefins. DIOP (H. B. Kagan, 1971) is the first widely used chiral coligand. The enantioselective hydrogenation of a C–C double bond is central in the synthesis of L-DOPA on an industrial scale.

- 1972, *Heck reaction*. Richard F. Heck (University of Delaware, Newark) reported a palladium-catalyzed coupling of aryl and vinyl halides with olefins, which developed into one of the most important methods for C_{sp^2}–C_{sp^2} coupling.

- 1972–1979, *Metal-catalyzed C–C cross-couplings* of organohalides with organometallic compounds were established as standard methods in organic synthesis, including those catalyzed with nickel using Grignard reagents (M = Mg; M. Kumada, 1972) and catalyzed with palladium using organozinc (M = Zn; E.-i. Negishi, 1976–1977), organoboron (M = B; A. Suzuki, 1979) and organotin compounds (M = Sn; J. K. Stille, 1979).

$$\text{R-X} + [\text{M}]-\text{R}' \xrightarrow{[\text{Ni}] \text{ or } [\text{Pd}]} \text{R-R}' + [\text{M}]-\text{X}$$

- *1973, Geoffrey Wilkinson* and *Ernst Otto Fischer* (Technical University of Munich, Germany). Nobel Prize in Chemistry for their pioneering work, performed independently, on the chemistry of the organometallic, so-called sandwich compounds.
- *1977, Biphasic catalysis (SHOP)*. With his work on nickel-catalyzed ethene oligomerization in a liquid—liquid biphasic system, W. Keim (Technical University of Aachen, Germany) created the basis for the Shell Higher Olefin Process (SHOP).

$$\text{CH}_2=\text{CH}_2 \xrightarrow{[\text{Ni}]} \text{higher olefins}$$

- *1980, K. Barry Sharpless* (Scripps Research Institute, La Jolla, California) developed an asymmetric epoxidation of allyl alcohols.

$$\underset{R}{\overset{R}{>}}=\underset{R}{\overset{\text{OH}}{<}} \xrightarrow[t\text{-BuOOH}]{\text{Ti}(Oi\text{-Pr})_4 / (S,S)(-)\text{-dialkyl tartrate}} \text{epoxide}$$

- *1980, Hansjörg Sinn* and *Walter Kaminsky* (both University of Hamburg, Germany), in polymerizing ethene with metallocene catalysts and methylaluminoxane (MAO) as cocatalyst, induced activity corresponding to that of highly active enzymes.

$$\text{CH}_2=\text{CH}_2 \xrightarrow[\text{MAO}]{[\text{ZrMe}_2(\text{Cp})_2]} \text{polyethylene}$$

- *1982/1985, Walter Kaminsky* and *Hans-Herbert Brintzinger* (University of Konstanz, Germany) synthesized *ansa*-metallocenes (H.-H. B.), which, with MAO as cocatalyst, are highly active and productive polymerization catalysts for olefins. Structure-selectivity relationships for propene polymerization into iso- or syndiotactic polypropylene are elucidated using these "single-site catalysts."

C_2 symmetry   bzw. C_s symmetry

- *1988/1993, Richard R. Schrock* (MIT, Cambridge, Massachusetts) and *Robert H. Grubbs* (Caltech, Pasadena, California) showed the broad applicability of, respectively, alkylidene–molybdenum and –tungsten complexes (**I**, 1988) and alkylidene–ruthenium complexes (**IIa**, 1993; **IIb**, 1999) as single-component catalysts for olefin metathesis, and thus established it as a standard method in both organic synthesis and polymer chemistry.

- *1997, Jean-Marie Basset* (CNRS, Lyons, France) found in his studies on surface organometallic chemistry, which can be seen as a bridge between homogeneous and heterogeneous catalysis, a way to achieve the metathesis of alkanes with a tantalum hydride supported on silica.

- *1998, Walter Kohn* (University of California, Santa Barbara) and *John A. Pople* (Northwestern University, Evanston, Illinois). Nobel Prize in Chemistry for the development of density functional theory (DFT) and computational methods in quantum chemistry. The development of theory (especially the DFT method, starting in 1990) and computer technology enabled a full analysis of intermediate complexes and transition states in catalytic cycles. This has been an exceptionally valuable aid to understanding activity and selectivity of catalysts and to catalyst development and optimization.

- *2001, William S. Knowles, Ryoji Noyori* (Nagoya University, Japan), and *K. Barry Sharpless*. Nobel Prize in Chemistry for their work on chirally catalyzed hydrogenation (W. K; R. N.) and oxidation reactions (B. S.).

- *2003, Dmitry V. Yandulov* and *Richard R. Schrock* (both MIT) described a defined mononuclear dinitrogen–molybdenum(III) complex that catalyzes nitrogen fixation at room temperature and normal pressure, hence under physiological conditions.

$$N_2 + 6\,H^+ + 6\,e^- \xrightarrow[\text{heptane (24 °C, 1 bar)}]{[Mo]-N\equiv N} 2\,NH_3 \quad (64\%,\ TON = 6)$$

- *2005, Yves Chauvin, Robert H. Grubbs,* and *Richard R. Schrock*. Nobel Prize in Chemistry for the development of the metathesis method in organic synthesis.

- *2007, Gerhard Ertl* (Fritz-Haber-Institut der MPG, Berlin, Germany). Nobel Prize in Chemistry for his studies on chemical processes on solid surfaces. These created an important foundation for understanding heterogeneously catalyzed reactions on the atomic and molecular levels.

- *2010, Richard F. Heck, Ei-ichi Negishi* (Purdue University, West Lafayette, Indiana), and *Akira Suzuki* (Hokkaido University, Sapporo, Japan). Nobel Prize in Chemistry for their work on palladium-catalyzed cross-coupling in organic synthesis.

3
Elementary Steps in Organometallic Catalysis

3.1
Cleavage and Coordination of Ligands

Reaction Principle[1]

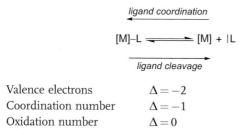

Valence electrons	$\Delta = -2$
Coordination number	$\Delta = -1$
Oxidation number	$\Delta = 0$

Ligand cleavage (ligand dissociation) from a metal generates coordinatively unsaturated complexes. In solution, the vacant coordination site may be occupied by a solvent molecule. If this is not counted in the electron balance for M and in the determination of the coordination number, these reactions reduce the number of electrons in the valence shell of M by two, and decrease the coordination number of M by one. The reverse reaction is called coordination or attachment of a ligand.

Ligand substitution reactions involve both ligand cleavage and attachment. They can proceed either by a dissociative mechanism (symbol: *D*), where ligand cleavage

---

[1] Those ligands that are not directly involved in the actual reaction will be indicated henceforth in square brackets. A "vacant" coordination site is occasionally indicated by a small square. In solution, "vacant" coordination sites are usually occupied by solvent molecules "s." When this needs to be emphasized, "s" is written as a ligand. From this point on, weakly coordinating solvent molecules are not considered when determining the coordination number and electron balance for M.

Fundamentals of Organometallic Catalysis. First Edition. Dirk Steinborn
Copyright © 2012 WILEY-VCH Verlag GmbH & Co. KGaA, Weinheim

occurs before attachment (reaction **a**; L_l, L_e = leaving ligand and entering ligand, respectively) or an associative mechanism (symbol: *A*), where ligand attachment occurs before cleavage (reaction **b**). Intermediates are complexes with a lower (*D*) or higher (*A*) coordination number than that of the reactant. In addition, a ligand substitution may be based on an interchange mechanism (symbol: *I*) in which no intermediate complex can be detected (reaction **c**).

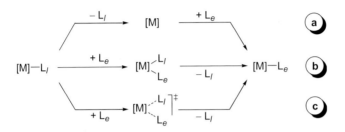

Ligand cleavages that occur in homogeneous catalysis are generally reversible. The coordinatively unsaturated complexes that are formed are frequently in a position to coordinate a substrate molecule (ligand attachment). This may be an indispensable precondition for catalysis, since it is associated with activation of the substrate. Examples are the formation of π complexes with unsaturated hydrocarbons (olefins, alkynes, dienes, etc.) and σ complexes with dihydrogen or hydrocarbons (H–H or C–H as ligand). In many cases, the cleavage of ligands from the precatalyst leads to the formation of the actual catalyst complex.

In homogeneous catalysis with transition metals, complexes with 16 and 14, but also with 12 valence electrons appear as intermediates. In precatalysts with 18 valence electrons, the catalytically active complex (16 *ve*) is generally not formed until ligand cleavage occurs.

In most cases, ligands L are cleaved off as Lewis bases. The electron pair from the M–L bond thus remains with the ligand. However, it is also possible for a ligand to be cleaved off as a Lewis acid. Since the bonding electron pair then remains with the metal, such reactions are associated with a change in the oxidation number (*ON*) of the metal ($\Delta ON = -2$; $\Delta ve = 0$). The most familiar examples are deprotonations of hydridometal complexes. Corresponding reverse reactions are protonations of metal bases:

$$[M]\text{–}H \rightleftharpoons \overline{[M]}^{\ominus} + H^{\oplus}$$

Examples
- In hydroformylation and hydrogenation of olefins, the catalysts are generated from carbonylhydridotris(triphenylphosphane)rhodium or from the Wilkinson complex by cleaving off PPh_3.

Precatalyst		Catalyst
[RhH(CO)(PPh$_3$)$_3$]	$\underset{+ \text{PPh}_3}{\overset{- \text{PPh}_3}{\rightleftharpoons}}$	[RhH(CO)(PPh$_3$)$_2$]
Rh^{I}; 18 ve		Rh^{I}; 16 ve
[RhCl(PPh$_3$)$_3$]	$\underset{+ \text{PPh}_3}{\overset{- \text{PPh}_3}{\rightleftharpoons}}$	[RhCl(PPh$_3$)$_2$]
Rh^{I}; 16 ve		Rh^{I}; 14 ve

- In many cases, carbonyl–hydrido complexes are strong acids (sometimes very strong) in protic solvents. An example is [CoH(CO)$_4$], whose pK_a is less than 2 in H$_2$O. By contrast, in MeCN, its pK_a is 8, so its acid strength is roughly equal to that of HCl in MeCN [23].

$$(CO)_4Co-H + H_2O \rightleftharpoons (CO)_4Co|^{\ominus} + H_3O^{\oplus}$$

Co^{I}; 18 ve Co$^{-\text{I}}$; 18 ve

Background: Classification of Ligands

Complex formation reactions can be regarded as Lewis acid-base reactions:

[M] + |L ⟶ [M]–L

According to the nature of the donor orbital of L, the following complex types can be differentiated:

Complex type	n complex	π complex	σ complex
Donor orbital of L	nonbonding	π MO	σ MO
Lewis structure	M–X	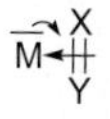 M⟵‖$^\text{X}_\text{Y}$	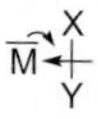 M⟵‖$^\text{X}_\text{Y}$
Orbital overlap$^{a)}$ (schematic)	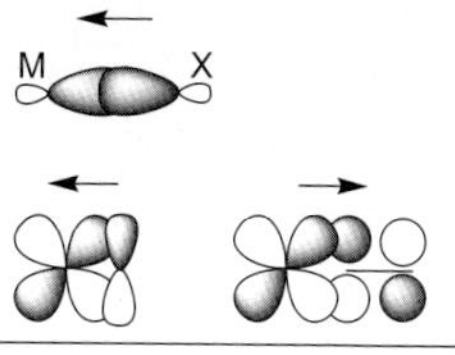	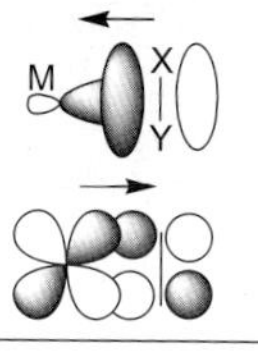	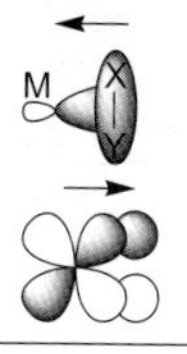

a) The arrows point from the occupied to the unoccupied orbital: ← = forward donation; → = back-donation.

In the case of the π and σ complexes (e.g., olefin and aromatic complexes or η^2-H$_2$ complexes), the metal–ligand bonding is strengthened by π back-donation, which transfers electron density from M into the π* or σ* orbitals of the ligands. n-Donor ligands can, in addition, be π donors (e.g., O^{2-}, F$^-$) or π acceptors (e.g., CO) [24, 25].

Note: σ complexes (e.g., dihydrogen complexes such as [W(η²-H₂)(CO)₃{P(i-Pr)₃}₂]) have only been synthesized recently. Previously, only n and π complexes needed to be differentiated from each other, and n complexes were designated σ complexes. This should be noted when the designation "n and σ complexes" is used; adding an explanation may be worthwhile. In accordance with general usage, the designation of simple ligands such as PR₃ and CO as σ-donor ligands or σ-donor/π-acceptor ligands remains unchanged in this book.

3.2
Oxidative Addition and Reductive Elimination

Reaction Principle

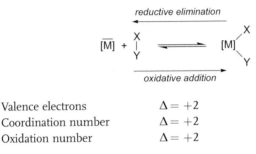

Valence electrons	Δ = +2
Coordination number	Δ = +2
Oxidation number	Δ = +2

During oxidative addition reactions, a substrate molecule X–Y is attached to a low-valent metal complex, breaking the X–Y bond. The metal's coordination number, oxidation number, and number of valence electrons are each increased by two. Structural and electronic preconditions for oxidative additions are that [M] be coordinatively and electronically unsaturated, have a nonbonding electron pair available, and be able to make a transition to an oxidation state higher by two. The reverse reaction is called reductive elimination.

Oxidative addition reactions are very frequently encountered for d^8 and d^{10} complexes where a transition occurs from a square-planar to an octahedral coordination geometry ($M^{II} \rightarrow M^{IV}$, M = Pd, Pt; $M^I \rightarrow M^{III}$, M = Rh, Ir) or from a linear/bent to a square-planar geometry ($M^0 \rightarrow M^{II}$, M = Pd, Pt; $M^I \rightarrow M^{III}$, M = Au). Typical substrates are dihydrogen H–H, halogens X–X (X = Cl, Br, I), hydrogen halides H–X, and halohydrocarbons R–X (R = alkyl, aryl, vinyl, alkynyl, etc.). With hydrocarbons and silanes, oxidative additions lead to activation of C–H and Si–H bonds, respectively, and sometimes also of C–C and Si–Si bonds.

Oxidative addition reactions display a great variety of mechanisms. They can proceed according to a radical or ionic mechanism. Bond formation (M–X/M–Y) and breaking (X–Y) can also occur synchronously (concerted mechanism). The initial step in the reaction may be coordination of the substrate as a σ complex or an agostic C–H···M interaction between substrate and [M], a special form of σ complex formation.

Oxidative addition reactions of R–X (R = alkyl) can also proceed as S_N2 reactions. These are then consecutive reactions (1 → 2 → 3).

$$L_{x-1}MI + C{-}X \longrightarrow [L_{x-1}M{-}C]^+ + X^- $$
$$\text{1} \qquad\qquad\qquad\qquad \text{2}$$

$$\longrightarrow L_{x-1}M\overset{C}{\underset{X}{\diagdown}}$$
$$\text{3}$$

$$L_xMI + C{-}X \longrightarrow [L_xM{-}C]^+ + X^- \xrightarrow{-L}$$
$$\text{1'} \qquad\qquad\qquad\qquad \text{2'}$$

Such reactions are, in principle, possible even for 18-ve complexes (1' → 2' → 3), where addition of X^- in the second step is accompanied by ligand cleavage. The reaction may also remain at the first stage at 2', in which case X^- will not be coordinated at all. Such reactions can also be called oxidative additions in an extended sense, but are more accurately considered as nucleophilic substitution of X^- by M or as electrophilic attack of R–X on the metal.

Background: Agostic C–H···M Interactions

Organometallic compounds with an electrophilic metal center are capable of forming weak interactions (with strength in the range of about 4–40 kJ/mol) using the bonding electron pair of a C–H bond. These interactions are called agostic C–H···M interactions. Depending on whether an α- or β-C–H bond is involved, one can speak more precisely of α- or β-agostic interactions. An agostic C–H···M interaction is a three-center/two-electron bond (3c–2e), which can be considered a σ complex formation (see Background, Section 3.1) with a C–H bond. Common ways of representing the formula are:

$$M{-\!\!\!\!-}\overset{H}{\underset{}{C}}\qquad \text{or}\qquad M\cdots\overset{H}{\underset{}{C}}\qquad \text{or}\qquad M{\diagup\!\!\diagdown}\overset{H}{C}.$$

Note the fundamental difference from normal hydrogen bonds (X–H–Y such as O–H–O, F–H–F, ...), which represent three-center/four-electron bonds (3c–4e). A more comprehensive definition for agostic C–H···M interaction, setting aside the bond description, includes any structural distortions of an organometallic moiety that bring an appended C–H bond into close proximity with the metal center, even those of a predominantly electrostatic nature, such as may be the case for C–H···Li interactions [26–28].

If instead of X–Y substrates (see above), substrates with an X=Y double bond are used, X and Y remain connected through a single bond after a successful oxidative addition:

$$[M] + \overset{X}{\underset{Y}{\|}} \rightleftharpoons [M]\overset{X}{\underset{Y}{<|}} \quad vs. \quad [M]\leftarrow\overset{X}{\underset{Y}{\|}} \quad\quad X=Y: CH_2=CH_2, O_2,...$$

$$\quad\quad\quad\quad\quad\quad\quad\quad\quad a \quad\quad\quad\quad\quad b$$

Whether it is correct to call this oxidative addition (formation of metallacyclopropane complexes, formula **a**) or whether the reaction is better understood as π-complex formation (formula **b**) must be decided on a case-by-case basis from spectroscopic and/or structural studies and/or quantum-chemical calculations. An analogous situation applies for reactions with alkynes, where the products can be classified as either metallacyclopropene complexes or as π-alkyne complexes.

Typical reductive eliminations of H–H, X–X, H–X, R–X correspond to the reversal of the oxidative addition reactions described previously. However, reductive eliminations with linkage of C–H or C–C bonds are also very important. The reverse reactions, oxidative additions of (unactivated) C–H and C–C bonds, proceed with much more difficulty. One of the great challenges of homogeneous catalysis is to incorporate such C–H and C–C activation into catalytic processes.

The rate of reductive elimination and oxidative addition reactions depends heavily on the electron structure of the central atom and on the ligand sphere. For example, for reductive elimination from square-planar d^8 complexes (M = Ni, Pd, Pt), a direct (**a**), a dissociative (**b**), or, in rare cases, an associative (**c**) course of reaction has been demonstrated. The designation is related to whether the actual reductive elimination step occurs directly (**a**) or only after previous ligand cleavage (**b**) or attachment (**c**) [29–31].

Background: Oxidation States of Metals in Olefin and Alkyne Complexes

The oxidation number relates to a conceptual model rather than a measurable quantity. In a compound A–B, it specifies the charges of A and B after heterolytic bond cleavage, after which the more electronegative partner contains the bonding electrons. This presupposes knowledge about the distribution of electrons. If, for example, the electron structure of A–B can be described by three resonance (canonical) structures **1a–1c**, and the electronegativity $\chi(B)$ is higher than $\chi(A)$, then the electrons should be assigned as indicated by the arc. Different oxidation numbers (ON) are then to be assigned to A and B according to whether the resonance formula **1a** or **1b/1c** is dominant.

3.2 Oxidative Addition and Reductive Elimination

	$\overset{\ominus}{A}\!\!-\!\!\overset{\oplus}{B}$ ⟷ $A\!\!=\!\!B$ ⟷ $\overset{\oplus}{A}\!\!-\!\!\overset{\ominus}{B}$	
ON(A)	$+n$	$+(n+2)$	$+(n+2)$
ON(B)	$-n$	$-(n+2)$	$-(n+2)$
	1a	1b	1c

To assign the oxidation number of M in metal complexes, it is assumed that the ligands, as nonmetallic derivatives, are generally more electronegative than M. Thus the electrons in M–L bonds are assigned to the ligands. It follows, therefore, that an appropriate assignment of oxidation number of M in a complex can only be made if one knows the electron structure, which can be derived from magnetic measurements, spectroscopic (e.g., ESR and Mößbauer spectroscopy) and structural experiments, as well as quantum-chemical calculations.

Olefin ligands are π donors and π^* acceptors, so for metal–olefin bonds, two bond components are of significance, the σ donation and the π back-donation (Dewar-Chatt-Duncanson model). The stronger the π back-donation, the greater the elongation (Δd) of the C=C bond in the complex and the more strongly the substituents on the olefin carbon atoms are bent back away from the metal (measured as angle α). When back-donation is very high, the complex is more appropriately described as a metallacyclopropane complex.

Within the framework of the concept of resonance, an η^2-olefin complex **2** is described by two resonance formulas **2a** and **2b**, which represent a complex with a neutral π-olefin ligand or an olefin dianion, respectively. This situation must be considered when assigning the oxidation number of the metal. Thus, the Pt^{II} complex K[PtCl$_3$(η^2-H$_2$C=CH$_2$)]·H$_2$O (Zeise's salt) is the classic example of a π-ethene complex (C–C 1.375 Å, $\alpha = 16°$; compare to C–C in uncoordinated ethene, 1.339 Å). In contrast, the C–C bond length in [Os(η^2-H$_2$C=CH$_2$)(CO)$_4$], is 1.49 Å; this is almost as long as in cyclopropane (1.512 Å), so it must be described as an Os^{II} complex.

Similarly, η^2-alkyne complexes **3** should be described as π-alkyne **3a** or metallacyclopropene **3b** complexes with a neutral or (formally) dianionic alkyne ligand. The comparison of C≡C bond lengths in [Pt(C$_6$F$_5$)$_2$(η^2-PhC≡CPh)$_2$] (C–C 1.20 Å) and [WCl$_2$(η^2-PhC≡CPh)(PMe$_3$)$_3$] (C–C 1.33 Å) with those in diphenylacetylene (1.21 Å) and 1,2-diphenylcyclopropene ($\approx$1.34 Å) shows that the description as a (π-diphenylacetylene)platinum(II) complex or a tungsta(IV)-cyclopropene complex is appropriate [24].

Examples

- The Vaska complex **4** is amenable to numerous oxidative addition reactions. H_2 is oxidatively added via an η^2-dihydrogen complex as an intermediate (concerted mechanism; **4** → **5** → **6**). The oxidative addition of methyl halides is completed in an S_N2 reaction (**4** → **7** → **8**). In this case, via **7** as intermediate, complex **8** is formed, and the two newly added ligands are *trans* coordinated [T4].

```
                                    H   H
                                     \ /
                     + H₂        Ph₃P—Ir—PPh₃              Ph₃P—Ir—PPh₃
                    ─────→          |    |         ───→      |    |
     Ph₃P—Ir—PPh₃               Cl   CO                      Cl   CO
        |  CO
        Cl                     5 (Ir¹; 18 ve)              6 (Irᴵᴵᴵ; 18 ve)

     4 (Ir¹; 16 ve)   + MeX            Me  CO  ⁺                 Me  CO
                      ─────→        Ph₃P—Ir—PPh₃  X⁻   ───→   Ph₃P—Ir—PPh₃
                                       |                          |
                                       Cl                         Cl  X

                                    7 (Irᴵᴵᴵ; 16 ve)            8 (Irᴵᴵᴵ; 18 ve)
```

- Oxidative addition of alkyl halides RX to [Pt(PPh₃)₃] takes place according to a radical mechanism (L = PPh₃):

$$[Pt^0L_3] \underset{+L}{\overset{-L}{\rightleftharpoons}} [Pt^0L_2] \xrightarrow[\text{(slow)}]{+RX} [Pt^IXL_2] + R\cdot \longrightarrow [Pt^{II}R(X)L_2]$$

The reaction occurs with homolytic cleavage of the R–X bond, which is initiated via the transfer of an electron (SET – single electron transfer) from Pt^0 into the σ^*-C–X orbital. With the transfer of X^-, the oxidation number and the coordination number of M are each increased by one. The radical pair thus formed recombines very rapidly, which raises the oxidation number and coordination number of M by an additional unit. Particularly, when the starting complex only permits the oxidation number and coordination number of M to be increased by one each, the result is a bimolecular oxidative addition, as is demonstrated by the following example:

$$[Co^{II}(CN)_5]^{3-} + RX \longrightarrow [Co^{III}X(CN)_5]^{3-} + R\cdot$$

$$[Co^{II}(CN)_5]^{3-} + R\cdot \longrightarrow [Co^{III}R(CN)_5]^{3-}$$

$$\overline{2\,[Co^{II}(CN)_5]^{3-} + RX \longrightarrow [Co^{III}X(CN)_5]^{3-} + [Co^{III}R(CN)_5]^{3-}}$$

Such reactions are also designated one-electron oxidative additions.

- The dinuclear platina-β-diketone[2] **9** reacts with donors L⌒L after cleaving the Pt–Cl–Pt bridges (**9** → **10**) under oxidative addition to acetylhydridoplatinum(IV)

---

[2] If the methine group =CH– in the enol form of a β-diketone is replaced by a metal complex fragment, a metalla-β-diketone is obtained. It should be regarded as a hydroxycarbene complex stabilized by an intramolecular hydrogen bond to an acyl ligand.

complexes (**10** → **11**), from which, in a reductive C–H elimination, acetaldehyde is cleaved off (**11** → **12**). With P⌒P donors such as dppe, reaction **9** → ... → **12** runs even at room temperature, while with N⌒N donors such as bpy, acetyl–hydrido complexes **11** are formed, which are not subject to reductive elimination to **12** in the solid state except above 140 °C [32].

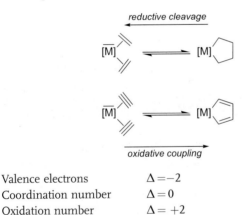

3.3
Oxidative Coupling and Reductive Cleavage

Reaction Principle

Valence electrons	$\Delta = -2$
Coordination number	$\Delta = 0$
Oxidation number	$\Delta = +2$

In oxidative coupling reactions, alkenes or alkynes form π complexes, which are converted to metallacycles with C–C bond linkage. In the reaction, the oxidation number of the metal increases by two. The reverse reaction is called reductive cleavage (of a C–C bond) or reductive decoupling.

Alkynes undergo oxidative coupling reactions more easily than alkenes. Other unsaturated substrates such as heteroolefins and heteroalkynes are also amenable to oxidative coupling reactions.

The oxidative coupling and reductive cleavage reactions described here are, in principle, cycloaddition or cycloreversion reactions, in which metallacycles are formed or cleaved, respectively. Related to these are [2 + 2] cycloadditions, in which a carbene-olefin complex is converted to a metallacyclobutane complex.

If the carbene ligand is counted as a neutral 2e donor, these reactions raise the oxidation number (ON) of M by two ($\Delta ON = +2$; $\Delta ve = -2$). Similarly, carbyne complexes and alkynes can react and form metallacyclobutadiene complexes.

Examples

- Equilibrium between a bis(ethene)nickel complex and a nickelacyclopentane complex, which undergoes a reductive C–C elimination reaction after cleaving off L (L = PPh$_3$) [33].

- Formation of an iridacyclopentene complex from a bis(ethene)iridium(I) complex via ligand substitution (**13** → **14**) and oxidative coupling (**14** → **15**). The reaction forming the iridacyclopentadiene complex (**15** → **16**) only runs when **15** contains tetrahydrofuran as a weakly binding ligand [34].

- Formation of an iridacyclopentadiene complex (L = PPh$_3$).

- Tebbe reagent **17** [35] is converted with olefins, in the presence of pyridine, via the carbene complex **18** (not isolated), to the stable titanacyclobutane complex **19** [36].

$$\text{Cp}_2\text{Ti}\overset{\overset{H_2}{\underset{}{C}}}{\underset{Cl}{\diagdown}}\text{AlMe}_2 \quad \xrightarrow[-\text{AlMe}_2\text{Cl(py)}]{+\text{py}} \quad \{\text{Cp}_2\text{Ti}=\text{CH}_2\} \quad \xrightarrow{+\text{H}_2\text{C}=\text{CH}t\text{-Bu}} \quad \text{Cp}_2\text{Ti}\overset{\overset{H_2}{\underset{H_2}{C}}}{\underset{C}{\diagdown}}\text{CH}t\text{-Bu}$$

    17   **18**   **19**

3.4 Olefin Insertion and β-Hydrogen Elimination

Reaction Principle

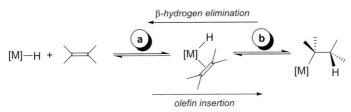

Reaction	a	b
Valence electrons	$\Delta = +2$	$\Delta = -2$
Coordination number	$\Delta = +1$	$\Delta = -1$
Oxidation number	$\Delta = 0$	$\Delta = 0$

Insertion of alkenes into M–H bonds (1,2–insertion[3]) and the reverse reaction, β-hydrogen elimination, are key steps in organometallic catalysis. They occur particularly easily with transition metals. Insertion involves coordination of an olefin to form an intermediary hydrido–olefin complex (reaction a) and the actual insertion reaction (reaction b). In the insertion step b, the number of valence electrons of the metal decreases by two and the coordination number decreases by one. The oxidation number of M does not change. For the reverse reactions, the β-hydrogen eliminations, the opposite is true. In general, from a stereochemical point of view, the reactions run uniformly as *cis* insertions (i.e., *syn* additions of [M] and H to olefins) and *cis* β-hydrogen eliminations (i.e., *syn* eliminations of [M] and H from metal alkyl compounds).

---

[3] In complexes of the type [M]–X (here: X = H), the term "1,n-insertion" refers, without any mechanistic implications, to reactions of the type [M]–X + Y → [M]–Y–X (Y = atom or group of atoms); n gives the number of atoms between M and X in the product. Elimination reactions are designated similarly; instead of the number, the Greek letters (1,1 = α; 1,2 = β; ...) frequently appear.

Agostic C_β–H $\cdots$ M interactions can be a step for initiating β-hydride eliminations. The transition state is cyclic and displays a coplanar M–C–C–H arrangement. If this cannot be achieved readily (which can be the case in cyclic systems), insertion and β-hydrogen elimination are hindered. This can be the reason for the stability of selected hydrido–olefin complexes and alkyl complexes with β-hydrogen atoms. In addition to stereoelectronic factors, the reaction course depends on many other factors. Hence, the stability of a hydrido–olefin complex can also be traced to an especially stable metal–olefin bond. Furthermore, β-hydrogen eliminations are generally hampered when the complex is coordinatively saturated.

Exercise 3.1
Although **1** and **2** are two isomeric *cis* hydrido–olefin complexes (L = phosphane), only **1** readily undergoes an insertion of the olefin into the Ir–H bond, which results in a very different thermal stability (T_{dec}: > −80 °C, **1**; > 20 °C, **2**). Explain this situation.

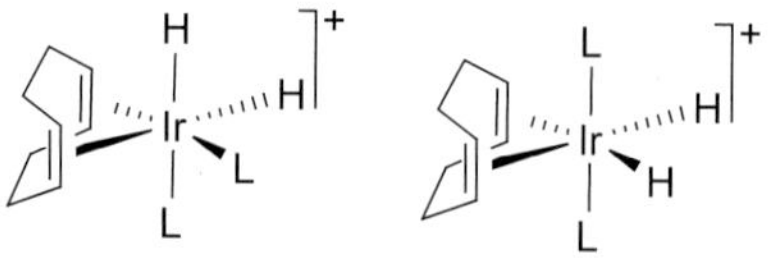

Olefins are also inserted easily into M–C bonds (**20** → **21** → **22**). The reverse reaction (**22** → **21**), a β-alkyl elimination, does not occur as frequently. If the β-C atom bears a hydrogen atom, a β-hydrogen elimination is favored (**22** → **23**).

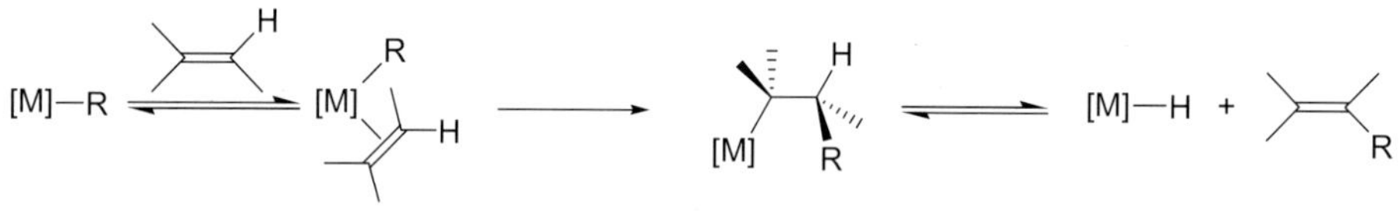

Insertion reactions of alkynes into M–H and M–C bonds proceed analogously, generally as *syn* additions:

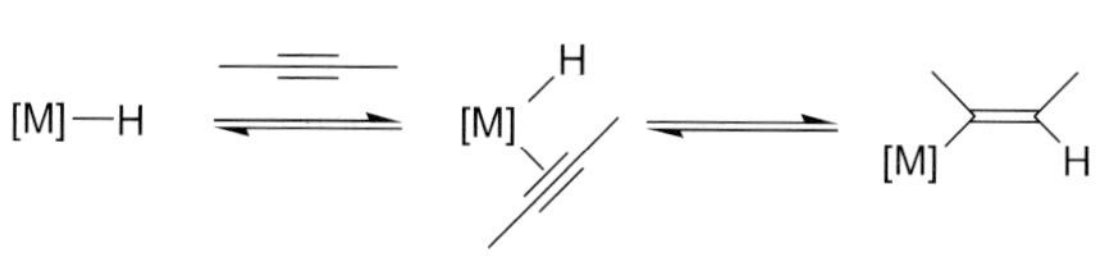

With respect to net transformation (addition of M–H or M–C to a double or triple bond), such insertion reactions are also called hydrometallation or carbometallation of olefins or alkynes. In addition to the insertion and elimination reactions discussed here, there are many other analogous reactions significant in homogeneous catalysis. Among these are insertion reactions with participation of heteroolefins (e.g., $R_2C=NR$) and in bonds other than M–C or M–H (e.g., M–OR, M–NR$_2$, or M–SiR$_3$ bonds).

3.4 Olefin Insertion and β-Hydrogen Elimination

Examples

- Formation of a cationic ethylplatinum complex via olefin coordination (24 → 25, s = solvent) and olefin insertion into a Pt–H bond (25 → 26) [37]. Single-crystal X-ray diffraction analysis of $[Pt(C_2H_5)\{(t\text{-}Bu)_2P(CH_2)_3P(t\text{-}Bu)_2\}][CB_{11}H_{12}]$, a complex of type 26 without coordinated solvent s, revealed a β-agostic M–C–C–H interaction.

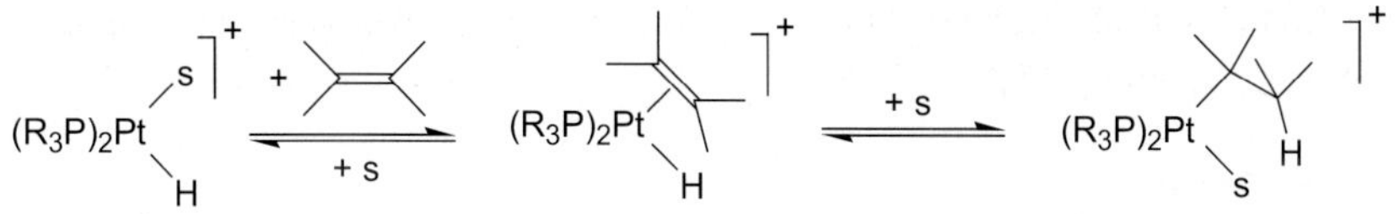

- Hydrozirconation of alkynes with Schwartz reagent [35] (28 → 29) is a *syn* addition. The following reaction with electrophiles E yields (E)-olefins (29 → 30) [38].

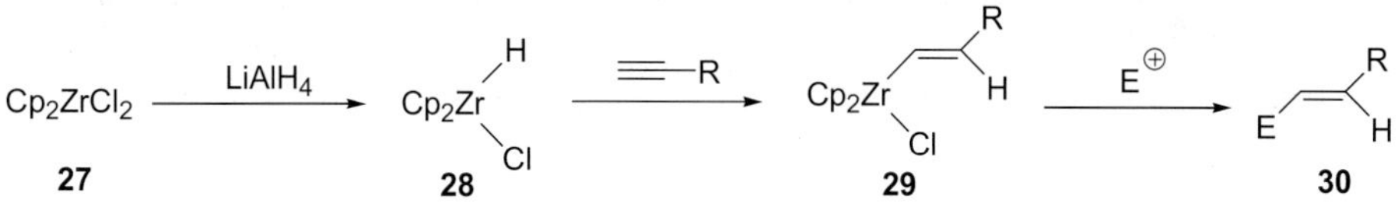

- Hydroalumination [39] (31 → 32) and carboalumination (32 → 33), which can be effectively Zr-catalyzed [40], are the key steps in Ziegler's Aufbau reaction. In this case, as in general, hydrometallation runs more easily than carbometallation.

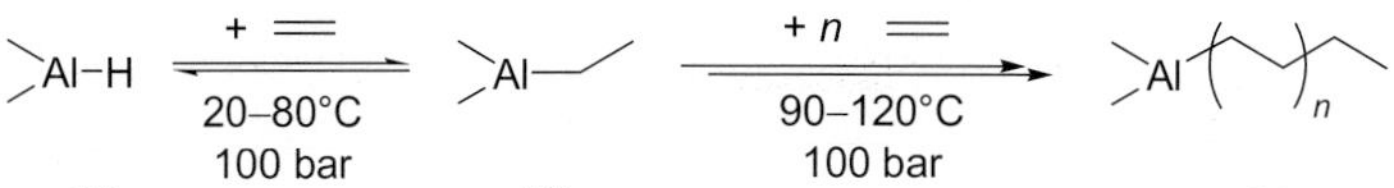

Exercise 3.2

In dialkyl complexes, a β-hydrogen elimination is frequently coupled with a reductive elimination. Formulate this reaction sequence for 1 (L = PR$_3$) and note that a vacant coordination site must first be generated by cleaving off L. Give the product ratio. Which products would you expect for a radical Pt–C bond cleavage?

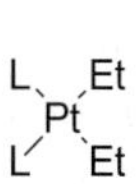

3.5
α-Hydrogen Elimination and Carbene Insertion Reactions

Reaction Principle

$$[M]-\underset{R}{\overset{H}{\underset{|}{C}}}- \;\rightleftharpoons\; [M]\overset{H}{\underset{C-}{\diagdown}} \;\rightleftharpoons\; [M]\overset{H}{\underset{C-}{=}}$$

← carbene insertion
→ α-hydrogen elimination

Valence electrons	$\Delta = +2$
Coordination number	$\Delta = +1$
Oxidation number	$\Delta = 0$

Transferring an α-hydrogen atom of an alkyl ligand to the central atom leads to the formation of a carbene–hydrido complex. If the carbene ligand is considered a neutral 2e donor, the oxidation number of M remains unchanged. The reverse reaction, an H transfer from the metal to the ligand, formally represents an insertion of a carbene into an M–H bond (1,1-insertion).

Agostic C_α–H $\cdots$ M interactions are the initial step of α-hydride elimination. Thus, high acceptor capability of the metal atom *and* donor capability for stabilizing the carbene ligand to be formed offer good conditions for a ready course of reaction. Steric overload on the metal due to bulky ligands can favor the hydride elimination.

For dialkylmetal complexes, α-hydride elimination is encountered particularly frequently, coupled with reductive C–H elimination (**34** → **35**).

$$[M]\diagdown_R^H \longrightarrow [M]=C\diagdown + RH \qquad [M]\diagdown_R^{C-H} \longrightarrow [M]\equiv C- + RH$$

34   **35**   **36**   **37**

This process does not always pass through an intermediate alkyl–carbene–hydrido complex; the α-hydrogen atom of an alkyl ligand can also be transferred directly onto the α-C atom of the other alkyl ligand (R). Analogously, carbyne complexes (**36** → **37**) can be generated from alkyl–carbene complexes via α-hydride elimination coupled with a reductive C–H elimination.

Examples
- The cationic hydrido(phosphorus ylide)tungsten(IV) complex **38** isomerizes thermally with formation of **41**. Quantum-chemical calculations suggest a hydrido(methylene)- (**39**) and a methyltungsten complex (**40**) as intermediates [41].

3.5 α-Hydrogen Elimination and Carbene Insertion Reactions

- Although α-hydride elimination normally runs slower than β-hydride elimination, the ethylbis(neopentyl)tantalum(V) complex **42** reacts with PMe$_3$ (L) with both α- and β-hydride elimination occurring. These are coupled with reductive elimination of neopentane, resulting in formation of, respectively, a neopentylidenetantalum complex **43** or an (η^2-ethene)tantalum complex **44**. The two tautomeric complexes **43** and **44** exist in equilibrium in solution [42].

- Tris(neopentyl)(neopentylidyne)tungsten (**45**) reacts with dmpe (Me$_2$PCH$_2$CH$_2$PMe$_2$) via α-hydride elimination and reductive elimination to form a neopentyl(neopentylidene)(neopentylidyne)tungsten(VI) complex **46** [43].

A single-crystal X-ray diffraction analysis shows **46** to be a tetragonal-pyramidal complex with bond lengths (a) W–CH$_2$CMe$_3$ (2.258(9) Å) > (b) W=CHCMe$_3$ (1.942(9) Å) > (c) W≡CCMe$_3$ (1.785(8) Å), which undoubtedly confirms an M–C multiple-bond character for **b** and **c**.

3.6
Addition of Nucleophiles and Heterolytic Fragmentation

Reaction Principle

Reaction	a	b
Valence electrons	$\Delta = +2$	$\Delta = 0$
Coordination number	$\Delta = +1$	$\Delta = 0$
Oxidation number	$\Delta = 0$	$\Delta = 0$

By coordinating an olefin to a metal (reaction **a**), the olefin can be activated so as to react with nucleophiles Nu in an intermolecular reaction to form 2-functionalized alkyl complexes (reaction **b**). In the actual addition step **b**, neither metal's number of valence electrons, nor its coordination number, nor its oxidation number changes. If the reaction is performed with a neutral nucleophile (Nu = NR_3, PR_3, SR_2, ...; NuH = NH_3, H_2O, R_2NH, RSH, ...), a cationic heteroatomic center forms. The center can then be easily deprotonated if the nucleophile is proton-acidic (NuH). Occasionally, these reactions are also called π–σ rearrangements since the π-bonded olefin ligand is transferred to a σ-bonded alkyl ligand. This designation is misleading, however, since these reactions are not isomerizations.

The reverse reactions should be classified as heterolytic fragmentation. If the cleavage of the nucleophile is induced by an electrophile (e.g., $E^+ = H^+$), they can also be called electrophilic abstraction (reaction **c**).

If the insertion of an olefin into an M–Nu bond (**d**) is compared with an intermolecular addition of nucleophiles (dealt with here) to a coordinated olefin (reaction **e**), the same products result, but the stereochemistry is different: Intermolecular additions of nucleophiles take place as *anti* additions and intramolecular insertion reactions as *syn* additions. In catalysis, the substitution of the olefin by the nucleophile (**f**) is generally undesired.

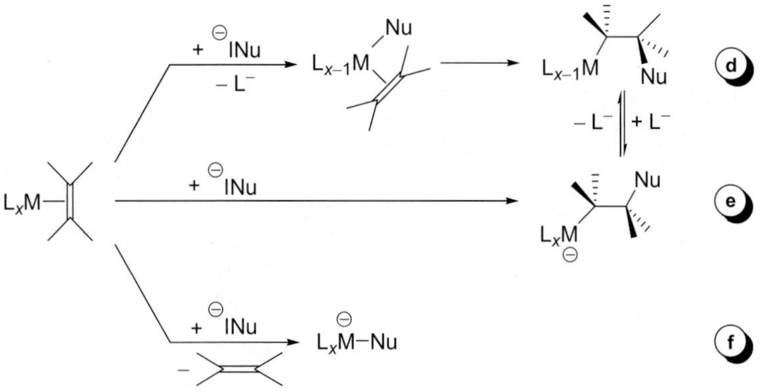

3.6 Addition of Nucleophiles and Heterolytic Fragmentation

In the same way that they can be added to olefins, nucleophiles can be added to alkynes. This forms 2-functionalized vinyl compounds that generally show a *trans* arrangement of M and Nu, as expected for an *anti* addition:

Background: Heterolytic Fragmentation (Grob Fragmentation)

Heterolytic fragmentations are 1,2-eliminations that run according to the following reaction scheme:

$$a\text{-}b\text{-}c\text{-}d\text{-}X \longrightarrow a\text{-}b^{+} + c{=}d + X^{-}$$

$a\text{-}b^{+}$ and X^{-} are called, respectively, electrofugal and nucleofugal groups. In contrast to normal 1,2-elimination ($H\text{-}c\text{-}d\text{-}X \rightarrow H^{+} + c{=}d + X^{-}$), where H^{+} is the electrofuge, the electrofuge here is a multiatomic group [10, 44, 45].

Examples
- Fragmentation of 3-hydroxypropyl tosylates:

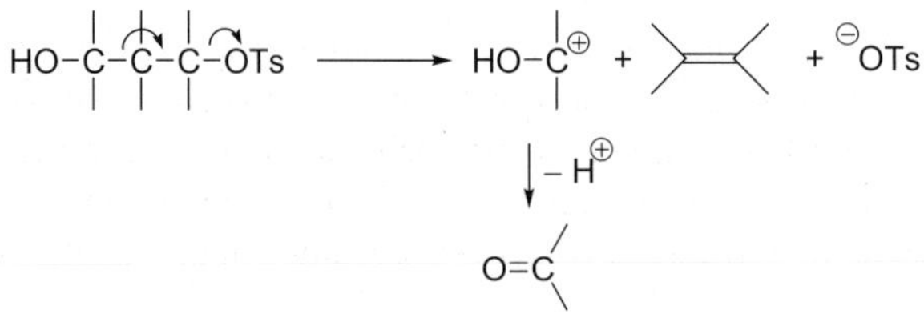

- Fragmentation of (2-ammonioethyl)tin compounds:

$$R_3Sn-CH_2-CH_2-\bar{N}R'_2 \xrightarrow{+H^{\oplus}} R_3Sn-CH_2-CH_2-\overset{\oplus}{N}HR'_2 \xrightarrow{T>100°C} R_3Sn^{\oplus} + H_2C=CH_2 + \bar{H}NR'_2$$

Examples
- The addition of a neutral nucleophile such as PPh_3 to a coordinated olefin leads to a 2-functionalized alkyl complex with a cationic heteroatomic center:

$$Cp(CO)_3W-\|\overset{+}{\|} \xrightarrow{+PPh_3} Cp(CO)_3W-CH_2-CH_2-\overset{\oplus}{P}Ph_3 \overset{+}{]}$$

- 2-Aminoethyl Grignard compounds **48** decompose with heterolytic fragmentation (**48** → **49**) at temperatures between −20 and −80 °C, depending on the steric and electronic properties of R [46].

$$R_2N-CH_2-CH_2-Br \xrightarrow[-78 \ldots -100\,°C]{+Mg} R_2N-CH_2-CH_2-MgBr \xrightarrow{-20 \ldots -80\,°C} Mg(NR_2)Br + H_2C=CH_2$$
$$\quad\;\;\, \mathbf{47} \qquad\qquad\qquad\qquad\qquad \mathbf{48} \qquad\qquad\qquad\qquad\qquad \mathbf{49}$$

The proven activation of magnesium in Grignard formation reactions with 1,2-dibromoethane is likewise caused by heterolytic fragmentation of the $BrCH_2CH_2MgBr$ formed as an intermediate.

- β-Haloalkyl esters and β-haloalkyl urethanes have found their way into peptide chemistry as protecting groups that can easily be cleaved off via heterolytic fragmentation after treatment with cobalt(I) phthalocyanine [47].

Ⓟ = peptide chain, X = halogen, H_2pc = phthalocyanine

Exercise 3.3
Which reaction do you expect when the cationic etheneiron(II) complex $[FeCp(CO)_2(\eta^2-H_2C=CH_2)]^+$ reacts with an excess of methylamine? How will the complex formed react with HCl?

3.7
Insertion and Extrusion of CO

Reaction Principle

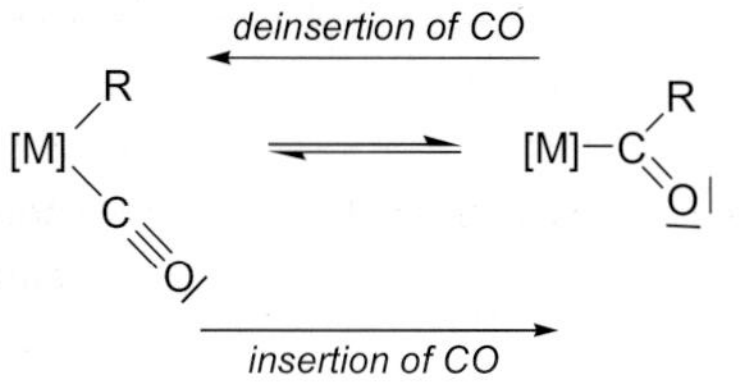

$$[M]\!-\!\!\overset{R}{\underset{C\equiv O}{|}} \quad \underset{\text{insertion of CO}}{\overset{\text{deinsertion of CO}}{\rightleftarrows}} \quad [M]\!-\!\!C\!\!\overset{R}{\underset{O|}{\diagdown}}$$

Valence electrons	$\Delta = -2$
Coordination number	$\Delta = -1$
Oxidation number	$\Delta = 0$

The insertion of CO into an M–C bond leads to an acyl complex. Such reactions proceed as a migration of the alkyl ligand to the carbon atom of a *cis* carbonyl ligand (migratory insertion; 1,1-insertion). The reverse reaction is called extrusion (deinsertion, elimination) of CO. The oxidation number of M remains unchanged by the CO insertion. The acyl complex primarily formed is electronically and coordinatively unsaturated. If the work is performed in an atmosphere of carbon monoxide, the now vacant coordination site is again occupied by CO.

Intermolecular addition of R^- to carbonyl complexes (**50** → **51**) is also well known. Protonation of the intermediate leads to hydroxycarbene complexes (**51** → **52**) (E. O. Fischer, 1964).

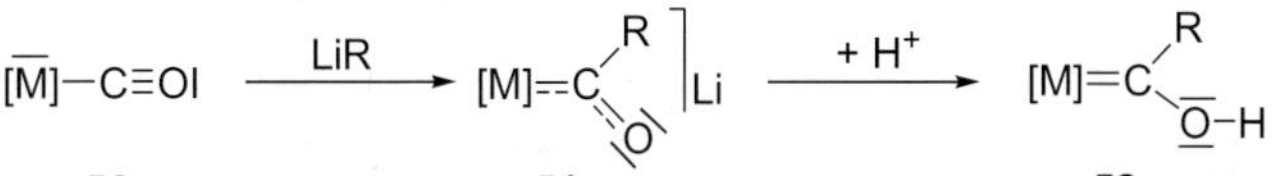

$$\overline{[M]}\!-\!C\!\equiv\!O| \quad \xrightarrow{\text{LiR}} \quad [M]\!=\!C\!\overset{R}{\underset{|O|}{\diagdown}}\,\text{Li} \quad \xrightarrow{+\,H^+} \quad [M]\!=\!C\!\overset{R}{\underset{O-H}{\diagdown}}$$

50   **51**   **52**

Although (anionic) acyl complexes (**51**) are also formed as intermediates, these reactions must not be classified as migratory insertion reactions. Rather, these reactions are intermolecular additions of a nucleophile to the (electrophilic) carbonyl C atom. The Hieber base reaction runs in a similar way:

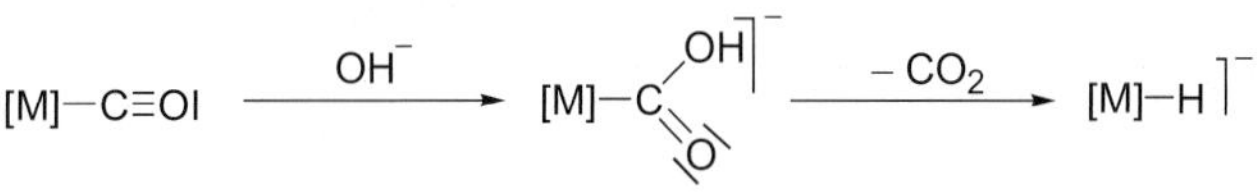

$$[M]\!-\!C\!\equiv\!O| \quad \xrightarrow{OH^-} \quad [M]\!-\!C\!\overset{OH}{\underset{|O|}{\diagdown}} \quad \xrightarrow{-\,CO_2} \quad [M]\!-\!H$$

Example
Evidence that the following reaction actually occurs as a migratory insertion was found via isotope labeling.

3 Elementary Steps in Organometallic Catalysis

$(CO)_4Mn-Me$ (C≡O) [18 ve] ⇌ $(CO)_4Mn$(–C(=O)–Me) [16 ve intermediate] ⇌ (+CO) $(CO)_4Mn$(CO)(–C(=O)–Me) [18 ve]

Exercise 3.4
Prove through suitable ^{13}C labeling for the previously shown reaction that (a) it runs intramolecularly and (b) a methyl migration takes place.

3.8
One-Electron Reduction and Oxidation

Reaction Principle

$$[M] + e^- \underset{\text{reduction}}{\overset{\text{oxidation}}{\rightleftharpoons}} [M]^{\ominus}$$

Valence electrons	$\Delta = +1$
Coordination number	$\Delta = 0$
Oxidation number	$\Delta = -1$

In a one-electron reduction or oxidation where the ligand sphere is retained, the reduced and oxidized complex differ only by one electron. If the orbital involved is a metal-centered molecular orbital, the oxidation number of the metal decreases or increases by one. Otherwise, reduction or oxidation of the ligand occurs.

Examples
- *Electron-variable complexes.* Structurally similar metal complexes that differ *only* in their number of electrons are called electron-variable. Electron variability occurs, for example, in phthalocyanine–metal complexes. The complexes **53a–f** (H_2pc = phthalocyanine) with iron as the central atom have been isolated. Starting from **53a**, the stepwise electron gain leads first to reduction of the central atom (**53b–d**) and then to reduction of the ligand (**53e, 53f**). Within the framework of LCAO-MO theory, the additional electrons are each accommodated in MO's for **53b–d**, which predominantly have metallic character (3d

orbitals of iron). For **53e/53f**, in contrast, electrons are accommodated in an MO that extends substantially over the ligand, and thus is "ligand-centered" [48].

	[FeBr(pc)]	[Fe(pc)]	Li[Fe(pc)]	Li$_2$[Fe(pc)]	Li$_3$[Fe(pc)]	Li$_4$[Fe(pc)][a]
ON(Fe)/ligand:	+3/pc^{2-}	+2/pc^{2-}	+1/pc^{2-}	0/pc^{2-}	0/pc^{3-}	0/pc^{4-}
	53a	53b	53c	53d	53e	53f

a) The anionic complexes crystallize as THF solvates.

- The ligand substitution $54 + L \rightarrow 58 + CO$, which does not readily occur for kinetically inert metal carbonyl complexes, can be induced by reduction ($54 \rightarrow 55$). For the kinetically labile 19-*ve* intermediate **55**, a ligand substitution is easily possible ($55 \rightarrow 56 \rightarrow 57$). If **57** can reduce complex **54**, catalytic quantities of reductant are sufficient. Otherwise, stoichiometric amounts would be necessary.

$$[M(CO)_6] \xrightarrow{+e} [M(CO)_6]^{\ominus\cdot} \xrightarrow[-CO]{} [M(CO)_5]^{\ominus\cdot} \xrightarrow{+L} [M(CO)_5L]^{\ominus\cdot} \longrightarrow [M(CO)_5L] + e$$

54 (18 ve)   **55** (19 ve)   **56** (17 ve)   **57** (19 ve)   **58** (18 ve)

4
Hydrogenation of Olefins

4.1
Introduction

The addition of dihydrogen H_2 to olefins (**1** → **3**) is a strongly exergonic reaction ($\Delta G^\ominus = -101$ kJ/mol for ethene), but the synchronous addition via a four-membered cyclic transition state **2** is symmetry-forbidden.

$$\underset{\mathbf{1}}{\diagup\!\!\!\!\!\diagdown\!\!\text{C=C}\!\!\diagdown\!\!\!\!\!\diagup} + H_2 \longrightarrow \underset{\mathbf{2}}{\left[\diagup\!\!\!\!\!\diagdown\!\!\text{C}\overset{H\text{-}\text{-}H}{\underset{\vdots\;\;\vdots}{=}}\text{C}\!\!\diagdown\!\!\!\!\!\diagup\right]^{\ddagger}} \longrightarrow \underset{\mathbf{3}}{-\overset{H}{\underset{|}{\text{C}}}-\overset{H}{\underset{|}{\text{C}}}-}$$

Figure 4.1 schematically represents the orbitals of H_2 and C_2H_4 that determine their reactivity. In addition, the two possible HOMO–LUMO interactions for olefin hydrogenation with a cyclic transition state without catalyst are shown. In both cases, due to symmetry considerations, they result in an overlap integral $S = 0$. Thus, no bond formation can take place in this way.

The method of choice is to carry out a catalytic reaction. Heterogeneous metal catalysts (e.g., Ni) for olefin hydrogenation have long been known (P. Sabatier, University of Toulouse; Nobel Prize 1912). In the mid-1960s, G. Wilkinson (Imperial College, London; Nobel Prize 1973) discovered that [RhCl(PPh$_3$)$_3$] at room temperature and normal pressure is a homogeneous catalyst for olefin hydrogenation. It is now called the Wilkinson catalyst.

Exercise 4.1
Can H_2 be added to an olefin in a radical chain reaction under standard conditions? Base your analysis on the following mean bond dissociation enthalpies (in kJ/mol): H–H 436, C–C 348, C=C 612, C–H 412.

Fundamentals of Organometallic Catalysis. First Edition. Dirk Steinborn
Copyright © 2012 WILEY-VCH Verlag GmbH & Co. KGaA, Weinheim

4 Hydrogenation of Olefins

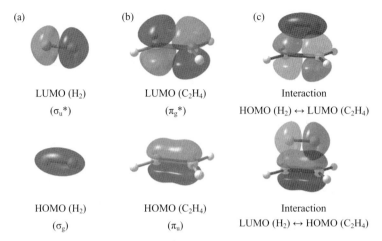

Figure 4.1 Schematic representation of the reactivity-determining orbitals of (a) dihydrogen and (b) ethene, and (c) the HOMO–LUMO interactions associated with the synchronous addition of H_2 to C_2H_4.

4.2
The Wilkinson Catalyst

4.2.1
Principles

The Wilkinson complex 4 is easy to synthesize from $[RhCl_3(H_2O)_3]$ and triphenylphosphane in ethanol.

$$[RhCl_3(H_2O)_3] + 4\ PPh_3 \xrightarrow{\text{EtOH}} \underset{4}{[RhCl(PPh_3)_3]} + Ph_3PO + 2\ HCl + 2\ H_2O$$

It is a precatalyst for hydrogenation of olefins. The catalytically active species is the 14-*ve* complex $[RhCl(PPh_3)_2]$ (5), which dimerizes very easily into complex 6.

$$\underset{4}{[RhCl(PPh_3)_3]} \underset{+\ PPh_3}{\overset{-\ PPh_3}{\rightleftharpoons}} \underset{5}{[RhCl(PPh_3)_2]} \rightleftharpoons \underset{6}{1/2\ [(Ph_3P)_2Rh\overset{Cl}{\underset{Cl}{\diamond}}Rh(PPh_3)_2]}$$

Hydrogenation with the Wilkinson catalyst is usually carried out at room temperature in a hydrogen atmosphere under normal pressure. Aromatics, alcohols, acetone, or ethers can be used as solvents. The hydrogenation rate depends heavily on the olefin (R = alkyl):

Terminal olefins react faster than inner olefins. Olefins with very bulky substituents can only be hydrogenated very slowly, and ethene itself not at all, probably because they are coordinated too weakly or too tightly (see below). Functional groups such as Ph, COOR, CONR$_2$, CN, OR are tolerated, but not CHO or COCl groups, which are decarbonylated.

The activity of the Wilkinson catalyst can be increased by suitable structural variation. With the more strongly basic phosphanes P(C$_6$H$_4$-p-X)$_3$ (X = Me, OMe) as coligands, the activity more than doubles, while with very strongly basic alkylphosphanes, the hydrogenation activity disappears entirely. Apparently, a moderate increase in basicity (in comparison with PPh$_3$) facilitates the insertion reaction, while phosphane ligands that are too strongly basic can no longer be sufficiently easily cleaved off (see below).

Cationic complexes [RhL$_2$s$_2$]$^+$ (9; L$_2$ = chelating phosphane ligand such as Ph$_2$P-(CH$_2$)$_n$PPh$_2$, n = 2, 3; s = solvent such as MeOH) possess a hydrogenation activity higher by up to a factor of 100 than that of the Wilkinson complex. They are easily accessible from cationic norbornadiene complexes 8 (or analogous COD complexes) via hydrogenation of the diene according to the following scheme.

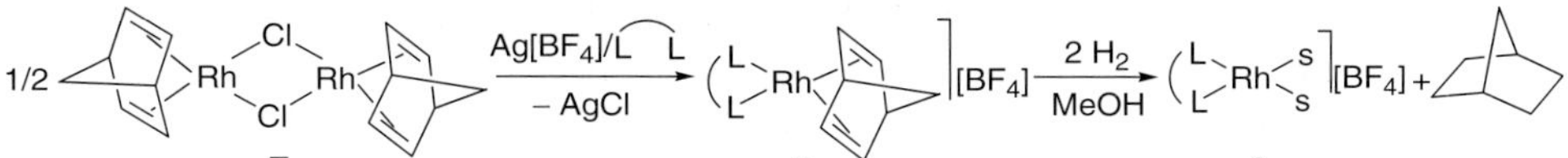

There is a broad palette of other hydrogenation catalysts of late transition metals (e.g., [RuH(Cl)(PPh$_3$)$_3$], [CoH(CN)$_5$]$^{3-}$) and of early transition metals and lanthanoids (e.g., [{LnH(η^5-C$_5$Me$_5$)$_2$}$_2$], Ln = La, Nd, Lu, ...). Just like olefins, other unsaturated compounds such as alkynes, dienes and arenes can be hydrogenated.

4.2.2
Mechanism of Olefin Hydrogenation

Olefin hydrogenation with the Wilkinson complex proceeds via the "dissociative hydride mechanism" represented in Figure 4.2 (shaded gray). This designation makes it clear that first a triphenylphosphane ligand is cleaved off and then hydrogen is attached. The following reaction steps are to be examined in detail:

4 → 5: *Ligand cleavage/attachment.* From the precatalyst [RhCl(PPh$_3$)$_3$] (4), the catalytically active species [RhCl(PPh$_3$)$_2$] (5) is formed by cleaving off a PPh$_3$ ligand. In benzene, the equilibrium is virtually entirely on the side of 4 ($K < 10^{-4}$ mol/l).

5 → 10: *Oxidative addition/reductive elimination.* Oxidative addition of H$_2$ yields a coordinatively unsaturated *cis*-dihydridorhodium(III) complex 10. The reaction is reversible; the reverse reaction is a reductive elimination of H$_2$.

10 → 11: *Ligand attachment/cleavage.* Coordination of the olefin leads to the dihydrido(olefin)rhodium(III) complex 11, which is saturated electronically (18 *ve*) and coordinatively (CN = 6).

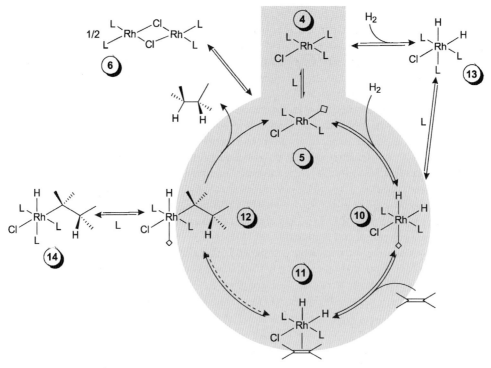

Figure 4.2 Hydride mechanism of olefin hydrogenation with the Wilkinson complex (L = PPh$_3$). The fundamental dissociative mechanism is shaded gray. The small squares indicate "vacant" coordination sites occupied by solvent molecules, at least in polar solvents (see Section 3.1).

11 → 12: *Insertion/β-H elimination*. Insertion of the coordinated olefin into the cis- Rh–H bond yields an ethyl(hydrido)rhodium(III) complex **12**. In the case of hydrogenation of cyclohexene, this is the rate-determining step. In principle, the reaction is reversible. However, the reverse reaction, a β-H elimination, is insignificant under conventional reaction conditions. Nor does double-bond isomerization occur as a side reaction. The insertion primarily forms a *trans* alkyl–hydrido complex, which then isomerizes to the *cis* alkyl–hydrido complex **12**.
12 → 5: *Reductive elimination*. In a reductive C–H elimination the alkane is cleaved off, where the starting complex **5** is re-formed. This reaction is irreversible; the reverse reaction – oxidative addition of an unactivated C–H bond – does not occur.

Exercise 4.2
Regarding the insertion step **11 → 12**, explain the statements that (a) its irreversibility allows us to conclude that no double-bond isomerization occurs, and that (b) a *trans* alkyl–hydrido complex is first formed.

Several side reactions must now be noted. In principle, the coordinatively unsaturated complexes are in equilibrium with the corresponding coordinatively saturated, catalytically inactive complexes (**10** ⇌ **13**, **12** ⇌ **14**). In addition, **5** occurs in equilibrium with the chloro-bridged dimer **6** ($K \geq 10^6$ l/mol). The forming of complexes **6**, **13**, and **14** reduces the concentration of the catalytically active complexes, so the activity of the catalyst is decreased.

The equilibrium constant of the coordinatively unsaturated dihydridorhodium(III) complex **13** with the dihydrido–olefin complex **11** (**13** ⇌ **11**) can be determined to be about 10^{-4}. Complex **13** can also be formed directly from the precatalyst [RhCl(PPh$_3$)$_3$] (**4**) by oxidative addition of H$_2$ ("associative hydride mechanism"). Kinetic studies show, however, that the coordinatively unsaturated complex **5** reacts at least 10^4 times faster with H$_2$ than with the precatalyst **4** (**5** → **10** vs. **4** → **13**). Thus, the coordinatively unsaturated complex [RhCl(PPh$_3$)$_2$] (**5**) determines the reaction mechanism and also the reaction rate, although it only exists in low concentration [49].

Quantum-chemical calculations have been used to determine the energies of intermediate complexes and transition states for a model system in the gas phase (Figure 4.3). The formation of the dihydrido complex (**5** → **10**) occurs via a dihydrogen complex and is not associated with any substantial activation barrier. The same applies to the subsequent coordination of ethene (**10** → **11**). Insertion and subsequent isomerization of the *trans* into a *cis* alkyl–hydrido complex (represented as step **11** → **12** in Figure 4.3) is exothermic and rate-determining (E_A about 85 kJ/mol). The subsequent reductive elimination (**12** → **5**) is nearly thermoneutral. Comparison of the calculated activation barriers (**12** → **5** vs. **12** → **11**) shows that the olefin insertion is irreversible. Although these quantum-chemical calculations are simpler than the newer options available today, they reflect important experimental results correctly. In particular, they explain significant causes of high catalyst activity, namely that the full reaction is broken into a sequence of partial reactions with very (or at least sufficiently) low activation barriers. Furthermore, the partial reactions lead, energet-

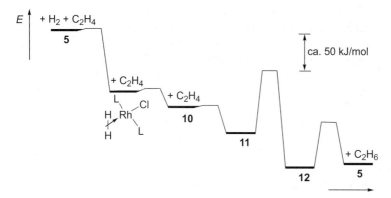

Figure 4.3 Energy profile diagram for the hydrogenation of ethene with [RhCl(PH$_3$)$_2$] as model catalyst (simplified, after Koga and Morokuma [50]). The numbering of substances corresponds to that in Figure 4.2 (L=PH$_3$).

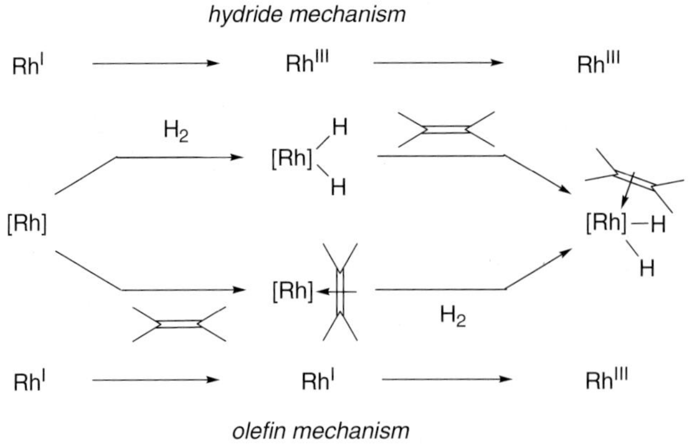

4 Hydrogenation of Olefins

Figure 4.4 Hydride mechanism vs. olefin mechanism.

ically speaking, "downhill," and no intermediate is thermodynamically stable enough to constitute a "thermodynamic trap" from which the reaction cannot proceed.

A distinction is made between the hydride and olefin mechanism. With the hydride mechanism, first the oxidative addition of H_2 and then the coordination of the olefin occurs, as previously explained. With the olefin mechanism, the opposite is true, as is shown schematically in Figure 4.4. The cationic complexes $[RhL_2s_2]^+$ catalyze the hydrogenation of olefins via the olefin mechanism. This means that the coordination of the olefin (with displacement of the solvent s) occurs first, followed by the oxidative addition of the H_2, which is rate-determining.

4.3
Enantioselective Hydrogenation

4.3.1
Principles

Hydrogenation with the Wilkinson catalyst proceeds stereoselectively as a *syn* addition. If the olefin to be hydrogenated is prochiral, the *syn* addition of H_2 leads to equal proportions of both enantiomers. In the following scheme, this is represented with the prochiral (Z)-α-acetamidocinnamic acid ester (**15**) as an example. *syn* addition of H_2 to the *Re* face (front-side attack) yields the (S)-enantiomer (S)-**16**, whereas addition to the *Si* face (back-side attack), yields the (R)-enantiomer (R)-**16**.

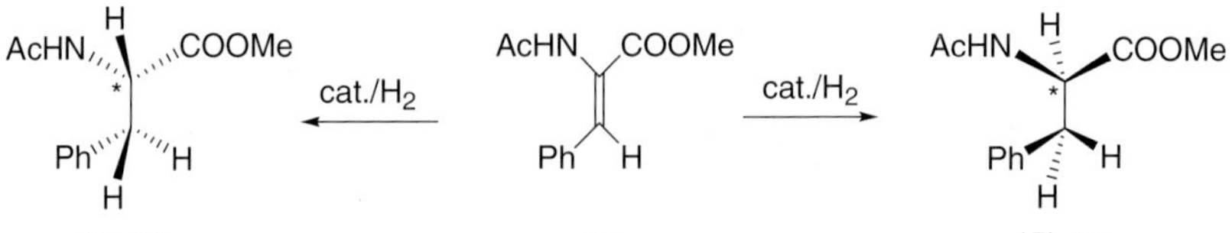

Background: Prostereogenicity, Prostereogenic Faces

A compound with a trigonal-planar carbon atom that bears three different substituents is prochiral, since the addition of a fourth substituent (different from those already present) to this C atom leads to a chiral molecule. The two faces of the prochiral compound are mirror-symmetric (enantiotopic). They are designated as *Re* (from Latin *rectus*) and *Si* (from Latin *sinister*). To determine *Re* and *Si*, the CIP rules (Cahn-Ingold-Prelog) are used to assign the priority of the substituents. If the priority of the substituents decreases clockwise, one is looking at the *Re* face of the molecule. If the molecule is observed from the *Si* face, the priority increases in the clockwise direction [10, 51].

Example: In **1**, the priority decreases in the order $NHAc > COOMe > =CH_2$, so the viewer is looking at the *Re* face.

The amino acid L-DOPA **20** is an effective medication for treating Parkinson's disease, since it eliminates the deficiency of dopamine in the brain (**20** → **21**). L-DOPA is now produced on an industrial scale in a process much less expensive than the one by which it was first synthesized (Hoffmann-LaRoche). In the older process, the palladium-catalyzed hydrogenation of a cinnamic acid amide **17** produced a racemic mixture of amino acids **18**, from which L-DOPA **20** was obtained via racemate separation and deprotection.

Enantioselective homogenous hydrogenation catalysts have led to a substantial simplification of the process. With a rhodium catalyst, **17'** is converted in a single step directly into the desired L-form **19'** with an enantiomeric excess of about 95% (Monsanto) [52].

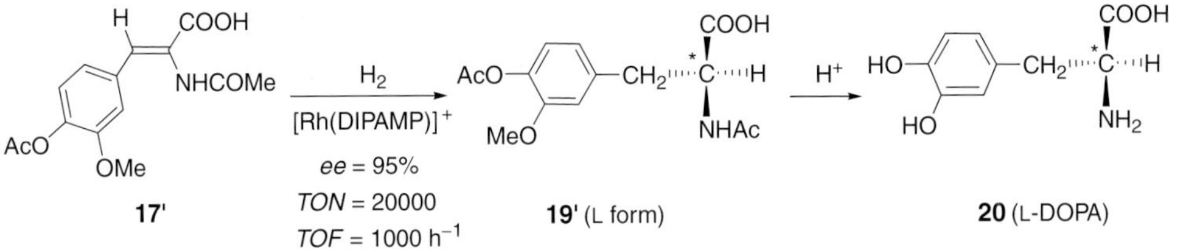

[Scheme: AcO/H-substituted styrene with NHCOMe → H₂, [Rh(DIPAMP)]⁺ → 19' (L form) → H⁺ → 20 (L-DOPA)]

17' → ee = 95%, TON = 20000, TOF = 1000 h⁻¹ → **19'** (L form) → **20** (L-DOPA)

Exercise 4.3
For amino acids, how are (a) the D- and L-configuration and, alternatively, (b) the R- and S-configuration determined? Does L-DOPA correspond to the R- or the S-configuration?

When a rhodium complex with a chiral P-ligand is used as catalyst, prochiral olefins can be hydrogenated enantioselectively. The coordination of a prochiral olefin with the Re or Si face to the chiral catalyst yields diastereomers that differ in stability and reactivity.

Enantioselective hydrogenation of styrene derivatives was first performed in 1968 by L. Horner and, independently, by W. S. Knowles (Nobel Prize in 2001 together with R. Noyori and K. B. Sharpless) with [RhCl(P*PhMePr)₃] (P*PhMePr = (S)-(+)-methylphenylpropylphosphane) as catalyst, but ee values only up to 15% were achieved. ee values of over 90% (sometimes over 99%) can be reached with cationic rhodium complexes [Rh(L₂*)s₂]⁺. L₂* are chiral bis(phosphane), bis(phosphinite) or bis(phospholane) ligands. Examples are given in Figure 4.5.

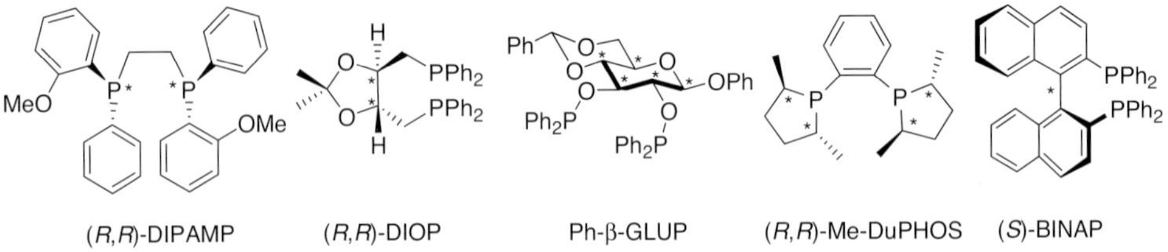

[Structures: (R,R)-DIPAMP, (R,R)-DIOP, Ph-β-GLUP, (R,R)-Me-DuPHOS, (S)-BINAP]

Figure 4.5 Examples of widely used chiral P,P-ligands in asymmetric catalysis.

The first P,P-chelating ligand (DIOP) with efficient behavior was developed in 1971 by K. B. Kagan. This also made it clear that the chirality of the ligand L₂* does not need to be located at the phosphorus atoms (DIPAMP). The carbon atoms of the "backbone" (DIOP, GLUP, DuPHOS) can also be chiral. In BINAP the chirality is caused by atropisomerism (helical chirality). No general relationship can be recognized between, on the one hand, the position and type of chirality of the ligand, and on the other, the ee values that can be achieved. Rather, the shape and conformational stability of the "coordination pocket" for the prochiral olefin are decisive factors. Both the shape and stability of the pocket are critically influenced by the size (5- or 7-membered) of the rhodadiphosphacycle. The selectivities achieved for the hydrogenation of **22/22'** into (R)- or (S)-**23/23'** with various cationic rhodium(I) catalyst complexes [Rh(L⌒L)]⁺ are given as an example [T11]:

4.3 Enantioselective Hydrogenation

				product	ee	product	ee
		L L					
		(R,R)-DIOP		(R)-**23**	85%	(R)-**23'**	73%
		(R,R)-DIPAMP		(S)-**23**	96%	(S)-**23'**	94%
		(S,S)-Et-DuPHOS		(S)-**23**	99%	(S)-**23'**	99%

22 (R = Ph), **22'** (R = H) → **23** (R = Ph), **23'** (R = H)

Many of the chiral P,P-ligands contain PAr$_2$ substituents, so the shape of the coordination pocket for the substrate built by the rhodadiphosphacycle is determined by the four aryl groups (Figure 4.6). If we are looking at the coordination pocket as an approaching substrate would, two aryl groups are seen edge-on and two are seen face-on ("edge-face" arrangement). The coordination pocket is chiral, in many cases C_2-symmetric, and can be broken into four quadrants, of which two are easier for the substrate to reach and two are more difficult. A prochiral olefin can be coordinated with its *Re* or *Si* face to the chiral Rh center, so two diastereomers can be formed. One of them is thermodynamically more stable, namely the one in which the bulky substituents protrude into the more readily accessible quadrants of the coordination pocket (Figure 4.6).

The catalysis proceeds according to the "olefin mechanism." First, the coordination of the olefin occurs as described, and then the oxidative addition of H$_2$, which is rate-determining and irreversible. It has been shown (see below) that the diastereomer that is thermodynamically more stable, and thus present in higher concentration, is less reactive. Thus, the diastereomer that is present in lower concentration, but is significantly more reactive, is the one that predominantly reacts. This "kinetically controlled enantioselectivity" is called an "anti-lock-and-key relationship," a reference to the "lock-and-key relationship" term (E. Fischer, 1894) that characterizes the reactivity of enzymes.

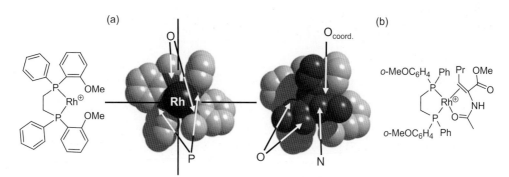

Figure 4.6 Structure of a catalyst–substrate complex [Rh(DIPAMP)(PrHC=C(COOMe)–NHAc)]$^+$. (a) Coordination pocket ([Rh(DIPAMP)]$^+$) for the prochiral olefin viewed along the C_2 axis with the typical "edge-face" arrangement of the four aryl substituents (Ph in edge arrangement and o-MeOC$_6$H$_4$ in face arrangement); the four quadrants are indicated by dashes. (b) Catalyst–substrate complex in the same projection. It is clear that the prochiral olefin ligand (shaded dark gray) "fits" well into the chiral coordination pocket, resulting in a high stability for the complex. That would not be the case for coordination to the other side of the prochiral double bond. This diastereomer is more reactive and determines the enantioselectivity (after McCulloch, Halpern and Landis [53]).

Although the course of reaction depends in a highly complex manner on stabilities and reactivities of all intermediates and transition states involved in the reaction, *ee* values of >99% can be achieved. This corresponds to energy differences of only a few kJ/mol between the transition states in the two reaction pathways.[1] They are of the same order of magnitude as the barrier to rotation around the C–C bond in ethane, which shows how subtly stability and reactivity in such homogeneous catalysts are attuned.

Enantioselective rhodium-catalyzed hydrogenation is not necessarily kinetically controlled. If instead of the α-acylaminoacrylates RCH=C(NHCOMe)–CO$_2$R' that have been considered so far, the (Z)-β-acylaminoacrylates **24** with [Rh{(S,S)-DIPAMP}(MeOH)$_2$][BF$_4$] (**25**) are hydrogenated in MeOH, the reaction follows the "lock-and-key principle": The main diasteromer present in solution of the catalyst-substrate complex *Re*-**26** (Rh is symbolized by a circle; the DIPAMP ligand is not shown) leads to the main diastereomer (S)-**27**. From the NMR-spectroscopically determined product ratios of the catalyst-substrate complex c_{maj}/c_{min} (about 9/1) and the *ee* values (38–60%), it can be calculated that the *minor* diastereomer *Si*-**26** only reacts about three times faster (and not 10^2–10^3 times faster as with the α-acylaminoacrylates) than the *major* diastereomer *Re*-**26**. That is not enough for the reaction pathway **24** → *Si*-**26** → (R)-**27** to gain the upper hand [54].

Ar	Re-26 : Si-26 (c_{maj} : c_{min})	(S)-27-selectivity % ee	k_{maj}/k_{min}
Ph	90 : 10	50	0.33
p-ClC$_6$H$_4$	88 : 12	38	0.30
p-MeC$_6$H$_4$	91 : 9	60	0.39

4.3.2
Applications and Examples

4.3.2.1 Applications for Asymmetric Hydrogenation
Enantioselective catalytic hydrogenation reactions play an important role, especially in the synthesis of pharmacological and agricultural chemicals. In addition to the

---
1) A calculation is performed in Exercise 4.6.

L-DOPA synthesis mentioned initially, two additional examples of industrial syntheses can be introduced.

- *Naproxen* (**29**) is one of the most important anti-inflammatory agents. According to R. Noyori's method, the (*S*)-enantiomer is obtained with 92% yield and 97% *ee* by hydrogenation of the aryl-substituted acrylic acid **28** with a ruthenium(II)-BINAP catalyst (12 h, r.t., 135 bar) [55]. The (*R*)-enantiomer is harmful to the liver.

BINAP-ruthenium complexes are excellent catalysts for asymmetric hydrogenation of a large number of functionalized olefins. They catalyze according to the monohydride mechanism, by which the olefin is inserted into a monohydride species (**30** → **31**) and the alkyl compound thus formed undergoes hydrogenolysis, without passing through a dihydride intermediate (**31** → **32**) [56].

- *Metolachlor* (**35**) (Ciba-Geigy/Syngenta) is one of the most important herbicides in the world. The synthesis is based on the enantioselective hydrogenation of the imine **33**. An iridium catalyst with a chiral ferrocenyldiphosphane ligand (R = Ph; R′ = 3,5-xylyl), which exhibits impressive productivity (*TON* = 10^6) and activity (*TOF* > 200 000 h^{-1}), is used [57].

4.3.2.2 Combinatorial Catalysis

A combinatorial approach (see Background) may be the method of choice for finding a suitable catalyst for a particular problem very efficiently. To demonstrate the approach, an example of automated parallel synthesis[2] of a chiral ligand library is

---

2) In the parallel synthesis, all reactions are performed – in an automated process – in separate vessels, so each ligand is accessible individually in a distinct vessel. This is the critical difference from the pool/split method (see Background), in which the components of the library are only present in a mixture.

introduced: Treatment of chlorophosphite **36**, derived from (R)-2,2'-binaphthol, with primary or secondary amines **37**, leads, in the presence of NEt$_3$, to the phosphoramidites **38**.[3] The smooth course of reaction enabled the direct further processing of the reaction solutions after filtering off [NEt$_3$H]Cl. They were reacted with **39**, in a reactor system designed to run the most reactions possible in parallel, to yield the catalyst that catalyzes the hydrogenation of the prochiral olefin **40** into the product **41**. Yield and *ee* values were determined using capillary GC with use of a chiral column. The devices were set up so that within two days, it was possible to produce 96 phosphoramidites **38** and, after Rh complex formation, react them with three different prochiral olefins.

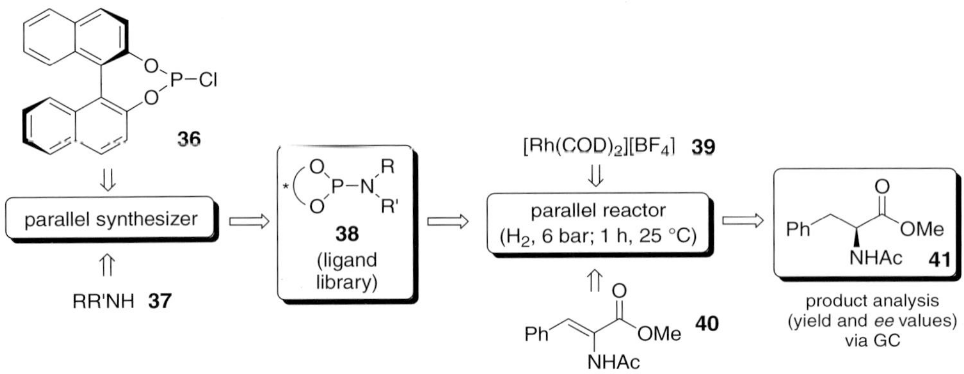

Alternatively, the ligand libraries can be assembled from the large palette of chiral ligands available. Such high-throughput catalyst testing is becoming increasingly important, especially for industrially relevant problems [58].

Background: Combinatorial Catalysis and High-Throughput Screening

The source of combinatorial chemistry can be found in the solid-phase synthesis of peptides (R. B. Merrifield, Nobel Prize in Chemistry, 1984). The pool/split procedure (division of the solid support into multiple portions of equal size, subsequent reaction of each of these portions individually with a single amino acid, and pooling of the individual portions into a single batch, as well as the multiple repetition of the divide, couple and recombine process) allowed the generation of a combinatorial library

---

3) Phosphoramidites have emerged as a highly versatile class of chiral ligands in asymmetric catalysis [59]. Note that, in contrast to previously discussed bidentate ligands, these are *monodentate* chiral ligands and their rhodium complexes catalyze the hydrogenation of prochiral olefins with excellent *ee* values [60].

with a very large number of peptides with a very small number of chemical steps [61].

Combinatorial methods are based on the idea to synthesize and screen large libraries of chemical entities as fast as possible using special techniques, rather than performing the task sequentially in a traditional and thus time-consuming manner (cited from [62]). In the case of combinatorial homogeneous catalysis, the libraries are created from ligands (which are converted into the catalysts by reaction with a metal complex) or directly from (pre-)catalysts, which then are subjected to a catalytic reaction in a parallelized and miniaturized screening, where, in general, the effect of the solvent, the temperature, the pressure, and other parameters are studied. For the requisite analysis, automated systems for the handling of samples for GC and HPLC can be considered, but also tests based on fluorescence and IR-thermographic assays, where "hot spots" indicate an exothermic reaction. GC and HPLC with chiral columns or in combination with circular dichroism (CD) spectroscopy can be used to determine ee values to evaluate enantioselective catalysts. The large amount of resulting data fundamentally requires effective computer-supported data collection and evaluation.

The combinatorial approach *complements rather than competes* with conventional catalytic research, so one approach will never replace the other. The design of libraries *that make sense* is a decisive key to success [63, 64].

4.3.2.3 Nonlinear Effects

Clearly, asymmetric synthesis requires a chiral auxiliary. Let us imagine that obtaining it in an enantiomerically pure form, would be, for example, too labor-intensive or expensive. Instead, we merely use an enantiomer-enriched product with an enantiomeric excess ee_{aux}. When the two enantiomers of the auxiliary in the catalyst act independently, it can be expected that the ee value of the product (ee_{prod}) is proportional to that of the auxiliary:

$$ee_{prod}(\%) = ee_{max} \cdot ee_{aux} \cdot 100.$$

The proportionality constant is the ee value (ee_{max}) achieved with use of the enantiomerically pure auxiliary. If this proportionality does not apply, a nonlinear effect (NLE) is present. It can be either positive ((+)-NLE) or negative ((−)-NLE) (Figure 4.7). There can be many causes of nonlinear behavior. If, for example, the catalyst complex is of type [ML*$_2$] (the same applies if the catalyst complex is aggregated into [{ML*}$_2$]), then three different catalysts exist, namely the two homochiral catalysts [ML$^R_2$] and [ML$^S_2$] and the heterochiral [ML^{R}L^S]. Each homochiral catalyst produces the corresponding enantiomeric product, while the heterochiral species catalyzes the formation of the racemic product. Its

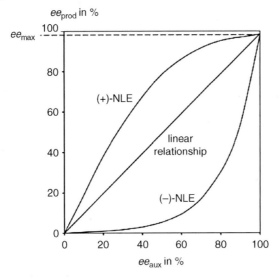

Figure 4.7 Nonlinear behavior in enantioselective catalysis. Positive and negative nonlinear effects are also called "chiral amplification" ((+)-NLE) or "chiral depletion" ((−)-NLE) (after Kagan in [9], p. 206).

activity can differ from that of the homochiral catalysts. This causes nonlinear behavior [65].

Exercise 4.4
- In the hydrogenation of itaconic acid dimethyl ester (prop-2-ene-1,2-dicarboxylic dimethyl ester) with [RhL$_2$(cod)][BF$_4$], nonlinear behavior has been found. The experiments showed that the catalyst complexes [RhL$^R_2$]$^+$/[RhL$^S_2$]$^+$ and [RhL^{R}L^S]$^+$ have the same thermodynamic stability (they are in equilibrium with each other) and in addition, the heterochiral combination is catalytically inactive. (This is a special case!) Formulate the reaction equation. Do you expect a positive or negative NLE? Justify your answer.
- Recently, it has been shown that effective catalysts can be obtained from mixtures of two monodentate ligands L^a and L^b and a transition metal. If, in the transition state for the reaction, two ligands L are bound to M and a rapid exchange of ligands occurs, there are three structurally different catalysts, the two homo-combinations **1a** and **1b**, and the hetero-combination **2** [62].

$$[M(L^a)_2] + [M(L^b)_2] \rightleftharpoons 2\,[ML^aL^b]$$
$$\text{1a} \quad\quad \text{1b} \quad\quad\quad\quad\quad \text{2}$$

This makes it possible to generate high catalyst diversity without creating new ligands. How many hetero-combinations can be produced with a library of 50 monodentate ligands?

- For the asymmetric hydrogenation of an olefin (not specified here) with $[Rh(COD)_2][BF_4] + 1\,L^a + 1\,L^b$ (L=monodentate chiral (enantiopure) phosphonite), with the homo-combinations $[Rh(L^a)_2]^+$ and $[Rh(L^b)_2]^+$, ee values of 76 and 13%, respectively, were found, while the hetero-combination $[RhL^aL^b]^+$ yielded 96% ee (conversion for each was 100%). What statements can be made about the composition of the catalyst mixture, the selectivity, and the activity?

4.3.3
Kinetically Controlled Enantioselectivity — A Closer Look

It was just shown that the type of coordination of the prochiral olefin (*Re* or *Si* face) determines which enantiomer is formed in enantioselective hydrogenation reactions.

Exercise 4.5
Which coordination (*Re* vs. *Si* face) of (*Z*)-α-acetamidocinnamic acid esters leads to L-DOPA upon hydrogenation?

(*Z*)-α-Acetamidocinnamic acid esters **42** react with cationic rhodium complexes $[Rh(L_2^*)s_2]^+$ (**43**), that have chiral *P*-ligands L_2^* coordinated, to form diastereomeric (acetamidocinnamic acid ester)rhodium complexes **44a** and **44b**, respectively. Coordination to the *Si* face yields the "D-diastereomer" **44a** and to the *Re* face, the "L-diastereomer" **44b**. **44a** reacts, after H_2 addition, via **45a**, forming D-DOPA, and **44b** via **45b**, forming L-DOPA. For the enantioselectivity of the hydrogenation, two effects are decisive:

- *Stability of the diastereomers (thermodynamic analysis)*. For the L-DOPA synthesis, the "D-diastereomer" **44a** is the thermodynamically more stable product and is thus formed in excess. The favoring of the *Si* face coordination to Rh is caused by the shape of the coordination pocket.
- *Reactivity of the diastereomers (kinetic analysis)*. The oxidative addition of H_2 is the rate-determining step. It runs faster for the diastereomer **44b** than for **44a**. One reason is that the rhodium atom in the diasteromer **44a** is sterically more shielded and the coordinated ligand must undergo larger structural changes on the way to the dihydrido complex than that in **44b**.

4 Hydrogenation of Olefins

[Structures 42 (viewed from the Si face), 43, 42 (viewed from the Re face), 44a, 44b, 45a, 45b leading to D-DOPA and L-DOPA]

The kinetics determine the course of reaction, and enantiomeric excesses of up to 99% in L-DOPA can be achieved. This is called kinetically controlled enantioselectivity. In Figure 4.8, both reaction pathways are represented in a reaction profile diagram.

The two diastereomers [M$_L$] and [M$_D$] are in equilibrium: [M$_L$]⇌[M$_D$]. The difference in thermodynamic stability between the two diastereomers is reflected in

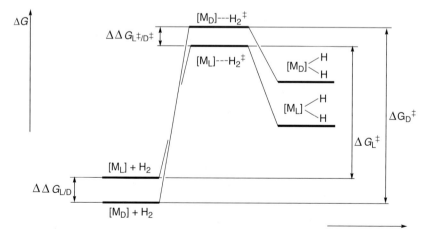

Figure 4.8 Reaction profile diagram for the formation of L- and D-DOPA (not to scale; explanation of symbols in the text).

the difference in Gibbs free energies $\Delta\Delta G_{L/D}$. The equilibrium constant $K_{L/D}$ is calculated from $\Delta\Delta G_{L/D} = -RT \ln K_{L/D}$. Provided there is rapid establishment of equilibrium between the two diastereomers, then the difference between the overall barrier heights of the two reaction paths, $\Delta\Delta G_{L\ddagger/D\ddagger}$, determines the reaction rates of the two diastereomers, that is, the enantioselectivty. The ratio of the reaction rates is *not* (!) determined by the difference between the Gibbs free energies of activation $\Delta G_L^\ddagger - \Delta G_D^\ddagger$. The Curtin-Hammett principle gives the theoretical description of this situation.

Background: The Curtin-Hammett Principle

Two conformers, **A** and **B**, of a reactant are in equilibrium with each other and interconvert with each other significantly faster than they react to form the products P_A and P_B, resulting in the reaction profile depicted.

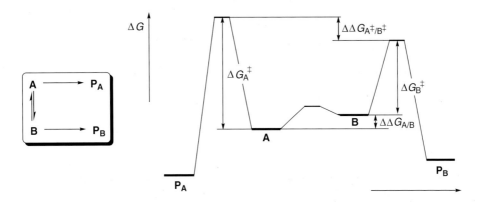

$\Delta\Delta G_{A/B}$	Difference between the Gibbs free energies of the conformers
$\Delta\Delta G_{A\ddagger/B\ddagger}$	Difference between Gibbs free energies of the transition states (determining the differences in reactivity)
$\Delta G_A^\ddagger$, $\Delta G_B^\ddagger$	Gibbs free energies of activation for reactions $A \rightarrow P_A$ and $B \rightarrow P_B$, respectively

The Curtin-Hammett principle states that the position of the equilibrium between the conformers of the reagent c_A/c_B does not determine the product composition c_{P_A}/c_{P_B}. The product composition is determined by the difference between the Gibbs free energies of the two transition states of the reactions $A \rightarrow P_A$ and $B \rightarrow P_B$. That must not be confused with the difference between the two Gibbs free energies of activation ($\Delta G_A^\ddagger - \Delta G_B^\ddagger$).

The Curtin-Hammett principle is the basis for understanding that a conformer only present in low concentration (here **B**) can determine the product when it leads to the transition state with the lowest Gibbs free energy [10, 66, 67].

Exercise 4.6

- Justify the statement that the difference between the Gibbs free energies of the two transition states $\Delta\Delta G_{L\ddagger/D\ddagger} = \Delta G_{L\ddagger} - \Delta G_{D\ddagger}$ and not the difference between the Gibbs free energies of activation $\Delta G_L^{\ddagger} - \Delta G_D^{\ddagger}$ determines the ratio of the reaction rates.
- Calculate the Gibbs free energy differences between the two transition states $\Delta\Delta G_{L\ddagger/D\ddagger}$ for which enantiomeric excesses of 90, 99, and 99.9% are achieved ($T = 298$ K).

Quantum-chemical DFT calculations (without consideration of the influence of a solvent) for the hydrogenation of the enamide **46** with $[\text{Rh}(\text{Me-DuPHOS})]^+$ (Me-DuPHOS: see Figure 4.5) as catalyst show that the (R)-enantiomer **47** is formed (Figure 4.9). The catalyst–substrate complex is diastereomeric: Re face (front-side) coordination of **46** to Rh yields the major diastereomer **48a** and Si face (back-side) coordination yields the minor diastereomer **48b**. The difference between the Gibbs free energies at 298 K is calculated as 15 kJ/mol. This corresponds to a concentration ratio of about 500 : 1. For the subsequent reaction of **48a/48b** with H_2, the energetically favorable pathway is the approach parallel to the C–Rh–P axis, and, in both cases, from the side on which the terminal C atom (C_β) of the coordinated double bond is located (see diagrams **49a** and **49b**). The total Gibbs free energy of activation for the less reactive diastereomer (**48a** → **49a** → ...) was determined to be 85 kJ/mol and for the more reactive diastereomer (**48b** → **49b** → ...) only 52 kJ/mol. This corresponds to a difference in the Gibbs free energies of the transition states of 18 kJ/

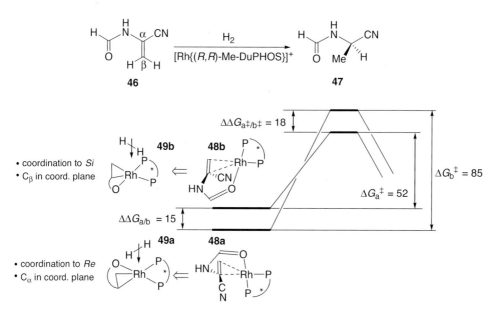

Figure 4.9 Quantum-chemical calculation of the enantioselective hydrogenation of enamides (all values in kJ/mol for 298 K; greatly abbreviated from Feldgus and Landis [68]).

mol. This quantity, according to the Curtin-Hammett principle, determines the reactivity difference between the two diastereomers. Thus an enantiomeric excess $ee = 99.9\%$ is calculated for the (R)-enantiomer **47**.

In Figure 4.10, the structures of the less reactive (**48a**) and the more reactive (**48b**) diastereomers are shown. The view of the molecular fragment [Rh(Me-DuPHOS)]$^+$ perpendicular to the coordination plane (Figure 4.10, a) makes clear that the two methyl groups of the phospholane groups at the upper left and lower right extend into the coordination pocket of the substrate and can sterically hinder its coordination. The coordination to the *Re* face is not hindered and in **48a**, the C_α atom of the double bond lies in the coordination plane [RhP$_2$OC] (**48a'**). In contrast, when the coordination occurs to the *Si* face, the interaction with a methyl group of the Me-DuPHOS ligand (see double arrow in Figure 4.10) forces the C_β atom of the double bond into the coordination plane, which requires energy (**48b'**). In both cases, H$_2$ is added from above. For **48b** – compare with view **48b'** in Figure 4.10 – this only necessitates small changes in the conformation of the enamide ligands. In **48a**, by contrast, C_β from the C=C double bond is not yet in the coordination plane and must be forced into it, shifting the CN substituent into the sterically hindered quadrant (lower right). All in all, this causes a substantially higher activation barrier for **48a** than for **48b** and explains the lower reactivity of **48a** with respect to H$_2$.

On the basis of a detailed understanding of the mechanism, enantioselective hydrogenation reactions of unsaturated substrates have developed into a "flagship" of asymmetric catalysis. They now belong to the standard repertoire of chemical

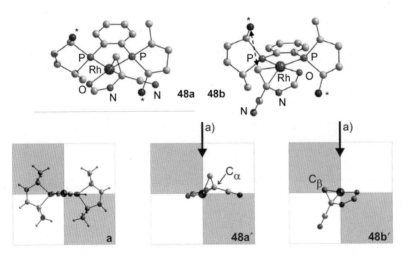

Figure 4.10 Molecular models for the catalyst-substrate diastereomers **48a** and **48b** (H atoms are not shown). The two methyl groups of the Me-DuPHOS ligands that extend into the coordination pocket of the enamide are marked with a star and drawn in dark gray. (a) Molecular fragment [Rh(Me-DuPHOS)]$^+$. (**48a'/48b'**) Enamide ligands in **48a** and **48b** (view along the coordination plane). The sterically hindered quadrants are highlighted in gray (after Feldgus and Landis [68]).
a) The arrow indicates the direction of addition of H$_2$.

synthesis in both academic and applied research and have also found broad industrial use [69, 70].

Exercise 4.7

Suppose we have an enantioselective catalytic reaction in which two diastereomeric product complexes are formed from two diastereomeric reactant complexes (**1** → **2** or **1'** → **2'**); see reaction diagrams **a–c**. Let us assume that this reaction step determines the enantioselectivity. Let us also assume that there is equilibrium between the two diastereomeric reactants (**1**/**1'**) and the Curtin-Hammett principle applies in each case. Thus, the enantiomer ratio is determined in each case by the difference of the Gibbs free energies of the transition states $\Delta\Delta G_{ts}$. Using the Hammond postulate for each reaction diagram, discuss the expected enantiomer ratio for the products.

4.4
Dihydrogen Complexes and H$_2$ Activation

4.4.1
Dihydrogen Complexes

Dihydrogen can coordinate to a metal without the H–H bond being cleaved. Such η^2-dihydrogen complexes [M(η^2-H$_2$)L$_x$] (**50**) are intermediates in concerted oxidative addition of H$_2$ to dihydrido complexes **51**:

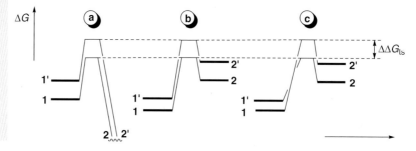

Dihydrogen complexes are σ complexes (compare Background, Section 3.1). Due to the exceptionally weak donor strength of the σ molecular orbital of H$_2$, substantial π back-donation is required to obtain stable complexes. Thus, for the bond – as for π-olefin complexes – two bond components are significant (Figure 4.11):

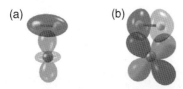

Figure 4.11 σ donation (a) and π back-donation (b) in η^2-dihydrogen complexes.

- *σ donation.* Transfer of electron density from the σ-bonding molecular orbital of the H_2 into an empty orbital with σ-symmetry (s, p_z, d_{z^2}; z-axis directed toward the H_2 ligand) of the metal. The d_{z^2} orbital of the metal and the σ orbital of the H_2 molecule are depicted.
- *π back-donation.* Transfer of electron density from a metal d orbital with π-symmetry (d_{xz} or d_{yz}) into the antibonding molecular orbital σ* of H_2.

Both bond components lead to a weakening of the H–H bond. If the back-donation is too weak, the η^2-H_2 complex is not stable. If the back-donation is too strong, too much electron density is transferred into the σ* H–H orbital, resulting in cleavage of the H–H bond. This means that the oxidative addition reaction of H_2 has completed and a dihydridometal complex with two conventional 2c–2e bonds has been formed.

Consequently, the coordination of H_2 to a metal leads to lengthening of the H–H bond (typical values: 0.8–0.9 Å in comparison to 0.75 Å in the H_2 molecule) and to lower wavenumbers ν_{HH} (typical values: 2500–3100 cm^{-1} in comparison to 4000 cm^{-1} in the H_2 molecule).

The synthesis of η^2-H_2 complexes occurs by reaction of a suitable precursor complex with molecular hydrogen (**a**) or by protonation of a hydridometal complex (**b**). In both cases, however, the classical dihydrido complex can also be obtained (**a'**/**b'**). A tautomeric equilibrium can exist between the two complexes in solution (**c**).

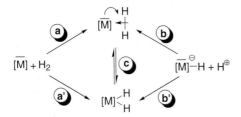

The following reaction equations demonstrate the two synthesis paths for dihydrogen complexes [71].

[W(CO)$_3${P(i-Pr)$_3$}$_2$] + H_2 ⇌ [W(η^2-H_2)(CO)$_3${P(i-Pr)$_3$}$_2$] ⇌ [W(H)$_2$(CO)$_3${P(i-Pr)$_3$}$_2$]

52 **53** (85%) **54** (15%)
W^0 (d^6, 16 ve) W^0 (d^6, 18 ve) W^{II} (d^4, 18 ve)

[Ru(H)$_2$(dppm)$_2$] ⇌ (+ ROH, – RO$^-$) [RuH(η^2-H_2)(dppm)$_2$]$^+$ →(+ RO$^-$, – H_2) [RuH(OR)(dppm)$_2$]

55 **56**
Ru^{II} (d^6, 18 ve) Ru^{II} (d^6, 18 ve)

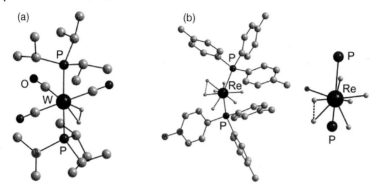

Figure 4.12 (a) Structure of [W(η^2-H$_2$)(CO)$_3${P(i-Pr)$_3$}$_2$] (**53**). (b) Structure of [ReH$_5$(η^2-H$_2$){P(p-Tol)$_3$}$_2$] (**58**) and Re coordination; the stretched H–H bond is shown with dashes. The H atoms of the phosphane ligands are omitted for clarity.

The tungsten complex **52** (16 ve) is stabilized by an agostic C–H $\cdots$ W interaction. In the dihydrogen complex **53**, this is broken in favor of coordination of H$_2$ (Figure 4.12). In particular, coordinatively unsaturated d^6 complexes are capable of forming η^2-H$_2$ complexes. On the other hand, when d^8 complexes react with H$_2$, oxidative addition is favored, producing classical d^6 dihydrido complexes. Both are apparently associated with the particularly high ligand field stabilization energy of d^6 complexes. Thus, [Cr(η^2-H$_2$)(CO)$_5$] (Cr^0, d^6) is a dihydrogen complex and [Fe(H)$_2$(CO)$_4$] (Fe^{II}, d^6) is a classical dihydrido complex.

The coordination of dihydrogen can – as a reversal of synthesis by protonation – be associated with a sharp increase in the acidity of H$_2$:

$$[M]\!\!-\!\!\genfrac{}{}{0pt}{}{H}{H} \;\rightleftharpoons\; [M]\!-\!H^{\ominus} + H^{\oplus}$$

$$[Os(\eta^2\text{-H}_2)(dppe)_2(MeCN)]^{2+} \;\rightleftharpoons\; [Os(H)(dppe)_2(MeCN)]^{+} + H^{+}$$

$$\begin{array}{cc} \textbf{57a} & \textbf{57b} \\ \text{Os}^{II}\,(d^6,\,18\text{ ve}) & \text{Os}^{II}\,(d^6,\,18\text{ ve}) \end{array}$$

Thus, the dihydrogen osmium complex **57a** is a very strong acid, with p$K_a = -2$ [72]. Such reactions correspond to a heterolytic bond cleavage of H$_2$. This is an important activation reaction for many hydrogenation catalysts. It has also been demonstrated for H$_2$-bonding hydrogenases, whereby η^2-H$_2$ complexes may play a role as well.

It is not always easy to distinguish between "nonclassical" η^2-H$_2$ complexes and "classical" dihydrido complexes. In general, in single-crystal X-ray diffraction analyses, it is difficult to localize hydrogen atoms in the immediate proximity of heavy metals to a sufficient degree of precision. Neutron diffraction experiments, on the other hand, require very large single crystals. In η^2-HD complexes, NMR-spectroscopic H–D coupling constants can be measured, from which the bonding situation can also be deduced.

4.4 Dihydrogen Complexes and H_2 Activation

In addition to the η^2-H_2 complexes discussed here (H–H 0.8–0.9 Å), there are also those with stretched H–H bonds (1.1–1.6 Å). H–H distances >1.7 Å indicate classical dihydrido complexes. In the complex [ReH$_7${P(p-Tol)$_3$}$_2$] (Figure 4.12), two of the seven H ligands show a separation of only 1.357(7) Å, so an H_2 complex with a stretched H–H bond is present. It should thus be written as [ReH$_5$(η^2-H_2){P(p-Tol)$_3$}$_2$] (**58**; Re^V, d^2, 18 ve).

The trigonal-bipyramidal unsaturated [Pt$_3$Re$_2$] metal cluster **59** is able to activate H_2 through multiple oxidative addition steps. In **59**, three equivalents of H_2 are added successively at room temperature (**59** → **59a** → **59b** → **59c**; in **59a–c** only the cluster core is shown) without CO or P(t-Bu)$_3$ being cleaved off. Under UV-vis radiation (**59c** → **59b** → **59a**), two H_2 molecules are cleaved off again by reductive elimination. This reversibility makes such complexes interesting, apart from their catalytic properties, as potential hydrogen storage systems [73].

4.4.2 Activation of Dihydrogen

A critical step in olefin hydrogenation is the activation of H_2. There are a number of possibilities:

- **Oxidative Addition of H_2**

$$[M] + H_2 \rightleftharpoons [M]\begin{matrix}H\\H\end{matrix}$$

Oxidative addition of H_2 results in a dihydrido complex. A σ-H_2 complex can appear as an intermediate. A change of two in the oxidation number of the intermediates is characteristic of the catalytic cycle. Examples are catalysts of the Wilkinson complex type and a large number of other complexes of late transition metals.

- **σ-Bond Metathesis of H_2**

A σ-alkylmetal complex (formed by insertion of an olefin into an M–H bond) reacts with H_2 via a four-membered cyclic transition state, representing a σ-bond metathesis with cleavage of the alkane from the complex (compare Section 7.4). The catalytic cycle completes with insertion of the olefin into the hydridometal complex formed. In the course of the catalysis, no change occurs in the oxidation number of the metal. Examples are cyclopentadienyllanthanoid complexes such as $[\{LuH(\eta^5\text{-}C_5Me_5)_2\}_2]$ (Lu^{III}) and other complexes of early transition metals.

- **Homolytic Cleavage of H_2**

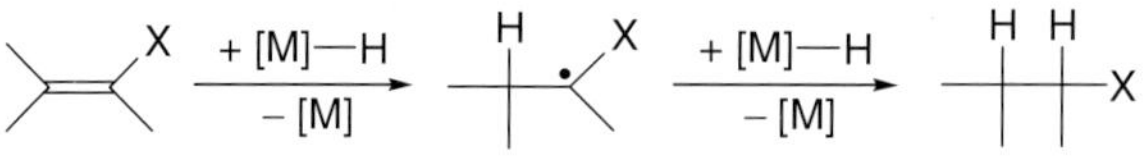

Homoyltic cleavage of H_2 proceeds via a radical mechanism. A hydridometal complex transfers, stepwise, two hydrogen atoms onto an olefin, usually functionalized. The regeneration of the hydrido metal complex occurs through homolytic bond cleavage of dihydrogen according to the equation $2\,[M] + H_2 \rightarrow 2\,[M]\text{–}H$. The intermediates characteristically undergo a change of one in their oxidation number. A familiar example is $[M]\text{–}H = [CoH(CN)_5]^{3-}$ (Co^{III}, 18 ve) and $[M] = [Co(CN)_5]^{3-}$ (Co^{II}, 17 ve).

- **Heterolytic Cleavage of H_2**

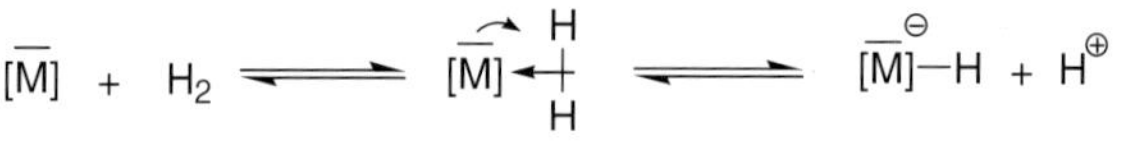

Heterolytic bond cleavage of H_2, which can proceed via a dihydrogen complex as intermediate, produces a hydridometal complex and a proton, which must be accepted by a suitable proton acceptor. An example is the heterolytic cleavage of H_2 on a ruthenium(II) complex (d^6) with an amine as proton acceptor ($P\frown P$ = dppe) [71]:

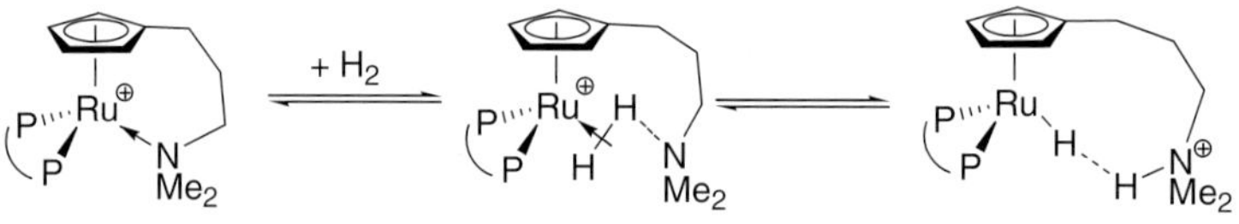

Such complexes are especially suitable for hydrogenating polar bonds $^{\delta+}A{=}B^{\delta-}$ (e.g., C=O groups in ketones), where the hydridic H atom binds to A and the protic H atom to B. In the course of the catalysis, no change occurs in the oxidation number of the metal.

Heterolytic bond activation of H_2 is possible even without participation of transition metals. The Lewis acid **60** does not react with sterically encumbered phosphanes such as **61**. Hence, no Lewis acid-base adduct is formed; **60** and **61** thus form a so-called "frustrated Lewis pair." However, it reacts under normal conditions with H_2, forming a phosphonium hydridoborate **62**. For the intramolecular variant **63/63'** this reaction is even reversible. **63'** is able to reduce imines and nitriles catalytically in the presence of H_2 to amines [74].

$$(t\text{-Bu})_3P + B(C_6F_5)_3 \xrightarrow{H_2 \text{ (1 bar)}, 25\,°C} [(t\text{-Bu})_3PH][HB(C_6F_5)_3]$$
$$\quad\; 61 \qquad\quad 60 \qquad\qquad\qquad\qquad\qquad 62$$

Compound **63**: $(Mes)_2P$–[tetrafluorophenylene]–$B(C_6F_5)_2$

$$63 \;\; \underset{-H_2}{\overset{+H_2 \text{ (1 bar)}, \text{ toluene (100\,°C)}}{\rightleftharpoons}} \;\; 63'$$

Compound **63'**: $(Mes)_2P(H)^{\oplus}$–[tetrafluorophenylene]–$B(H)^{\ominus}(C_6F_5)_2$

The activation of H_2 by heterolysis plays a large role in heterogenous catalysis as well. Hence, at an elevated temperature and high pressure, ZnO catalyzes the formation of methanol from CO and H_2. It is assumed that H^- is bound to zinc and H^+ is bound to oxygen on the ZnO surface.

- **Activation of H_2 in Hydrogenases**

With hydrogenases, heterolytic cleavage of H_2 generally takes place. Hydrogenases are enzymes that catalyze the consumption or formation of dihydrogen according to the following equation:

$$H_2 \;\; \underset{b}{\overset{a}{\underset{\longleftarrow}{\overset{\text{hydrogenase}}{\longrightarrow}}}} \;\; 2H^+ + 2e^-$$

64: Fe$_4$S$_4$ cluster linked to Fe$_2$ unit with CysS, SCys, Cys, CN, CO, and NH ligands.

Reaction **a** is coupled to a reduction of physiological electron acceptors and **b** is coupled to oxidation of physiological electron donors. Most hydrogenases are metalloenzymes (Fe/Ni–S or Fe–S clusters). The active center of a hydrogenase that contains a Fe$_4$S$_4$ cluster and a Fe$_2$ unit is shown in **64** (SCys = binding sites of the protein). In the reduced form, a hydrido complex (L = H) exists, and in the oxidized form, an aqua complex (L = H_2O) exists. The terminal iron atom is probably the binding site for H_2 [75].

4.5
Transfer Hydrogenation

Transfer hydrogenation indicates a reaction for which the source of hydrogen is not H_2; rather, hydrogen is transferred under the catalytic influence of metal complexes from an organic substrate DH_2 ("hydrogen donor") onto another organic substrate A ("hydrogen acceptor"):

$$DH_2 + A \xrightleftharpoons{[M]} AH_2 + D$$

Favored hydrogen donors are alcohols, aldehydes, formic acid, and amines, which can also be used directly as solvents. Hydrogen acceptors are often ketones and aldehydes, but also nitriles, amines, and activated olefins. Reduction of aldehydes and ketones to primary and secondary alcohols, respectively, in isopropanol as solvent, under the influence of catalytic quantities of $Al(O\text{-}i\text{-}Pr)_3$ is a transfer hydrogenation reaction that has long been known (Meerwein-Ponndorf-Verley-Reduktion, 1925).

For hydrogen transfer, using the following reaction as an example, two general paths are to be considered:

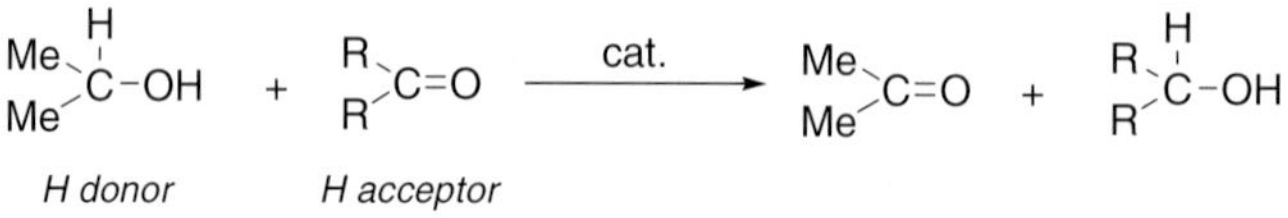

$$\underset{\text{H donor}}{\underset{Me}{Me}}\!\!\!>\!\!\!\underset{|}{\overset{H}{C}}\!\!-OH \;+\; \underset{\text{H acceptor}}{\underset{R}{R}}\!\!\!>\!\!\!C\!=\!O \;\xrightarrow{\text{cat.}}\; \underset{Me}{\underset{Me}{}}\!\!\!>\!\!\!C\!=\!O \;+\; \underset{R}{\underset{R}{}}\!\!\!>\!\!\!\underset{|}{\overset{H}{C}}\!\!-OH$$

- *Direct hydrogen transfer.* The α-C–H hydrogen atom is transferred in a concerted reaction via a six-membered cyclic transition state (**65**) in which both the H donor and the H acceptor are coordinated to the metal. Metal hydrides do not appear as intermediates. The catalysts are typically main-group metal compounds. The Meerwein-Ponndorf-Verley reduction is an example.
- *Metal hydride mechanisms.* In H-transfer reactions in which metal monohydrides appear as intermediates, the α-C–H hydrogen atom of the hydrogen donor is transferred to the metal, which can be considered to be a β-hydride elimination[4] (**66** → **67**). The transfer of the OH proton from the H donor to the H acceptor ($[M]\text{-}OCHR_2 + Me_2CHOH \rightarrow [M]\text{-}OCHMe_2 + R_2CHOH$) occurs without participation of metal hydride intermediates. If metal dihydrides appear as intermediates, both the OH proton and the α-C–H hydrogen atom of the hydrogen donor are transferred onto the metal, the former via an oxidative addition reaction (**68** → **69**) and the latter via a β-hydride elimination (**69** → **70**).

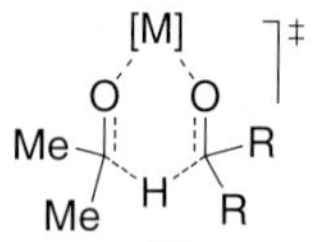

---

[4] The C1 atom (α-C atom) in alcohols is in β-position in alcoholate complexes [M]–O–CH<.

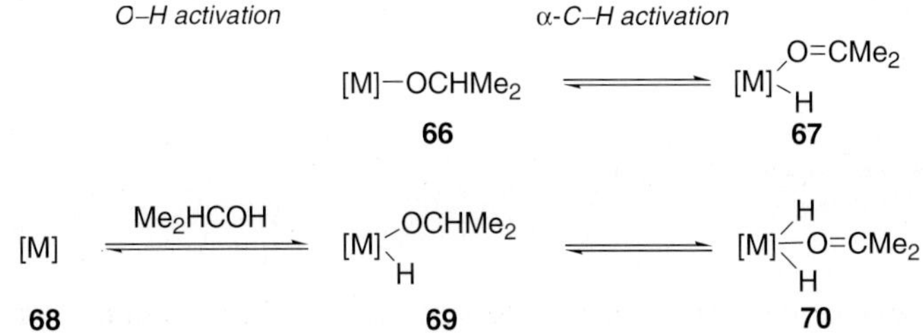

O–H activation
α-C–H activation
[M]–OCHMe2
O=CMe2
[M]
H
Me2HCOH
[M]
[M]
OCHMe2
H
H
[M]
O=CMe2
H
66 67
68 69 70

As catalysts, Ru, Rh, and Ir complexes are particularly noteworthy. For example, the Wilkinson complex is a precatalyst which forms a monohydride intermediate in transfer hydrogenation reactions.

Asymmetric H-transfer reactions with prochiral ketones and imines to chiral alcohols and amines, respectively, are of special importance. In the following reaction scheme, the enantioselective reduction of aromatic ketones using isopropanol is shown. The precatalysts are arene–ruthenium complexes [RuCl{(*S*,*S*)-(YCHPh–CHPhNH$_2$)}(η^6-arene)] (Y = O, NTs), which react in the presence of isoproponol and an alkaline base, forming **71**. Complex **71** is bifunctional. Both hydrogen atoms are simultaneously transferred onto the ketone, as indicated in the transition state **73**. The cycle is completed with the reaction of the coordinatively unsaturated complex **72** with isopropanol, forming **71**. This is an example where substrate–metal coordination is not a necessary precondition for a homogeneously catalyzed hydrogenation [76].

R'n
H
Ru
H2N
Y
Ph
Ph
R
Ph
O
R
Ph
H
O
H
Me2CO
Me2HCOH
R'n
Ru
HN
Y
Ph
Ph
Ph
R
H
O
H
NH
Ru
R'n
Y
Ph
Ph
‡

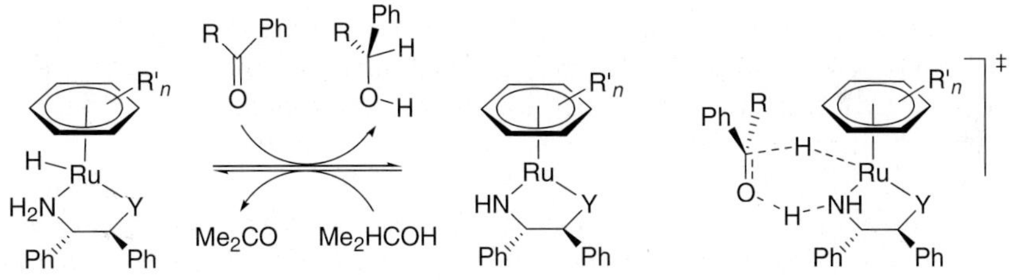

71 (Ru^{II}, 18 *ve*) **72** (Ru^{II}, 16 *ve*) **73**

5
Hydroformylation of Olefins and Fischer-Tropsch Synthesis

5.1
Cobalt Catalysts

In 1938 at Ruhrchemie, Otto Roehlen discovered and brought to technical maturity the reaction of ethene with synthesis gas (CO/H_2), forming propionaldehyde in the presence of a heterogeneous cobalt–thorium catalyst. He was at the time performing experiments under the conditions of the Fischer-Tropsch synthesis (synthesis of hydrocarbons from CO/H_2) in order to obtain oxygen-containing compounds as main products, not merely side products. When terminal olefins are used as substrates, either *n*-aldehydes or isoaldehydes – the latter being mostly undesired – are formed.

$$R-CH=CH_2 \xrightarrow{CO/H_2, \text{ cat.}} R-CH(H)-CH_2-CHO + R-CH(CHO)-CH_3$$

This process is designated the hydroformylation of olefins, since the reaction corresponds formally to the addition of an H atom and a formyl group (–CHO) to an olefin double bond. The designation "oxo synthesis" is also common, though not completely correct.[1] Hydroformylation of olefins is exergonic ($\Delta G^{\ominus} = -65$ kJ/mol for propene). It proceeds in the sense of a *syn* addition to the double bond.

At the end of the 1940s, it was shown that when using the heterogeneous catalyst mentioned above, dicobaltoctacarbonyl is first formed, which is soluble in the reaction mixture, so the reaction is homogeneously catalyzed. Fundamental experiments on the mechanism can be traced back to R. F. Heck and D. S. Breslow at the beginning of the 1960s. Starting from [$Co_2(CO)_8$], under exposure of H_2, the

---

[1] Immediately after the discovery of the hydroformylation of ethene, it was "... initially assumed that it must be merely a question of further development of processing methods until someday any olefins could be converted into either aldehydes or ketones, as desired – and thus generally into *oxo compounds*. In anticipation of this expected later development, the patent department of Ruhrchemie introduced, in due course, the abbreviated name '*oxo synthesis*' for the new reaction. However, it has not been possible to confirm the original assumption.... However, the name '*oxo synthesis*' has spread as a catchphrase due to its brevity and terseness, and it has been impossible to retire it from usage, despite the opposing facts and all the efforts made to argue against it." (cited from O. Roelen [77]).

Fundamentals of Organometallic Catalysis. First Edition. Dirk Steinborn
Copyright © 2012 WILEY-VCH Verlag GmbH & Co. KGaA, Weinheim

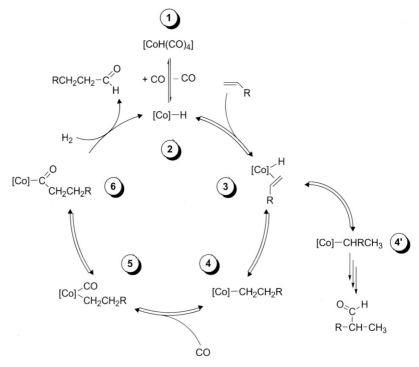

Figure 5.1 Mechanism (simplified) of hydroformylation of olefins with [CoH(CO)$_4$] as precatalyst ([Co] = Co(CO)$_3$).

precatalyst [CoH(CO)$_4$] is first formed. Cleavage of CO from the precatalyst leads to formation of the catalytically active complex [CoH(CO)$_3$]. The mechanism of catalysis is given in Figure 5.1. The following individual reaction steps can be named:

1 → 2: *Ligand cleavage/attachment.* Formation of the catalytically active complex (Co^I, 16 ve) from the precatalyst (Co^I, 18 ve).

2 → 3: *Ligand attachment/cleavage.* Olefin activation by formation of a coordinatively saturated π-olefin complex (Co^I, 18 ve).

3 → 4: *Insertion/β-H elimination.* Formation of an n-alkylcobalt(I) complex (16 ve) by Co–C1 bond linkage.

4 → 5: *Ligand attachment/cleavage.* Coordination of CO leads to an alkyltetracarbonylcobalt(I) complex (18 ve).

5 → 6: *CO insertion/deinsertion.* Migratory CO insertion creates an acylcobalt(I) complex (16 ve).

6 → 2: *σ-bond metathesis (oxidative addition/reductive elimination).* The n-aldehyde could be formed in a σ-bond metathesis, without passing through a cobalt(III) intermediate. However, H$_2$ can also react in an oxidative addition reaction, forming an acyldihydridocobalt(III) complex [CoH$_2$(COCH$_2$CH$_2$R)(CO)$_3$] (7, 18 ve), which

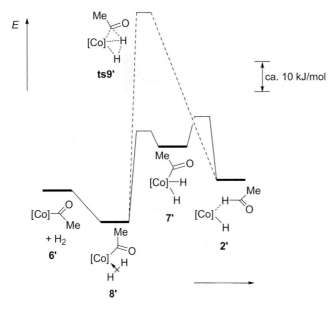

Figure 5.2 Energy profile for the aldehyde formation from [Co]–COR' + H_2 (R' = Me; **6'** → **2'**) as model for the catalyst in Figure 5.1 (R' = CH_2CH_2R; **6** → **2**) via oxidative addition/reductive elimination and σ-bond metathesis, respectively ([Co] = $Co(CO)_3$) (simplified, after Solà and Ziegler [78]).

undergoes a reductive C–H elimination with cleavage of the aldehyde. This reaction step is irreversible and probably also rate-determining.

DFT calculations on an acetyl model complex, [Co(COMe)(CO)$_3$] (**6'**), show that in the gas phase, the pathway via oxidative addition (**6'** → **8'** → **7'** → **2'**) is energetically favored over σ-bond metathesis (**6'** → **8'** → ts9' → **2'**) (Figure 5.2). In both cases, a σ-dihydrogen complex **8'** is first formed in an exothermic reaction from **6'** and H_2. The primary reaction product is complex **2'**, to which acetaldehyde is bound via an agostic C–H ⋯ Co interaction.

The regioselectivity of the reaction (n-aldehyde vs. isoaldehyde) is controlled via olefin insertion. Formation of a Co–C1 bond (**3** → **4**) leads to n-aldehydes, while the linkage of a Co–C2 bond (**3** → **4'**) opens the reaction channel to formation of the isoaldehydes. There are indications that the acyl complex [Co]–$COCH_2CH_2R$ (**6**) might isomerize to [Co]–COCHMeR, which would represent an additional option for formation of isoaldehydes [79, 80].

Addition of phosphanes leads to a considerable increase in the n/iso ratio. This is caused, at least partly, by the steric strain of the phosphane ligand, which hinders the formation of secondary alkyl- and acylcobalt complexes relative to the analogous primary alkyl/acyl complexes. The reaction mechanism when phosphane-modified cobalt precatalysts such as [CoH(CO)$_3${P(n-Bu)$_3$}] are used, is in principle analogous to the phosphane-free system.

5 Hydroformylation of Olefins and Fischer-Tropsch Synthesis

Table 5.1 Catalyst systems and process parameters in industrially used hydroformylation processes (after [T6]).

Precatalyst	[CoH(CO)$_4$]	[CoH(CO)$_3${P(n-Bu)$_3$}]	[RhH(CO)(PPh$_3$)$_3$][a]
Catalyst	[CoH(CO)$_3$]	[CoH(CO)$_2${P(n-Bu)$_3$}]	[RhH(CO)(PPh$_3$)$_2$]
p (in bar)	200–300	50–100	7–25
T (in °C)	140–180	180–200	90–125
C$_4$ selectivity (in %)	82–85	>85	>90
n/iso ratio	4/1	9/1	19/1

a) In the presence of an up to 500-fold excess of PPh$_3$.

The most important catalysts are unmodified cobalt carbonyl catalysts and phosphane-modified cobalt and rhodium catalysts. Typical process parameters are compiled in Table 5.1. In general, the cobalt-containing systems require carbon monoxide under high pressure to suppress the decomposition of the catalyst into Co and carbon monoxide. The higher thermal stability of the phosphane-modified cobalt catalysts, which permits somewhat lower pressures, comes at the price of lower catalytic activity, which in turn requires higher reaction temperatures. These catalysts already possess considerable hydrogenation activity and the processes can be carried out so that the alcohols, rather than the aldehydes, are obtained directly. That offers an advantage for hydroformylation of higher olefins: The thermally sensitive, but higher-boiling, aldehydes can no longer be separated by distillation from the catalyst system, but the less sensitive alcohols can. Particularly for short-chain olefins, however, the favored alcohol synthesis is a two-stage process, a Rh-catalyzed aldehyde synthesis followed by a separate hydrogenation step.

Exercise 5.1
Give reasons for the higher thermal stability of the phosphane-modified cobalt catalysts compared to those that are not phosphane-modified. How does phosphane substitution change the acidity of [CoH(CO)$_4$]?

5.2
Phosphane-Modified Rhodium Catalysts

G. Wilkinson, back in the middle of the 1960s, discovered the hydroformylation activity of phosphane-modified rhodium catalysts, which are now the most broadly used. Rhodium catalysts possess higher catalytic activity by about a factor of 1000 than cobalt catalysts [81]. In addition, phosphane-modified rhodium catalysts are distinguished by a high n/iso ratio of aldehydes formed (Table 5.1). Selectivities of greater than 90% are achieved in aldehyde formation. Side reactions that appear include hydrogenation of aldehydes to form alcohols and of olefins to form saturated hydrocarbons, as well as condensation reactions (e.g., aldol reactions) of aldehydes.

Rhodium catalysts lead in general to fewer side products than cobalt catalysts, but require careful purification of the reactants. The higher stability of the rhodium complexes and the lower reaction temperatures result in longer catalyst lifetimes. Thus, despite the higher precious metal price, when setting up a new plant for hydroformylation of lower olefins, rhodium catalysts are favored.

The fundamental mechanism starting from [RhH(CO)(PPh$_3$)$_3$] (**10**) as precatalyst is shown in Figure 5.3. This is a "dissociative mechanism" since cleavage of a PPh$_3$ ligand precedes the olefin coordination. The following individual reaction steps are similar to the cobalt-catalyzed reaction, apart from the product formation step:

10 → 11: *Ligand cleavage/attachment.* The coordinatively saturated trigonal-bipyramidal precatalyst is converted by cleavage of phosphane into the catalytically active square-planar rhodium(I) complex (16 ve).

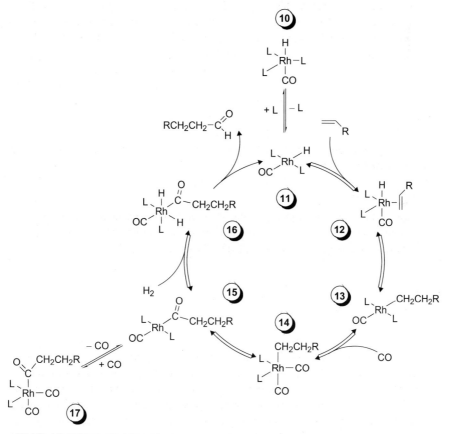

Figure 5.3 (Simplified) mechanism for hydroformylation of olefins with [RhH(CO)(PPh$_3$)$_3$] (**10**) as precatalyst (L=PPh$_3$).

11 → 12: *Ligand attachment/cleavage.* Olefin activation by π-complex formation, yielding a coordinatively saturated complex (Rh^I, 18 ve).

12 → 13: *Insertion/β-H elimination.* Formation of an n-alkylrhodium(I) complex (16 ve) by Rh–C1 bond linkage.

13 → 14: *Ligand attachment/cleavage.* Coordination of CO leads to an alkyldicarbonylbis(phosphane)rhodium(I) complex (18 ve).

14 → 15: *CO insertion/deinsertion.* Migratory CO insertion creates an acylrhodium(I) complex (16 ve).

15 → 16: *Oxidative addition/reductive elimination.* Oxidative addition of H_2 leads to an electronically and coordinatively saturated rhodium(III) complex. This step is probably rate-determining.

16 → 11: *Reductive elimination.* In a reductive C–H elimination reaction the aldehyde is cleaved off, which causes the starting complex **11** to be regenerated. This reaction is irreversible.

15 → 17: *Ligand attachment/cleavage.* CO coordination yields an electronically saturated Rh^I complex, which is not capable of oxidative addition of H_2 until cleavage of CO occurs. Thus, **17** is a reservoir (resting state) for a catalytically active complex, and its formation lowers the catalytic activity.

The phosphane-modified rhodium catalyst [RhH(CO)(PPh$_3$)$_3$] (**10**), also used in industry, has been studied well from a mechanistic point of view. Under hydroformylation conditions, it is in equilibrium with a series of other complexes:

[RhH(CO)(PPh$_3$)$_3$]
10 (18 ve)

$+ PPh_3$ ↕ $- PPh_3$

[RhH(CO)(PPh$_3$)$_2$] ⇌ (+ CO / − CO) [RhH(CO)$_2$(PPh$_3$)$_2$] ⇌ (− PPh$_3$ / + PPh$_3$) [RhH(CO)$_2$(PPh$_3$)]

11 (16 ve)　　　　**18** (18 ve)　　　　**19** (16 ve)

↓　　　　　　　　　　　　　　　　　　　↓

n-aldehydes　　　　　　　　　　　　isoaldehydes

The two 16-ve complexes **11** and **19** catalyze the oxo synthesis. The mechanism with the bis(triphenylphosphane) catalyst complex **11** has already been described. The dicarbonylrhodium complex **19** catalyzes hydroformylation in a similar way. While complex **11** forms linear (unbranched) aldehydes with high selectivity, complex **19** favors the formation of isoaldehydes. A key reason is that the high steric load of the two PPh$_3$ ligands in the bis(triphenylphosphane) complex **11** highly favors the insertion of the olefin into the Rh–H bond to form a primary alkyl ligand (Rh–C1 bond formation) rather than the insertion leading to a secondary alkyl ligand (Rh–C2 bond formation). By contrast, with the dicarbonylrhodium complex **19**, the insertion with Rh–C2 bond formation seems to dominate, creating branched aldehydes.

The position of the equilibrium **11** ⇌ **19** is strongly dependent on the CO pressure and phosphane concentration. The formation of **11** is favored by low CO pressure and high phosphane concentration, but the latter lowers the catalyst activity. Under industrial conditions for butyraldehyde synthesis using a 100–200-fold excess of PPh_3, the formation of the *n*-aldehydes dominates.

Quantum-chemical calculations give a glimpse into the course of the hydroformylation of ethene with $[RhH(CO)_2(PH_3)_2]$ (**20**) as model catalyst (Figure 5.4). The overall reaction is exergonic. No intermediate is thermodynamically too stable, and no partial reaction shows an exceptionally high activation barrier. To understand solvent effects at least to some degree, ethene has been selected as a model solvent. The coordinatively unsaturated (square-planar) complexes (**21**, **23**, **25**, **27**), in particular, are substantially stabilized through solvation (here: $H_2C=CH_2$ coordination), while the Gibbs free energies of the transition states change only slightly. This leads to substantial changes in the activation barriers of the individual reaction steps. Thus, this example demonstrates that only limited conclusions about the course of reaction in solution can be drawn from quantum-chemical calculations from the reaction course in the gas phase.

With ethanol as the solvent, $[Rh(acac)(CO)_2]/PEt_3$ catalyzes a hydrocarbonylation of olefins, leading directly to alcohols, without aldehydes appearing as intermediates:

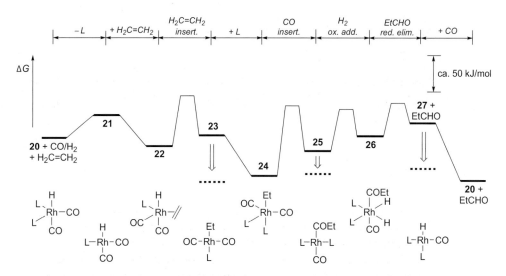

Figure 5.4 Quantum-chemically calculated profile of the Gibbs free energy ΔG ($T = 298$ K) for the hydroformylation of ethene with $[RhH(CO)_2(PH_3)_2]$ (**20**) as model catalyst (L = PH_3). The dashed lines take solvent effects ($H_2C=CH_2$) into account (simplified, after Matsubara and Morokuma [82]).

5 Hydroformylation of Olefins and Fischer-Tropsch Synthesis

$$R-CH=CH_2 + CO + 2 H_2 \xrightarrow[125\ °C]{[Rh]} R-CH_2-CH_2OH$$

For insight into the reaction, we start with the intermediate **15** (compare to Figure 5.3). With L = PPh$_3$, hydroformylation occurs with formation of aldehydes (reaction path **a**), as previously discussed. With PEt$_3$ as coligand, protonation of the acyl oxygen atom could occur, which would form a cationic hydroxycarbene complex **15'**, opening the reaction path **b** to formation of alcohols. The solvent (e.g., ethanol) is assumed to be the proton source.

$$R\text{-}CH_2\text{-}CHO \xleftarrow{H_2, \, L=PPh_3} \underset{\mathbf{15}}{OC-\overset{L}{\underset{L}{Rh}}-\overset{O}{\underset{}{C}}-CH_2R} \xrightleftharpoons[-H^+]{+H^+} \underset{\mathbf{15'}}{\left[OC-\overset{L}{\underset{L}{Rh}}=\overset{OH}{\underset{}{C}}-CH_2R\right]^+} \xrightarrow{H_2, \, L=PEt_3} R\text{-}CH_2\text{-}CH_2OH$$

(a) ... (b) ...

Exercise 5.2
- Formulate a possible mechanism for hydrocarbonylation of alkenes, starting from **15'** with oxidative addition of H$_2$, followed by migration of H from Rh onto the carbene ligand, and finally leading to **11** (Figure 5.3).
- What causes the different reactions of the catalysts with PPh$_3$ and PEt$_3$ ligands? With P(i-Pr)$_3$ as ligand, the formation of aldehydes predominates again. Give an explanation.

5.3
Enantioselective Hydroformylation

The use of rhodium catalysts with optically active phosphane ligands enables enantioselective hydroformylation of prochiral olefins. If the olefins are terminal, like styrene, only Markovnikov addition produces a chiral aldehyde **28**, while the unbranched aldehyde **29** is achiral. Enolizable optically active aldehydes can undergo relatively rapid racemization, which must be considered when carrying out the reaction.

$$PhCH=CH_2 \xrightarrow{CO/H_2, \, cat.} (S)\text{-}\mathbf{28} + (R)\text{-}\mathbf{28} + \mathbf{29}$$

The stereochemistry of the aldehyde formed in the reaction is determined by the olefin coordination (Re vs. Si), since the subsequent steps proceed in a stereochemically uniform manner, namely via a *cis* olefin insertion and a migratory

5.3 Enantioselective Hydroformylation

CO insertion (**12** → ... → **15**, Figure 5.3), so *syn* addition of H/CHO to the olefin occurs.

An efficient precatalyst for asymmetric hydroformylation of olefins like styrene is the complex [Rh(acac){(R,S)-BINAPHOS}] (**30**), which contains a phosphane–phosphite chelating ligand (**33**) for chiral induction. In the presence of CO/H$_2$ at normal pressure and room temperature, a dicarbonyl–hydrido complex **31** is formed, which can coordinate the olefin after cleaving off a CO ligand (**31** → **32**) [83].

As shown in **32**, the C_1-symmetric (R,S)-BINAPHOS (**33**) coordinates with the phosphite *P* atom in an apical position (*trans* to the hydrido ligand) and with the phosphane *P* atom in an equatorial position (*ae* coordination). The reverse coordination and *ee* coordination are not significant. The hydrido ligand is likewise apically coordinated. The overall result is a very stable configuration, which is *one* precondition for the very high enantiomeric excess that is to be achieved. Nevertheless, due to the trigonal-bipyramidal structure of the complex, the reasons for the stereodifferentiation are more complex than for square-planar complexes, which we have encountered in enantioselective hydrogenation with cationic Rh^I complexes (see Section 4.3.3).

For an explanation, we start, for the sake of simplicity, from a C_2-symmetric chiral P⌒P ligand, which only coordinates apical-equatorially (*ae*). The hydrido ligand occupies the other apical position. The precatalyst is shown in Figure 5.5. The two CO ligands (CO^1 and CO^2) are diastereotopic. There are two different coordination pockets for the olefin. One is created by cleaving off CO^1 (reaction path **a/b**) and the other by cleaving off CO^2 (reaction path **c/d**). Thus, substitution of CO by a prochiral olefin (here PhHC=CH$_2$) leads to four different catalyst–substrate complexes, since the olefin can coordinate with either the *Re* or *Si* face (**34**$_{re}$/**34**$_{si}$ and **35**$_{re}$/**35**$_{si}$). Another four catalyst–substrate complexes, in which the phenyl group points upwards, need not be considered here, since in the subsequent insertion reaction they yield a [Rh]–CH$_2$CH$_2$Ph complex and thus open the reaction channel to the unbranched (achiral) aldehyde. Now the C–C double bond is inserted into the Rh–H bond. Hence, via one of the four transition states **34′**$_{re}$/**34′**$_{si}$ and **35′**$_{re}$/**35′**$_{si}$, a [Rh]–CHPh–CH$_3$ complex is formed. The subsequent reaction steps proceed in a stereochemically consistent

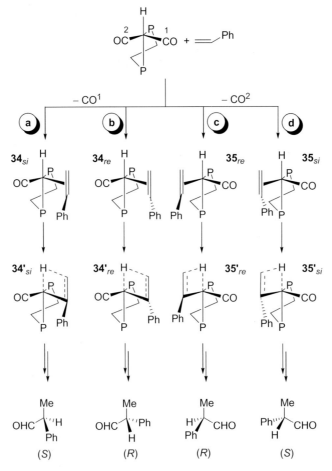

Figure 5.5 Stereodifferentiation in trigonal-bipyramidal rhodium complexes with an ae-coordinated C_2-symmetric P⌢P ligand (shown schematically) and an apically coordinated hydrido ligand (based on Gleich and Herrmann [84]).

manner. Thus, coordination of the *Si* face of the olefin gives rise to the formation of the S-configured aldehyde and vice versa (*Re* coordination ↔ (R)-aldehyde).

This makes it possible to formulate preconditions necessary for high *ee* values [84]:

- *Requirement of synchronous asymmetric induction.* If both reaction paths (cleavage of CO^1 and cleavage of CO^2) are significant, they must lead to the same reaction product ((R)- or (S)-aldehyde). Thus, reactions **a** and **d** in Figure 5.5 or reactions **b** and **c** must occur. If reactions **a** and **c** or reactions **b** and **d** were favored, the stereoselectivities would be at least partially compensated.
- *Requirement of preferred asymmetric induction.* The formation of a single favored stable ligand arrangement, so that only one reaction channel (Figure 5.5, **a/b** or **c/d**)

is opened, fosters high *ee* values. Equatorial-equatorial coordination of the P⌒P ligand should also be considered, though that has not been done here. Within certain limits, the preference for one coordination (*ae* vs. *ee*) can be predicted and controlled by the "bite angle" of the chelating ligand. For large bite angles, an *ee* coordination is favored (see Background, Section 5.4.1).

- Each of the two reaction channels in Figure 5.5 (cleavage of CO^1 (**a/b**) vs. CO^2 (**c/d**)) must favor one reaction with sufficient selectivity (either **a** or **b** in the first case, and either **c** or **d** in the second). This reaction control follows thermodynamic and kinetic regularities similar to those that have been discussed for enantioselective olefin hydrogenation on square-planar Rh complexes.

Thus, it is basically understood why many C_2-symmetric P⌒P ligands, which lead to high *ee* values for homogeneous hydrogenation on square-planar complexes, cause only poor stereodifferentiation, or none at all, for asymmetric hydroformylation. This also explains why high *ee* values in asymmetric hydroformylation were first achieved with (square-planar) platinum complexes, although platinum catalysts are in many cases clearly inferior to rhodium catalysts with respect to activity and regioselectivity (*n*/iso) in hydroformylation. An example of a platinum-catalyzed hydroformylation with (*R,R*)-DBD-DIOP (**36**) as coligand is offered below (chemoselectivity: 80% aldehydes; regioselectivity branched/linear: 77/23). DIOP itself, the "parent compound" for the C_2-symmetric bis(phosphane) ligands with chirality in the carbon backbone, only yields *ee* values of 26%, and furthermore, the branched aldehyde is only the minor product (branched/linear: 23/77) [85].

For both platinum and rhodium systems, there are now many chiral ligands with which *ee* values of over 90% for asymmetric hydroformylation are achieved. For rhodium complexes, in addition to the already mentioned phosphane–phosphite chelating ligand (*R,S*)-BINAPHOS (**33**), bis(phosphite) ligands of type **37** should be cited [86].

Optically active aldehydes and thus enantioselective hydroformylation reactions have a broad potential for applications. This is shown, for example, by the production of amino acids via Strecker synthesis and the generation of 2-arylpropionic acids via the reaction sequence $ArHC=CH_2 + CO/H_2 \rightarrow$ (S)-ArC*HMe–CHO $\rightarrow$ (S)-ArC*HMe–CO$_2$H, which are important anti-inflammatory agents (Ibuprofen: Ar = 4-isobutylphenyl; Naproxen: Ar = 6-methoxynaphth-2-yl).

5.4
Significance of Hydroformylation and Outlook

After the polymerization reactions of olefins and dienes, hydroformylation is the most important organometallic catalytic reaction in terms of quantity and value in the chemical industry. In 2008, the worldwide capacity for production of aldehydes via hydroformylation was about $12 \cdot 10^6$ metric tons. Of this, about 75% was the hydroformylation of propene and about 20% that of C_4–C_{12} olefins. About half of the butyraldehyde is processed via aldol condensation and subsequent hydrogenation into 2-ethylhexan-1-ol. This is further processed into dioctyl phthalate (bis(2-ethylhexyl) phthalate), used as a plasticizer for PVC. About a quarter of the butyraldehyde is hydrogenated into butanol, which is used as a solvent and serves as the starting material for synthesis of various esters. Industrial significance has been achieved almost exclusively for hydroformylation of terminal olefins.

As a rule of thumb for unfunctionalized olefins (R = alkyl), terminal linear olefins (**38**) are hydroformylated more easily than internal linear olefins (**39**). The same applies to branched olefins (**40/41**). Since the attachment of the formyl group with formation of a quaternary C atom is not very likely, hydroformylation of the olefins **40** and **41** are quite regioselective and hydroformylation of olefins of type **42** is generally not possible.

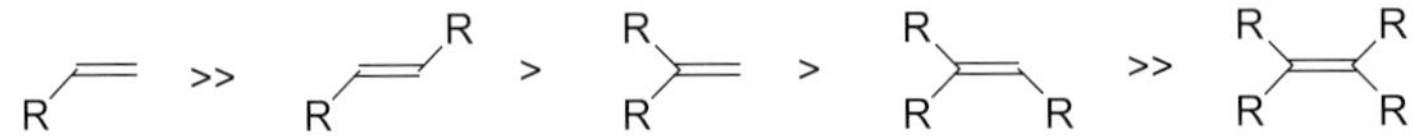

Hydroformylation catalysts can also catalyze double-bond isomerization. This phenomenon is used to selectively produce n-aldehydes from mixtures of linear inner olefins, which are created in the Shell Higher Olefin Process (SHOP).

Exercise 5.3
Which properties must a hydroformylation catalyst have if it is to favor production of n-aldehydes from a mixture of linear olefins with inner and terminal double bonds? Base your discussion on the following reaction scheme:

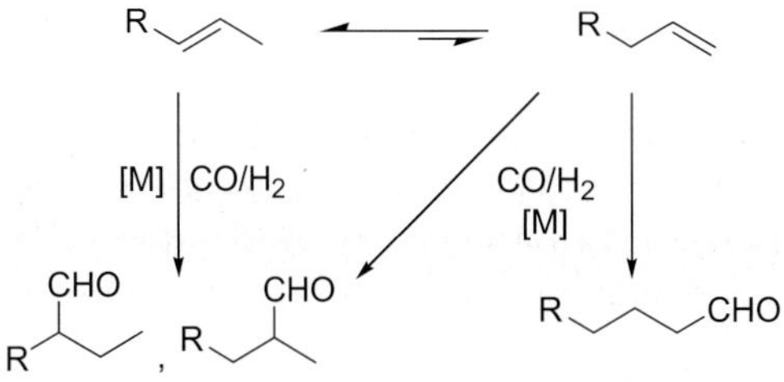

The following examples demonstrate new developments and additional industrial applications of hydroformylation.

5.4.1
Diphosphites as Ligands

If bulky diphosphites of type **43** (where the unsubstituted parent compound is shown) are used instead of triphenylphosphane as coligand in Rh-catalyzed hydroformylation reactions, the process is significantly improved. With careful cleaning of propene and synthesis gas (CO/H_2), n-butyraldehyde (n/iso ratio 30/1) is obtained with high selectivity (99%). The activity of the catalyst is so high that a recycle stream can be dispensed with, which significantly simplifies the technology [81].

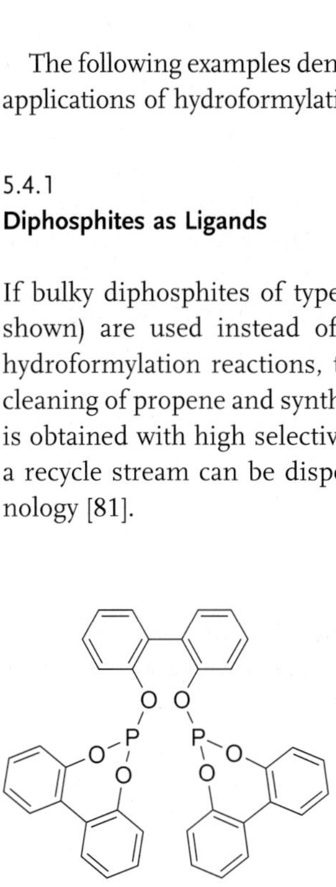

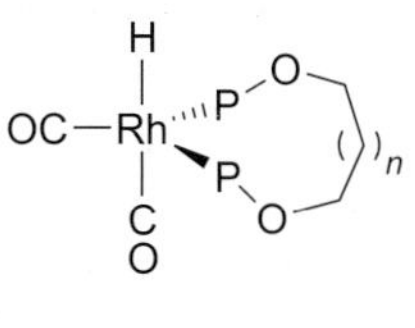

For these catalyst properties, a bis-equatorial (*ee*) coordination (complex type **44**; ligand structure shown schematically) seems to be a necessary (but not sufficient) precondition. In contrast, complexes of type **45** with an apical-equatorial (*ae*) chelating ligand coordination give worse n/iso ratios. Chelating ligands with large "bite angles" tend toward bis-equatorial coordination (**44**) and those with a smaller bite toward *ae* coordination (**45**). This enables targeted catalyst development [87, 88].

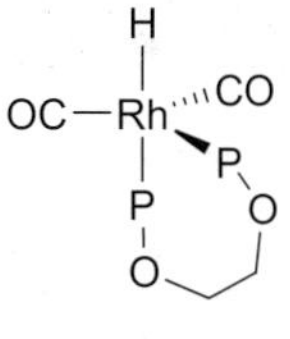

Background: The "Bite" of P,P-Chelating Ligands

Complexes with P,P-chelating ligands permit fewer isomers and are sterically less flexible than those with two monodentate P-ligands. Furthermore, coordination sites are more reliably blocked by the chelating ligand than by monophosphanes. All of this opens the possibility of better control over the regioselectivity and stereoselectivity of homogeneously catalyzed reactions. The effect of the chelating ligand on the catalyst properties can be influenced by the "bite" of the ligand, where steric (such as ligand–ligand/ligand–substrate interactions or fixing of conformations or configurations) and/or electronic effects (influence on orbital energies) can be critical.

Bis(phosphane) ligands can be synthesized with broad variation in the "bite angle" P–M–P (β_n). Since the bite angles depend substantially on the M–P bond lengths, a normalized M–P distance of 2.315 Å has been used. They are either obtained from quantum-chemical calculations or taken from single-crystal structure data. Examples are given in the following table [87, 89, 90].

Ligand	β_n[a] (in °)	Ligand	β_n[a] (in °)
1, n=1 (dppm)	72	5 (BISBI)	113 (92–155)
1, n=2 (dppe)	84 (70–95)	6 (TRANSPHOS)	111
1, n=3 (dppp)	91	7 (NORPHOS)	123 (110–145)
1, n=4 (dppb)	98	8 (DPEphos)	102 (86–120)
2 (DIOP)[b]	102 (90–120)	9 (DBFphos)	131 (117–147)
3 (BINAP)[b]	92	10 (Xantphos)[c]	112 (97–135)
4 (Me-DuPHOS)[b]	83	11 (DPPF)[c]	96

a) The numbers in parentheses give the flexibility range (obtained from molecular mechanics calculations) that is defined as the accessible range of bite angles within less than 12.6 kJ/mol excess strain energy starting from β_n.
b) For formula, see Figure 4.5, Section 4.3.1.
c) For formula, see Section 12.4.2.2.

For steric and electronic effects of monodentate P-ligands, see Background in Section 10.4.2.2.

5.4.2
Biphasic Catalysis

Rhodium complexes with sulfonated arylphosphane ligands (e.g., [RhH(CO){P(C_6H_4-m-SO_3Na)$_3$}$_3$]) are water-soluble. This makes a continuously operated biphasic process possible. The catalyst remains in the aqueous phase. The aldehyde, which is immiscible with water, forms the organic phase and is easily isolated by phase separation. This process (Ruhrchemie/Rhône-Poulenc) has been used since 1984 on a large scale for synthesis of butyraldehyde and for hydroformylation of other short-chain linear α-olefins.

This process is an excellent application of biphasic catalysis. At relatively low pressure (about 40–60 bar) and low temperature (110–130 °C), propene is converted with high selectivity into C_4 aldehydes (99%) and with high regioselectivity into n-butyraldehyde (n/iso about 20/1). Additional advantages are the very low Rh losses ($<10^{-9}$ g Rh/kg PrCHO) and very simple technology meeting high ecological standards, with cost savings of about 10% relative to the conventional process. A disadvantage is that the process is not suitable for higher olefins, since these are not sufficiently soluble in water. About 10% of the C_4 and C_5 aldehydes are produced using the Ruhrchemie/Rhône-Poulenc process.

Exercise 5.4

Rhodium complexes with triphenylphosphane ligands **1**, which are functionalized with weakly basic amidine groups ([Rh(acac)(CO)$_2$]/**1** = 1/50), catalyze hydroformylation. The catalyst is soluble both in toluene and in CO_2-saturated water (why?). On this basis, develop a process for hydroformylation of a hydrophobic and of a hydrophilic olefin (oct-1-ene, allyl alcohol), that takes advantage of pH-dependent solubility for catalyst separation.

Ionic liquids (see Background) can be used as alternative solvents. In particular, they dissolve ionic catalyst complexes well, while the organic products (here aldehydes) form a second phase, resulting in a liquid-liquid biphasic reaction. A fast catalytic reaction, low solubility of the reactants in the ionic liquids, and mass transfer limitations can cause the ionic liquid to become impoverished in reactant and the reaction to take place primarily at the interphase or in the diffusion layer rather than in the bulk solvent. This makes it attractive to carry out a heterogenized variant of the reaction. The ionic liquid in which the catalyst is dissolved is immobilized in a thin film on the surface of a highly porous solid support material with large surface area (supported *ionic liquid-phase* (SILP) catalysis) [91]. Now, the preconditions have been met for a continuous gas-phase reaction that runs as a gas-liquid biphasic reaction. An example is the reaction of propene and butene with a rhodium catalyst that has coordinated sulfonated Xantphos (L) [92]:

	R	Me	Et
TOF (in h^{-1})		308	647
n/iso		18/1	41/1

Reaction scheme: R-CH=CH₂ (2 bar) + CO/H$_2$ (1/1, 10 bar); 120 °C, SiO$_2$ → R-CH$_2$-CHO + R-CH(CHO)-

catalyst [Rh(acac)(CO)$_2$] + L (1 / 10)

L = xanthene ligand with NaO$_3$S, SO$_3$Na substituents and Ph$_2$P, PPh$_2$ groups

ionic liquid: N-methylimidazolium cation with octyl-OSO$_3^-$ anion

Background: Ionic Liquids

Nonaqueous ionic liquids are salts that melt below 100 °C, but are typically already fluid at room temperature. They consist of bulky organic cations (see the examples 1–5; R, ..., R''' = alkyl) and mostly weakly coordinating anions such as $[BF_4]^-$, $[PF_6]^-$, $[SbF_6]^-$, $[AlCl_4]^-$, $CF_3CO_2^-$, $CF_3SO_3^-$, RSO_3^-, $ROSO_3^-$.

Structures 1–5: ammonium, phosphonium, imidazolium, pyridinium, sulfonium cations.

A wide variation in physical and chemical properties can be obtained by assembling cations and anions in various combinations. As expected for salts, they display a high ionic conductivity and have a negligible vapor pressure below their decomposition temperature. In ionic liquids, high polarity can be associated with low nucleophilicity and they can – for example, with tetrafluoroborate or triflate anions – be stable against hydrolysis. Ionic liquids are an attractive alternative to conventional organic solvents due to their solubility properties for organic and inorganic compounds and a defined miscibility (or immiscibility, as the case may be) with other liquids, in addition to, in general, high environmental friendliness and nonflammability.

Ionic liquids have already found an opening as a new kind of solvent in organometallic catalysis. Depending on the structure of the cation and anion, they display very different Lewis basicities and acidities. This permits a targeted influence on catalyst–solvent interactions, which enables optimization of catalytic processes with regard to selectivity, activity, and stability. Functional groups can be incorporated into the anion or the cation of ionic liquids to exploit their specific interactions for catalytic applications, for which the term "task-specific ionic liquids" was coined. A multiphase reaction can often be carried out with ionic liquids, so they represent suitable solvents for biphasic catalysis. However, the need for great quantities of an ionic liquid may be a disadvantage for industrial applications [93–96].

5.4.3
Synthesis of Vitamin A

Vitamin A (**52**) is produced in a quantity of several thousand tons a year. The synthesis from 1,2-diacetoxy-but-3-ene (**46**) (BASF) or from 1,4-diacetoxy-but-2-ene (**47**) (Hoffmann-La Roche) includes a hydroformylation step. Special reaction conditions and use of an unmodified rhodium catalyst in the BASF synthesis guarantee the (normally undesired) formation of the branched aldehyde (**46** → **48**), while in the hydroformylation **47** → **49**, there is no problem regarding regioselectivity. Conversion of **48** or **49** into **50** and subsequent Wittig reaction ultimately produces the esterified vitamin A (**51**), where the arrow points to the newly formed C–C bond [M6].

5.4.4
Carbon Dioxide as Alternative to CO

The substitution of toxic by less toxic reactants is one of the ongoing challenges in chemistry and the chemical industry. For hydroformylation reactions, there is the possibility of replacing the highly toxic carbon monoxide by the nontoxic, abundant, and also inexpensive carbon dioxide when the hydroformylation reaction is coupled with a reduction of CO_2 by H_2 [92].

Such reactions proceed in two steps: First, CO_2/H_2 is converted to CO/H_2O via the reverse water-gas shift (Section 6.5.1). This is followed by the actual olefin hydroformylation, though it generally leads to the alcohols as main products due to the hydrogenation activity of the catalysts used. The conversion of hex-1-ene to heptanol with a ruthenium carbonyl as catalyst in an ionic liquid as solvent can be cited as an example. It is advantageous that CO_2 is highly soluble in ionic liquids. The hydrogenation activity of the catalyst used leads to formation of hexane as a side product.

$$n\text{-Bu-CH=CH}_2 \xrightarrow[\text{[Ru}_3\text{(CO)}_{12}\text{]}]{\text{CO}_2/\text{H}_2 \text{ (1/1, 80 bar); 160 °C}} n\text{-Bu-CH}_2\text{-OH} + n\text{-Bu-CH(OH)-}n\text{-Bu} + n\text{-Bu-H} + H_2O$$

in: [bmim][Cl] / [(CF$_3$SO$_2$)$_2$N]

(82%) TOF = 16 h^{-1} (8%)

5.4.5
Combinatorial and Supramolecular Catalysis

The hydroformylation of terminal olefins such as oct-1-ene is catalyzed by [Rh(acac)(CO)$_2$] in the presence of a bidentate P,P-ligand (L⌒L'). Both halves of the chelating ligand L⌒L' can now be linked by hydrogen bonds, instead of covalent bonds, as is the case with conventional chelating ligands. One example is the catalyst complex **53** (D–H/D'–H = proton donor; A/A' = proton acceptor), in which a linkage analogous to Watson-Crick base pairing between adenine (D–H = N–H, A = N) and thymine (D'–H = N–H, A' = O) occurs.

$$n\text{-Hex-CH=CH}_2 \xrightarrow[\text{toluene, 80 °C}]{\text{CO/H}_2 \text{ (1/1, 10 bar)}} n\text{-Hex-CH}_2\text{CH}_2\text{CHO} + n\text{-Hex-CH(CHO)-CH}_3$$

[Rh(acac)(CO)$_2$] + L / L'

Complex **53**: L and L' linked by D–H···A' and A···H–D' hydrogen bonds, both bearing Ph$_2$P groups coordinated to [Rh].

Complex **53'**: L has F$_3$C(O)C group with N–H···O and N···H–N hydrogen bonds to L' (benzothiazole-based), both bearing Ph$_2$P groups coordinated to [Rh].

In all, five ligands L and two ligands L' were introduced, resulting in a total of 10 different combinations of L⌒L'. It was shown that L and L' formed a self-organizing ligand system with which excellent regioselectivities (up to $n/iso > 99/1$ in the combination **53'**, $TOF = 3900 \text{ h}^{-1}$) were reached. Thus, an exemplary library of monodentate ligands that organized themselves into chelating ligands was designed. This is a new kind of combinatorial approach to libraries of chelating ligands, which is simpler from a synthetic perspective [97].

The efficiency of enzymatic catalysis, one of the great examples of homogeneous catalysis, depends in part on what is generally very high substrate selectivity, for which the protein component is responsible. In homogeneous catalysis, it can be modeled with ligands L, which, in addition to the ligating atom, have other binding sites available for a substrate. An example is the ligand **54**, which has bound an acylguanidine group in its periphery. This enables molecular recognition of carboxylic acid functionality via hydrogen bonds; see the schematic depiction of the presumed transition state (**55**).

In the 1/1 mixture of **56a/57a**, **56a** is converted with high substrate selectivity (10:1), where **56b** is formed as the main product with high regioselectivity (50:1). The hydroformylation of the doubly unsaturated carboxylic acid **58a** produces **58b** as the main product, which serves as further evidence of the directing effect of the acylguanidine group in L (**54**). This is one example how the formation of a supramolecular catalyst complex ([Rh(acac)(CO)$_2$] + **54**) can substantially increase the selectivity of homogeneously catalyzed reactions [98].

5.5
The Fischer-Tropsch Synthesis

In 1913, A. Mittasch and C. Schneider (BASF) succeeded in converting synthesis gas (CO/H$_2$) in the presence of iron oxide catalysts at elevated pressure and temperature to a mixture of higher hydrocarbons and oxygen-containing compounds (alcohols, acids, esters, etc.). Continuing along this line, F. Fischer and H. Tropsch (KWI – today MPI – for Coal Research in Mülheim/Ruhr, Germany) developed a high-pressure synthesis (10–15 MPa, 400 °C) in 1922, using alkalized iron contact catalysts. In 1925, they succeeded in performing a synthesis at normal pressure in which hydrocarbons appeared as the main products. In 1936, the first large industrial plant at Ruhrchemie with a Kieselguhr-supported Co–ThO$_2$–MgO catalyst went into operation. This indirect coal liquefaction (coal → synthesis gas (CO/H$_2$) → hydrocarbons/oxygenates) achieved great industrial significance, but then subsided due to the rise of petrochemistry. The growing scarcity of petroleum resources and the possibility of producing synthesis gas from biomass could lead in the future to renewed interest in Fischer-Tropsch synthesis.

Although the Fischer-Tropsch synthesis is in the domain of heterogeneous catalysis, we will deal with it here, since the organometallic chemistry has contributed substantially to our understanding of mechanism. Furthermore, it is an instructive example of the mechanistic complexity of heterogeneously catalyzed reactions. We will limit ourselves to the "hydrogenolysis" of CO into hydrocarbons according to the following equation:

$$n\text{ CO} + 2n\text{ H}_2 \xrightarrow{\text{cat.}} -(\text{CH}_2)_n- + n\text{ H}_2\text{O}$$

$$\left[\diagup\!\!\!\diagdown_n \diagup\!\!\!\diagdown, \text{CH}_4, \diagup\!\!\!\diagdown_n \diagup\!\!\!\diagdown \right]$$

Fischer-Tropsch reactions are not very selective and produce a wide palette of hydrocarbons. Typical C-numbers n are 1–35. The process can be optimized so as to obtain up to 40% gasoline ($n = 5$–11), but due to the high proportion of linear hydrocarbons, its octane number is very low.

Primary products of Fischer-Tropsch synthesis are n-alkenes and methane. Alkanes are predominantly formed by subsequent hydrogenation. In addition, isomerizations and cyclizations occur. Although the formation of hydrocarbons from carbon monoxide and hydrogen is strongly exergonic ($\Delta G^\ominus = -151$ kJ/mol for CH$_4$; $\Delta G^\ominus = -783$ kJ/mol for C$_8$H$_{18(g)}$; $\Delta G^\ominus = -695$ kJ/mol for C$_8$H$_{16(g)}$), it requires a catalyst due to the kinetically inert reactants.

The substrate activation proceeds by dissociative chemisorption (see Figure 14.4, Section 14.2.2). That means that H$_2$ is adsorbed on the catalyst surface and the H–H bond is cleaved with formation of surface-bound H atoms (metal hydrides) {H$_{(s)}$}.[2] CO is activated via coordination to metal atoms on the catalyst surface. C–O cleavage and hydrogenation then follow. At this point the following species are sequentially formed: carbide {C$_{(s)}$} (**59**), methylidyne {CH$_{(s)}$} (**60**), methylene {CH$_{2(s)}$} (**61**) and methyl {CH$_{3(s)}$} (**62**). Each species is bound to metal surface atoms. The C$_1$ component **62** can also be released, forming methane, an inherent product of Fischer-Tropsch synthesis.

All these surface-bound organic groups are well known as ligands in metal complexes. Many methylene and methylidyne ligands bound to a metal with a double ([M]=CH$_2$, **61′**) or triple ([M]≡CH, **60′**) bond, respectively, can be found in

---

[2] The index "(s)" (from "surface") refers to the binding to the metal surface atoms.

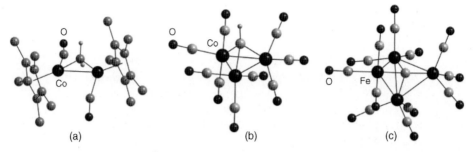

Figure 5.6 Structures of [{Co(η^5-C$_5$Me$_5$)(CO)}$_2$(μ-CH$_2$)] (H atoms of the C$_5$Me$_5$ ligands are not shown) (a), [{Co(CO)$_3$}$_3$(μ_3-CH)] (b) and the anion in [NMe$_3$(CH$_2$Ph)]$_2$[{Fe(CO)$_3$}$_4$(μ_4-C)] (c).

Schrock-type alkylidene or alkylidyne complexes, especially for electron-deficient transition metals in high oxidation states. In numerous other complexes, they appear as bridging ligands μ_2-CH$_2$ (**61″**) or μ_3-CH (**60″**). In terminal ligands, the C atom can be considered to be sp^2- (**61′**) or sp-hybridized (**60′**), while in the bridging ligands (**61″/60″**), sp^3-hybridized C atoms are assumed. The examples in Figure 5.6a/b can be seen as model complexes for the bonding of ligands of types **61″** and **60″**.

Aside from long-known clusters with interstitial C atoms (example: [Rh$_6$(μ_6-C)(CO)$_{13}$]$^{2-}$), which, however, cannot be regarded as model complexes for surface-bound C atoms, those with terminal carbido ligands [M]≡C| (example: [Ru(≡C)-Cl$_2$(PCy$_3$)$_2$], Ru≡C 1.632(6) Å [99]) have also been described. Complexes with bridging carbido ligands of type [M]=C=[M] (example: [LFe=C=FeL], H$_2$L = 5,10,15,20-tetraphenylporphine) and [M]≡C–[M] (example: [(PCy$_3$)$_2$Cl$_2$-Ru≡C–PdCl$_2$(SMe$_2$)]) are known and the structure in Figure 5.6c with a μ_4-C ligand shows a coordination pattern that is, in principle, also conceivable on metal surfaces.

5.5.1
Mechanism

In Fischer-Tropsch synthesis, hydrocarbon formation occurs by oligomerization of surface-bound methylene (**61**) or methylidyne species (**60**). The following mechanisms are to be considered [100–102]:

- *The alkyl mechanism.* According to the alkyl mechanism, oligomerization starts with C–C coupling between {CH$_{2(s)}$} (**61**) and {CH$_{3(s)}$} (**62**) (see **a**) and then takes place according to **b** (R = growing alkyl chain). One option for terminating the chain is β-hydride elimination, which releases an α-olefin (**c1**). Alternatively, with participation of a metal hydride, a bimolecular reductive elimination can occur, which results in the primary formation of an alkane (**c2**). Under Fischer-Tropsch conditions, olefins are generally hydrogenated to form alkanes, so nothing about the mechanism can be deduced from the experimentally determined ratio of the two products. It is likely that initial formation of the olefins predominates.

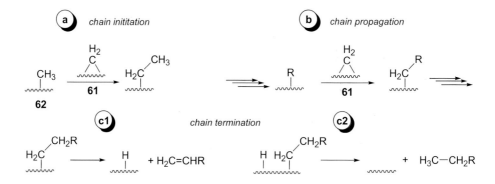

- The alkenyl mechanism. According to the alkenyl mechanism, the oligomerization is triggered by a surface-bound unsaturated C_2 component, namely a vinyl {CH=CH$_{2(s)}$} (63) created by the reaction between two C_1 components (60 + 61) (a). The oligomerization is based on an insertion of 61 into a metal–alkenyl bond, forming an allyl that in turn is converted by 1,3-H shift into an alkenyl (b). Chain termination can occur with the participation of a metal hydride through reductive elimination (c). Primary products are α-olefins that can be further hydrogenated, forming alkanes.

- **The alkylidene mechanism.** First, a surface-bound vinyl **63** is generated as in the alkenyl mechanism, which is followed by an H addition to the β-C atom, yielding an ethylidene species **64 (a)**. The chain growth takes place through reaction of **64'** with a methylidyne **60**, forming a vinyl, which is converted into a surface-bound alkylidene via H addition to the β-vinyl C atom **(b)**. Similarly to the alkenyl mechanism, H addition to the α-vinyl C atom results in a reductive elimination, causing chain termination **(c)**.

Particularly useful experiments for clarifying the mechanism have taken place on metal surfaces loaded with methylene groups by decomposition of diazomethane. In addition, experiments have been performed with deuterated and ^{13}C-labeled compounds, and with reaction mixtures to which ethene has been added in a targeted way. Nevertheless, some of the mechanisms that have been mentioned remain speculative, and not all experimental findings can be easily explained. The most energetically demanding is C–C coupling between two sp^3-hybridized C atoms, so the alkyl mechanism is less likely.

At least in hydrocarbon synthesis, cleavage of the C–O bond occurs in a very early stage of the reaction and the hydroxycarbene mechanism, which was earlier favored, no longer seems to play a role. It is based on the formation of a surface-bound hydroxycarbene **66** as a central intermediate, created by successive addition of $\{H_{(s)}\}$ to coordinated CO via a formyl intermediate **65**.

In contrast to the previously described mechanisms, here the oligomer chain grows before the C–O bonds have fully broken. It is not certain to what extent hydroxycarbene intermediates play a role in the formation of oxygen-containing products.

6
Carbonylation of Methanol and Water-Gas Shift Reaction

6.1
Principles

Acetic acid is one of the most important intermediate products in the chemical industry. In 2009, the worldwide production was about $8 \cdot 10^6$ metric tons. It is mainly used in the synthesis of polyvinyl acetate and cellulose acetate.

The most important industrial process for production of acetic acid is the carbonylation of methanol, about 80% of which is based on:

$$\text{MeOH} + \text{CO} \xrightarrow{\text{cat.}} \text{MeCOOH}$$

Under standard conditions, the reaction is thermodynamically feasible ($\Delta G^\ominus = -86$ kJ/mol), but requires a catalyst.

Acetic acid is also produced by direct oxidation of saturated hydrocarbons (especially butane):

$$\text{MeCH}_2\text{CH}_2\text{Me} \xrightarrow[\text{cat.}]{+ 5/2\ O_2} 2\ \text{MeCOOH} + H_2O$$

The reaction proceeds according to a radical mechanism and can be catalyzed by Mn, Co, Ni or Cr compounds (150–200 °C, 5–6 MPa). The oxidation of actetaldehyde by oxygen (60–80 °C, 0.3–1 MPa), forming acetic acid, is also radically catalyzed; transition metal compounds (Mn, Co, Cu) are the catalysts.

$$\text{MeCHO} \xrightarrow[\text{cat.}]{+ 1/2\ O_2} \text{MeCOOH}$$

The biotechnological production of 4–12% aqueous acetic acid via fermentation has been known for more than 5000 years. The worldwide production of 10% acetic acid by fermentation amounts to about $2 \cdot 10^6$ metric tons per year.

$$\text{EtOH} \xrightarrow[\text{acetobacter}]{+ O_2 / -H_2O} \text{MeCOOH}$$

The first industrial process for carbonylation of methanol was developed at BASF (Ludwigshafen, Germany) at the end of the 1950s. In 1960, a plant for the BASF acetic

Fundamentals of Organometallic Catalysis. First Edition. Dirk Steinborn
Copyright © 2012 WILEY-VCH Verlag GmbH & Co. KGaA, Weinheim

acid high-pressure process (250 °C, 70 MPa) went into operation. An especially corrosion-resistant material, a Ni–Mo/Cr alloy ("Hastelloy"), was required for the construction of the reactor. In the BASF process, CoI_2 was used for the *in situ* generation of $[Co_2(CO)_8]$ (**1**) and HI. Under the reaction conditions, **1** reacts to form $[CoH(CO)_4]$ (**2**), which can be regarded as a water-gas shift reaction. The actual catalyst is yielded by deprotonation of **2**.

$$2\ CoI_2 + 2\ H_2O + 10\ CO \longrightarrow \underset{\mathbf{1}}{[Co_2(CO)_8]} + 4\ HI + 2\ CO_2$$

$$\underset{\mathbf{1}}{[Co_2(CO)_8]} + H_2O + CO \longrightarrow \underset{\mathbf{2}}{2\ [CoH(CO)_4]} + CO_2$$

The selectivity of acetic acid is about 90% based on methanol and about 70% based on CO (i.e., about 10% of the MeOH and 30% of the CO are consumed by side reactions). In 1968, Monsanto (St. Louis, Missouri) worked out a rhodium-catalyzed process that runs at significantly milder reaction conditions (150–200 °C, 3–6 MPa) and also displays a significantly higher selectivity (about 99% for MeOH and 90% for CO). The BASF process could not compete and is today only of historical interest. A substantial improvement of the Monsanto process was achieved in 1995–96 by BP Chemicals (Hull, England) through use of an iridium complex as catalyst in the Cativa process. Important parameters for processes of acetic acid production via carbonylation of methanol are collected in Table 6.1.

The low selectivity with respect to CO in the BASF and Monsanto processes can be traced to, respectively, the Co- or Rh-catalyzed water-gas shift reaction ($CO + H_2O \rightleftharpoons CO_2 + H_2$, see Figure 6.3). Hence, synthesis gas (CO/H_2) is formed, reacting under the influence of transition metal catalysts as a methylene equivalent ($CO + 2\ H_2 \rightarrow$ "$-CH_2-$" $+ H_2O$), thus forming oxygen-containing C_2 and C_3 compounds (**4**–**6**) as side-products:

$$\underset{\mathbf{3}}{MeOH + CO + H_2} \xrightarrow[-H_2O]{[M]} \underset{\mathbf{4}}{MeCHO} \xrightarrow[[M]]{+H_2} \underset{\mathbf{5}}{MeCH_2OH} \xrightarrow[[M]]{+CO} \underset{\mathbf{6}}{MeCH_2COOH}$$

The hydrocarbonylation (homologation) of methanol to form ethanol (**3** → **5**) can be Co-, Rh- or Rh/Ru-catalyzed so as to make ethanol the main product.

Table 6.1 Important process parameters for Co-, Rh- and Ir-catalyzed carbonylation of methanol, forming acetic acid (compiled from [103, 104, M6, M14]).

	Co	Rh	Ir
Industrial introduction	1960 (BASF)	1970 (Monsanto)	1995 (BP Chemicals)
T (in °C)	250	150–200	180
p (in bar)	600–700	30–60	30–40
Selectivity for MeOH (in %)	90	99	99.5
Selectivity for CO (in %)	70	90	>94
Important side-products	CH_4, CO_2, EtOH, MeCHO, EtCOOH	CH_4, CO_2, H_2, EtCOOH	negligible

6.2
The Monsanto Process

In the Monsanto process, the catalyst system consists of a rhodium(III) halide and a cocatalyst containing iodine (e.g., HI/H_2O). Under the reaction conditions, a square-planar dicarbonyldiiodorhodate(I) complex 7 is formed as a precatalyst, and methyl iodide is formed from MeOH/HI:

$$RhI_3 + 3\ CO + H_2O \longrightarrow [RhI_2(CO)_2]^- + I^- + 2H^+ + CO_2$$
$$7$$

$$MeOH + HI \rightleftharpoons MeI + H_2O$$

The mechanism of catalysis is shown in Figure 6.1. There are two cycles to be distinguished, namely the "rhodium cycle," the true organometallic-catalyzed reaction, and the "iodide cycle," which does not include any metal-catalyzed reactions. The rhodium-catalyzed formation of acetyl iodide from methyl iodide and CO is completed in the following steps:

7 → 8: *Oxidative addition/reductive elimination.* Oxidative addition of methyl iodide to the Rh^I complex 7 (16 ve) yields a coordinatively and electronically saturated methylrhodium(III) complex (18 ve). Under typical reaction conditions (with higher water content), this reaction is rate-determining. The reverse reaction, although possible in principle, does not play a role.

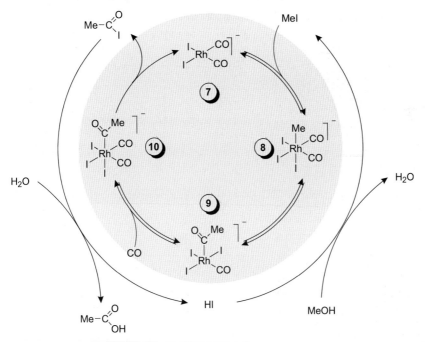

Figure 6.1 Reaction mechanism of rhodium-catalyzed carbonylation of methanol consisting of the rhodium cycle (shaded gray) and the iodide cycle.

8 → 9: *CO insertion/deinsertion.* Migration of the methyl ligand to a *cis* CO ligand (insertion of CO into the Rh–C bond) yields a coordinatively unsaturated (16 *ve*) acetylrhodium(III) complex **9**, which is a dimer in the solid state. The reaction is reversible.

9 → 10: *Ligand attachment/cleavage.* Addition of CO leads to an electronically (18 *ve*) and coordinatively saturated rhodium(III) complex.

10 → 7: *Reductive elimination.* In a reductive elimination, acetyl iodide is cleaved off, which restores the catalyst complex $[RhI_2(CO)_2]^-$ (**7**). Although this reaction is reversible in principle, the equilibrium lies practically entirely to the right. The acetyl iodide that has been cleaved off is rapidly (and irreversibly) converted to HOAc and HI [105].

The reaction profile obtained from kinetic data for the oxidative addition of MeI (**7 → 8**) and the subsequent migratory CO insertion (**8 → 9**) is shown in Figure 6.2. It can thus be concluded that in these model studies the oxidative addition is rate-determining and the methylrhodate(III) complex **8** is unstable with respect to both the reductive elimination (**8 → 7**) and the migratory CO insertion (**8 → 9**). As a result, the equilibrium concentration of **8** is very low, but it was possible to identify **8** through both IR and NMR spectroscopy in reaction mixtures $[RhI_2(CO)_2]^-$ (**7**)/MeI.

Exercise 6.1

Suggest a way to accelerate the rate-determining oxidative addition (**7 → 8**) in rhodium-catalyzed methanol carbonylation [106].

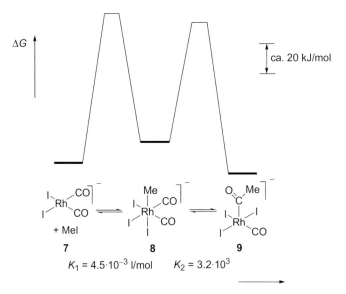

Figure 6.2 Profile of the Gibbs free energy of oxidative addition and CO insertion (**7 → 8 → 9**) during methanol carbonylation (35 °C, in CH_2Cl_2/MeI) (after Maitlis [105]).

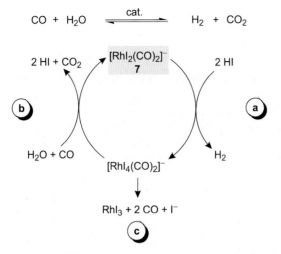

Figure 6.3 Water-gas shift reaction and its mechanism (schematic) with $[RhI_2(CO)_2]^-$ (7) as catalyst. Complex 7, which catalyzes both the water-gas shift reaction and methanol carbonylation, is shaded gray.

The Monsanto process is carried out in polar solvents (acetic acid/water). Special demands are placed on the reactor materials due to the highly corrosive acidic iodide-containing reaction solutions. Acetic acid is formed with a selectivity for methanol of about 99%. A disadvantage of the process is that rhodium complexes – among them $[RhI_2(CO)_2]^-$ (7), the complex active in methanol carbonylation – also catalyze the water-gas shift reaction. This leads to a decrease in selectivity with respect to CO. The water-gas shift reaction is based on two (complex) reactions (Figure 6.3):

- Reduction of H^+ to form H_2 and formation of a dicarbonyltetraiododrhodate(III) complex (a).
- Oxidation of CO to form CO_2, whereupon 7 is regenerated (b).

In addition, the Rh^{III} complex $[RhI_4(CO)_2]^-$ also tends toward decomposition with precipitation of RhI_3 (c).

The iodide cycle in the Monsanto process is shown in Figure 6.4. The core is the rhodium-catalyzed formation of acetyl iodide from methyl iodide and carbon monoxide (a). In the classical process, cycle I runs as follows: Methyl iodide is formed by reaction of MeOH with HI (b), and acetyl iodide is hydrolyzed to form acetic acid (c). In the reaction mixture, sufficient water is present to ensure that cycle II, the formation of methyl acetate from acetyl iodide and methanol according to d, as well as the reaction of methyl acetate with HI to form MeI and acetic acid (e), only plays a minor role.

Operating the process with a low quantity of water as in the Hoechst-Celanese technology causes cycle II to run to an increasing degree. Acetyl iodide then reacts with methanol (rather than with water) (d, Figure 6.4). The acetic acid ester thus formed is then available for reaction with HI, which leads to the formation of acetic

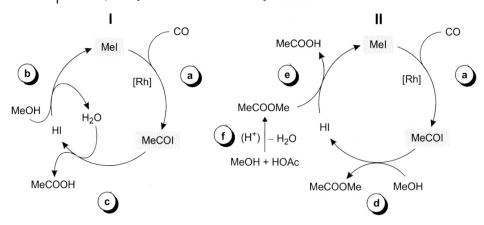

Figure 6.4 Iodide cycle in the classical Monsanto process (cycle I) and in the Hoechst-Celanese process (cycle II). The reactant and the product of the rhodium-catalyzed cycles are each shaded gray.

acid and MeI (e). At a high concentration of methyl acetate, which, under the reaction conditions, is also formed via acid-catalyzed esterification of HOAc (reaction f), the concentration of HI is low. This reduces the rate of reaction **a** (Figure 6.3) in the cycle of the water-gas shift reaction. Thus, the water-gas shift reaction as a whole is suppressed, resulting in a substantial increase in the selectivity with respect to CO during methanol carbonylation. However, the low concentration of HI would favor the formation of insoluble RhI_3 (see reaction **c**, Figure 6.3). Thus, iodides such as LiI, $[NR_4]I$ or $[PR_4]I$ are added to the catalyst system. An additional effect of the iodides, namely their function as promoters, will later be discussed in the context of acetic anhydride synthesis.

6.3
Synthesis of Acetic Anhydride

The rhodium-catalyzed carbonylation of methyl acetate produces acetic anhydride:

$$\text{MeCOOMe} + \text{CO} \xrightarrow[\text{(HI)}]{\text{[Rh]}} (\text{MeCO})_2\text{O}$$

The methyl acetate required is produced beforehand by acid-catalyzed esterification from acetic acid and methanol. In acetic anhydride synthesis, use is made of a Monsanto catalysator ([Rh]/MeI/CO) to which an iodide is added as promoter (Figure 6.5). The rhodium-catalyzed reaction (a) corresponds to acetic acid synthesis. Without the addition of iodides, the reaction proceeds according to cycle III. Methyl iodide is generated in a reaction between MeCOOMe and HI (b). Acetyl iodide and acetic acid react in an equilibrium reaction, forming acetic anhydride and hydrogen iodide (c).

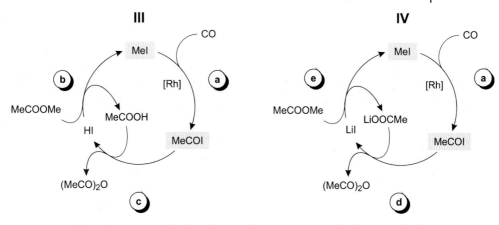

Figure 6.5 Iodide cycle in acetic anhydride synthesis without promoters ("acid cycle" **III**) and with LiI as promoter ("salt cycle" **IV**).

In contrast to the acetic acid synthesis via the Monsanto process, the reaction system here contains no water. This leads to a relatively long induction period, and furthermore, the reaction still runs comparatively slowly. The following improvements have been achieved:

- The existence of an induction period is due to the fact that during catalyst formation, no suitable reductant ($Rh^{III} \rightarrow Rh^I$) is present. Introducing hydrogen to the reaction system shortens the induction period. Furthermore, H_2 is also able to reduce inactive Rh^{III} complexes formed during the course of the reaction.
- When iodides are added as promoters (LiI, [NR$_4$]I, [PR$_4$]I, …) the reaction proceeds according to cycle **IV** (Figure 6.5). Now, methyl iodide is formed in the reaction of MeCOOMe with LiI (e). This generates lithium acetate, which reacts with acetyl iodide, forming acetic anhydride (d). The latter comes with a crucial advantage: The equilibrium of reaction **d** – in contrast to the analogous reaction **c** in the "acid cycle" **III** – lies to the right. The same effect can be achieved when other iodide salts are employed as promoters.

Exercise 6.2
Despite the similarities of the catalyst systems for the acetic anhydride and acetic acid syntheses, no induction period occurs in the Monsanto acetic acid process. Which reaction in the acetic acid process causes the reduction of Rh^{III} to Rh^I?

In sum, acetic anhydride is obtained from CO and methanol according to the following overall scheme. Both C_1 building blocks can be produced cost-effectively from coal, natural gas, or petroleum.

$$MeOH \xrightarrow[\text{[Rh]}]{CO} MeCOOH \xrightarrow[-H_2O\ (H^+)]{MeOH} MeCOOMe \xrightarrow[\text{[Rh]}]{CO} (MeCO)_2O$$

A process variant is based on co-carbonylation of methanol and methyl acetate, forming a mixture of acetic anhydride and acetic acid.

Exercise 6.3
What is the course of reaction upon which the co-carbonylation of a (nonaqueous) mixture of methanol and methyl acetate is based?

Exercise 6.4
Oxidative carbonylation of methanol. Due to its environmental friendliness and very low toxicity, dimethyl carbonate has increasingly established itself as a substitute for highly toxic carbonylation and methylation agents such as phosgene and dimethyl sulfate. More than 100 000 metric tons a year are produced, and among other uses, it is employed for synthesis of polycarbonates. One method of synthesis is the oxidative carbonylation of methanol, catalyzed by CuCl, that is, the reaction of methanol with CO in the presence of O_2. What are the disadvantages of a "conventional" (non-catalytic) industrial synthesis of dimethyl carbonate? Formulate the reaction equation of the oxidative carbonylation of MeOH and a possible reaction mechanism. Although it has not been proven in detail, mixed-valence copper complexes of type **1** seem to play a role as intermediates.

6.4
The Cativa Process

A substantial improvement of the Monsanto process was achieved in 1996 by BP Chemicals (Hull, England) through the introduction of the corresponding iridium complex as catalyst (Cativa process). The elementary steps of the rhodium- and iridium-catalyzed reactions are similar, but differ in their relative rates. This has decisive consequences for the overall course of the catalysis. The oxidative addition of MeI on $[MI_2(CO)_2]^-$ (M = Rh, Ir) assumes a key role.

Kinetic studies and quantum-chemical calculations confirm that the oxidative addition of methyl iodide to $[MI_2(CO)_2]^-$ (M = Rh, **7**; Ir, **7'**; 16 ve, d^8), with formation of a methylmetal(III) complex (M = Rh, **8**; Ir, **8'**; 18 ve, low-spin d^6) proceeds according to the S_N2 mechanism (Figure 6.6). The square-planar M^I complex **7/7'** reacts as a nucleophile; the nucleophilic center is the doubly-occupied d_{z^2} orbital. In the transition state **ts**, the M–C bond has been formed and the C–I bond has been broken. Thus, the energy necessary to split the C–I bond is partially compensated by the formation of the M–C bond. Since the bond dissociation enthalpy of the Ir–C bond is now greater than that of the Rh–C bond, however, the activation barrier for the Ir complex is smaller than for the Rh complex (Figure 6.6). As a result, the formation of the rhodium complex is an endergonic process, while formation of the iridium complex is exergonic.

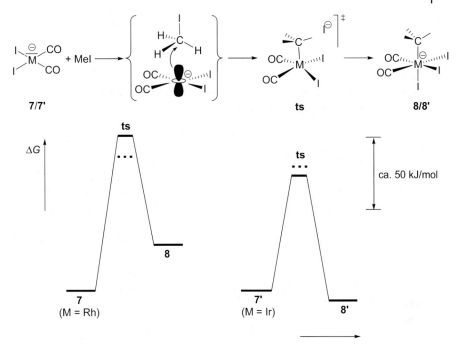

Figure 6.6 Mechanism of nucleophilic addition of MeI to $[MI_2(CO)_2]^-$ (M = Rh, **7**; Ir, **7'**). ΔG has been calculated for 298 K with MeOH as solvent. The dashed lines give experimentally determined values (for M = Rh in MeOH and for M = Ir in CH_2Cl_2) (simplified from Cheong and Ziegler [107]).

Table 6.2 gives the results of kinetic measurements for the addition of alkyl iodides RI to $[MI_2(CO)_2]^-$ (**7/7'**). As is typical for organic S_N2 reactions, MeI reacts much faster than the other alkyl iodides. From the last row in Table 6.2, it can be concluded that the iridium complex $[IrI_2(CO)_2]^-$ generally oxidatively adds alkyl iodides faster by a factor of more than two orders of magnitude than the analogous rhodium complex.

Thus, the oxidative addition of methyl iodide to $[IrI_2(CO)_2]^-$ runs about 150 times faster than to $[RhI_2(CO)_2]^-$, and is no longer rate-determining in iridium-catalyzed methanol carbonylation. On the other hand, however, the next reaction step, the

Table 6.2 Relative rates for the reactions of $[MI_2(CO)_2]^-$ (M = Rh, **7**; Ir, **7'**) with different alkyl iodides RI. The rate for MeI, which is used here as a standard, is set equal to 1000 (after Ellis and Maitlis [108]).

M	solvent/T (in °C)	MeI	EtI	n-PrI	i-PrI
Rh	RI/80	1000	3	1.7	4
Ir	CH_2Cl_2/30	1000	2.3	0.75	
a)		1000	33	13	0.8
k_{Ir}/k_{Rh} b)		150	220	140	

a) Typical relative rates for organic S_N2 reactions.
b) Approximate ratio of reaction rates (extrapolated to 80 °C).

migratory CO insertion (8/8′ → 10/10′), is slower by several orders of magnitude for complex **8** (M = Rh) than for complex **8′** (M = Ir) in aprotic solvents.

$$[\text{Me–M}(\text{I})(\text{CO})_2]^- \; + \; CO \;\rightleftharpoons\; [\text{Me–C(O)–M}(\text{I})(\text{CO})_2]^-$$

8/8′ → 10/10′

Acceleration of the reaction results from the following [109]:

- In the presence of protic solvents such as methanol, dissociative substitution of I⁻ by CO (8′ → 11′ → 12′) is substantially accelerated.
- Complex **12′** undergoes a significantly faster CO insertion (12′ → 13′) than complex **8′**.

8′ →(−I⁻)→ 11′ →(+CO)→ 12′ → 13′

Exercise 6.5
Explain (a) the reaction-accelerating effect of MeOH and (b) the difference in reactivity for complexes **12′** and **8′**.

On this basis, a simplified mechanism for the Cativa process is given in Figure 6.7. The following individual reaction steps can be named:

7′ → 11′ → 8′: *Oxidative addition/reductive elimination.* Oxidative addition of methyl iodide to the Ir^I complex (16 ve) yields a coordinatively and electronically saturated methyliridate(III) complex (**8′**, 18 ve) in a rapid reaction according to the S$_N$2 mechanism via a coordinatively unsaturated Ir^{III} complex (**11′**, 16 ve). Probably, however, **8′** is an off-loop complex and represents the resting state [109].

11′ → 12′: *Ligand addition.* Addition of CO to the intermediate complex **11′** yields a neutral iridium(III) complex.

12′ → 13′: *CO insertion.* CO insertion produces a coordinatively unsaturated (16 ve) iridium(III) complex.

13′ → 7′: *Reductive elimination.* In a reductive elimination, acetyl iodide is cleaved off, which restores the catalyst complex [IrI$_2$(CO)$_2$]⁻ (**7′**) by addition of I⁻. Analogously to the Monsanto process, water rapidly and irreversibly converts the acetyl iodide that has been cleaved off into HOAc and HI.

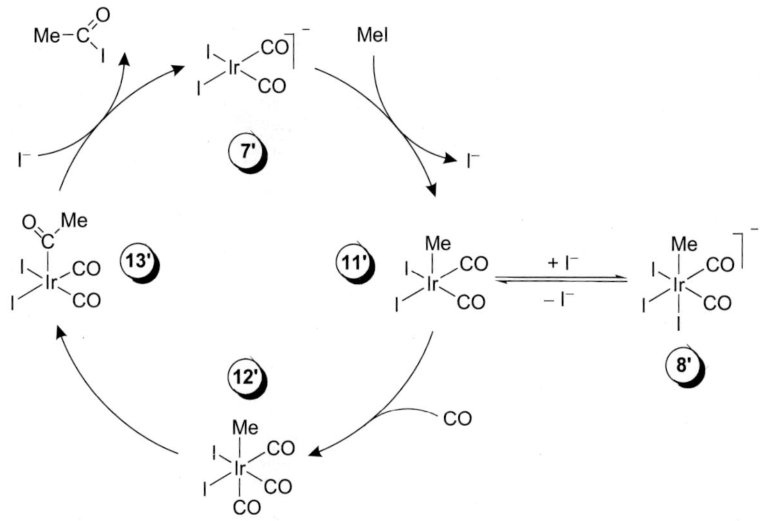

Figure 6.7 Mechanism of the Cativa process. The "iodide cycle" (formation of MeI from MeOH and hydrolysis of MeCOI to form MeCOOH) corresponds to that depicted in Figure 6.1 and is not shown.

Although all reactions are reversible in principle, the reverse reactions are insignificant at the conditions under which the reaction is carried out in industry. In accord with very rapid oxidative addition of MeI, the reaction rate is given by: $r \sim c_{Ir} \cdot p_{CO} \cdot c_{I^-}^{-1}$. The inverse first order in the iodide concentration of the reaction rate shows the inhibiting effect of iodide ions, which can be traced back to the equilibrium $11' \rightleftharpoons 8'$. Without promoters, the carbonylation of the anionic complex $8'$ is rate-determining ($8' \rightarrow 10'$) [103].

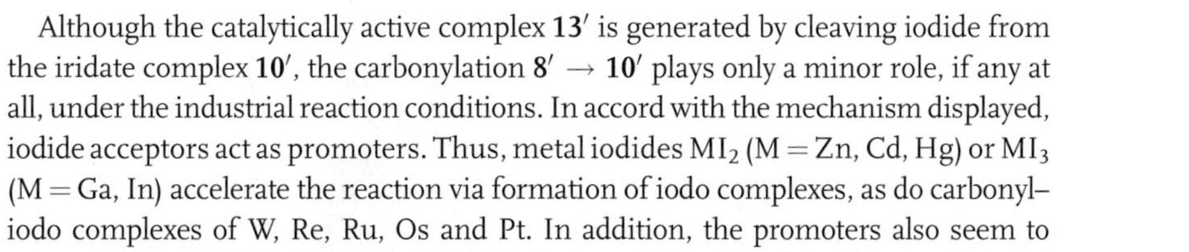

Although the catalytically active complex $13'$ is generated by cleaving iodide from the iridate complex $10'$, the carbonylation $8' \rightarrow 10'$ plays only a minor role, if any at all, under the industrial reaction conditions. In accord with the mechanism displayed, iodide acceptors act as promoters. Thus, metal iodides MI_2 (M = Zn, Cd, Hg) or MI_3 (M = Ga, In) accelerate the reaction via formation of iodo complexes, as do carbonyl–iodo complexes of W, Re, Ru, Os and Pt. In addition, the promoters also seem to suppress the formation of inactive iodide-rich complexes such as $[IrI_4(CO)_2]^-$ and $[IrI_3(CO)_3]$. In the Cativa process, a carbonyl(iodo)ruthenium complex is used as promoter.

6.5
Water-Gas Shift Reaction and Carbon Monoxide Dehydrogenases

6.5.1
Water-Gas Shift Reaction

Carbon monoxide and water react in a weakly exothermic ($\Delta H^\ominus = -41$ kJ/mol) and exergonic ($\Delta G^\ominus = -29$ kJ/mol) reaction, forming carbon dioxide and hydrogen:

$$CO + H_2O_{(g)} \xrightleftharpoons{cat.} H_2 + CO_2$$

This reaction is called the water-gas shift reaction[1] but is also known as carbon monoxide conversion. The equilibrium is called water-gas (shift) equilibrium. The equilibrium constant K has the value 1.0 at 830 °C. Since this is an exothermic reaction, it should be run at the lowest possible temperatures in order to shift the equilibrium to the right. Establishment of the equilibrium occurs very slowly under these conditions and can be accelerated by heterogeneous catalysts. In this regard, a distinction is made between high-temperature shift ($T = 350$–380 °C, residual CO content 3–4% by volume) by catalysts containing Fe/Cr or Cr/Mo, and low-temperature shift ($T = 200$–250 °C, residual CO content <0.3% by volume) by catalysts containing Cu/Zn.

Homogeneous catalysts are also able to catalyze the water-gas shift reaction, but in many cases, it is an undesired side reaction. This affects reactions where water is present and CO is used as substrate. Examples are the Monsanto process (a), the hydrocarboxylation of olefins (b) and Fischer-Tropsch synthesis (c).

$$MeOH + CO \xrightarrow[HI/H_2O]{[Rh]} MeCOOH \quad (a)$$

$$H_2C=CH_2 + CO + H_2O \xrightarrow{[Co], [Pd], ...} MeCH_2COOH \quad (b)$$

$$CO + 2\,H_2 \xrightarrow{cat.} 1/n\,{-}(CH_2)_n{-} + H_2O \quad (c)$$

The mechanism of the metal-catalyzed water-gas shift reaction is shown in Figure 6.8. The following individual reaction steps can be named:

14 → 15: *Hieber base reaction.* Nucleophilic attack by OH⁻ or water on a carbonyl carbon atom, followed by deprotonation in the case of water. An unstable hydroxycarbonyl complex is formed. Isotope labeling ([M]–CO + ^{18}OH⁻ ⇌ [M]–C^{18}O + OH⁻) confirms the reversibility of the reaction.

---

[1] In coal-based chemical industries, water and coal were used to produce a mixture of CO and H_2 (C + $H_2O \rightleftharpoons CO + H_2$), so-called "water gas," followed by CO conversion (CO + $H_2O \rightleftharpoons H_2 + CO_2$), hence the name "water-gas shift reaction."

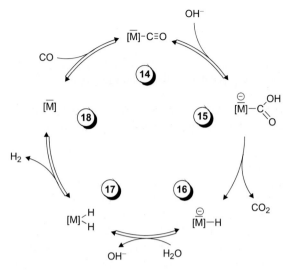

Figure 6.8 Principal mechanism of the water-gas shift reaction catalyzed by transition metal complexes.

15 → 16: *β-Hydrogen elimination.* Decarboxylation of the hydroxycarbonyl complex with cleavage of CO_2 and formation of a hydridometal complex. The reverse reaction, an insertion of CO_2 into an M–H bond, has also been described. However, it leads predominantly to formato complexes [M]–OC(O)H and not to complexes of type **15**.

16 → 17: *Protonation/deprotonation.* Protonation of a metal base with formation of a dihydridometal complex. The reverse reaction represents a deprotonation of a dihydrido complex.

17 → 18: *Reductive elimination/oxidative addition.* Reductive elimination of H_2. This reaction step is reversible.

18 → 14: *Ligand attachment/cleavage.* Coordination of CO completes the catalytic cycle.

The essential steps of the mechanism have been demonstrated for [Fe(CO)$_5$] (**14′**). However, the catalytic activity of **14′** is low, and in an acidic medium, **14′** is catalytically inactive. This is because of the conflicting pH requirements for the two partial reactions **14 → 15** and **16 → 17**. Firstly, the reactivity of iron pentacarbonyl (**14′**) towards OH^- is low, and it decreases for mononuclear metal carbonyls in this order:

$$[Os(CO)_5] > [Ru(CO)_5] \gg [Fe(CO)_5]$$

Secondly, the acid strength of [FeH$_2$(CO)$_4$] (**17′**, $pK_a = 4.4$ in water) is comparable to that of acetic acid and as a result, the concentration of **17′** in alkaline solutions is low. Ruthenium carbonyls have a catalytic activity higher by up to 3–4 orders of magnitude than iron pentacarbonyl [110].

6.5.2
Carbon Monoxide Dehydrogenases

Enzymes that catalyze the reversible oxidation of carbon monoxide to carbon dioxide according to the following equation are called carbon monoxide dehydrogenases (CODH).[2]

$$CO + H_2O \xrightleftharpoons{CODH} CO_2 + 2\,H^+ + 2\,e^-$$

Carbon monoxide dehydrogenases are enzymes that contain iron–sulfur clusters [Fe$_4$S$_4$] and nickel–iron–sulfur clusters [NiFe$_4$S$_4$], on which the oxidation of CO takes place. Fe$_4$S$_4$ clusters are formed surprisingly easily under "abiotic" conditions, as the following example shows:

$$4\,Fe^{3+} + 14\,RS^- + 4\,S \longrightarrow [Fe_4S_4(SR)_4]^{2-} + 5\,RSSR$$

The complexes formed are electron-variable, meaning that they can undergo multistep oxidation or reduction without undergoing changes of the cluster structure:

$$[Fe_4S_4(SR)_4]^{4-} \underset{}{\overset{-e^-}{\rightleftharpoons}} [Fe_4S_4(SR)_4]^{3-} \underset{}{\overset{-e^-}{\rightleftharpoons}} [Fe_4S_4(SR)_4]^{2-} \underset{}{\overset{-e^-}{\rightleftharpoons}} [Fe_4S_4(SR)_4]^{-}$$

(4 Fe^{II})   (3 Fe^{II} + Fe^{III})   (2 Fe^{II} + 2 Fe^{III})   (Fe^{II} + 3 Fe^{III})

The complexes display a heterocubane structure; for example, see the structure in Figure 6.9. The cluster core is obtained by alternately filling the vertices of a distorted cube with iron and sulfur atoms or by interpenetrating a Fe$_4$ tetrahedron and a S$_4$ tetrahedron in such a way as to produce a (distorted) cube.

The structure of an Ni–Fe–S cluster within a carbon monoxide dehydrogenase and its incorporation into the protein matrix (Cys = cysteine, His = histidine) is shown schematically in **19**. The enzymatic carbon monoxide conversion occurs

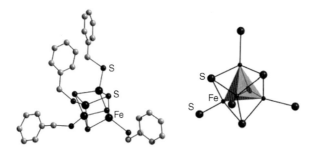

Figure 6.9 Structure of the anion of [NEt$_4$]$_3$[Fe$_4$(μ_3-S)$_4$(SCH$_2$Ph)$_4$]·DMF (left) and the Fe$_4$S$_4$ cluster core with the four S atoms of the terminal benzylthio ligands (right). The Fe$_4$ tetrahedron is shaded gray.

---

2) Although CO does not contain hydrogen, enzymes that catalyze the oxidation of CO according to the cited equation are called "dehydrogenases" in biochemical usage.

6.5 Water-Gas Shift Reaction and Carbon Monoxide Dehydrogenases

on the Ni–X–Fe unit, highlighted in bold. If nickel and iron are connected by a sulfide bridge (**19**, X = S), it probably must be cleaved to activate the enzyme. The elementary steps seem to be very similar to those of the abiotic reaction: Water bound to **19** (X = H_2O, charges are ignored) is deprotonated (**19** → **20**). Coordination of CO to nickel yields a carbonylnickel and a hydroxoiron species in close proximity (**20** → **21**). Then a nucleophilic attack by the hydroxo ligand on the carbonyl carbon atom occurs (Hieber base reaction) and complex **22** is formed by deprotonation of the hydroxy group. The cleavage of CO_2 leaves behind a reduced cluster **23**. Oxidation involving Fe_4S_4 clusters and coordination of water conclude the catalytic cycle (**23** → **19**). This completes the CODH reaction mentioned at the beginning [111, 112].

7
Metathesis

7.1
Metathesis of Olefins

7.1.1
Introduction

The metathesis (Greek: μετάθεσις = transfer, transposition, exchange of places) of olefins is a catalytic reaction in which a redistribution of alkylidene groups, with accompanying cleavage and re-formation of double bonds, takes place according to the following equation:

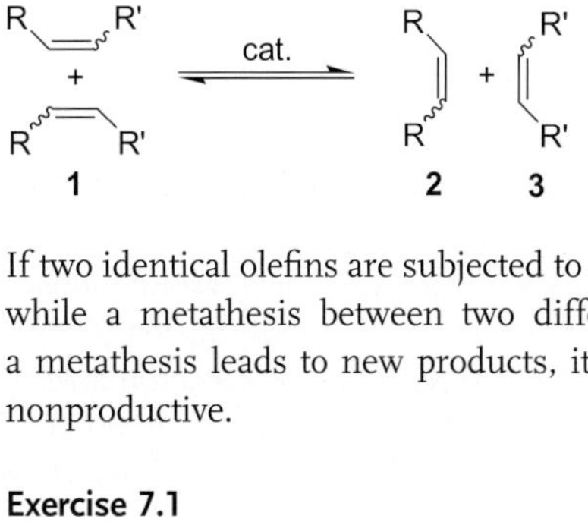

$$\underset{1}{\overset{R}{\underset{R}{\diagdown}}\!\!=\!\!\overset{R'}{\underset{R'}{\diagup}}} \;+\; \overset{\text{cat.}}{\rightleftarrows} \;\;\underset{2}{\overset{R}{\underset{R}{\diagdown}}\!\!=\!\!\overset{R'}{\underset{R'}{\diagdown}}} \;+\; \underset{3}{\overset{R'}{\underset{R'}{\diagup}}\!\!=\!\!\overset{R'}{\underset{R'}{\diagdown}}}$$

If two identical olefins are subjected to a metathesis, this is called homometathesis, while a metathesis between two different olefins is called cross-metathesis. If a metathesis leads to new products, it is called productive; otherwise, it is called nonproductive.

Exercise 7.1
Give the reaction products of the homometathesis of symmetrically substituted olefins RHC=CHR and $R_2C=CR_2$. Which products are created by the cross-metathesis of two different unsymmetrically substituted olefins RHC=CHR' and R"HC=CHR"'?

At the end of the 1950s, the metathesis of acyclic olefins (**1**, R = Me, R' = H) by heterogeneous catalysts such as MoO_3 (reduced with $Al(i\text{-}Bu)_3$) on Al_2O_3 was described in several patents, with the first, in 1957, by E. F. Peters and B. L. Evering (Standard Oil Co. of Indiana). An industrial process for metathesis of propene, which was available in surplus at that time, into ethene and butene had already been implemented in 1966 (R. L. Banks, G. C. Bailey; "Phillips triolefin process"). In 1966, N. Calderon had reported the first homogeneous catalyst system (WCl_6/$AlEtCl_2$/EtOH) (**1**: R = Et, R' = Me), which proved highly active and achieved equilibrium at room temperature within a few seconds.[1] For their contributions to the development of metathesis as a useful method in organic synthesis, Y. Chauvin, R. H. Grubbs, and R. R. Schrock were honored in 2005 with the Nobel Prize in Chemistry.

In metathesis the number of bonds does not change. Since in general the strength of the broken and newly formed double bonds differs only insignificantly ($\Delta H \approx 0$), the reaction is entropically driven: From $\Delta G = \Delta H - T \Delta S$, it follows that when $\Delta H = 0$, $\Delta G = -T \Delta S$. If olefin **1** is subjected to metathesis, an equilibrium of 50 mol% of the unsymmetrically substituted olefin **1** and 25 mol% each of the symmetrically substituted olefins **2** and **3** is established. Thus, in these cases the conversion for olefin metathesis cannot be greater than 50%.

Exercise 7.2

Explain why olefin metathesis cannot proceed without a catalyst in a concerted reaction according to the following equation:

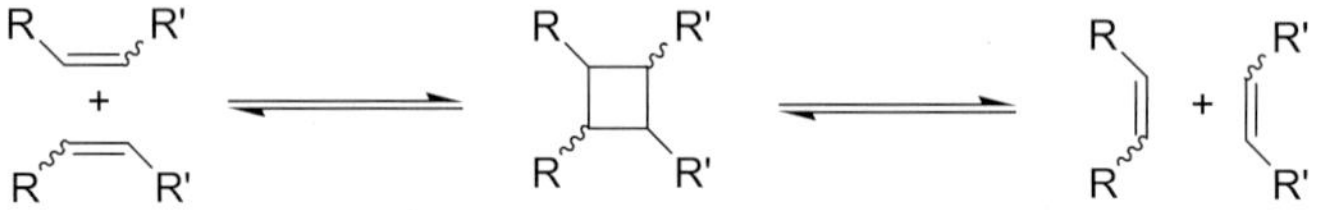

7.1.2
Mechanism

The catalysis of the olefin metathesis proceeds according to the so-called carbene mechanism (Y. Chauvin, J.-L. Hérisson, 1971), in which alkylidene group redistribution occurs via metallacyclobutane complex **5**. It is in equilibrium with two carbene transition metal complexes **4** and **6**, which have an olefin coordinated in the cis position, so [2 + 2] cycloaddition and cycloreversion occur in alternation. Cleavage of the olefin ligand (**4** → **4'**; **6** → **6'**) and addition of another olefin at **4'**/**6'** now ensure the conversion. All partial reactions are reversible, so the result of the catalysis is a composition corresponding to the thermodynamic equilibria.

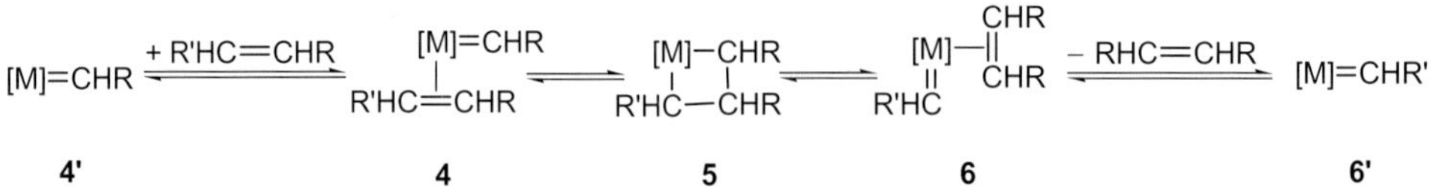

[M]=CHR + R'HC=CHR → [M]=CHR / R'HC=CHR ⇌ [M]–CHR / R'HC–CHR ⇌ [M]=CHR / R'HC=CHR – RHC=CHR → [M]=CHR'

4'   **4**   **5**   **6**   **6'**

---

1) A *cis-trans* isomerization of double bonds also takes place, leading to the establishment of the thermodynamically favored *cis-trans* composition of the olefins.

By decomposing the intermediates, the catalyst can be deactivated, for example, by reductive elimination of cyclopropane from the metallacyclobutane complex **5**, which ultimately represents a transfer of the carbene onto the olefin (**5** → **7**). The reaction of two carbene complexes can lead to olefin formation (**4'** → **8**).

$$[M] \underset{\text{5}}{\overset{\text{H R R H}}{\underset{\text{R' H H R}}{\bigsqcup}}} \rightleftharpoons [\overline{M}] + \underset{\text{7}}{\overset{\text{H R R H}}{\underset{\text{R' H H R}}{\bigsqcup}}}$$

$$2\,[M]{=}CHR \quad \longrightarrow \quad 2\,[\overline{M}] + RHC{=}CHR$$

    4'   **8**

Exercise 7.3

Originally, another mechanism was postulated for olefin metathesis, according to which the olefins were pairwise coordinated to the metal and then the rearrangement of the alkylidene groups took place, perhaps via a cyclobutadiene-like transition state:

$$\begin{array}{c} R'HC \quad CHR' \\ \|\text{-[M]-}\| \\ RHC \quad CHR \end{array} \rightleftharpoons \left[\begin{array}{c} R'HC{-}CHR' \\ / \, [M] \, / \\ RHC{-}CHR \end{array}\right]^{\ddagger} \rightleftharpoons \begin{array}{c} R'HC{=}CHR' \\ [M] \\ RHC{=}CHR \end{array}$$

An experiment that excluded such a pairwise mechanism in favor of the non-pairwise Chauvin mechanism was the cross-metathesis of a cycloolefin (cyclooctene) with two symmetric acyclic olefins (but-2-ene, oct-3-ene), leading to C12, C14, C16, and C6 products:

cyclooctene + MeHC=CHMe + PrHC=CHPr →[cat.] ring(=CHMe, =CHMe) [C12] + ring(=CHMe, =CHPr) [C14] + ring(=CHPr, =CHPr) [C16] + MeHC=CHPr [C6]

The product ratios C14:C12 and C14:C16 were determined as a function of time. Which ratios do you expect from extrapolation to time $t=0$ for the pairwise and for the non-pairwise (Chauvin) mechanism?

7.1.3
Catalysts

Starting with precatalysts that do not contain a carbene ligand, the carbene ligand will be created in the course of catalyst formation. In typical first-generation homogeneous metathesis catalysts (WCl_6/$EtAlCl_2$/EtOH; MCl_n/SnR_4 with M = W, n = 6 or M = Re, Mo, n = 5; $WOCl_4$/$AlEtCl_2$), the carbene functionality is formed by double alkylation and subsequent α-hydride elimination/reductive elimination, as is demonstrated by the reaction of a metal chloride with $SnMe_4$ (**9** → **10** → **11**).

$$[M]\overset{Cl}{\underset{Cl}{<}} \xrightarrow[-2\ SnMe_3Cl]{+2\ SnMe_4} [M]\overset{CH_3}{\underset{CH_3}{<}} \xrightarrow{-CH_4} [M]=CH_2$$

$$\quad\ \ \text{9} \qquad\qquad\qquad\qquad\quad \text{10} \qquad\qquad\qquad \text{11}$$

Tebbe reagent **12**, generated by reacting titanocene dichloride with aluminum trimethyl, reacts with bases such as pyridine, forming a methylenetitanium complex (**12** → **13**). Alternatively, Cp$_2$TiCl$_2$ can be doubly methylated with methyllithium. The dimethyl compound **14** thus formed undergoes thermal decomposition by α-hydride elimination, coupled with reductive elimination of methane (**14** → **13**) [113]. The methylenetitanium complex **13**, which is not stable in its pure form, is a typical Schrock carbene complex, and is active in metathesis.

[Reaction scheme showing Cp$_2$TiCl$_2$ converting via +2 AlMe$_3$, −AlMe$_2$Cl, −CH$_4$ to complex **12** (Cp$_2$Ti(μ-CH$_2$)(μ-Cl)AlMe$_2$), and via +2 LiMe, −2 LiCl to Cp$_2$Ti(CH$_3$)$_2$ (**14**), both leading to Cp$_2$Ti=CH$_2$ (**13**) by +py, −AlClMe$_2$(py) or Δ, −CH$_4$ respectively.]

If carbene complexes such as [W(=CHt-Bu)Br$_2$(OR)$_2$] are used as precatalysts for the olefin metathesis, activation by Lewis acids (AlBr$_3$, GaBr$_3$, …) is necessary. However, there are also carbene complexes that are catalytically active without a cocatalyst (single-component catalysts). Examples are the Schrock catalysts and first- and second-generation Grubbs catalysts (Figure 7.1). The activity and selectivity of these well-defined single-component catalysts in metathesis reactions can be fine-tuned by targeted variation of their coligands, which has led to a deeper understanding of catalysis [114, 115].

For the Schrock catalysts **A**, the o,o' substitution of the arylimido ligands increases the stability by hindering bimolecular decomposition reactions and the formation of catalytically inactive μ-imido complexes. Electron-withdrawing alkoxo ligands destabilize the metallacyclobutane intermediate and thus lead to a higher catalytic activity. Grubbs catalysts **B** and **C** are square-pyramidal 16-*ve* Ru^{II} complexes with the carbene ligand in the apical position. They are converted by cleavage of a PCy$_3$ ligand into the catalytically active form [Ru(=CHPh)Cl$_2$L] (L = PCy$_3$, NHC; NHC = *N*-heterocylic carbene). The cleavage of the phosphane ligand is favored by its bulkiness. The 14-*ve* complex formed is stabilized by the pronounced σ-donor strength of the PCy$_3$ (**B**) or NHC ligand (**C**).

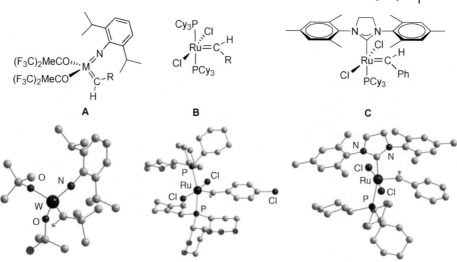

Figure 7.1 Single-component homogeneous catalysts for olefin metathesis. (**A**) Schrock catalysts (M = Mo, R = CMe$_2$Ph, "Schrock's catalyst"; M = W, R = t-Bu), structure of [W(=CHt-Bu){N{2,6-(i-Pr)$_2$C$_6$H$_3$}}(Ot-Bu)$_2$]. (**B**) First-generation Grubbs catalysts (R = Ph, CH=CPh$_2$), structure of [Ru{=CH(C$_6$H$_4$-p-Cl)}Cl$_2$(PCy$_3$)$_2$]. (**C**) Second-generation Grubbs catalysts, structure of [Ru(=CHPh)Cl$_2$(PCy$_3$)(Mes$_2$Imid)] (Mes$_2$Imid = 1,3-dimesitylimidazolidin-2-ylidene). (Prototypical catalysts for each are shown; many other catalytically active complexes can be derived by changing the ligand sphere. For the molecular structures, only the H atoms on the carbene carbon atoms are shown).

In many cases, heterogeneous metathesis catalysts (such as MoO$_3$, WoO$_3$ or Re$_2$O$_7$ on oxidic supports Al$_2$O$_3$/SiO$_2$) are favored in industry, while the Schrock and Grubbs catalysts are the catalysts of choice in organic synthesis. The Schrock alkylidene–molybdenum and –tungsten complexes (**A**) are air- and moisture-sensitive, due to the pronounced oxophilicity of Mo and W. As a result of their high reactivity, they have a low tolerance for functional groups, but they are highly active catalysts. In contrast, the first-generation Grubbs catalysts (**B**) are less active, but are distinguished by a high tolerance for functional groups (–COOH, –OH, –CHO, –COR, –NH$_2$, etc.). The second-generation Grubbs catalysts (**C**) are high in both catalytic activity and productivity[2] and also have a pronounced tolerance for functional groups. The high stability of Grubbs catalysts even enable metathesis to be achieved in water. Different strategies (e.g., the incorporation of polar anchor groups into the catalyst complex) are being developed to make the ruthenium complexes compatible with aqueous media [36, 116].

---

2) Activity and productivity can be greater by 1–2 orders of magnitude than for first-generation Grubbs catalysts. In the metathesis of oct-1-ene with [Ru(=CHPh)Cl$_2$(PCy$_3$)(Ar$_2$Imid)] (Ar$_2$Imid = 1,3-bis(2,6-diisopropylphenyl)imidazolidin-2-ylidene), effective turnover numbers of 6.4·10^5 mol octene/mol Ru (22 °C) and turnover frequencies of 2.3·10^5 min^{-1} (60 °C) were determined [117].

In catalyst optimization for a given synthesis problem, in addition to tolerance for functional groups, control over selectivity (e.g., homo- vs. heterocoupling in cross-metathesis, E/Z ratio for the olefins formed) may be of primary concern for obtaining the desired product with a high yield. In addition, double-bond isomerizations are potential (generally disruptive) side reactions in olefin metathesis. On the other hand, the hydrido(carbonyl)ruthenium complex **15**, obtained by reaction of the Grubbs catalyst **C** (Figure 7.1) with the silyl enol ether $H_2C=CHOSiMe_3$ has proven to be a very selective catalyst for the isomerization of a double bond from a terminal position to the adjacent position. This transformation ($H_2C=CHCH_2R \rightarrow H_3C-CH=CHR$), whose selectivity is apparently due to steric reasons, has been applied especially in the synthesis of natural products [118].

Exercise 7.4
- Two olefins $RCH=CH_2$ (**1**) and $R'CH=CH_2$ (**2**) are subjected to a cross-metathesis with a nonselective catalyst. (Assume that the conversion is complete and ethene is removed from the equilibrium.) What yield of $RCH=CHR'$ (**3**) is achieved if 1 or 10 mol, respectively, of **1** reacts with 1 mol of **2**? What quantities of undesired homocoupling products are produced?
- Suggest a mechanism for the formation of the hydrido(carbonyl)ruthenium complex **15** from **C** (Figure 7.1) and $H_2C=CHOSiMe_3$.

Background: Stable Carbenes as Ligands
Carbenes, neutral derivatives of divalent carbon, can form many metal complexes, which can be divided into the following three types:

- *Fischer carbene complexes* such as **1**, which are characteristic of transition metals in low oxidation states and O- or N-functionalized electrophilic carbene ligands.
- *Schrock carbene complexes* such as **2**, which are characteristic of transition metals in high oxidation states and unfunctionalized nucleophilic carbene ligands.
- *Main and transition metal complexes with stable N-heterocyclic carbene ligands* (NHC = N-heterocyclic carbene), whose reactivities are fundamentally different from those of the Fischer and Schrock carbene complexes.

The stability of carbenes depends markedly on the substituents on the carbene carbon atom. Particularly stable carbenes can be derived from imidazole derivatives (NHCs). The stability of NHCs (**3–6**) is based on the π-donor and σ-acceptor strength of the two vicinal substituents of the carbene carbon atom, which results in both a reduction in its electrophilicity via p_π–p_π interaction and its nucleophilicity via the –I effect.

NHCs are exceptionally strong σ-donor ligands (σ-donor strength: NHCs > P(alkyl)$_3$!) with a very low – generally negligible – π-acceptor effect. As a result, they form strong σ-M–C single bonds. NHC ligands are relatively inert with respect to nucleophiles and electrophiles and exert a comparatively high *trans* influence. With regard to their use as coligands in homogeneous catalysts, it is important to be able to control the accessibility of the coordination sites next to the carbene ligand via the bulkiness of the substituents R on the nitrogen atoms.

Examples: Electron-rich, saturated 1,3-R$_2$-imidazolidin-2-ylidenes (**3**) with the resonance structures (**3a–3c**), 6π-aromatic 1,3-R$_2$-imidazolin-2-ylidenes (**4**), a bidentate NHC ligand (**5**), and an N,S-heterocyclic carbene (**6**). For each case the basic structural element is shown, which can be further functionalized by substitution at C4 and C5. Noncyclic carbene ligands doubly functionalized with heteroatoms, such as **7**, can have ligand properties similar to those of NHCs (non-NHC = non-N-heterocyclic carbene) [119–124].

Nomenclature: Since carbenes of type **3** are formally created by cleavage of two H atoms from the same ring member from 1,3-R$_2$-imidazolidines (**3'**), they should be called 1,3-R$_2$-imidazolidin-2-ylidenes. In a similar way, the formal cleavage of two H atoms from 1,3-R$_2$-imidazolines (**4'**) leads to 1,3-R$_2$-imidazolin-2-ylidenes (**4**).

Also common, although we will not use them, are the designations as 1,3-R$_2$-imidazol-2-ylidenes (**4**) and 1,3-R$_2$-4,5-dihydroimidazol-2-ylidenes (**3**).

7.1.4
Mechanism – A Closer Look

For the Grubbs catalysts **B** (L = PCy$_3$) and **C** (L = NHC ligand), both kinetic experiments and quantum-chemical calculations have shown that first a phosphane ligand is cleaved off and then the olefin is coordinated to the 14-*ve* complex formed (**16** → **17** → **18**; "dissociative mechanism") [125, 126]. A direct addition of the olefin with formation of an 18-*ve* complex (**16** → **20**; "associative mechanism") can be ruled out. From the *cis*-carbene(olefin)ruthenium(II) complex **18**, a ruthenacyclobutane complex (**18** → **19**) forms through [2 + 2] cycloaddition. This raises the oxidation state of the ruthenium atom by two ($Ru^{II} \to Ru^{IV}$).

Now, the reaction cascade runs in the opposite sequence, except that cycloreversion occurs with formation of **18'** (**19** → **18'**). The difference in activity between the two types of Grubbs catalysts (**B** vs. **C**) shows that the cycloaddition with formation of ruthenacyclobutane is facilitated by the strong σ-donor NHC ligand.

For a quantum-chemical analysis of first- and second-generation Grubbs catalysts, we start with the model complexes **21** with, respectively, L = PMe$_3$ and L = 1,3-dimethylimidazolidin-2-ylidene (Me$_2$Imid) (Figure 7.2). The nonproductive homometathesis of ethene runs via the reaction sequence **21** → ... → **24**. For the complexes **23**, four conformations with different arrangements of the carbene and olefin ligands are possible (**23a–23d**). First, the most thermodynamically stable conformers **23a** are formed. In order to run cycloaddition easily, two conditions must be fulfilled, namely a coplanar arrangement of the four atoms involved (Ru=C, C=C) and a parallel orientation of the two π systems. These are only encountered in the conformations **23b**, though it has not been possible to localize them as a minimum structure. The transition states **23b'** have the same alkene and carbene conformation as the hypothetical structures **23b**.

The critical difference between the first- and second-generation catalysts is the lower energy of **23b'** and **24** with L = Me$_2$Imid compared to the analogous complexes with L = PMe$_3$. This can be traced to the fact that the donor strength of the NHC ligand is greater than that of PMe$_3$. The result is a stabilization of both the Ru^{IV} complex **24** (L = Me$_2$Imid) and the Ru^{II} transition state **23b'** (L = Me$_2$Imid) via a stronger back-donation of Ru^{II} to the methylene ligand. Furthermore, in the actual catalyst (L = 1,3-dimesitylimidazolidin-2-ylidene) with the "inactive" carbene conformations **23a/23d**, steric interactions occur between the hydrogen atom of the carbene ligand and a mesityl substituent, forcing an "active" carbene orientation

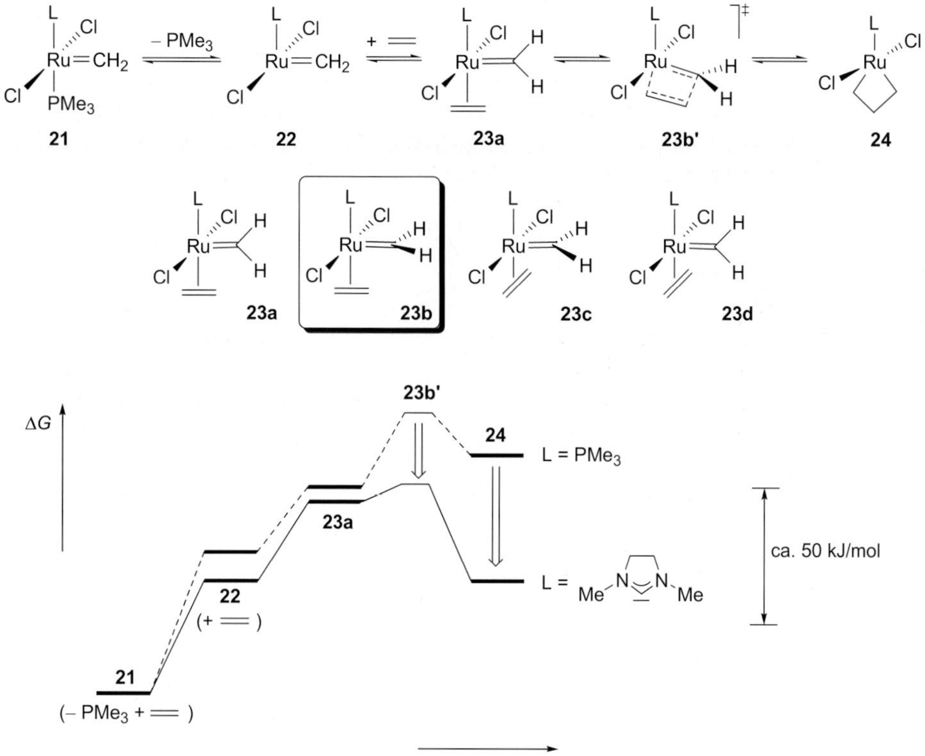

Figure 7.2 Reaction course and profile of the Gibbs free energy of the nonproductive homometathesis of ethene with first-generation **21** ... **24** (L = PMe$_3$) and second-generation Grubbs model catalysts **21** ... **24** (L = 1,3-dimethylimidazolidin-2-ylidene). The four possible conformations of the olefin and carbene ligands in complexes **23** are shown in the middle (shortened, after Straub [127]).

as in **23b/23c**. These seem to be the essential reasons for the difference in activity between the two types of catalysts.

7.1.5
Metathesis of Cycloalkenes

Cycloalkene metathesis generates unsaturated polymers called polyalkenamers (**25** → **26**). Bicyclic olefins and cycloalkadienes react in a similar fashion.

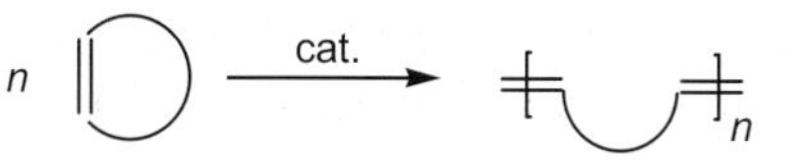

The course of this reaction is reflected by the term "ring-opening metathesis polymerization" and the acronym "ROMP." The reaction course (cleavage and new formation of double bonds) also reveals that the polymerization retains all double bonds of the monomer. Ring-opening metathesis polymerization of cycloalkenes is "enthalpy-driven"; the driving force is the loss of ring strain in the monomer. Thus, ROM polymerization reactions of highly strained cycloalkenes such as norbornene are irreversible, and in principle, complete conversion of the monomers can be achieved. This is significantly different from the reversible metathesis of acyclic olefins.

Exercise 7.5
Particularly in less strained ring systems, a thermodynamic equilibrium can be established between the open-chain polymer, the monomer, and the cyclic oligomers by involvement of inner double bonds of the growing polymer chain in the metathesis (back-biting). Formulate such a reaction. What consequences do corresponding intermolecular reactions have?

On the other hand, unstrained cycloolefins such as cyclohexene do not polymerize under ROMP conditions: The lack of ring strain ($\Delta H = 0$) and the negative reaction entropy ($\Delta S < 0$) result in $\Delta G > 0$. However, this has the consequence that polymers with tetramethylene units ($-(CH_2)_4-$) between two double bonds cleave off cyclohexene under metathesis conditions. An example is the metathesis of cyclodeca-1,5-diene following the reaction sequence **27** → **28** → **29**.

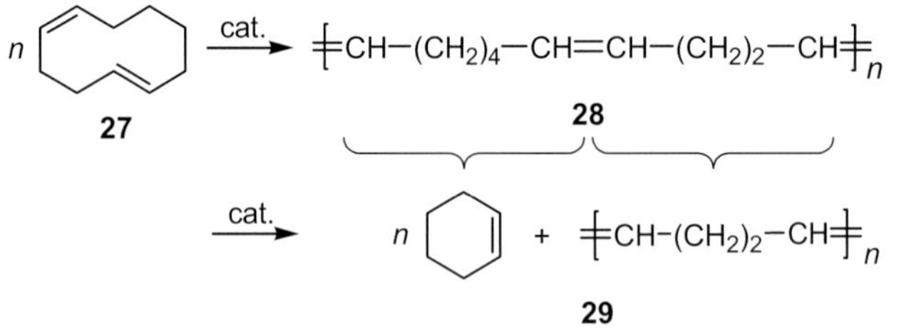

$$n \; \text{(27)} \xrightarrow{\text{cat.}} \;\; \text{⁅}CH-(CH_2)_4-CH=CH-(CH_2)_2-CH\text{⁆}_n$$
$$\text{28}$$

$$\xrightarrow{\text{cat.}} \;\; n \;\text{(cyclohexene)} \; + \; \text{⁅}CH-(CH_2)_2-CH\text{⁆}_n$$
$$\text{29}$$

The cross-metathesis of cycloolefins with ethene generates acyclic terminal dienes (**30** → **31**).

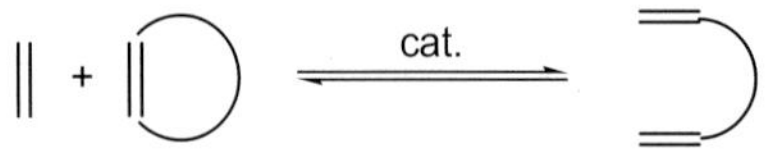

$$\| \; + \; \bigcirc \;\; \xrightleftharpoons{\text{cat.}} \;\; \text{(diene)}$$
$$\text{30} \qquad\qquad\qquad \text{31}$$

Such a reaction is called ring-opening metathesis (ROM). The driving force is, as for ROMP, the release of ring strain in the cycloolefins. The reverse reaction (RCM: ring-closing metathesis) can be verified when the ethene formed is removed from equilibrium. The reaction **30** → **31** also shows that for ROM polymerization, the addition of small quantities of an acyclic olefin induces chain transfer, and thus the chain length can be controlled.

ROMP reactions can, in principle, be designed so as to proceed as living polymerizations.[3] Thus, polymers with precisely controlled molecular weight and a narrow molar mass distribution can be produced. As a terminating reaction with Schrock catalysts, one should consider the reaction with aldehydes or ketones, which proceeds as a Wittig reaction (32 → 33; M = Mo, W). For Grubbs catalysts, a reaction with vinyl ethers leads to stable carbene–ruthenium complexes, which do not start further chains (34 → 35).

$$[M]=C\diagup_{P} \xrightarrow[-[M]=O]{+RR'C=O} \underset{R'}{\overset{R}{\diagdown}}C=C\diagup_{P} \qquad [Ru]=C\diagup_{P} \xrightarrow[-[Ru]=CHOR]{+H_2C=CHOR} H_2C=C\diagup_{P}$$

$$\quad\;\; 32 \qquad\qquad\qquad 33 \qquad\qquad 34 \qquad\qquad\qquad\qquad 35$$

Such reactions can also be used to produce polymers with desired end-group functionality; for example, R in 33 could be a luminescent substituent. Living polymerization also enables the targeted synthesis of block copolymers.

Exercise 7.6
Produce polyacetylene from the tricyclodecatrienes 1 (R = H, CF_3, CO_2Me) and from benzvalene 2, a (nonplanar) valence isomer of benzene. What double-bond selectivity do you expect for the ROM polymerization of 1?

The first ring-opening polymerization of a cycloolefin (norbornene) occurred with a Ziegler catalyst system in 1954. In 1967–1968, the relation between ring-opening polymerization and metathesis reactions became clear. Today, ring-opening metathesis polymerization finds many industrial applications. The first commercially produced polymer was polynorbornene with predominant *trans* structure of the double bonds (Norsorex®; CdF-Chimie, 1976), which was obtained by ruthenium-catalyzed ($RuCl_3$/HCl in butanol) ROM polymerization of norbornene (36 → 37). Due to its very high molar mass ($>3 \cdot 10^6$ g/mol; $P > 31\,000$) it cannot be melted ($T_{dec.} > 200\,°C$; $T_g = 37\,°C$). With plasticizers, useful elastomers are obtained. Hydrogenation leads to an amorphous, colorless, transparent polymer (37 → 38) with a very high glass transition temperature ($T_g = 140\,°C$); it belongs to the class of cyclic olefin polymers and is suitable for optical applications (lenses, prisms, ...) (Zeonex®) [128].

---

[3] Essential criteria are that chain transfer and termination reactions be practically insignificant and the number of active centers remain constant for the duration of the polymerization. The monomer reacts completely and when more is added, the chain growth continues. In these cases, only one polymer chain is formed per catalyst molecule, so it would be more accurate to refer to an initiator molecule. However, we will continue to speak of catalysts, referring to the linking of monomers rather than the formation of polymer chains.

Cyclooctene is converted with a catalyst system derived from $WCl_6/AlEtCl_2/EtOH$ in hexane, forming a polyoctenamer of high purity (>99.5%) (Vestenamer®, Degussa). The polymer consists of linear (75%, $M > 10^5$ g/mol) and cyclic (25%) macromolecules (see Exercise 7.5). The crystallinity depends heavily on the microstructure (*cis/trans* ratio of double bonds) and can be controlled by the polymerization conditions. Polyoctenamers have rubber-elastic properties and are added to other types of rubber to improve their properties. ROM polymerization of *endo*-dicyclopentadiene (DCPD), an excess product from naphtha crackers, can be performed in such a way that only the more highly strained double bonds react, but also such that cross-linking via the other double bonds occurs (see Exercise 7.7). A thermosetting plastic of high mechanical strength is obtained. The ROM polymerization runs so quickly that molded products (for automotive construction, sports equipment, etc.) can be produced via reaction injection molding (RIM) technology. This is *in situ* polymerization of the monomer (DCPD) directly in the molded form.

Exercise 7.7
Which of the two double bonds in DCPD is more reactive? Give an explanation. Formulate the equation for the ROM polymerization of DCPD under the condition that only the more reactive double bond reacts. Which structural elements can form when both double bonds react?

7.1.6
Metathesis of Acyclic Dienes

Acyclic terminal dienes can react in an intramolecular metathesis reaction with formation of cycloolefins and ethene (39 → 40; RCM: ring-closing metathesis; ROM: ring-opening metathesis) or in an intermolecular metathesis reaction cleaving off ethene, forming a polymer (39 → 41; ADMET: acyclic diene metathesis).

7.1 Metathesis of Olefins

RCM and ADMET reactions are in principle reversible for all steps, so the reaction can only generate the composition corresponding to thermodynamic equilibrium. Since the ethene formed can easily be removed from the reaction mixture, however, complete conversion can easily be achieved. Intramolecular and intermolecular metathesis reactions of dienes are parallel reactions, catalyzed by the same catalysts. The ratio (RCM vs. ADMET) with which they proceed can be influenced to a certain extent. Thus, for instance, dilute solutions and formation of unstrained five- and six-memebered rings favor the intramolecular RCM reactions. On the other hand, energetically demanding conformational changes to bring the two reacting double bonds in the substrate in close proximity disfavor RCM reactions.

Dienes with inner double bonds react similarly, but at a lower rate. However, it remains difficult to synthesize tetrasubstituted olefins (42 → 43; R, R′ ≠ H) by ring-closing metathesis (RCM).

This problem can be circumvented by relay ring-closing metathesis (relay RCM) [129]. This is based on the observation that dienes **42** with R = H can be converted in an RCM reaction with Grubbs catalysts. This means that a double bond –CR′=CH$_2$ is *intra*molecularly accessible to the catalyst. If a spacer is now used to link one of the two (non-reactive) double bonds in **42** to a terminal double bond and two RCM reactions are now performed, the desired products **43** are obtained. An example is the reaction (**44** → **45** → **46**), which is not catalyzed by Grubbs catalysts **B** (see Figure 7.1). However, if **44′** is now used as the starting point, the first step results in the formation of the alkylidene complex **44″** and two consecutive RCM reactions (**44″** → **45** → **46**) lead to the desired product with cleavage of cyclopentene.

The reverse reaction of acyclic diene metathesis (ADMET) polymerization leads to acyclic dienes and is known as ADMET depolymerization. ADMET depolymerization is interesting when the unsaturated polymers can be obtained straightforwardly (and not via an ADMET polymerization). This is the case for *cis*- and *trans*- 1,4-polybutadienes which, in the presence of ethene, are broken down into hexa-1,5-diene (**48** → **49**) via oligomeric intermediates (**47** → **48**).

7.1.7
Enantioselective Metathesis

Over the course of the catalytic metathesis cycle using Schrock's catalyst, the coligands (the imido ligand and the two alkoxo ligands) are not cleaved from the catalyst (Figure 7.1, **A**). Thus, the substitution of the two achiral alkoxo ligands by a chiral bis(aryloxo/alkoxo) ligand or by two monodentate ligands (a chiral aryloxo ligand and a pyrrolide ligand) offers a good possibility to design catalysts for asymmetric metathesis. Examples are the molybdenum complexes **50a–d** (Hoveyda-Schrock catalysts) [130–132].

a) The complex **50a** contains an additional THF ligand

They have proven their usefulness in asymmetric ring-closing and ring-opening metathesis (ARCM/AROM) reactions as shown by the formation of the dihydrofuran compound **52** from **51** and the ring-opening of the *meso* norbornene compound **53** with styrene, forming **54** (TBS = *tert*-butyldimethylsilyl). Enantioselective olefin metatheses have found their way into natural product synthesis [133].

7.2 Metathesis of Alkynes

Alkynes can undergo metathesis in essentially the same fashion as alkenes, based on an exchange of alkylidyne groups:

$$R{\equiv}R' + R{\equiv}R' \xrightleftharpoons{cat.} \begin{array}{c} R \\ ||| \\ R \end{array} + \begin{array}{c} R' \\ ||| \\ R' \end{array}$$

A. Mortreux and M. Blanchard reported on the first alkyne metathesis (R = p-MeC$_6$H$_4$, R' = Ph) in 1974 (catalyst: [Mo(CO)$_6$]/resorcinol (**55**); 160 °C). The alkylidyne–tungsten complex [W($\equiv$Ct-Bu)(Ot-Bu)$_3$] (**56**) catalyzes this reaction at room temperature (R. R. Schrock, 1981). Polymerization and cyclotrimerization reactions are particularly likely to appear as side reactions for metathesis of terminal alkynes. The mechanism of alkyne metathesis seems to be similar to that of olefin metathesis: Catalytically active intermediates are alkyne–carbyne complexes (**57/59**), which can interconvert via a metallacyclobutadiene complex **58**. Only when the alkyne ligands in **57** and **59** are substituted by another alkyne, productive metathesis does take place.

$$\underset{\mathbf{57}}{\begin{array}{c}[M]{\equiv}CR \\ | \\ R'C{\equiv}CR\end{array}} \rightleftharpoons \underset{\mathbf{58}}{\left\{\begin{array}{c}[M]{=}CR \\ | \quad | \\ R'C{=}CR\end{array} \leftrightarrow \begin{array}{c}[M]{-}CR \\ || \quad || \\ R'C{-}CR\end{array}\right\}} \rightleftharpoons \underset{\mathbf{59}}{\begin{array}{c}[M]{-}CR \\ ||| \\ CR'\end{array}}$$

Tungstacyclobutadiene complexes of type **58** ([M] = W(OAr)$_3$) have essentially been isolated and also structurally characterized (Ar = 2,6-diisopropylphenyl, R = R' = Et). They have proven to be metathetically active.

With the Mortreux catalyst **55**, other phenols were later also used as cocatalysts, which allowed for optimization with regard to selectivity and/or activity for any particular problem. Thus, for example, Mo(CO)$_6$/4-chlorophenol catalyzes the dimerization of ArC$\equiv$CMe at 130–150 °C with high yield, liberating butyne. Mortreux catalysts are very robust, and need not be used under inert conditions with scrupulously cleaned solvents. However, due to the drastic reaction conditions, their use is essentially limited to nonfunctionalized alkynes. The alkylidyne complex **56** and metathetically active amido(alkylidyne)molybdenum complexes [Mo($\equiv$CR)-{N(t-Bu)(3,5-Me$_2$C$_6$H$_3$)}$_3$] (R = H, Me, Et, ...) display a high tolerance for functional groups (ketones, acetals, esters, ethers, sulfones, urethanes, ...). They also behave strictly chemoselectively toward π systems: Olefinic double bonds are not attacked! This substantially expands the synthetic potential of olefin and alkyne metathesis.

An additional highly active single-component catalyst for alkyne metathesis is the alkylidyne(imidazolin-2-iminato)tungsten complex **60**. At room temperature it very effectively catalyzes the metathesis of PhC≡CMe, forming tolane and but-2-yne [134, 135].

Exercise 7.8
- $(t\text{-BuO})_3W≡W(Ot\text{-Bu})_3$ reacts with symmetric alkynes RC≡CR (R=Me, Et, Pr, ...), forming alkylidyne complexes $(t\text{-BuO})_3W≡CR$ (**56′**), but, apparently for steric reasons, not with $t\text{-BuC}≡Ct\text{-Bu}$ to form **56**. However, the corresponding reaction with an excess of $t\text{-BuC}≡CMe$ produces $(t\text{-BuO})_3W≡Ct\text{-Bu}$ (**56**) exclusively, provided it is performed under a vacuum. Which reaction course do you assume is occurring?
- Analyze complex **60**, comparing it to the Schrock catalysts **A** in Figure 7.1 with respect to the oxidation state and the net valence electron count of the central atom and the electronic properties of the ligands. Describe the nature of the N-bound coligand in **60** and explain why this is a strong σ donor.

Cycloalkynes of medium ring size (12- to 28-membered rings) can be obtained by ring-closing metathesis of acyclic diynes (**61** → **62**; RCAM: ring-closing alkyne metathesis) using the dilution principle, where in many cases the alkylidyne–tungsten complex **56** is used as catalyst. It is preferable to use dialkynes **61** with R = Me, Et so that the butene or hexene formed can easily be removed from the reaction mixture under vacuum. A subsequent partial hydrogenation of the triple bond with the Lindlar catalyst (a Pd–CaCO$_3$ catalyst modified with lead) selectively yields the (Z)-cycloolefin (**62** → (Z)-**63**). Catalytic hydrosilylation of **62** and cleavage of the silyl group with fluorides offer an approach to the (E)-cycloolefin (**62** → (E)-**63**). These two-step reactions (**61** → **62** → **63**) are in many cases to be preferred to the direct synthesis of **63** by olefin metathesis of a diolefin, since the diastereoselectivity ((Z)- vs. (E)-cycloolefin) can only be controlled to a limited extent, if at all [136].

The metathesis of cycloalkynes leads via a ROMP reaction to polyalkynamers. The polymerization of the strained cyclooctyne to the polyoctynamer is shown as an example (**64** → **65**). It can also be accessed from an acyclic diyne, the dodeca-2,10-diyne, via an acyclic diyne metathesis polymerization (**66** → **65**; ADIMET: acyclic diyne metathesis) [137].

ADIMET reactions are gaining increased interest for the synthesis of alkyne-bridged polymers with special optical and electric properties. Among these are the polymers **67** and **68**, called poly(p-phenylene-ethynylene)s (PPEs) (**67**) or generally as poly-(arylene-ethynylene)s (PAEs). **69** is an example of a hybrid polymer (PPE + PPV; PPV = poly(p-phenylene-vinylene)) [138].

Exercise 7.9

Nitrido- (**1**) and alkylidyne–tungsten complexes (**2**) are mutually interconvertible. If X is an alcoholate ligand with low donor strength (X = OCMe(CF$_3$)$_2$, OCMe$_2$CF$_3$), then an alkyne–nitrile cross-metathesis is catalyzed, which surprisingly leads to a symmetric alkyne Ar–C≡C–Ar (in addition to low quantities of Ar–C≡C–Et):

$$X_3W\equiv N \underset{+\ RC\equiv N}{\overset{+\ RC\equiv CR}{\rightleftharpoons}} X_3W\equiv CR$$
$$\mathbf{1} \qquad\qquad\qquad \mathbf{2}$$

$$2\ Ar-C\equiv N\ +\ Et-C\equiv C-Et \xrightarrow[\text{toluene (95 °C)}]{\mathbf{1}\ (5\ \text{mol\%})} Ar-C\equiv C-Ar\ +\ 2\ Et-C\equiv N$$

It has been shown that the reaction runs similarly to alkyne metathesis. Formulate the relevant catalytic cycles. Which intermediates are involved?

7.3 Enyne Metathesis

Grubbs catalysts can also be used to perform enyne metathesis, forming conjugated dienes. Formally, the olefin's two alkylidene groups are added to the triple bond of the alkyne so as to transform it into the C_{sp^2}–C_{sp^2} single bond of the conjugated diene. Enyne metathesis can be carried out intramolecularly as ring-closing metathesis

(70 → 71; RCEYM: ring-closing enyne metathesis), intermolecularly as cross-metathesis (72 → 73) or as tandem metathesis (74 → 76). In tandem metathesis, two ring-closing metatheses (74 → 75: RCEYM; 75 → 76: RCM) run consecutively.

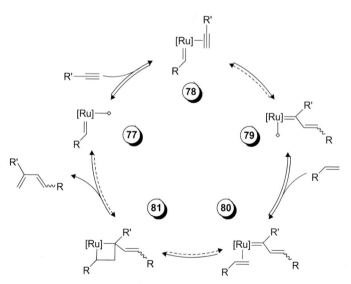

A possible mechanism for an enyne metathesis with a Grubbs catalyst is offered in the example of an intermolecular enyne metathesis shown in simplified form in Figure 7.3. The following reaction steps occur:

77 → 78: Similarly to olefin metathesis, the 14-*ve* alkylidene complex **77** is the starting point. Coordination of the alkyne leads to the alkyne–alkylidene complex **78**.

78 → 79: Insertion of the alkyne into the Ru=C bond yields a vinylcarbene complex **79**. Apparently, no ruthenacyclobutene appears as an intermediate.

Figure 7.3 Mechanism of the intermolecular enyne metathesis with Grubbs catalysts ([Ru] = $RuCl_2(R_2Imid)$), simplified, after Diver [139] as well as Lippstreu and Straub [140].

In the case of metathesis of HC≡CH/H₂C=CH₂, however, quantum-chemical calculations show that a ruthenacyclobutene-like transition state **78‡** occurs.

79 → 80 → 81 → 77: The further reaction (olefin coordination, cycloaddition, cycloreversion) corresponds to the "normal" course of olefin metathesis, just starting from a vinylcarbene complex. It is not clear which reaction step is rate-determining, and this also seems to depend heavily on the nature of the olefin and alkyne.

7.4
σ-Bond Metathesis

σ-Bond metathesis is understood to refer to concerted reactions of type **82 → 84**, which proceed via four-center transition states **83**.

$$\begin{array}{c} A-B \\ + \\ C-D \end{array} \rightleftharpoons \left[\begin{array}{c} A\text{-}\text{-}B \\ | \quad | \\ C\text{-}\text{-}D \end{array}\right]^{\ddagger} \rightleftharpoons \begin{array}{c} A \quad B \\ | \; + \; | \\ C \quad D \end{array}$$

82   **83**   **84**

Here, the reactions that take place with participation of metals are of interest. They occur with activation of H–H (**a**), C–H (**b**) and C–C bonds (**c**) and correspond to hydrogenolysis (**a**) or alkanolysis (**b/c**), respectively, of M–C bonds.

(a) H–H + [M]–C⟨ ⇌ [H----H / [M]---C⟨]‡ ⇌ H–[M] + H–C⟨

(b) ⟩C–H + [M]–C⟨ ⇌ [⟩C----H / [M]---C⟨]‡ ⇌ ⟩C–[M] + H–C⟨

(c) ⟩C–C⟨ + [M]–C⟨ ⇌ [⟩C---C⟨ / [M]---C⟨]‡ ⇌ ⟩C–[M] + ⟩C–C⟨

For early transition metals with a d⁰ electron configuration, the following sequence applies for the tendency to form the four-membered transition states: **a > b ≫ c** [141]. If the σ-bond metathesis is preceded by σ-complex formation, as has been demonstrated for late transition metals, this is called σ-complex-assisted metathesis (σ-CAM). As an example, a ligand metathesis is introduced that proceeds via discrete σ complexes as intermediates. These complexes are able to undergo dynamic rearrangements:

$$[M]{-}Si{\lesssim} + H{-}H \rightleftharpoons [M]{\cdots}H_2{\cdots}Si \rightleftharpoons [M]{\cdots}H{\cdots}Si{-}H \rightleftharpoons [M]{-}H + {\geq}Si{-}H \quad (d)$$

Both σ-H_2 (see Section 4.4.1) and σ-H–SiR_3 complexes (see Section 12.3.2.4) have been isolated and characterized. No change occurs in the oxidation state of the metal in the σ-CAM mechanism [142].

An alternative mechanism (a′) is possible if the metal complex is capable of oxidative addition of H_2:

$$\frac{H-H}{[M]-C{\lesssim}} \rightleftharpoons [M]{\overset{H}{\underset{C}{\lesssim}}}H \rightleftharpoons [M]{-}H + H{-}C \quad (a')$$

The reaction sequence in a′, oxidative addition (possibly preceded by σ-H_2 complex formation) and reductive C–H elimination, generates the same products as in a. Here the oxidation state of M changes; this is the critical difference from the σ-CAM mechanism.

As with alkanes, σ-bond metathesis proceeds with silanes (d/d′).

$$\frac{H-H}{[M]-Si{\lesssim}} \rightleftharpoons \left[{[M]{\cdots}Si{\lesssim}}^{H----H}\right]^{\ddagger} \rightleftharpoons [M]{-}H + H{-}Si \quad (d')$$

Thus, catalytic dehydrocoupling of silanes (**85 → 86**) is possible. Dihydrosilanes R_2SiH_2 are thus converted into polysilanes **87**, which are of interest due to their special electric and optical properties, as well as the option of pyrolysis (R = Me) to form silicon carbide fibers. Intermediates can be silylmetal and hydridometal compounds.

$$\underset{\mathbf{85}}{\overset{{\geq}Si{-}H}{\underset{{\geq}Si{-}H}{+}}} \quad \overset{cat.}{\rightleftharpoons} \quad \underset{\mathbf{86}}{\overset{Si}{\underset{Si}{\overset{H}{\underset{H}{+}}}}} \quad \underset{\mathbf{87}}{H{-}{\left[Si\overset{R}{\underset{R}{|}}\right]}_n{-}H}$$

Exercise 7.10

Justify the statement that an alternative mechanism to σ-bond metathesis of silanes based on oxidative addition of Si–H or H–H bonds to hydrido- or silylmetal complexes is not possible for d^0 complexes.

σ-Bond metathesis can be the basis for hydrogenolysis of polyethylene into oligomers or short-chain alkanes. Thus, zirconium hydrides can be grafted to the surface of dehydroxylated silica-alumina, displaying a structure as in **88** (where the wavy line indicates the solid surface). In addition, aluminum hydride or silicon dihydride groups are bound to the surface in close proximity to the zirconium center. The grafted hydridozirconium complex 88 (symbol: $[Zr]_s-H$) catalyzes the polymerization of ethene into polyethylene (89 → 90) even at room temperature. At 150 °C, in the presence of H_2, depolymerization occurs all the way to production of short-chain alkanes (90 → 91). Hydrogenolytic polymer degradation also occurs when 88 reacts directly with polyethylene in the presence of H_2.

$$n \, \text{==} \xrightarrow[RT]{88} \left(\bigwedge\right)_n \xrightarrow[88]{H_2, 150\,°C} \text{short-chain alkanes}$$

89   90   91   88

The first step in the hydrogenolysis of the polymer (88 → 92, **P** and **P'** symbolize the polymer chain) is probably a σ-bond metathesis with C–H activation according to reaction **a** (see above). A surface-grafted secondary alkyl complex 92 and a hydrogen molecule are thus formed.

88   92

Exercise 7.11
Suggest a reaction mechanism for the hydrogenolytic polymer degradation 90 → 91 starting with complex 92. Consider a β-alkyl elimination (Section 3.4), the microscopic reverse of the olefin insertion into a σ-alkyl–metal bond. Think about what could be the driving force for the β-alkyl transfer, which by itself is thermodynamically unfavorable.

7.5 Metathesis of Alkanes

7.5.1 Principles

The formal analogy with olefin and alkyne metathesis is the catalysis of a reaction that converts an alkane of medium chain length (93) into an alkane with lower and higher C-number (94/95) in equal parts (alkane metathesis).

$$\begin{array}{c} Me(CH_2)_n-(CH_2)_m Me \\ + \\ Me(CH_2)_n-(CH_2)_m Me \end{array} \xrightleftharpoons{cat.} \begin{bmatrix} Me(CH_2)_n \\ Me(CH_2)_n \end{bmatrix} + \begin{bmatrix} (CH_2)_m Me \\ (CH_2)_m Me \end{bmatrix}$$

93   94   95

Alkane metathesis is fundamentally less selective than olefin and alkyne metathesis and, since all C–C bonds are included in the reaction to at least some degree, produces a variety of alkanes with lower and higher C-numbers than the starting alkane. Alkane metathesis could only be achieved in recent years, and then only in special cases (J.-M. Basset, 1997). Hydridometal complexes grafted to the surface of silica have been used as precatalysts, particularly a hydridotantalum(III) complex **96**,[4] which is accessible in a well-defined way by grafting [Ta(=CHt-Bu)(CH$_2$$t$-Bu)$_3$] onto partially dehydroxylated silica and then performing hydrogenolysis. Silica thermally treated at 700 °C (SiO$_{2\text{-}(700)}$) has isolated surface silanol groups. On such a support, a monografted surface complex is first created and then treated with hydrogen to form the bis(siloxy)-hydridotantalum surface complex **96**. If the silica used was partially dehydroxylated at a lower temperature (SiO$_{2\text{-}(200)}$), a bisgrafted surface complex is created, which likewise yields **96** upon hydrogenolysis. Grafted surface complexes as in **96** are subsequently abbreviated as "[Ta]$_s$" (**96**: [Ta]$_s$–H) [143].

Complexes such as **96** are extremely electrophilic (8 ve!) hydridotantalum(III) complexes (d^2 electron configuration). The selectivity achieved with such catalysts ([Ta]$_s$–H, 150 °C) in ethane, propane, and n-butane metathesis is shown in Figure 7.4. The diagram of butane metathesis, in particular, shows that while the main products are the adjacent homologous alkanes (propane, pentanes), other homologues are formed to a great extent. In propane metathesis, a hydridotungsten complex grafted on Al$_2$O$_3$ has shown to be about twice as active as **96** [144].

7.5.2
Mechanism

The grafted hydridotantalum complex **96** reacts with the alkane[5] in a σ-bond metathesis forming an ethyltantalum complex **97**:

---

4) The silanol groups from which the bis(siloxy) complexes are formed need not be located on neighboring Si–O–Si-bridged silicon atoms. Our scheme depicts it that way for the sake of simplicity. Moreover, the fact that hydrogenolysis of the organo ligands can form surface Si–H moieties besides **96** is not considered here.

5) For the sake of clarity in the following discussion, we write formulas for ethane, since no regioisomers occur.

7.5 Metathesis of Alkanes

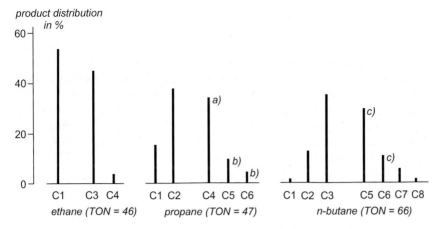

Figure 7.4 Selectivity (at 3% turnover) of ethane, propane, and n-butane metathesis with [Ta]$_s$–H (**96**) as precatalyst (150 °C; 50–80 h). The starting alkane, present in excess, is not shown. The turnover numbers (TON) are given in moles of alkane per moles of tantalum (after Vidal and Basset [145]). a) n/i ratio about 4. b) n/i ratio about 2. c) n/i ratio about 9.

$$[Ta]_s \overset{H}{|} + \overset{H}{|}CH_2-CH_3 \;\rightleftharpoons\; [Ta]_s\text{-}\overset{H---H}{|}CH_2\text{-}CH_3 \;\rightleftharpoons\; [Ta]_s\overset{H-H}{-}CH_2\text{-}CH_3$$

96   **97**

Exercise 7.12

It was originally thought that the alkane metathesis starting from **97** as catalyst complex completed via σ-bond metathesis. Formulate this with ethane as substrate. Is a reaction course via oxidative addition and reductive elimination possible in principle?

Careful time-dependent product analysis for propane metathesis[5] has demonstrated that olefins and hydrogen are primary products of metathesis. To account for this, the following mechanism, which has been increasingly supported by experiments, has been postulated [141, 146]:

97 → 98 or 97 → 99 → 96: α/β-*Hydrogen elimination*. The grafted alkyltantalum complexes **97** are subject to α- and β-H elimination, forming hydrido–carbene complexes (**98**) or hydrido complexes **96** and olefins, respectively.

$$\begin{array}{c}
\textbf{98}\\
[Ta]_s\!\!\diagup\!\!\overset{CHCH_3}{\underset{H}{}}\\
\nearrow\\
[Ta]_s\text{–}CH_2CH_3\\
\textbf{97}\\
\searrow\\
[Ta]_s\!\!\diagup\!\!\overset{H_2C}{\underset{H}{\diagdown CH_2}} \;\rightleftharpoons\; [Ta]_s\text{–}H + H_2C{=}CH_2\\
\textbf{99} \hspace{3cm} \textbf{96}
\end{array}$$

98 → 100 → 101 → 100′ → 98′: π-bond metathesis. Via π-bond metathesis, olefins of different chain lengths are formed in the usual way, except that the catalytically active complexes are not only alkylidene, but hydrido–alkylidene complexes **98/98′**.

$$98 \quad [Ta]_s \!\!\begin{array}{c} \diagup CHCH_3 \\ \diagdown H \end{array} \qquad\qquad [Ta]_s \!\!\begin{array}{c} \diagup CH_2 \\ \diagdown H \end{array} \quad 98'$$

$$\big\Updownarrow H_2C{=}CH_2 \qquad\qquad\qquad\qquad \big\Updownarrow H_2C{=}CHCH_3$$

$$\begin{array}{c} CH_2{=}CH_2 \\ | \\ [Ta]_s{=}CHCH_3 \\ | \\ H \end{array} \rightleftharpoons \begin{array}{c} H_2C{-}CH_2 \\ |\ \ \ | \\ [Ta]_s{-}CHCH_3 \\ | \\ H \end{array} \rightleftharpoons \begin{array}{c} CH_2 \\ \| \\ CH_2 \\ [Ta]_s{-}\| \\ | \ \ CHCH_3 \\ H \end{array}$$

$$\quad\quad\ 100 \qquad\qquad\qquad\qquad 101 \qquad\qquad\qquad\qquad 100'$$

98′ → 97′ → 96: Carbene insertion and σ-bond metathesis. Transfer of hydrogen from the metal onto the carbene ligand forms a methyltantalum complex from which methane is formed by hydrogenolysis.

$$[Ta]_s\!\!\begin{array}{c} \diagup CH_2 \\ \diagdown H \end{array} \rightleftharpoons [Ta]_s{-}CH_3 \xrightarrow{H_2} \left[\begin{array}{c} H{-}{-}{-}H \\ \vdots\quad\vdots \\ [Ta]_s{-}{-}CH_3 \end{array}\right]^{\ddagger} \rightleftharpoons \begin{array}{c} H \\ | \\ [Ta]_s \end{array} + CH_4$$

$$\quad 98' \qquad\qquad\quad 97' \qquad\qquad\qquad\qquad\qquad\qquad 96$$

96 → 99′ → 97″ → 96: Olefin insertion and σ-bond metathesis. Insertion of the olefin into the Ta–H bond generates an alkyltantalum complex **97″**, from which propane is liberated via hydrogenolysis, with regeneration of **96**.

$$[Ta]_s{-}H \xrightleftharpoons{H_2C=CHCH_3} \begin{array}{c} H_2C \\ \diagup\ \ \diagdown CHCH_3 \\ [Ta]_s \\ \diagdown H \end{array} \rightleftharpoons [Ta]_s{-}CH_2CH_2CH_3 \xrightarrow{H_2} \left[\begin{array}{c} H{-}{-}{-}H \\ \vdots\quad\vdots \\ [Ta]_s{-}{-}CH_2CH_2CH_3 \end{array}\right]^{\ddagger}$$

$$96 \qquad\qquad 99' \qquad\qquad\qquad 97''$$

$$\qquad\qquad\qquad\qquad\qquad\qquad 96\ \ \begin{array}{c} H \\ | \\ [Ta]_s \end{array} + CH_3CH_2CH_3$$

It should be emphasized that according to this mechanism, the critical steps in alkane metathesis, namely C–C bond cleavage and formation, are analogous to olefin metathesis, and hence occur through π-bond metathesis. However, DFT calculations on model complexes show that in olefin complexes **100/100′**, the carbene ligand cannot be arranged *trans* to the hydrido ligand.[6] In the tetrahedral starting

---

6) It is energetically unfavorable for two ligands to have to share the same metal atomic orbitals for bonding with the metal. This general principle is not only seen in the *trans* influence of ligands, but is also expressed by the fact that for two strongly bonding σ ligands (such as H$^-$ and CH$_2^{2-}$), the *cis* arrangement is generally energetically favored over the *trans*. The same is true for two strong π ligands.

complex **98m**, the bidentate binding of the tantalum to an SiO$_2$ surface is modeled by a bidentate bis(siloxy) ligand, formed by double deprotonation of disilicic acid (H$_6$Si$_2$O$_7$). The distortion of the ligand sphere into *cis*-**98m** costs only about 6 kJ/mol, but forming *trans*-**98m** costs more than 300 kJ/mol, so the ethene complex *trans*-**100m** is ruled out as an intermediate.

This causes, in turn, a somewhat more complex reaction course of π-bond metathesis starting from a hydrido–alkylidene complex (Figure 7.5). The [2 + 2] cycloaddition **100m** → **101m** leads to a hydridotantalacyclobutane complex, in which the hydrido ligand is positioned (approximately) *trans* to a CH$_2$R ligand. Both are strong σ donors, so the energy of the complex is relatively high. By reorganization of the ligand sphere (**101m** → **102m**), a significantly more stable isomer is obtained in which the tantalum displays a strongly distorted trigonal-bipyramidal structure such that the H ligand is no longer located *trans* to an organyl ligand. **102m** is also the most energetically stable intermediate. Now, probably via a turnstile mechanism, a ligand rearrangement (**102m** → **102m'**) occurs, which is followed by the cycloreversion **102m'** → **101m'** → **100m'**.

During the catalysis, the oxidation state of the grafted tantalum switches from Ta^{III} (**96, 97, 99**) to Ta^V (**98, 100, 101**). The reactivity of the grafted hydrido- (**96**) and alkyltantalum(III) complexes (**97**) is unique, and there are no molecular complexes with comparable reactivity. This can probably be traced to the fact that the complexes are electronically extremely unsaturated (8 ve!) and that the exceptionally electrophilic Ta atom is not shielded by bulky ligands. Nevertheless, the grafting to the surface apparently is enough to prevent dimerization reactions, which probably would lead to deactivation and also to decomposition.

7.5.3
Alkane Metathesis Via Tandem Reactions

Another method for perfoming alkane metathesis has been found in a tandem reaction of catalytic alkane dehydrogenation and olefin metathesis with two catalysts

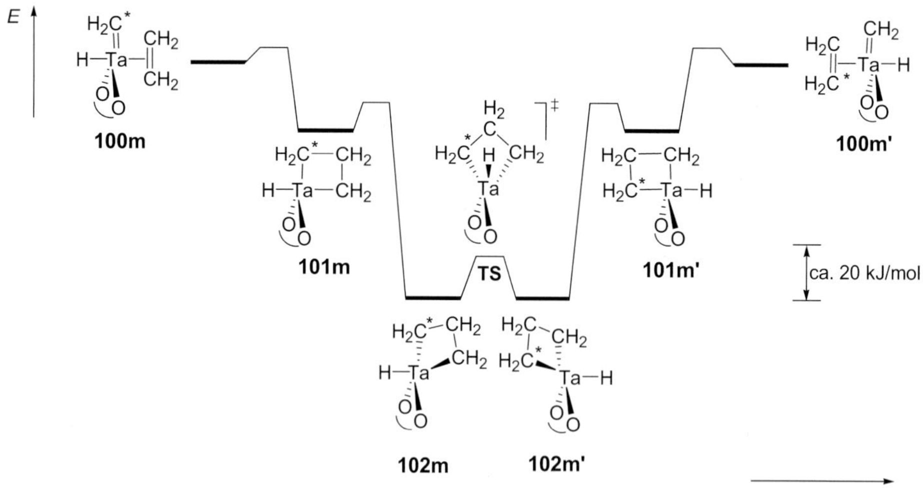

Figure 7.5 Energy profile diagram for a π-bond metathesis in the alkane metathesis using the example of the conversion of a hydrido(methylene)tantalum complex **100m** (O⌒O = bis(siloxy) ligand). In order to make it easier to follow the reaction course for this nonproductive metathesis, a C atom is marked by a star (simplified, after Schinzel and Chermette [147]).

working independently of each other: Transfer dehydrogenation of an alkane RCH_2CH_3 to form a terminal alkene, catalyzed by [M]/[M]H_2 (**103** → **104**), is followed by olefin metathesis catalyzed by [M'] (**104** → **105**). Then the reaction is completed by transfer hydrogenation of the olefin metathesis products to form RCH_2CH_2R and H_3CCH_3 (**105** → **106**), catalyzed by [M]H_2/[M] [148].

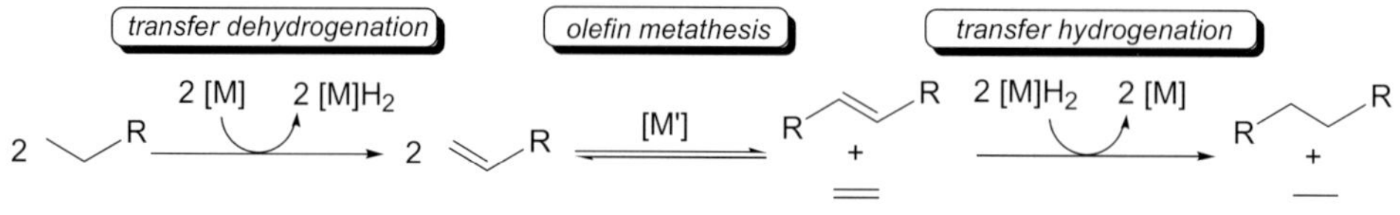

Pincer iridium complexes of type **107** (R = t-Bu, i-Pr; X = CH_2, O) have proven themselves as catalysts for transfer dehydrogenation of n-alkanes into α-olefins (**108** → **108'**), where olefins **109'** such as tert-butylethene, dec-1-ene and norbornene (**109'** → **109**) have found use as H acceptors. The terminal olefin **108'** formed is, under the catalytic influence of **107**, also partially isomerized into inner olefins.

$$RH_2C-CH_3 + R'HC=CH_2 \xrightarrow[150\ °C]{107} RHC=CH_2 + R'H_2C-CH_3$$

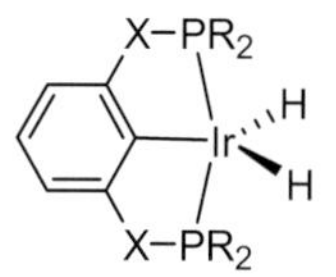

108   **109'**   **108'**   **109**

7.5 Metathesis of Alkanes

If an alkane (model substrate: *n*-hexane, **103** with R = *n*-Bu) is allowed to react in the presence of the transfer dehydrogenation catalyst **107**, which catalyzes the reactions **103** → **104** and **105** → **106**, *and* a metathesis catalyst (Schrock catalyst), a conversion to alkanes **106**, as described above, occurs (125 °C, 24 h). Since **107** also catalyzes double-bond isomerization, a broad palette of alkanes, mainly C_2–C_5 and C_7–C_{10} alkanes, is created, with turnover numbers of about 135.

Exercise 7.13
Suggest a mechanism for the iridium-catalyzed reaction **108** + **109'** → **108'** + **109**.

While the system just described consists of a combination of two independently acting catalysts – a dehydrogenation/hydrogenation catalyst (Ir) and a metathesis catalyst (Mo) – a surface-grafted Schrock-type imidomolybdenum complex **110** unites these two functionalities in a single complex. The propane metathesis produces *n*-butane and ethane with good selectivity (>90%) [149]:

$$2 \; \wedge \xrightarrow[150 \,°C, \, 120 \, h]{\textbf{110}} \; - \; + \; \wedge \; + \; i\text{-}C_4, C_5, C_6$$

(56%)   (35%)   (< 10%)

(degree of conversion: 10%, TON = 55)

110: *t*-BuCH$_2$, Mo(=N-2,6-*i*-Pr$_2$C$_6$H$_3$)(CH*t*-Bu)(O–Si(O)$_3$) (silica-grafted)

The experiments confirm that transfer dehydrogenation/hydrogenation converts alkanes into olefins and vice versa and that the C-number redistribution is completed via olefin metathesis. Similar results have been found for *n*-butane metathesis, while **110** does not catalyze ethane metathesis. This is a significant difference from the previously described grafted hydridotantalum complex, although in both cases, π-bond metathesis is of importance for C–C cleavage and formation reactions.

The catalytic activation and functionalization of C–H and C–C bonds of simple alkanes is one of the great challenges of homogeneous catalysis.

Background: Organometallic Pincer Complexes
Organometallic pincer ligands are special types of chelating ligands that typically bind with two neutral two-electron ligator atoms on opposite sides of a metal and have a σ-M–C bond between them, thus exhibiting a meridional κY,κC,κY coordination. The general formula **1** shows that organometallic pincer complexes are special organometallic inner complexes, typically with two five-membered metallacycles. Very often, the central structural unit is a 2,6-disubstituted aryl group, so an M–C$_{sp^2}$ bond results (**2**). Corresponding complexes with an alkyl backbone have also been described (**3**); the M–C bond experiences additional stabilization through its incorporation into a metallacyclic system. P-, N- and S-donors (YR$_n$ = PR$_2$, NR$_2$, SR; R = alkyl, aryl) are especially to be considered as donor groups.

Organometallic pincer ligands have proven themselves as coligands (control ligands) in organometallic catalysis. The electronic properties of M can be influenced in a targeted way by the ligand structure (systematic modification of the skeletal structure *and* variation of YR_n, R'). Likewise, the coordination pocket for the substrate and additional ligands (see arrow in **1**) can be varied according to goal. Functionalization with chiral substituents (e.g., R in YR_n) of asymmetrically substituted benzyl C atoms in **2** lead to chiral (mostly C_2-symmetric) pincer ligands, which allow asymmetric catalysis [150, 151].

8
Oligomerization of Olefins

8.1
Ziegler Growth Reaction

Studies by Karl Ziegler and co-workers at the MPI für Kohlenforschung (MPI for Coal Research, Mülheim/Ruhr) beginning in mid-1949 showed that aluminum triethyl reacts with ethene (90–120 °C, 100 bar) with insertion into the Al–C bonds to form long-chain aluminum alkyls (**1** → **2**). Ziegler named this reaction "Aufbaureaktion" (Aufbau reaction, growth reaction). At higher temperatures (320 °C), β-hydrogen elimination occurs with formation of α-olefins and aluminum hydride (**2** → **3**). The latter reacts at temperatures as low as 20–80 °C with ethene, with insertion into the Al–H bonds, forming aluminum triethyl (**3** → **1**).

$$\text{Al-H} \underset{\substack{20-80°C \\ 100\,\text{bar}}}{\overset{+ \text{H}_2\text{C=CH}_2}{\rightleftharpoons}} \text{Al-Et} \underset{\substack{90-120°C \\ 100\,\text{bar}}}{\overset{+ n\,\text{H}_2\text{C=CH}_2}{\longrightarrow}} \text{Al-(CH}_2\text{-CH}_2\text{)}_n\text{-Et} \underset{\substack{320°C \\ 10\,\text{bar}}}{\rightleftharpoons} \text{H}_2\text{C=CH-(CH}_2\text{-CH}_2\text{)}_{n-1}\text{-Et} + \text{Al-H}$$

$$\mathbf{1} \qquad\qquad \mathbf{2} \qquad\qquad \mathbf{3}$$

An equilibrium exists between trialkyl- and dialkyl(hydrido)aluminum compounds such as **2** and **3** (AlR$_3$ ⇌ R$_2$Al–H + olefin). This causes the Aufbau reaction (**1** → **2**) to be accompanied by a displacement reaction,

$$\text{Al-(CH}_2\text{-CH}_2\text{)}_n\text{-Et} + \text{H}_2\text{C=CH}_2 \rightleftharpoons \text{Al-Et} + \text{H}_2\text{C=CH-(CH}_2\text{-CH}_2\text{)}_{n-1}\text{-Et},$$

$$\mathbf{2} \qquad\qquad\qquad \mathbf{1}$$

which proceeds via hydridoaluminum intermediates. The α-olefins formed in this way can now insert themselves (instead of ethene) into an Al–H bond. This leads to branched olefins, which are also observed as side products. The positions of the equilibria of the displacement reactions allow to deduce that the affinity of olefins for insertion into an Al–H bond decreases in the sequence H$_2$C=CH$_2$ > H$_2$C=CHR > H$_2$C=CR$_2$. The Aufbau reaction is the basis for a catalytic synthesis from ethene of α-olefins with chain lengths C$_{10}$–C$_{18}$, which in turn are used as the starting compounds for synthesis of fatty alcohols.

Fundamentals of Organometallic Catalysis. First Edition. Dirk Steinborn
Copyright © 2012 WILEY-VCH Verlag GmbH & Co. KGaA, Weinheim

At that time, the Aufbau reaction was coupled with an oxidation to form aluminum alcoholates (2 → 4) and their hydrolysis (4 → 5). Fatty alcohols were thus obtained (ALFOL process); however, this process consumed stoichiometric quantities of AlEt$_3$. Therefore, this process is now of only historical interest.

$$\text{>Al-(CH}_2\text{-CH}_2)_n\text{-Et} \xrightarrow{+ 1/2\, O_2} \text{>Al-O-(CH}_2\text{-CH}_2)_n\text{-Et} \xrightarrow[-\,>\text{Al-OH}]{+ H_2O} \text{HO-(CH}_2\text{-CH}_2)_n\text{-Et}$$

$$\mathbf{2} \qquad\qquad\qquad \mathbf{4} \qquad\qquad\qquad \mathbf{5} \quad n = 4\text{--}8$$

From a scientific point of view, it is interesting that the Aufbau reaction can be performed such that the termination reaction 2 → 3 is practically insignificant. At room temperature and an ethene pressure of 20–100 bar, a markedly high-molecular (molar masses up to $9 \cdot 10^6$ g/mol), highly linear polyethylene (melting point 140–143 °C) is obtained. However, the reaction runs extremely slowly [152].

In the 1960s, the Gulf Oil Company (Texas, USA) and a few years later, the Ethyl Corporation (Texas, USA) developed industrial processes for synthesis of α-olefins, based on the Aufbau reaction, which are still being run today. The Gulf process is a one-step procedure: The oligomerization of ethene is carried out with catalytic quantities of AlEt$_3$ in hexane as solvent at about 200 °C and 250 bar. Growth *and* displacement reactions run simultaneously in one reactor. C$_4$–C$_{30}$ α-olefins (Gulftene®) are produced that only contain about 1% paraffin hydrocarbons and a few percent of branched olefins as side products.

The Ethyl process permits better control over the chain length of the olefins. It is a two-step process. In the first process step, as in the Gulf process, ethene is oligomerized with catalytic quantities of AlEt$_3$ (160–175 °C, 130–270 bar). After hydrolytic decomposition of the catalyst, the product is fractionally distilled and the desired fraction of α-olefins (RHC=CH$_2$, R = C$_{10}$–C$_{16}$) is introduced for use. In a second process step, the shorter-chain olefins (R'HC=CH$_2$, R' = C$_2$–C$_8$) are reacted in the "transalkylation reactor" (300 °C, 100 bar) with a stoichiometric quantity of aluminum alkyls with longer-chain alkyl groups R (6). A displacement reaction occurs (6 → 7) in which, with formation of aluminum alkyls with shorter-chain alkyl group R' (7), α-olefins of the desired chain length are cleaved off.

$$\text{Al(CH}_2\text{CH}_2\text{R)}_3 + 3\ \text{R'CH=CH}_2 \longrightarrow \text{Al(CH}_2\text{CH}_2\text{R')}_3 + 3\ \text{RCH=CH}_2$$

$$\mathbf{6} \qquad\qquad\qquad\qquad\qquad\qquad\qquad \mathbf{7}$$

$$n\ \text{H}_2\text{C=CH}_2$$

After separation of these olefins via distillation, the aluminum alkyls 7 are subjected to the Aufbau reaction with ethene (7 → 6, 100 °C, 200 bar) and the aluminum alkyls 6 formed are introduced once more into the transalkylation reactor. The Gulf process has the advantage that ethene can be converted to olefins of the desired chain length with 95% selectivity. However, the process has the disadvantage that substantial quantities of branched α-olefins and inner olefins are formed. Thus,

8.2
Nickel Effect and Nickel-Catalyzed Dimerization of Ethene

for instance, the C_{14}/C_{16} and the C_{16}/C_{18} fractions contain only about 76 or 63% linear α-olefins, respectively.

In experiments on the growth reaction (1953) with aluminum triethyl in a PhD project in Ziegler's group at the MPI für Kohlenforschung, only butenes had been formed, rather than long-chain alkyl aluminum compounds. It turned out, that traces of nickel compounds had been left after the V2A autoclave used in these experiments had been cleaned. Further investigations then showed that traces of acetylene contained in the technical-grade ethene used in these experiments were also needed to stabilize the catalyst. This effect of nickel on the Aufbau reaction is called "nickel effect" [153, 154].

The catalysis of the ethene dimerization proceeds by insertion of ethene into an Al–C bond (Aufbau reaction: $R_2AlEt + H_2C=CH_2 \rightarrow R_2AlBu$) followed by a nickel-catalyzed cleavage of butene. This occurs as a "transalkylation" on a nickel(0) complex 8, which has coordinated ethene and also – probably via a multi-center bond – the α-C atom from a butyl group of an aluminum butyl species. Reorganization then takes place, probably by an electrocyclic reaction (9), forming a butenenickel(0) complex with a coordinated aluminum ethyl species (10). Ligand substitution (butene/ethene) and insertion of ethene into the Al–C bond (10 → 8) complete the catalytic cycle.

Diorganyl magnesium complexes of Ni^0 (obtained by reaction of $[Ni(\eta^2\text{-}H_2C=CH_2)_3]$ with MgR_2) can be regarded as model complexes for the intermediates 8/10 (Figure 8.1). As with 8/10, these are olefinnickel(0) complexes that have a μ-alkyl ligand bound.

Nickel(II) compounds also catalyze the dimerization of ethene without being reduced to nickel(0). Aluminum alkyls are used as cocatalysts in this process. The mechanism is shown in Figure 8.2. Starting with a nickel(II) hydride species, coordination and insertion of the ethene into the Ni–H bond (11 → 12 → 13) occurs as well as further olefin coordination (13 → 14). A butylnickel complex

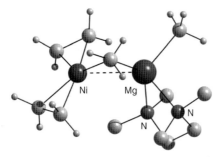

Figure 8.1 Molecular structure of [Ni(η^2-C$_2$H$_4$)$_2$(μ-Me)MgMe(tmeda)] (tmeda = N,N,N',N'-tetramethylethylenediamine). The H atoms of tmeda have been omitted for clarity.

(**14** → **15**) forms as a result of insertion into the Ni–C bond. This complex undergoes a β-hydride elimination, whereupon but-1-ene cleaves off (**15** → **16** → **11**). Thus, for every two insertion steps, one β-H elimination reaction occurs. In principle, all reaction steps are reversible except for the insertion into the Ni–C bond (**14** → **15**), since **15** is subject to a very rapid β-H elimination (**15** → **16** → **11**).

Organometallic-catalyzed polymerization of olefins (see Section 9.2.1) is based on the same elementary steps as the dimerization of ethene according to Figure 8.2, except that the ratio of the rates for the chain-growth reaction (k_p, **14** → **15**) and the β-H elimination, a chain termination reaction without catalyst deactivation (k_t, **15** → **16**), is fundamentally different for the two reactions. For very rapid olefin insertion (k_p ↑) and slow β-H elimination (k_t ↓) many insertion steps occur before

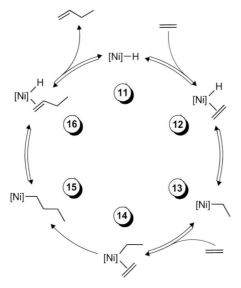

Figure 8.2 Reaction scheme for the nickel(II)-catalyzed dimerization of ethene.

chain termination takes place and polymerization results. The opposite is true for oligomerization. Thus, there is a fundamental relationship between oligomerization and polymerization catalysts. They may be mutually interconvertible by structural variation and/or changes to the reaction conditions. Thus, for example, even typical metallocene catalysts for the polymerization of olefins such as $[Zr(Cp')_2Cl_2]$/MAO (MAO – methylaluminoxane, Section 9.4.1) can oligomerize them under altered reaction conditions such as a higher temperature and a lower excess of cocatalyst [155].

During nickel-catalyzed ethene dimerization, a double-bond isomerization to form but-2-ene has also been observed: The *n*-butylnickel complex **15** can isomerize via β-hydride elimination (**15** → **16**) and reinsertion to form the *sec*-butylnickel (1-methylpropylnickel) complex **17**, which can then undergo a β-hydride elimination with formation of but-2-ene (**17** → **18**).

In catalyst generation, that is, the formation of a coordinatively unsaturated hydrido- or alkylnickel(II) complex, the aluminum alkyl (in general, an ethylaluminum compound) serves two purposes as cocatalyst:

- *Alkylation* (**a**). If one assumes that a nickel halide is present as the precatalyst, an ethylnickel species is formed in a metathesis-like (double displacement) reaction.
- *Lewis acidity* (**b**). A Lewis acid-base interaction between the Lewis acidic aluminum compound ($AlEt_3$ or $AlEt_2Cl$) and a chloro ligand can be a precondition for the formation of a coordinatively unsaturated nickel complex, capable of coordinating ethene.

The mechanism of olefin dimerization has been studied more closely with the dinuclear allylnickel bromide complex **19** as precatalyst in the presence of a phosphane PR$_3$ and of AlEtCl$_2$ as cocatalyst ("Wilke catalyst"; G. Wilke, B. Bogdanović, MPI für Kohlenforschung). First, a reaction forming a mononuclear allyl(phosphane)nickel complex **20** occurs. Coordination of ethene and insertion into the η^3-allyl bond yields a pent-4-enyl complex **21**, which cleaves off penta-1,4-diene in a β-H elimination, thus generating the catalyst complex **11** (see Figure 8.2: [Ni]–H = NiH(PR$_3$)(AlEtCl$_2$Br)). As precatalysts, cationic allyl(phosphane)nickel complexes of type **20** such as [Ni(C$_3$H$_5$)(PR$_3$)][AlCl$_4$] can also be used from the outset.

The catalytic cycle corresponds to the one in Figure 8.2 with [Ni]–H = NiH(PR$_3$)(AlEtCl$_2$Br). The coordination of the anion is noteworthy, as [AlEtCl$_2$Br]$^-$ probably serves as a hemilabile ligand (see Background, Section 8.4), which is coordinated monodentately (**I**, in the intermediates **12**, **14** and **16** in Figure 8.2) or bidentately (**II**, in the intermediates **11**, **13** and **15** in Figure 8.2). The bidentate coordination contributes to the stabilization of the coordinatively unsaturated intermediates.

(X = Cl, Br)

The highest selectivity is achieved with PR$_3$ = PMe$_3$ (C$_4$ > 98%). Using more sterically demanding phosphane ligands leads to an increased occurrence of higher oligomers. In fact, with P(t-Bu)$_3$ as ligand ($c_{Ni}/c_{P(t-Bu)3}$ = 1/4), polyethylene is obtained. Sterically demanding groups thus hinder the β-H elimination to an increasing extent [156].

Nickel-catalyzed formation of butene from ethene can also run according to another reaction mechanism. From preparative and kinetic experiments, it is known that the bis(ethene)nickel(0) complex **22** (L = PPh$_3$) in the presence of PPh$_3$ is in equilibrium with the oxidative coupling product **23** (**22** → **23**), which is converted by successive cleavage of PPh$_3$ into the nickelacyclopentane complexes **24** and **25**. These decompose via reductive C–C elimination, forming cyclobutane (**24** → **26**), or, via coupled β-H and reductive C–H elimination, forming butene (**25** → **26**). Accordingly, complex **23** in isolated form catalyzes the dimerization of ethene to form butene and cyclobutane without requiring a cocatalyst, but its activity is low [33].

8.2 Nickel Effect and Nickel-Catalyzed Dimerization

$$L_2Ni \underset{}{\overset{+L}{\rightleftharpoons}} L_3Ni \underset{}{\overset{-L}{\rightleftharpoons}} L_2Ni \underset{}{\overset{-L}{\rightleftharpoons}} LNi$$

22   23   24   25

$+2 \rightleftharpoons (+L)$

$+ LNi/L_2Ni$

26

The Wilke catalyst is also suited to the dimerization of propene, a process in which the structure of the dimers can be controlled to a high degree via the phosphanes. PMe$_3$ generates about 80% methylpentenes (head-to-tail linkage)[1] and about 10% each of n-hexenes (tail-to-tail linkage) and 2,3-dimethylbutenes (head-to-head linkage). With more sterically demanding phosphanes, the highly branched dimers (2,3-dimethylbutenes) are increasingly formed; for example, with P(i-Pr)$_2$(t-Bu), up to about 80%. The catalysts display exceptional activity: With [Ni(η^3-C$_3$H$_5$)(PCy$_3$)]/AlEtCl$_2$ in chlorobenzene, turnover frequencies (TOF) of about $2.1 \cdot 10^8$ mol propene/(mol Ni·h) are achieved (experimental values from −55 to −75 °C extrapolated to 25 °C), so about 150 metric tons of product/(g Ni·h) are formed. This corresponds to the activity of highly active enzymes[2] [157].

The so-called non-regioselective dimerization without addition of phosphane produces about 20% n-hexenes, 75% methylpentenes, and 5% dimethylbutenes. The activity of the phosphane-free catalyst amounts to about 1/15 of the value given above. It is of industrial significance (Dimersol process from IFP) [M6]. The hexenes formed are added to gasoline to raise its octane number. In an analogous way, n-butenes are dimerized into octenes.

Exercise 8.1
State the reaction products of (non-regioselective) propene dimerization. Neglect subsequent double-bond isomerization reactions.

Ionic liquids (see Background, Section 5.4.2) consisting of 3-dialkylimidazolium cations and chloroaluminate anions have proven to be excellent solvents for the nickel dimerization catalysts in the Dimersol process. The latently Lewis-acidic ionic liquid acts also as cocatalyst, since in mixtures of an imidazolium chloride [3]Cl with excess AlCl$_3$ ([3]Cl : AlCl$_3$ < 1), Lewis-acidic higher aggregated chloroaluminates ([AlCl$_4$]$^-$ → [Al$_2$Cl$_7$]$^-$ → [Al$_3$Cl$_{10}$]$^-$) are formed to an increasing extent.

---

[1] For vinyl and vinylidene monomers H$_2$C=CHR and H$_2$C=CRR', the more highly substituted end is called the head (H) and the CH$_2$ group is called the tail (T).

[2] The activity of enzymes can be specified by the number of substrate molecules that are converted by an enzyme molecule per second under optimal conditions. In biochemistry this is usually called "turnover number," but should be called here "turnover frequency." The turnover frequencies of many enzymes lie between 15 and 150 s^{-1} (20–38 °C); extremely high turnover frequencies are found in carbonic anhydrase ($6 \cdot 10^5$ s^{-1}) and catalase ($8 \cdot 10^4$ s^{-1}). The turnover frequency of the catalysts mentioned above is found to be $6 \cdot 10^4$ s^{-1}.

Due to the almost complete immiscibility of the hexenes (in the C_3 dimerization) and the octenes (in the C_4 dimerization) with the ionic liquid, biphasic catalysis occurs, and the separation of product is simply achieved by phase separation. The corresponding process, whose industrial feasibility has been demonstrated, is called the Difasol process (IFP). Higher dimer selectivity is achieved than in the Dimersol process; a significantly higher reaction rate and a lower consumption of the catalyst results in markedly higher space–time yields [M7].

For the dimerization of ethene to form but-1-ene, a soluble titanium-containing catalyst system ($Ti(OR)_4/AlR'_3$; R, R' = alkyl; 50 °C, 25 bar) is also used (Alphabutol process from IFP). Very pure but-1-ene is obtained, since practically no isomerization to but-2-ene (<0.01%) occurs. Trimers (C_6, about 7%) and higher (C_{8+}, <1%) hydrocarbons are separated via simple distillation [158].

Exercise 8.2

A trihydridotungsten complex **1** grafted on aluminum oxide catalyzes the direct conversion of ethene to propene with high selectivity (>95%):

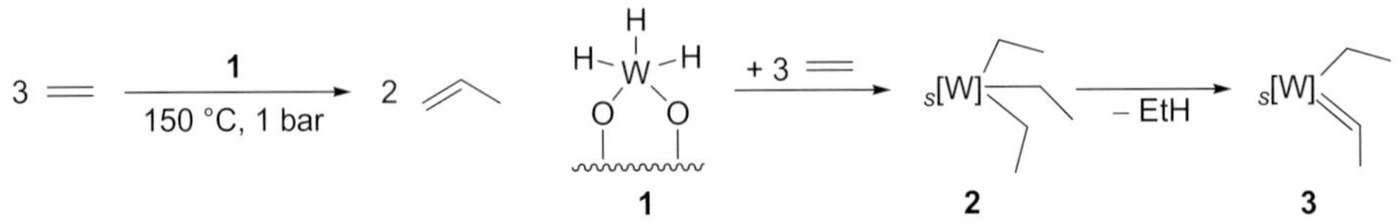

$$3 = \quad \xrightarrow[150\,°C,\,1\,bar]{1} \quad 2 \diagup\diagdown \quad \xrightarrow{+3=} \quad _s[W] \quad \xrightarrow{-EtH} \quad _s[W]$$

$$\quad\quad\quad\quad\quad\quad\quad\quad\quad\quad 1 \quad\quad\quad\quad\quad\quad 2 \quad\quad\quad\quad\quad\quad 3$$

Detailed experiments argue for **1** being a trifunctional single-site catalyst, which catalyzes the dimerization of ethene into but-1-ene, the double-bond isomerization of but-1-ene and the ethene/but-2-ene metathesis. The actual catalyst is formed from the triethyl complex **2** by α-hydride elimination, coupled with reductive C–H elimination (**2 → 3**). Formulate a probable reaction mechanism for all three partial reactions.

8.3
Trimerization of Ethene

Selective trimerization of ethene has gained increasing significance due to the increased demand for hex-1-ene as a comonomer in the polymerization of ethene to form LLDPE (see Section 9.2.4). Mainly chromium- and titanium-containing catalyst systems are used to carry out the selective trimerization, the mechanism of which is shown in Figure 8.3. The following individual reaction steps can be named:

27 → 28: *Ligand attachment/cleavage.* Coordination of ethene to a metal complex in low oxidation state leads to a bis(η^2-ethene) complex.

28 → 29: *Oxidative coupling/reductive cleavage.* Oxidative coupling yields a metallacyclopentane complex while raising the oxidation state of M by two.

29 → 30 → 31: *Ligand attachment/insertion.* Coordination of ethene followed by its insertion into an M–C bond leads to a metallacycloheptane complex.

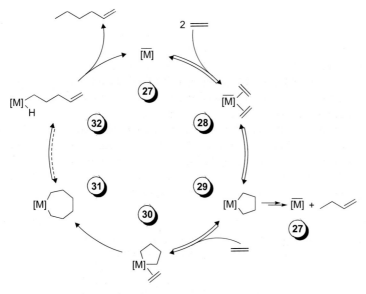

Figure 8.3 Mechanism for selective trimerization of ethene.

31 → 32 → 27: β-H elimination/reductive elimination. Hex-1-ene is formed by transfer of a β-hydrogen atom onto the metal followed by a reductive C–H elimination. There may not necessarily be an alkyl–hydrido intermediate **32**. The H atom can also be transferred directly – assisted by agostic C–H $\cdots$ M interactions – onto the C atom.

29 → 27: β-H elimination/reductive elimination. Similar to the previously described trimerization, ethene dimerization is also possible, where the metallacyclopentane complex **29** decomposes via β-H and reductive C–H elimination to form butene and the catalyst complex **27**.

Exercise 8.3
Experimental investigations suggest a similar mechanism for the Alphabutol process (Section 8.2). Formulate it. The main components of the C_6 fraction are 2-ethylbut-1-ene and 3-methylpent-1-ene. How do you explain their formation?

If trichlorotitanium(IV) compounds with a functionalized cyclopentadienyl ligand such as **33** are used as precatalysts and methylaluminoxanes (MAO, see Section 9.4.1) as cocatalysts, high activity (>12 000 mol C_6/(mol Ti·h)) and high selectivity (97%, of which 86% is hex-1-ene and 14% C_{10} olefins by inclusion of hexene in trimerization) with respect to trimerization are achieved [159].

The mechanism of this reaction is well-studied. Precatalyst **33** reacts with MAO in the presence of ethene, with cleavage of oligomeric olefins and alkanes, to form the actual catalyst, a cationic titanium(II) complex **34**, which corresponds to **27** in Figure 8.3 with $[M] = [Ti(\eta^5\text{-}C_5H_4CMe_2Ph)]^+$. This is a complex with a hemilabile (2-phenylprop-2-yl)cyclopentadienyl ligand. The η^5-cyclopentadienyl group is strongly coordinated to the titanium. Quantum-chemical calculations show that in the coordinatively unsaturated intermediates **27**, **28**, **29** and **31** (Figure 8.3), additional π interactions exist between the aromatic group and the titanium. The highly unsaturated complex **27** has a very strongly coordinated η^6-bound benzene ligand. As expected, the π interactions in the titanium(II) complex **28** are stronger than in the titanium(IV) complex **29**. This is precisely what is required, since after successful oxidative coupling (**28** → **29**), the coordination site for the new ethene ligand (**29** → **30**) becomes more easily accessible [160].

In addition, the calculations show that the intermediate **32** ($[M] = [Ti(\eta^5\text{-}C_5H_4CMe_2Ph)]^+$) is not part of the reaction path; instead, a concerted H transfer and reductive elimination takes place. In contrast, the conformatively less flexible titanacyclopentane complex **29** decomposes markedly slower with butene formation (**29** → **27**) by consecutive β-H and reductive C–H elimination. This is a significant reason for the high selectivity with regard to trimerization.

Many chromium-containing catalyst systems have been described that are characterized by high activity, selectivity, and thermal stability. There is evidence for a change in oxidation state between Cr^I and Cr^{III}, as well as between Cr^{II} and Cr^{IV} [161, 162]. Experiments on model systems including the synthesis and structural characterization of chromacycloheptane complexes (such as **35**) and their thermal decomposition into hex-1-ene (which in the case of **35** is generated along with butene/ethene as main product) show that the mechanism illustrated in Figure 8.3 applies.

At Phillips Petroleum, a catalyst system has been developed ($Cr[O_2CCH(Et)Bu]_3$, 2,5-dimethylpyrrole, $AlEt_2Cl/AlEt_3$ in toluene) that produces hex-1-ene (93% hex-1-ene, <1% other hexenes, 5% decenes) with high activity (about 10^5 mol hex-1-ene/(mol Cr·h); 115 °C, 100 bar) and selectivity. In hexachloroethane, with $AlEt_3$ as co-catalyst, activity of $2.3 \cdot 10^6$ mol hex-1-ene/(mol Cr·h) (105 °C, 50 bar) can be achieved. In 2003, Chevron-Phillips put into operation an industrial plant for synthesis of hex-1-ene (50 000 metric tons/year) in Qatar [159].

Exercise 8.4

$[CrCl_3(THF)_3]$ and the bis(diarylphosphino)amine **1'** (Ar = 2-methoxyphenyl) activated by MAO form a highly active and very selective catalyst system for the trimerization of ethene. If, by contrast, the compound **1'** is used with Ar = Ph, the tetramerization of ethene is catalyzed. Oct-1-ene is formed as the main product. The most important side products are hex-1-ene as well as **2'** and **3'** in a 1 : 1 ratio. Give a possible mechanism for the tetramerization of ethene and the formation of the C_6 hydrocarbons.

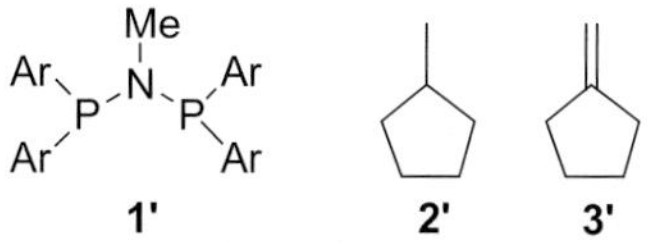

Background: Hemilabile Ligands

Hemilabile ligands are flexidentate chelating ligands bound to the central atom with varying denticity in the course of a catalytic cycle. This process is reversible:

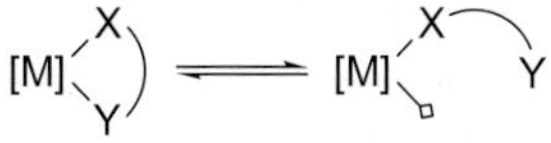

Hemilabile ligands can vacate coordination sites if they are needed in the catalytic cycle, and hence *temporarily*, for example, for the coordination of a substrate molecule or a β-H elimination. The alternative, namely that these coordination sites be *permanently* available, leads to coordinatively unsaturated, less stable complexes. In addition, the reaction center is electronically affected by the change in the coordination mode of the hemilabile ligand, which can promote the activity and/or selectivity of the reaction.

Hemilabile ligands generally offer one substitutionally inert and one substitutionally labile donor group. Substitutionally labile groups can be *n*- and π-donor ligands, as shown by the examples **1–3** and **4**. An important role is also played by σ-donor ligands, namely CH groups, coordinated via agostic C–H···M interactions (**5**). In special cases, the product bound to the metal can also be a hemilabile ligand. Thus, for example, in polymerizations/copolymerizations an *n*-donor atom (**6**) or a β-CH group (**7**) may appear as the substitutionally labile group in the growing polymer chain **P**.

(The substitutionally labile groups are shown at the bottom; see arrow.) Hemilabile ligands also play a major role in coordination chemistry [163–165].

8.4
Shell Higher Olefin and α-Sablin Processes

8.4.1
The Shell Higher Olefin Process (SHOP)

The Shell Higher Olefin Process (SHOP) is an important industrial application of olefin oligomerization and olefin metathesis. It is a process for producing linear α-olefins (C_{12}–C_{18}) from ethene. This can be coupled with hydroformylation so as to obtain the corresponding aldehydes or alcohols. From the flowchart in Figure 8.4, it can be seen that the Shell Higher Olefin Process is a combination of oligomerization, isomerization, and olefin metathesis. The plant consists of the following individual units:

a) *Oligomerization reactor.* The oligomerization of ethene to form linear α-olefins is nickel-catalyzed (solvent: butane-1,4-diol).
b) *Phase separator.* The oligomer mixture is immiscible with the solvent in which the catalyst is dissolved, and is removed by simple phase separation.
c) *Distillation column.* The oligomer mixture is separated into the desired olefin fractions.
d) *Isomerization reactor.* The α-olefins with too high a C-number are subjected to a double-bond isomerization, whereupon olefins with internal double bonds are formed. α-Olefins with lower C-number may be added as necessary.
e) *Metathesis reactor.* The metathesis of the mixture of inner olefins leads to an olefin mixture that comes closer to the desired C-number distribution.

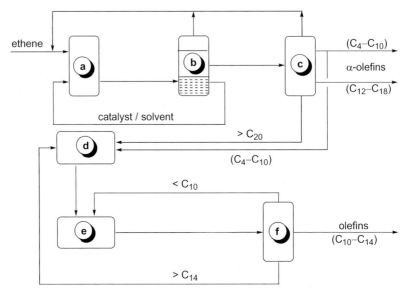

Figure 8.4 Flowchart for the Shell Higher Olefin Process (SHOP) (after A. Behr "Hydrocarbons" in [M 14]).

f) *Distillation column.* Fractional distillation of the olefin mixture combined with recycling of olefins of undesired chain lengths, which are fed to the metathesis reactor **e** (short-chain olefins) or to the isomerization reactor **d** (long-chain olefins).

- *Oligomerization.* The oligomerization of ethene is performed in the liquid phase with nickel catalysts in polar solvents such as butane-1,4-diol at 80–120 °C and 7–14 MPa. The olefins formed are immiscible with the solvent, so the catalyst, dissolved in butanediol, is removed from the product by simple phase separation (biphasic catalysis). Linear olefins of C_4 to C_{30+} with an unusually high linearity (about 99%) and only a low proportion of inner olefins (1–2%) are obtained. Nickel complexes of type $[NiR(P\frown O)(PR'_3)]$ ($P\frown O$ = anionic P,O-chelating ligand) are used as precatalysts. The mechanism for oligomerization reaction is shown in the following scheme with $[NiPh(Ph_2PCH_2COO\text{-}\kappa O,\kappa P)(PPh_3)]$ (**36**) as precatalyst.

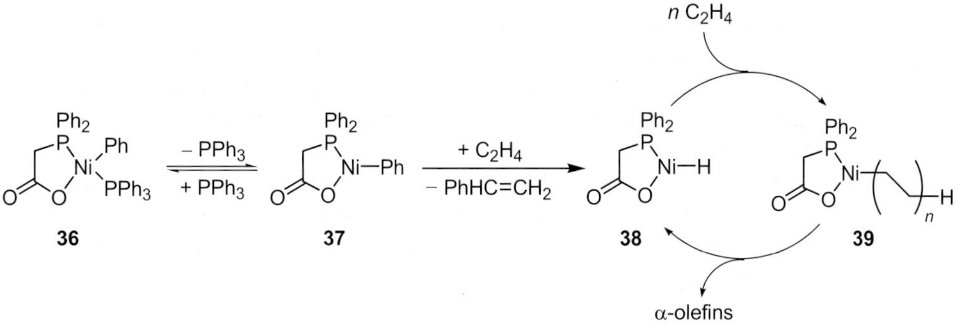

Cleavage of triphenylphosphane from the complex generates a coordinatively unsaturated (14 ve) phenylnickel(II) species (**36** → **37**). This enables the insertion of ethene into the Ni–C bond with formation of a 2-phenylethyl complex, which forms the actual catalyst, a hydridonickel(II) complex (**37** → **38**), in a β-hydride elimination. Multiple insertion of ethene into the Ni–H or Ni–C bond gives a nickel complex with an oligomeric alkyl ligand (**38** → **39**). Via β-hydride elimination, the α-olefin is cleaved off, re-forming the hydrido complex (**39** → **38**) [166].

- *Isomerization.* The ethene oligomerization is followed by a distillation. Olefins with the desired C-number (e.g., C_{12}–C_{18}) are separated. Those with too low or too high a C-number are subjected to a double-bond isomerization. This occurs in the liquid phase with magnesium oxide catalysts (80–140 °C, 0.3–2 MPa). About 90% olefins with inner double bonds are obtained.

- *Olefin metathesis.* The isomerization is followed by an olefin metathesis, which is generally performed with heterogeneous rhenium or molybdenum catalysts. A mixture of (even- and odd-numbered) inner olefins with a completely new chain length distribution is obtained. The desired fraction (e.g., C_{10}–C_{14}) is separated by distillation. The longer-chained olefins (C_{14+}) are fed into the isomerization reactor. The short-chain olefins are fed into the metathesis reactor. The high proportion of shorter-chain olefins in the metathesis reaction results in a shift of the double bond toward the end of the chain. Isomerization and metathesis produce linear inner olefins of the desired chain length, which can be converted directly into n-aldehydes via hydroformylation (see Exercise 5.3 in Section 5.4).

The metathesis reaction ensures complete utilization of the material: Olefins with too low or too high a C-number are fed repeatedly into the circulation for the metathesis reactor. α-Olefins must undergo double-bond isomerization beforehand. The same is true for long-chain olefins with double bonds near the end of the chain. An interesting variant of the Shell Higher Olefin Process is the metathesis of the high-boiling portion (C_{20+} with inner double bonds) with ethene, which leads directly to α-olefins of the desired chain length. The worldwide production capacity of SHOP plants amounts to $1.2 \cdot 10^6$ metric tons/year of linear α- and inner olefins (2006).

Exercise 8.5
Remember that in olefin metathesis the position of the double bond in the olefin starting materials and their concentration ratios have a decisive effect on the C-number distribution of the product mixture. (For simplicity, we indicate alkylidene groups here simply by the number of C atoms.)
- Assume that these isomeric linear olefins $C_{20}H_{40}$ are present: (a) C_9H_{19}–CH=CH–C_9H_{19} (abbreviation: C10=C10), (b) $C_{14}H_{29}$–CH=CH–C_4H_9 (abbreviation: C15=C5) and (c) $C_{18}H_{37}$–CH=CH_2 (abbreviation: C19=C1). They are each subjected to metathesis with an equimolar quantity of but-2-ene (abbreviation: C2=C2). Give the equilibrium composition for each.
- What is the product distribution if C_9H_{19}–CH=CH–C_9H_{19} (abbreviation: C10=C10) is subjected to a metathesis with but-2-ene (abbreviation: C2=C2) in molar ratio 1 : 9?

8.4.2
α-Sablin Process

Linde and SABIC (Saudi Arabia) have recently developed the α-Sablin process for production of ethene oligomers [167, 168]. The process is based on the fact that zirconium(IV) carboxylates $Zr(O_2CR)_4$ (R = C_3–C_7 alkyl) with aluminum alkyls as cocatalysts have proven to be very good catalysts for ethene oligomerization. Thus, this is a Ziegler catalyst system, for which – in contrast to typical polymerization catalysts – β-H elimination displays a relatively low activation barrier, so ethene oligomers are obtained (see Section 8.2).

The oligomerization is performed in toluene as solvent at 20–30 bar and 60–100 °C. The product ratio can be controlled easily via the Zr/Al ratio, since with an increased amount of cocatalyst, the proportion of C_4–C_{10} olefins rises. α-Olefins C_4–C_{20+} are formed very selectively, so after decomposition of the catalyst by addition of H_2O or ROH, a fractional distillation is sufficient to separate the product. In 2006, the first commercial plant (150 000 metric tons/year) in Saudi Arabia was put into operation.

8.4.3
Use of Linear α-Olefins

Linear α-olefins (LAOs) are used in a wide range of applications. Those with shorter chains (C_4–C_8) are mainly used as comonomers in the production of polyethylene (LLDPE) and of other polymers. LAOs of medium chain length (C_8–C_{12}) are used as starting materials for production of synthetic lubricants. LAOs of chain length C_{12}–C_{18} are further processed to produce detergents, while the long-chain (C_{18+}) LAOs are made into plasticizers or used directly as lubricants and drilling fluids. The worldwide production of α-olefins amounts to about $2.8 \cdot 10^6$ metric tons/year (2006). About half are used as comonomers in the production of polymers, 1/4 in the production of detergents and 15% for the production of lubricants.

9
Polymerization of Olefins

9.1
Introduction

Polyolefins are among the most important synthetic polymers. In the 1920s, Hermann Staudinger (University of Freiburg, Germany; Nobel Prize 1953) developed fundamental concepts on the structure of macromolecules. Experiments on sodium- and radical-initiated polymerization of dienes, starting in 1910, led in the 1930s to the large-scale industrial production of synthetic rubber. The first industrial process for radical polymerization of vinyl chloride, vinyl acetate, and styrene had already been developed in 1930.

There are four fundamental mechanisms by which the polymerization of olefins can take place. The name assigned to each is derived from the nature of the reactive intermediate ("chain carrier"), as can be recognized from the propagation steps (**P** = growing polymer chain) listed below.

- *Radical polymerization.* Radical polymerization is started by radical initiators (e.g., dibenzoyl peroxide).

- *Cationic polymerization.* Initiators for cationic polymerization are Brønsted acids (e.g., H_2SO_4) or Lewis acids (e.g., BF_3) in the presence of traces of water or alcohol.

Fundamentals of Organometallic Catalysis. First Edition. Dirk Steinborn
Copyright © 2012 WILEY-VCH Verlag GmbH & Co. KGaA, Weinheim

- *Anionic polymerization.* Initiators of anionic polymerization of dienes and vinyl compounds with acceptor substituents are bases (e.g., sodium amide, lithium alkyls).

- *Coordination (metal-catalyzed) polymerization.* Organotransition metal complexes generated from a transition metal compound and a cocatalyst (e.g., $TiCl_4/AlEt_3$) function as catalysts.

In the following, we will take a closer look at the coordination polymerization of ethene and propene.[1] The radical polymerization of ethene was not developed until the second half of the 1930s, at ICI (Imperial Chemical Industries, UK), where the first small-scale industrial plant was put into operation in 1939. By this point, the polymerization of vinyl compounds with electron-withdrawing substituents had already been industrialized. Radical polymerization of propene to form industrially useful polymers proved unfeasible.

9.2
Ethene Polymerization

9.2.1
Ziegler Catalysts

The nickel effect (see Section 8.2) prompted Karl Ziegler to investigate the catalytic effect of transition metal compounds with aluminum alkyls. Thus, it was discovered in the autumn of 1953 that the combination of $TiCl_4$ and $AlEt_3$ in gasoline, at room temperature and normal pressure, polymerizes ethene into polyethylene of high crystallinity (HDPE: high-density polyethylene). The significance of this discovery becomes apparent when one considers that polyethylene previously could be

---

1) Here, we call polymers by their generally familiar source-based names, which are derived from the monomer and sometimes are also trade names. For simple polymers, we make no use of the structure-based names, which are based on the names of constitutional repeating units. Example: $-(CH_2-CH_2)_n-$ is called "polyethene" or "polyethylene," but not "poly(methylene)."

obtained only by radical polymerization at 200 °C and a pressure of about 1000–2000 bar [169]. Soon after (1954), Giulio Natta (Institute of Technology, Milan, Italy) found that Ziegler catalysts converted propene into completely new types of polymers through stereoregular polymerization. The discovery of organometallic mixed catalysts[2] and the evidence of stereoselective polymerization with these catalysts mark the beginning of modern plastics production. In 1963, K. Ziegler and G. Natta were honored with the Nobel Prize in Chemistry for these ground-breaking chemical discoveries.

The polymerization occurs on the transition metal (titanium). The growing polymer chain is σ coordinated to titanium (**1**, **P** = growing polymer chain). The chain growth occurs through a rapid sequence (in comparison with termination reactions) of olefin coordination (**1** → **2**) and insertion of the coordinated olefin into the Ti–C bond (**2** → **1'**).

The insertion (**2** → **1'**) is a *syn* addition. According to P. Cossee and E. J. Arlman (1964), a *cis* migration (migratory insertion) of the σ-bound polymer chain to the coordinated olefin occurs via a four-center transition state **3**. This is pivotal: the polymerization is completed at two coordination sites, at which, alternately, the monomer and the polymer chain are bound.

The most important chain termination reaction[3] is the β-hydride elimination, by which a hydride from the polymer chain is transferred onto the titanium (**1** → **4**) or the coordinated olefin (**2** → **5**). In either case, the catalyst center is not deactivated, since insertion of ethene into the Ti–H or Ti–C bond allows a new chain to start. In contrast, a homolytic cleavage of the metal–carbon bond (**1** → **6**) leads to catalyst deactivation. The first two termination reactions produce polymers with olefinic end groups. If the homolytic cleavage results in stabilization of the carbon radical through disproportionation (H transfer), polymers are formed with saturated and unsaturated end groups in a 1:1 ratio.

---

2) Ziegler himself called the catalysts "organometallic mixed catalysts" ("metallorganische Misch-katalysatoren") or "Mülheim catalysts." In the literature, they are called Ziegler or Ziegler-Natta catalysts. In the broadest sense, this is understood to refer to a combination of a transition metal compound (preferably a halide) with a main group metal alkyl, aryl, or hydride, which is able to polymerize ethene or α-olefins.

3) Here and in the following, chain termination reactions are all steps that cause release of the polymer chain from the active catalyst. They can proceed with complete destruction of the catalytically active species and thus lead to (irreversible) deactivation of the catalyst. Alternatively, a new active catalyst molecule can be generated. Such termination reactions without deactivation of the catalyst are also called chain transfer reactions.

In the course of catalyst formation, a coordinatively unsaturated σ-organotitanium compound is formed. The cocatalyst, an alkylaluminum compound, exhibits an alkylating and a Lewis acid function, as in the nickel-catalyzed dimerization of ethene. Titanium-containing catalyst systems can contain Ti^{III} or Ti^{IV} in their catalytically active form. For the classical Ziegler catalyst system ($TiCl_4$/$AlEt_3$ in gasoline), reduction of $TiCl_4$ (**7**) occurs first via $TiEtCl_3$ (**8**) as intermediate. $TiCl_3$ (**9**) forms, and precipitates as a fibrous solid substance.

$$TiCl_4 \xrightarrow[- AlEt_2Cl]{+ AlEt_3} TiEtCl_3 \longrightarrow \{TiCl_3\}_s + 1/2\ (C_2H_4 + C_2H_6)$$
$$\quad\ \ \textbf{7} \qquad\qquad\qquad\qquad \textbf{8} \qquad\qquad\qquad \textbf{9}$$

The cocatalyst $AlEt_3$ (or alternatively $AlCl_xEt_{3-x}$, $x = 0, 1, 2$) ethylates $TiCl_3$ on the surface (alkylating function) on the one hand, and, on the other hand, enables ethene coordination (Lewis acid function) by formation of ethylchloroaluminate anions $[AlCl_{x+1}Et_{3-x}]^-$. Thus, this is heterogeneous catalysis with typical organometallic species on the surface. However, since it displays the characteristic features of homogeneous catalysis (see Section 2.1), and there are analogous homogeneous catalyst systems, it makes sense to deal with the process within the framework of homogeneous catalysis. At the end of the 1950s (D. S. Breslow and N. R. Newburg, as well as G. Natta and P. Pino), a homogeneous system was found ($[TiCl_2Cp_2]$/$AlEt_2Cl$). It was of relatively low activity, but served a valuable purpose in clarifying the mechanism. In this case, the tetravalent titanium is catalytically active; reduction to Ti^{III} leads to deactivation of the catalyst. The generation of the catalyst involves an alkylation of titanocene dichloride via the cocatalyst (**10** → **11**) and the formation of a 1:1 complex by means of the Lewis acidity of the aluminum compound (**11** → **12**).

Under the influence of additional cocatalyst, the ethylchloroaluminate anion [AlEtCl$_3$]$^-$ can be displaced from **12** by ethene, with the polymerization proceeding as described above. Later, it was shown for analogous zirconocene systems that the actual polymerization-active species are cationic complexes of type [ZrCp$_2$P]$^+$ (**P** = growing polymer chain). Highly active catalyst systems are obtained when the anion is such a weak Lewis base that it does not hinder the coordination of the olefin to the metal center. Suitable anions are the perfluorated tetraphenylborate anion [B(C$_6$F$_5$)$_4$]$^-$ and methylated methylaluminoxanes [MAO–Me]$^-$, which have also achieved industrial significance for "metallocene catalysts."

In contrast to radical ethene polymerization, coordination polymerization produces strictly linear polymers. Branching occurs in side reactions when, rather than ethene, a polymer or oligomer with an unsaturated end group is inserted into the growing polymer chain. Classical Ziegler catalysts produce about 1.2 methyl branchings per 1000 C atoms. The chain length of the polymers can be governed through the ratio of insertion and β-H elimination (chain termination without deactivation) rates. Low temperatures and high ethene pressure favor chain growth, while higher temperatures and low ethene pressure lead to increased chain termination. To control molar mass, chain transfer agents such as hydrogen can be added. These are reagents that stop the chain growth without deactivating the catalyst, and thus are in a position to control the molar mass of the polymer efficiently. H$_2$ leads via hydrogenolysis of the M–C bond in the catalyst complex (**13** → **14**) to chain termination, followed by the reinitiation of a polymer chain on the hydridometal complex **14**.

[M]–P $\xrightarrow{+ H_2}$ [M]–H + P

 13   **14**

9.2.2
Mechanism – A Closer Look

The mechanism of the coordination polymerization of ethene (and other α-olefins) is more multifacted and detailed than presented above, so for deeper understanding, a few additional notes are necessary.

- *Agostic interactions.* The coordination of the growing polymer chain can be stabilized by a β-agostic C–H $\cdots$ M interaction (**15**, **16**), which upon insertion (**16** → **17** → **18** → **15'**) must be broken. In the insertion reaction itself, an α-agostic C–H $\cdots$ M interaction in the ground state (**17**) and/or in the transition state (**18**) may be of significance. This additional fixing of the conformation of the growing polymer chain during polymerization of α-olefins can result in an increase in selectivity, and can lower the activation barriers by stabilizing the transition state [170].

If the α-H atom were fully transferred onto the metal, a carbene–hydrido complex **19** would result (M. L. H. Green, J. J. Rooney, 1978). Starting from **19**, the insertion proceeds via a metallacyclobutane complex **20** as an intermediate. This is a critical difference from the "classical" mechanism (with or without α-agostic interaction), in which the metallacyclobutane-like structures **3** (see Section 9.2.1) and **18** are transition states [171]. The significance of the agostic interactions described here depends, in any particular case, on the nature of the catalyst, and is sometimes only known in a rudimentary way.

- *Chain termination reactions (without catalyst deactivation).* β-Agostic interactions in the complexes **15** and **16** can initiate a chain transfer via β-hydride transfer onto the metal (**15** → **21**) or onto the coordinated olefin (**16** → **23** with **22** as transition state), forming complexes in which the polymer chain is bound to the metal via an olefinic end group. Dissociative or associative exchange of the polymer with the monomer starts the growth of a new polymer chain (**21/23** → **24**). The growing polymer chain may be transferred onto the cocatalyst in a metal–metal exchange reaction ("transalkylation"). If the cocatalyst is an ethylaluminum compound, [M]–Et is formed, upon which a polymer chain grows again after olefin coordination (**15** → **24**).

- *Chain branching.* The reversible formation of hydrido olefin complexes **21** from **15** opens the way to methyl-branched polymers when an insertion with M–C2 bond formation (**21** → **25**) occurs instead of reinsertion with M–C1 bond formation (**21** → **15**). Multiple repetition of this reaction (**25** → **21′** → **25′** → ...) (chain running) produces higher branching.

- *Insertionless migration.* The polymerization is completed at two coordination sites, of which one is occupied by the monomer and the other by the growing polymer chain P_n. The "normal" course of reaction for a migratory insertion (**26** → **27** → **28** → **29** → **26′** → ...; in **26′** $n = n + 2$ applies) includes a migration of the polymer chain from one coordination site to the other. If the polymer chain migrates before the olefin is coordinated (**28** → **26′**; in **26′** $n = n + 1$ applies), the cycle **26** → **27** → **28** → **26′** → ... will proceed, which is called "insertionless migration" (back-skip) of the polymer chain.

This may be the case if the vacant coordination sites in **26** and **28** are structurally different. The driving force could be, for instance, that the coordination of the bulky polymer chain in **28** is energetically less favorable than in **26** [172].

9.2.3
Phillips Catalysts

In the mid-1950s, it was discovered at the Phillips Petroleum Company that chromium oxides on oxidic supports are polymerization catalysts for ethene. To this end, CrO_3 is supported on silica **30**; subsequent calcination in air yields the precatalyst, which contains Cr^{VI} (**31**). The activation of the catalyst occurs via reduction with ethene or CO (**31** → **32**). In the 1960s, at Union Carbide, another heterogeneous chromium-containing catalyst system **33** was developed by impregnating **30** with chromocene [173].

The precise composition of the catalysts is not known, so structures **31–33** are only schematic. In the (active) reduced form of the Phillips catalysts (**32**), most of the chromium is divalent. However, as merely less than 1% of the grafted chromium is catalytically active, the catalytically active chromium may have another oxidation state.

At temperatures as low as 70 °C, the grafted bis(neopentyl)chromium(IV) complex **34** liberates neopentane. The neopentylidenechromium(IV) complex formed (**34** → **35**) has been well characterized. It has been found to be a single-component polymerization catalyst for ethene, which in many properties (activity, behavior with respect to H_2, discrimination between ethene and higher α-olefins, microstructure of the polymer) is similar to the classical Phillips catalysts. Quantum-chemical calculations support the assertion that a cationic chromium(IV) complex **36** is the actual catalyst [174, 175].

Homogeneously catalyzed ethene polymerization has been reported for chromium(III) complexes. Thus, the chromium(III) complexes **37** (E = N, P; R = alkyl, aryl; X = Cl, Me) in the presence of methylaluminoxane (MAO) as cocatalyst are highly active catalysts for ethene polymerization. A cationic chromium(III) complex **38** is a catalytically active intermediate that has bound the growing polymer chain and the coordinated monomer as ligands [176, 177].

$$\underset{\underset{R_2}{37}}{\overset{}{\text{Cp-Cr(X)(X)-E-R}_2}} \xrightarrow[\text{(MAO)}]{n\ H_2C=CH_2} \underset{\underset{R_2}{38}}{\overset{}{[\text{Cp-Cr-E-R}_2\cdots P]^{\oplus}}}$$

37 **38**

9.2.4
Polymer Types and Process Specifications

The polymerization mechanism substantially determines the properties of the polymer. A radical polymerization of ethene leads to heavily branched polymers, which are amorphous and of low density (LDPE: low-density polyethylene; threshold value for 100% amorphous PE: 0.85 g/cm^3). Coordination polymerization yields linear polymers of high crystallinity and density (HDPE: high-density polyethylene; threshold value for 100% crystalline PE: 1.00 g/cm^3). The copolymerization of ethene with α-olefins such as butene, hexene, or octene (typically up to 10% comonomers) yields a linear polymer with short branches, which due to its low density is called LLDPE (linear low-density polyethylene). Characteristic properties of the different polymer types are compiled in Table 9.1.

Polyethylene is the most widely used plastic. It accounts for about 1/3 of the plastic produced worldwide. In 2009, the worldwide production of polyethylene came to about 68 million metric tons with a ratio of LDPE : HDPE : LLDPE of about 1 : 2 : 1. Ziegler and Phillips catalysts were used in a ratio of about 2 : 1. The significance of radical polymerization has steadily dropped; in 1983, the proportion of LDPE was still about 60%.

Table 9.2 lists typical process parameters for the industrial synthesis of HDPE and LDPE. The low-pressure process with Ziegler-Natta catalysts was implemented industrially for the first time in 1957 (Montecatini, Italy). It was originally performed in suspension with low-boiling hydrocarbons such as hexane as solvent. The polymer is insoluble under the process conditions and is separated together with the (insoluble) catalyst by filtration. Since catalyst residues can influence the aging of polyethylene, they must be removed.

In 1968, it was found that supporting classical Ziegler-Natta catalysts on magnesium chloride leads to highly active catalyst systems. First, TiCl$_4$ and MgCl$_2$ are milled together, and then the tetravalent titanium is reduced by the cocatalyst, an aluminum alkyl compound. An essential factor for the catalytic activity is the morphology of the support (MgCl$_2$), which should be prepared so that the catalyst grains are already fragmented in the initial stage of polymeri-

Table 9.1 Properties of three characteristic polyethylene types[a] (compiled from K. S. Whiteley "Polyolefins" in [M 14]).

PE type	LDPE	HDPE	LLDPE
Structure (basic sketch)			
Density (in g/cm^3)[b]	0.924 (0.91–0.94)	0.961 (0.94–0.97)	0.922 (0.91–0.94)
Crystallinity (in %)	40	67	40
Melting point (in °C)	110	131	122
Molar mass M_w (in g/mol)	200 000	136 300	158 100
Short branches[c]	23	1.2	26
Modulus of elasticity (in MPa)	240 (high elasticity)	885 (high stiffness)	199 (high elasticity)

a) LDPE: Repsol PE077/A; HDPE: Hoechst GD-4755; LLDPE: BP LL 0209.
b) Typical range in parentheses.
c) Number of methyl groups per 1000 C atoms.

zation and thus, all active centers offer easy access to the monomer. Activity and productivity of such catalysts are so high that an expensive removal of catalyst residues from the polymer could be eliminated, with enormous simplification of the polymerization plants. Such supported Ziegler-Natta systems are used to

Table 9.2 Comparison of characteristic process parameters in the coordination and radical polymerization of ethene (compiled from [178], Vol. 2 and K. S. Whiteley, "Polyolefins" in [M 14]).

Mechanism	coordination polymerization (low-pressure process)				radical polymerization (high-pressure process)
Process	suspension	suspension	solution[a]	gas phase[b]	solution[c]
Catalyst	Ziegler type (e.g., TiCl$_4$/AlEt$_3$)	Phillips type (e.g., CrO$_3$/silica)	Ziegler or Phillips type		0.05–0.1% oxygen[d]
p (in bar)	<15	30–50	50–150	10–20	1500–3000
T (in °C)	75–85	<110	100–300	90–120	200–300
Structure	high crystallinity, linear (unbranched)				highly branched[e]
Polymer type	HDPE (high-density PE)				LDPE (low-density PE)

a) The polymer may be dissolved in the solvent or present in a molten state.
b) Fluidized-bed process.
c) Under the process conditions, ethene is in the supercritical state in which the polymer is dissolved.
d) Initiator.
e) Random long-branching structure, with branches on branches.

polymerize ethene in a solution process and in a fluidized-bed process (i.e., without solvent).

The process from the Phillips Petroleum Company has the advantage that it does not require a cocatalyst, and according to market requirements, different polyethylene types can be produced. It uses somewhat higher pressures than the Ziegler process. If the process is performed with the solvent being a cycloparaffin in which monomer and polymer are soluble, the insoluble catalyst can easily be separated by filtration at the end of polymerization. In another variant, the catalyst is dispersed in a paraffin hydrocarbon and the polymer grows around the catalyst core. Due to the high activity (250 kg PE/(mol Cr·bar C_2H_4·h)), the catalyst does not need to be separated from the polymer. Finally, the process can be carried out without any solvent in a fluidized bed, whereupon a separation of the catalyst can be dispensed with.

9.3
Propene Polymerization

9.3.1
Regioselectivity and Stereoselectivity

Organometallic mixed catalysts also enable propene to be polymerized (G. Natta, 1954). This is significant because it is not possible to produce polypropylene with molar masses of industrial interest by radical polymerization, due to the high stability of the allyl radical.

Propene, like all other α-olefins, has an unsymmetrically substituted double bond (=CH_2 vs. =CHMe), and is thus prochiral. Thus, polymerization can form different regio- and stereoisomers:

- *Regioselectivity.* Polymerization is regioselective when head-to-tail linkage[4] (C2–C1 bond formation) occurs, leading to **39**. Alternating head-to-head (C2–C2) and tail-to-tail linkage (C1–C1) leading to polymer **40** is also regioselective, but cannot be achieved by direct synthesis.

- *Stereoselectivity.* The polymerization is stereoselective when within one polymer strand all stereocenters display the same (relative) configuration (...RRR... or ...SSS...) ("isotactic polypropylene," abbreviation: i-PP, **41**) or are of alternating

---

[4] For the definition of "head" (C2) and "tail" (C1) for vinyl monomers, see footnote 1 in Section 8.2. and the numbering of the C atoms indicated in **39** and **40**.

R- and S-configuration ("syndiotactic polypropylene," abbreviation: s-PP, **42**). In a non-stereoselective polymerization, atactic polypropylene (abbreviation: a-PP, **43**) is formed.

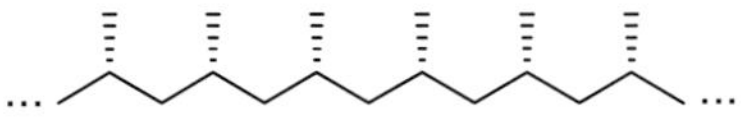

41

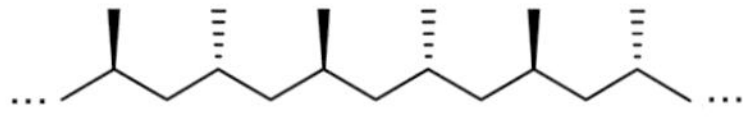

42

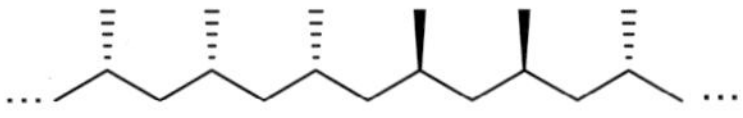

43

Exercise 9.1
Propose a way to synthesize H–H polypropylene (head-to-head polypropylene) **40**.

Background: Configuration of Polypropylene
Stereocenters in polymers are assigned *relative* configurations. The configuration of the stereocenter is determined at one end of the polymer chain, the configuration of the adjacent stereocenter is given relative to it, and so on. If all stereocenters have the same configuration, namely either (...RRRR...)$_{rel.}$ or (...SSSS...)$_{rel.}$, the polymer is called isotactic; if there is an alternation (...RSRSRS...)$_{rel.}$, it is called syndiotactic. To emphasize that this is *not* the absolute configuration, for which the priority of the substituents would have to be considered, we add here the index "rel." The difference between relative and absolute configuration is shown below, using the example of the isotactic pentamer of propene with an ethyl and a methyl end group:

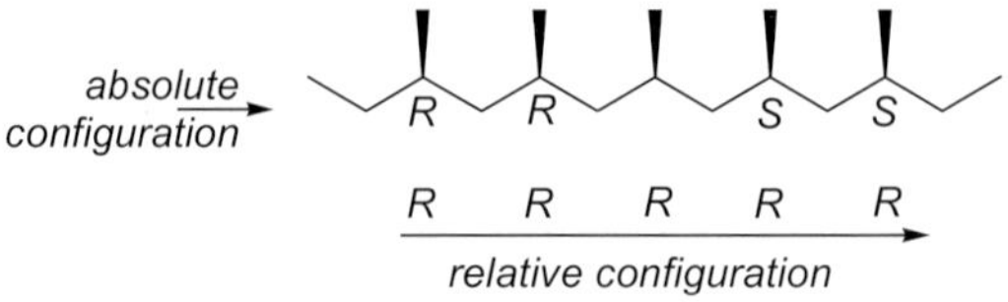

absolute configuration →   R   R   S   S

R   R   R   R   R
relative configuration →

In the example, the relative configuration is given based on the first chiral C atom at the left end of the chain. If one were to begin at the right end of the oligomer chain in determining the relative configuration, (...SSSS...)$_{rel.}$ would be written. If two adjacent stereocenters have the same relative configuration, this is the meso form, and is called an *m*-diad (**1**); otherwise, it is called a racemic diad (**2**, *r*-diad).

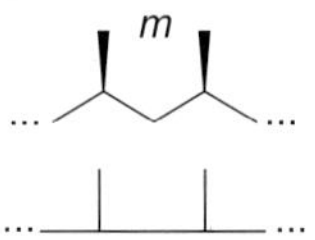

m   r

1 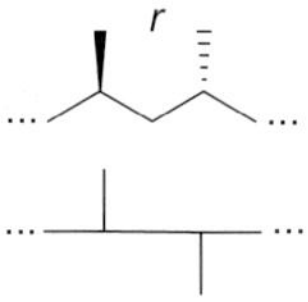   **2**

Thus, in *i*-PP there are only *m*-diads (...mmmmm...) and in *s*-PP there are only *r*-diads (...rrrrr...) [178–180].

9.3 Propene Polymerization

Background: Analysis of the Microstructure of Polypropylene

The microstructure of polypropylene can be reliably determined by NMR spectroscopy, since the ^{1}H and ^{13}C chemical shifts for the methyl groups are sensitive to the relative stereochemistry of the adjacent monomer units. Thus, ^{13}C NMR spectroscopy can detect differences of up to five monomer units on each side.

There are three different triads, which are represented by their stereoformulas and modified Fischer projections (mm = isotactic, rr = syndiotactic, and mr = heterotactic triads):

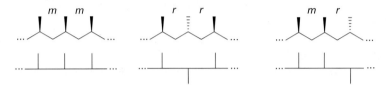

Generally, the pentads are included for analysis. There are ten different ones (with the chemical shift of the marked C atom given in ppm in parentheses; for details on the measurement conditions, see [181]:

$m\ m\,m\ m$	$m\ m\,m\ r$	$r\ m\,m\ r$	$m\ m\,r\ r$	$m\ m\,r\ m$
(21.78)	(21.55)	(21.33)	(21.01)	(20.85)
$r\ m\ r\ r$	$r\ m\,r\ m$	$r\ r\,r\ r$	$r\ r\,r\ m$	$m\ r\,r\ m$
(20.85)	(20.71)	(20.31)	(20.17)	(20.04)

Quantitative analysis of the NMR spectra of polypropylenes allows precise statements of their microstructure, from which deductions about the polymerization mechanism can be derived [178, 182].

The coordination polymerization of propene by Ziegler-Natta catalysts has proven itself to be fundamentally regioselective. This is achieved by exclusive M–C1 (**a**, "primary or 1,2-insertion") or exclusive M–C2 bond formation (**b**, "secondary or 2,1-insertion") in the insertion step.

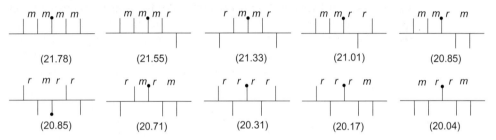

Exercise 9.2

Bis(phenolato) complexes of type **1** enable the oligomerization of hex-1-ene at room temperature in the presence of $B(C_6F_5)_3$ (1/1). The oligomerization proceeds regioselectively. With M = Ti, oligomers with vinylene end groups (ethene-1,2-diyl) are obtained and with M = Zr, Hf, oligomers with vinylidene end groups (ethene-1,1-diyl) are generated. Give the (probable) composition of the catalyst complex and the structure of the oligomers. From the nature of the end groups deduce the insertion type.

As a condition for stereoselective polymerization of propene, the coordination of the prochiral propene and the insertion step must run in a stereochemically consistent manner. The latter is fundamentally ensured, since the insertion proceeds as a *syn* addition. Thus, the coordination mode of the olefin to the metal is decisive for stereocontrol. This coordination can occur from the *Re* or *Si* face (see Background, Section 4.3.1). Binding of the *Re* face to the metal and subsequent primary insertion (*cis*!) generates an S-configured asymmetric C atom. If the *Si* face coordinates to the metal, primary insertion results in an asymmetric C atom with R-configuration (Figure 9.1). For secondary insertion (M–C2 bond formation), the opposite is true.

Isotactic polypropylene is obtained when the prochiral propene always coordinates with the same (prostereogenic) face, *Re* or *Si*, to the transition metal. Syndiotactic polypropylene results when the coordination alternates between the *Re* and *Si* face. A stereoselective polymerization clearly requires a chiral induction, which can originate from chain end stereocontrol, where the last formed asymmetric C atom on the growing polymer chain determines the stereochemistry of the next insertion step, and/or enantiomorphic site stereocontrol, where the active

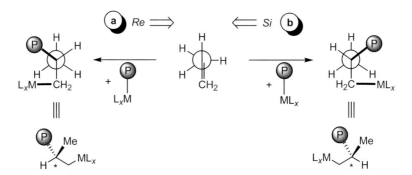

Figure 9.1 Newman projection of propene along the C2–C3 bond. Addition of L_xM to the *Re* face of propene and subsequent primary insertion leads to a C2 atom with S-configuration (a). Consequently, coordination to the *Si* face results in an (R)-C2 atom (b). The newly formed bonds M–C1 and C2–P (P = growing polymer chain) are shown in bold.

site of the catalyst complex is chiral and determines the stereochemistry of the insertion.

Exercise 9.3
Suppose that in the polymerization of propene to isotactic polypropylene, polymers form with the microstructure **1** or **2**. Consider how the type of correction of a stereoerror can allow one to make conclusions about the nature of the stereoregulation.

stereoblock polypropene (**1**)

isoblock polypropene (**2**)

Exercise 9.4
Draw the structure of iso- and syndiotactic polypropylene with one error each in stereocontrol by the chain end and by the catalytic center. Show the additional triads and pentads that appear due to the stereoerror and thus enable identification of the error by NMR spectroscopy.

9.3.2
Ziegler-Natta Catalysts

During polymerization of propene, the classical Ziegler catalyst, $TiCl_4/AlEt_3$, produces the isotactic polymer, albeit with low yield. The isotactic index[5] shows that the majority of the polymer formed is amorphous (atactic). However, if crystalline $TiCl_3$ is produced (e.g., by reaction of $TiCl_4$ with hydrogen or with aluminum to form $TiCl_3 \cdot 1/3\ AlCl_3$), which then is converted to the actual catalyst with an aluminum alkyl, the proportion of atactic polymer decreases to about 15%. Addition of Lewis bases such as ethers, esters or amines leads to an increase in activity and selectivity, so only 2–5% amorphous polypropylene accrues. The $MgCl_2$-supported catalysts used in ethene polymerization did not initially bring the desired success. Only when Lewis bases were added, a clear increase in activity and productivity as well as good stereoselectivities could be achieved.

Crystalline violet α-$TiCl_3$ is a modification used as precatalyst. The crystals contain edge-sharing $TiCl_6$ octahedra that form layers (BiI_3 type, Figure 9.2a). All chloro

---

[5] Isotactic polypropylene is insoluble in hydrocarbons, while the (comparatively low-molecular) atactic polymer that is also formed is hydrocarbon-soluble. The "isotactic index" indicates the percentage of the polymer that is insoluble in boiling heptane. Thus, it is a measure of the ratio in which the two polymer types have been formed, but not of the stereoregularity of the isotactic polymer.

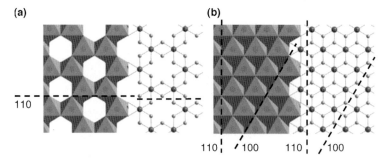

Figure 9.2 Edge-linked TiCl$_6$ and MgCl$_6$ octahedra in the layer structures of α-TiCl$_3$ (a) and MgCl$_2$ (b). View orientation: vertical onto the layer (001) from which the crystals are built. The dashed lines show position and structure of the crystal faces given by the Miller indices. (Structural images with the friendly permission of U. Müller, Inorganic Structural Chemistry, Wiley, Chichester, UK 2006.) To ensure clarity and comprehensibility of the related discussion, it is recommended to view the structures by means of a program for visualizing crystal structures commonly used. The necessary structural data are given with the solution to Exercise 9.5.

ligands are bridge-bonded (μ-Cl) and belong to two octahedra. This corresponds to the required stoichiometry: TiCl$_3$ = TiCl$_{6/2}$. Since the catalysis occurs on the crystal surface, the surface structure must be analyzed separately. Cleavage of crystals in the direction of the layer plane results in a (001) surface in which all Ti atoms are still surrounded in the same way by six μ-Cl ligands. However, a cut perpendicular to this plane so as to form a (110) surface now leads – for electroneutrality reasons – to coordinatively unsaturated Ti atoms ($CN = 5$) with four μ-Cl ligands and one terminal Cl ligand. (Figure 9.3a). These Ti atoms are catalytically active centers; the polymer chain can be built in the familiar way after alkylation (substitution [Ti]–Cl$_{terminal}$ → [Ti]–Et by means of the cocatalyst AlEt$_3$ or AlEt$_2$Cl) and coordination of propene to the vacant coordination site.

The two coordination sites at which the catalysis proceeds are not equivalent. One is oriented inwards and the other outwards. The coordination sites are chiral, so the principal precondition for stereoselective polymerization is fulfilled [183]. The exact mechanism, however, is not known.

MgCl$_2$ crystallizes in the CdCl$_2$ type with cubic closest packing of the chloride anions. Half of the octahedral holes are filled with Mg^{2+}, yielding a layered structure. The MgCl$_2$ structure is closely related to the TiCl$_3$ structure and is (formally) derived from it in that Ti^{3+} is substituted by Mg^{2+} and on the other hand the holes of the TiCl$_3$ structure are likewise filled with Mg^{2+} (Figure 9.2b). Thus, the MgCl$_6$ octahedra are linked with each other via six edges (μ$_3$-Cl). This corresponds to the stoichiometry: MgCl$_2$ = MgCl$_{6/3}$. Cleavage of crystals in the direction of the layer plane results in a (001) surface in which all Mg atoms are octahedrally surrounded by six chloride ions. However, Mg atoms of an (110) surface only have a coordination number of four, while those of an (100) surface have a coordination number of five (Figure 9.3b).

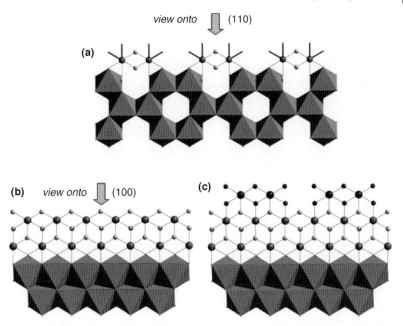

Figure 9.3 (a) View onto the (110) surface of the α-TiCl$_3$ structure. For the Ti atoms on the surface, two coordination sites are indicated by bold dashes. In TiCl$_3$ crystals, one site is occupied by a terminal Cl atom (CN(Ti$_{surface}$) = 5). In the catalyst complex, the growing polymer chain is coordinated to one of these coordination sites and the propene is coordinated to the other. (b) MgCl$_2$ structure with view onto the (100) surface (CN(Mg$_{surface}$) = 5). (c) MgCl$_2$ structure (same view orientation as b), covered with Ti$_2$Cl$_6$ (CN(Ti) = 5).

Exercise 9.5
Demonstrate that in the structure given for the crystal faces, the electroneutrality of the crystals is ensured in all three cases.

Deposition of TiCl$_4$ on a MgCl$_2$ support leads to an epitaxial coordination of mononuclear TiCl$_4$ and dinuclear Ti$_2$Cl$_8$ units on both (100) and (110) MgCl$_2$ surfaces. The subsequent reduction by the cocatalyst yields a surface covered with TiCl$_3$ or Ti$_2$Cl$_6$, respectively. The Ti$_2$Cl$_6$ species on a (100) MgCl$_2$ surface have a structure that is very similar to that of the (110) surface of TiCl$_3$ (Figure 9.3c). Apparently, this surface structure of the supported catalysts leads to stereoselective polymerization, while catalysis at other surface titanium centers is only slightly stereoselective, if at all.

In the pretreatment of MgCl$_2$ with Lewis bases, which are called "internal donors" in this context, the bases are coordinated to all Lewis acid centers on the MgCl$_2$ surface. The subsequent treatment with TiCl$_4$ now leads – with partial displacement of Lewis bases – to selective coverage of the (100) surface with Ti$_2$Cl$_8$ units. The reason for this selectivity is probably that the internal donors are more strongly bound to the stronger Lewis acidic Mg atoms of the (110) surface ($CN=4$!) and thus are more difficult to displace by $TiCl_4$ than from the weaker Lewis acidic centers on the (100) surface ($CN=5$!). However, the activation of the catalyst with aluminum alkyls now leads, in addition to the reduction $Ti^{IV} \rightarrow Ti^{III}$, to a partial displacement of the Lewis bases and also of the titanium halides. These can be deposited on the $MgCl_2$ surface again, but nonselectively at Mg surface atoms of any site, thus decreasing the selectivity of the polymerization. To prevent that, either very low concentrations of AlR_3 are used during polymerization, or additional Lewis base, the so-called external donor, is added during the polymerization process. In this way, highly active and selective catalysts have been successfully obtained, requiring no separation of either the catalyst or the a-PP from the isotactic polypropylene. Thus, $MgCl_2/TiCl_4$–$AlEt_3$ with diisobutyl phthalate as internal and alkoxysilanes as external donor produces 15000 kg PP/(mol Ti·MPa·h) with isotacticity of 97–98% [184].

9.3.3
Polymer Types and Process Specifications

In 2009, about 46 million metric tons of polypropylene were produced. Thus, polypropylene is one of the three most heavily used plastics (share: PE about 30%; PP about 20%; PVC about 14%). Industrially, by far the most important are isotactic polypropylene and its modifications through copolymerization. In comparison to HDPE, i-PP has a lower density and a higher melting range, but also a significantly higher glass transition temperature (Table 9.3). When industrial production of isotactic polypropylene began, the atactic polymer was only an undesired by-product. As a result of the improvement of the polymerization process, a-PP is no longer an unavoidable by-product, and today, it is produced directly to a small extent. Crystalline s-PP was first obtained by G. Natta from soluble catalyst systems such as $V(acac)_3/AlEt_2Cl$ or $VCl_4/AlEt_2Cl/PhOMe$.

The application properties of polypropylene can be controlled by means of the polymerization conditions (microstructure) and processing conditions (macrostructure). Copolymerization with other olefins is another powerful method to systematically influence the properties of polymers. Copolymers often display

Table 9.3 Physical-chemical properties of polypropylenes (compiled from [178], Vol. 2 and T. G. Heggs, "Polyolefins" in [M14]).

	i-PP	s-PP	a-PP
Density (in g/cm³)	0.91–0.94	0.88–0.93	0.85–0.89
Melting point (in °C)[a]	about 176 (160–165)	about 217 (140–150)	—
Glass temperature (in °C)[b]	−13...−35	−8	−5...−10
Solubility in hydrocarbons (20 °C)	—	moderate	high

a) Extrapolated to 100% iso- (α-form) or syndiotacticity. Typical range in parentheses.
b) For comparison, the glass temperature of PE is less than −100 °C.

properties that differ fundamentally from those of the homopolymers. For instance, ethene–propene copolymers are rubber-elastic (EPR: ethene–propene rubber).

9.4
Metallocene Catalysts

9.4.1
Cocatalysts and Anion Influence

Immediately after the discovery of heterogeneous Ziegler catalysts, the search for soluble catalyst systems began. Thus, as early as the end of the 1950s (D. S. Breslow; G. Natta), homogeneous catalyst systems for the polymerization of ethene with titanocene dichloride [$TiCl_2Cp_2$] and $AlEt_{3-x}Cl_x$ ($x = 0, 1$) as cocatalyst were found, and were valuable in illuminating the mechanism of polymerization. However, they were unsuitable for industrial applications due to their low activity. The discovery that metallocenes with methylaluminoxanes (MAO) as cocatalysts displayed excellent activity (H. Sinn, W. Kaminsky, 1980) was the starting point for the development of (modern) metallocene polymerization catalysts that found use in industry. These were preceded by observations that water – long considered "poison" for Ziegler-Natta catalysts – raises the polymerization rate of ethene for several systems (e.g., [$TiEt(Cl)Cp_2$]/$AlEtCl_2$) and transforms inactive systems like [$ZrMe_2Cp_2$]/$AlMe_3$ into surprisingly highly active catalysts [185].

In metallocene catalysts, catalytic centers are structurally uniform ("single-site catalysts"); they have been characterized in detail, thus providing exact knowledge of the mechanism of polymerization. This, in turn, was a precondition for the synthesis of "tailor-made" polymerization catalysts for the chemical industry.

Methylaluminoxanes (MAO) are created during the controlled partial hydrolysis of aluminum trimethyl. They are not structurally uniform. Rather, they are a mixture of several different oligomeric compounds which typically consist of 5–25 –O–Al(Me)– structural units. In addition, they usually contain $AlMe_3$. As represented by the basic structures shown below, the oligomers can be linear (**44**) or cyclic (**45**) or display a cage structure (**46**).

44 **45** **46**

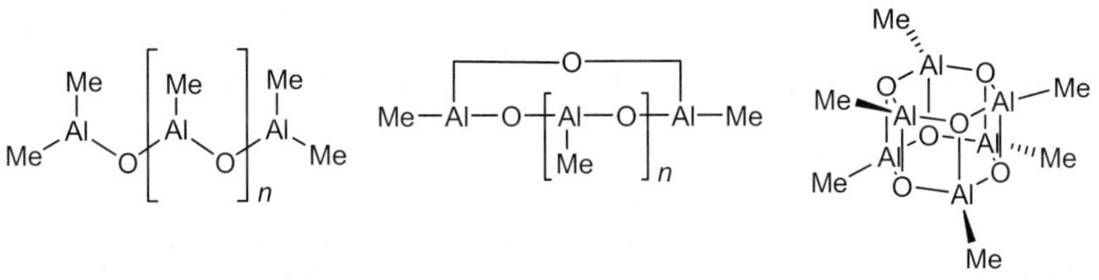

44 and **45** only have three-coordinate Al and μ_2-O ligands, while **46** has four-coordinate Al and μ_3-O ligands. Through combination of these structural elements, more complex two- and three-dimensional structures are formed.

In general, MAO is used in large excess (Al/M about 10^3–10^4). This causes "buffer" behavior because MAO scavenges impurities and thus protects the catalyst complex from decomposition; in some circumstances, deactivated catalysts are even reactivated. However, the two main functions of the cocatalyst are:

- **Methylating function.** Methylation of the metallocene dichloride (**47**, M = Group 4 metal; generally [ZrCl$_2$Cp$_2$] or a derivative) to form the corresponding dimethyl compound **48**:

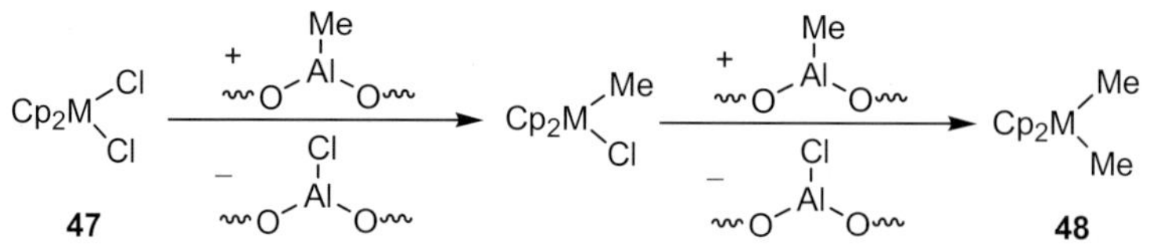

- **Lewis acid function.** Abstraction of a methyl anion from **48** with formation of the actual polymerization-active compound, a cation [MMeCp$_2$]$^+$ (**49**). The anion [MAO–Me]$^-$ thus formed coordinates so weakly, if at all, to the [MMeCp$_2$]$^+$ cation that it does not hinder the coordination of the olefin.

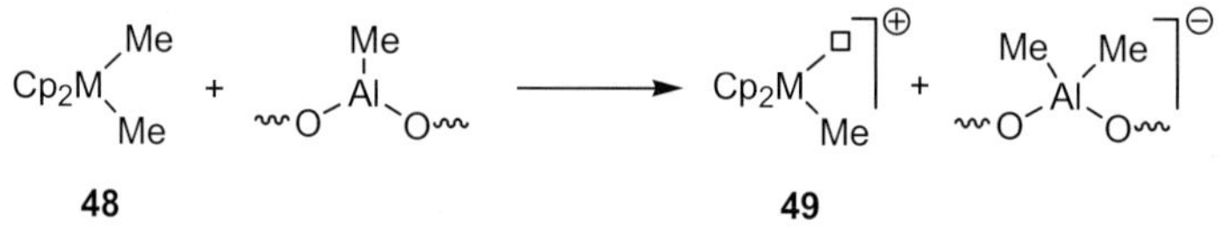

Thus, the actual catalytically active compound (**49**) has been formed, a cationic 14-ve alkyl metallocene complex with a weakly coordinating counter-ion. There are other formation paths as well [186]:

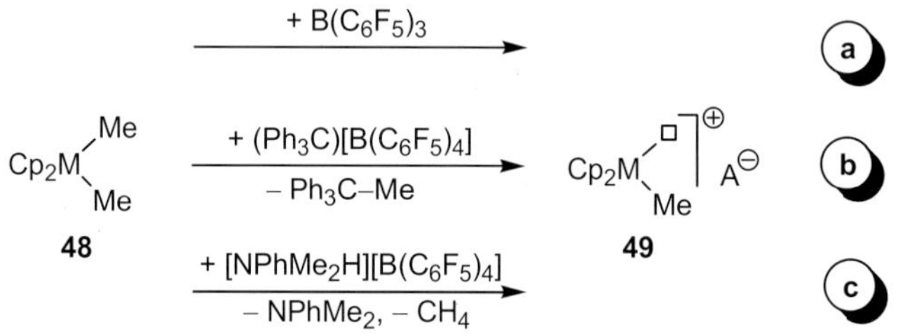

Analogous to the reaction with MAO, the reaction of **48** with neutral Lewis acids such as B(C$_6$F$_5$)$_3$ leads to abstraction of a methyl anion (**a**: A$^-$ = [BMe(C$_6$F$_5$)$_3$]$^-$). The trityl cation is suited to the demethylation of **48**, so its reaction with a trityl salt having a weakly coordinating anion results in complexes of type **49** (**b**: A$^-$ = [B(C$_6$F$_5$)$_4$]$^-$). In addition, protolytic cleavage of the M–C bond can be used to

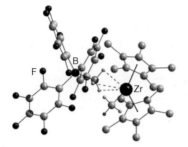

Figure 9.4 Molecular structure of [ZrMe(η^5-C$_5$Me$_5$)$_2$][BMe(C$_6$F$_5$)$_3$]. The H atoms on the C$_5$Me$_5$ ligands are not shown. The crystal contains contact ion pairs in which the Me group on the boron atom creates the contact to the cation. However, the Zr···C distance to the μ-methyl ligand is significantly longer than to the terminal methyl ligand (2.640(7) Å vs. 2.223(6) Å).

form **49** (c: A$^-$ = [B(C$_6$F$_5$)$_4$]$^-$). Figure 9.4 shows the structure of such a complex, which is, without further additions, a highly active homogeneous catalyst for ethene polymerization.

In toluene, a solvent often used for olefin polymerization, no coordinatively unsaturated cations [MMeCp$_2$]$^+$ (M = Ti, Zr) seem to exist (Figure 9.5). Quantum-chemical calculations with [BMe(C$_6$F$_5$)$_3$]$^-$ as counter-anion indicate that in toluene, contact ion pairs **49a** with μ-Me bridges between cations and anions are energetically the most stable. The insertion of an ethene molecule between cation and anion to form olefin-separated ion pairs **49b** has a low energy demand. The formation of ethene complexes in the form of separated solvated ions (**49c**) requires significantly more energy, and formation of the coordinatively unsaturated cations (**49d**) even more, by a substantial amount [187, 188].

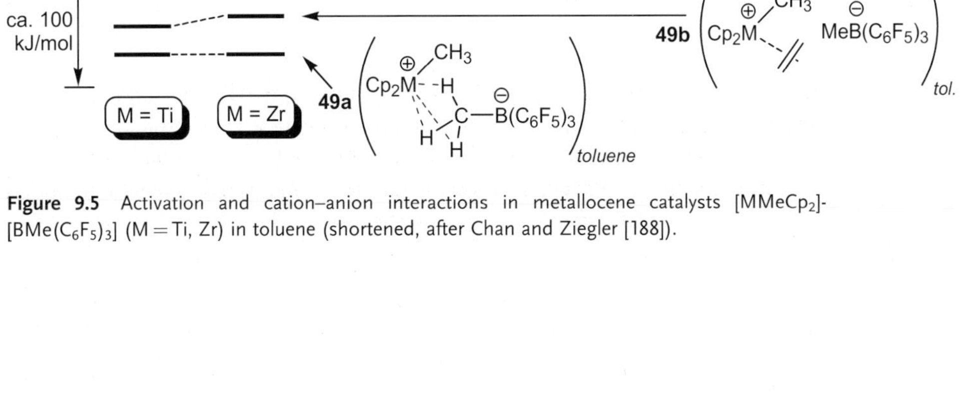

Figure 9.5 Activation and cation–anion interactions in metallocene catalysts [MMeCp$_2$][BMe(C$_6$F$_5$)$_3$] (M = Ti, Zr) in toluene (shortened, after Chan and Ziegler [188]).

The binding of the olefin and of the anion are competitive processes, and it is probable that olefin-separated ion pairs of type **49b** are important intermediates in the olefin polymerization with metallocene catalysts. Normally, the activation barriers for the insertion step for d^0 metals (including rare earth metals with d^0f^n electron configuration) are very low, provided that no larger reorganization of the conformation of the coordinated olefin and the growing polymer chain is involved [189, 190].

9.4.2
C_2- and C_s-Symmetric Metallocene Catalysts

9.4.2.1 Principles

If the catalyst is to exert a stereoregulating effect during propene polymerization, it must be chiral. Metallocene dichlorides $[MCl_2Cp_2]$ (M = Ti, Zr) are achiral. Even when unsymmetrically substituted cyclopentadienyls are used instead of cyclopentadienyl ligands, no chiral induction is possible, or at least none is effective, due to a low rotation barrier for the cyclopentadienyl ligands. Only with the synthesis of *ansa*-metallocenes (H. H. Brintzinger, 1982) was the foundation established for the development of metallocene catalysts that were suitable for stereoselective propene polymerization. The bridge (Latin: *ansa* = handle) that binds the two cyclopentadienyl ligands fixes the conformation of the complex so that the catalyst center can exert a stereoregulating function.

Exercise 9.6
Describe the structure of metallocene dichlorides $[MCl_2Cp_2]$ (M = Ti, Zr) and determine the symmetry group. Assume unhindered rotation of the cyclopentadienyl ligands around the M–Cp_{cg} axis (Cp_{cg} = center of the Cp ligand).

Figure 9.6 shows the structures of three *ansa*-zirconocene dichlorides. In all three complexes, the two π ligands are linked by dimethylsilyl bridges. From the parent compound **50**, which displays C_{2v} symmetry, the two most important systems for catalysis can be derived:

- C_2-*symmetric catalysts* (**51**). To each of the two cyclopentadienyl ligands on the parent compound, a benzene ring is fused such that a C_2-symmetric bis(η^5-indenyl) complex is formed.
- C_s-*symmetric catalysts* (**52**). Two benzene rings are fused to one of the two cyclopentadienyl ligands on the parent compound so as to form a C_s-symmetric η^5-fluorenyl–η^5-cyclopentadienyl complex.

The expressions "C_2-" and "C_s-symmetric catalyst" can be misleading: The actual catalyst complex, which contains, in addition to the η^5-bound ligands, the coordinated propene and the growing polymer chain, is always only C_1-symmetric. The notations C_2 and C_s refer to the symmetry relationship between the two "coordination sites" for the propene. In the "C_2-symmetric catalyst" they are

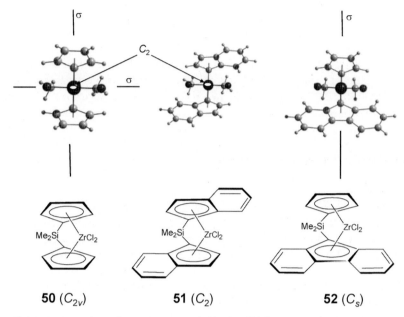

Figure 9.6 Structures of *ansa*-zirconocene dichlorides of different symmetry. Complex **51** is chiral; one of the two enantiomers is shown. The view of the molecular structures is along the Zr–Si direction so that the Si atoms are hidden by the Zr atoms. The symmetry elements (symmetry plane σ: ———; symmetry axis C_2: ◯) are indicated in the molecular structures.

identical (homotopic). They are interconverted by a C_2-symmetry operation such that the prochiral propene always coordinates with the same face (*Re* or *Si*) and isotactic polypropylene is formed. In the "C_s-symmetric catalyst" they are mirror-symmetrical (enantiotopic). They are interconverted by a σ-symmetry operation (σ = mirror plane) such that the prochiral propene coordinates to one coordination site with the *Re* face and to the other with the *Si* face and syndiotactic polypropylene is formed.

Background: Topic Relationships Between Molecular Fragments

To establish topic relationships between molecular fragments of the same atomic composition, symmetry criteria can be used: homotopic and enantiotopic fragments are symmetry-equivalent, while diastereotopic fragments are not. Homotopic fragments are mapped to each other by a rotation symmetry operation (C_n with $n \geq 2$) and enantiotopic fragments *only* by a reflection (σ), inversion (*i*), or rotation-reflection (improper rotation, S_n with $n > 2$) symmetry operation. To determine topic relationships, the following algorithm (after K. Mislow) can be applied [51, 191, 192].

```
                    ┌──────────────────────────────┐
                    │ molecular fragments having the│
                    │    same atomic composition   │
                    └──────────────┬───────────────┘
                                   ▼
                           ╭───────────────╮
            yes    ┌───────│   symmetry    │───────┐   no
                   │       │  equivalent?  │       │
                   ▼       ╰───────────────╯       ▼
           ╭───────────────╮              ╭─────────────╮
   yes ┌───│ sym. equiv. by│── no ┐  yes ─│    same     │── no
       │   │  Cₙ (n ≥ 2)?  │      │       │ constitution?│
       ▼   ╰───────────────╯      ▼       ╰─────────────╯
  ┌─────────┐              ┌──────────────┐  ▼            ▼
  │homotopic│              │ enantiotopic │ diastereotopic  const. heterotopic
  └─────────┘              └──────────────┘
```

9.4.2.2 Mechanism

The mechanism of the catalysis of propene polymerization with the C_2-symmetric precatalyst **51** is analogous to the one with [TiCl$_2$Cp$_2$]/AlEt$_2$Cl: The catalytically active complex is an *ansa*-bis(η^5-indenyl)zirconium(IV) cation (**53**, bridge not shown), to which the growing polymer chain and a monomer molecule (propene) are coordinated. The chain growth occurs by a primary *cis* insertion via migration of the polymer chain. The catalyst complex is formed again by coordination of propene to the vacant coordination site (**53** → **53'**), and an additional insertion and propene coordination occur (**53'** → **53''**).

In order to understand the nature of the stereocontrol, we consider the transition state **53ts** of the insertion reaction, which is characterized by a nearly planar metallacycle ZrC_3. Repulsive interactions force the following orientation:

- The growing polymer chain $-C_\beta HMeP$ points into the least sterically hindered sector of the metallocene-ligand backbone.
- The growing polymer chain $-C_\beta HMeP$ forces the entering propene into an orientation in which the two alkyl groups (Me and CHMeP) of the newly created C_2-C_α bond are in *anti* arrangement, hence on different sides of the ZrC_3 plane.

Model calculations show that transition states with other orientations have a higher energy and are irrelevant in the catalytic process. The transition state **53ts** is stabilized by an α-agostic $C_\alpha-H \cdots Zr$ interaction. However, it can only be formed with one of the two H atoms of the $C_\alpha H_2$ group. If the other H atom were involved in such an interaction, the $C_\beta HMeP$ group would have to rotate into a sterically hindered sector of the metallocene-ligand backbone.

The transition state **53ts** develops from a coordination of propene to the *Si* face, so an *R*-configured asymmetric C atom is formed (**53** → **53ts** → **53'**). Now, another propene molecule is coordinated, but to the other coordination site of the catalyst complex. The subsequent insertion step leads in turn to an asymmetric C atom with *R*-configuration. This is easy to understand, since both catalyst complexes are interconverted by a C_2-symmetry operation. In other words, the catalyst complex can distinguish between the two prochiral faces of the propene, and the coordination always occurs to the *Si* face. The result is isotactic polypropylene, in which all asymmetric C atoms display the same (relative) configuration. This induces a helical structure, since only a helix allows a repetition of monomer units with identically configured asymmetric C atoms.

Typically, the racemate of the chiral C_2-symmetric $[ZrCl_2\{(\eta^5\text{-Ind})_2SiMe_2\}]$ complex is used as precatalyst. The preceding discussion relates to the enantiomer shown in Figure 9.6; for the other enantiomer, an analogous discussion would apply. Each of the two enantiomers produces a *pseudo*chiral polymer.

Exercise 9.7
- Draw the other enantiomer of **53** and indicate to which face propene is coordinated.
- What result do you expect when an enantiomerically pure catalyst **53** is used and the reaction conditions (in particular, a low propene/metallocene ratio) are chosen so that an oligomerization of propene occurs?

The catalytically active complex **54** (Me_2Si bridge between Cp and fluorenyl ligand is not shown) is generated when starting from a C_s-symmetric precatalyst complex **52**. The course of the polymerization is shown in the following scheme.

In turn, the capability of the catalyst complex to coordinate propene in precisely one orientation is critical for stereoregulation. In the transition state **54ts**, the growing polymer chain –CHMeP points into the least sterically hindered sector of the metallocene (here: upwards) and is in an *anti* orientation relative to the methyl group of the reacting propene. From qualitative considerations, this was initially unexpected, since it points in the direction of the fluorenyl ligand. The opposite arrangement, where the methyl group points in the direction of the Cp ligand, but then is oriented *cis* to the –CHMeP group, is energetically less favorable. The weak direct interactions between the substituents of the coordinated olefin and the ligand backbone are of less importance in stereodifferentiation.

The transition state **54ts** coordinates propene to its *Re* face and forms an *S*-configured asymmetric C atom (**54** → **54ts** → **54'**). Due to the C_s-symmetry of the precatalyst, however, the two coordination pockets for propene are enantiotopic; that is, they interconvert via a reflection. This means that in the catalyst complex **54/54"** the prochiral propene coordinates with the *Re* face, and in the catalyst complex **54'**, it coordinates with the *Si* face. Therefore, the asymmetric C atoms of the growing polymer chain show alternating *S*- and *R*-configuration; that is, syndiotactic polypropylene is formed.

Table 9.4 gives a summary: Metallocene catalysts – due to the *migratory* insertion – have two coordination pockets available for the olefin. The catalyst complex (with two vacant coordination sites or, averaged over time, equally occupied coordination sites) is C_{2v}-, C_2- or C_s-symmetric. In C_{2v}-symmetric complexes, the coordination sites possess symmetry (C_s), from which it follows that they are nonselective. In all complexes mentioned, the two coordination sites are symmetry-equivalent, hence homotopic or enantiotopic. Thus, a relationship between the microstructure of the polymers and the structure of the catalyst is revealed.

Metallocene catalysts are extremely active! The "parent compound" [ZrCl$_2$Cp$_2$] with MAO as cocatalyst achieves an activity of 3600 kg PE/(mmol Zr·h) (95 °C, 8 bar) for ethene polymerization. Per zirconium atom, 13 polymer chains are produced per second. Every 0.03 ms, an ethene molecule is inserted into the growing polymer chain; the turnover frequency is $3 \cdot 10^4 \, \text{s}^{-1}$. This corresponds to the activity of highly

Table 9.4 Symmetry and symmetry relationships in metallocene catalysts.

Catalyst complex (stylized)	◇–M–◇	◇–M–◇	◇–M–◇
Symmetry of the catalyst complex	C_{2v}	C_2	C_s
Symmetry of the coordination sites	C_s (nonselective)	C_1 (enantioselect.)	C_1 (enantioselect.)
Symmetry relationship between the coordination sites	C_2, σ (homotopic)	C_2 (homotopic)	σ (enantiotopic)
Microstructure of the polymer[a]	atactic	isotactic	syndiotactic

a) Provided there is no stereochemical chain end control.

active enzymes (see footnote 2, Section 8.2). Even higher activity can be achieved by *ansa*-metallocenes with substituted cyclopentadienyl ligands. The productivity of metallocene catalysts is likewise exceptionally high; even after 100 h polymerization time, they are still highly active.

Propene polymerization also proceeds with high activity, but often lower by 1–2 orders of magnitude than for ethene polymerization. Stereoregularities of >99% and regioirregularities (regioerrors) of <1% are achieved [193].

9.4.3
Metallocene Catalysts with Diastereotopic Coordination Pockets

9.4.3.1 Principles

Up to now, metallocene catalysts that offer two symmetry-equivalent olefin coordination sites have been considered. Now, we abandon this restriction and allow diastereotopic coordination sites. Metallocenes of C_1-symmetry result. The structure of a corresponding catalyst pre-stage is shown in Figure 9.7. Since a coordination site for a prochiral olefin may or may not be capable of enantiotopic differentiation, the following three combinations are possible for the catalyst complexes that are to be discussed here: nonselective–nonselective, enantioselective–enantioselective, and nonselective–enantioselective.

In the stylized representation **a/a′**, a nonselective coordination site which is itself symmetric (σ)[6] is shown. In **b/b′**, an enantioselective coordination site is schematically shown, exhibiting preferential orientation for the growing polymer chain and thus also for the coordinated olefin. Examples of catalysts with two symmetry-equivalent enantioselective coordination sites are extensively discussed in the previous chapter. An example of a catalyst complex with two nonselective coordi-

---

6) The σ-symmetry ("eigensymmetry") of a coordination site in a metallocene catalyst is a sufficient, but not a necessary condition for nonselectivity. Such a symmetry of a coordination site must not be confused with the symmetry relation that exists between the two coordination sites in a metallocene catalyst.

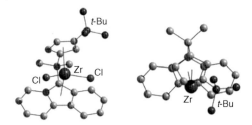

Figure 9.7 Structure of [ZrCl$_2${η^5-3-(t-Bu)C$_5$H$_3$–CMe$_2$–η^5-C$_{13}$H$_8$}], an *ansa*-zirconocene dichloride with diastereotopic coordination sites (without H atoms; the *tert*-butyl group is shaded dark gray). The top view (right; without the two chloro ligands) makes clear the differing degrees of steric shielding of the two coordination sites for the monomer and the growing polymer chain, respectively.

nation sites is the "parent compound" (**50** in Figure 9.6), from which the *ansa*-zirconocenes are derived. Metallocenes with one nonselective and one enantioselective coordination site are briefly dealt with below.

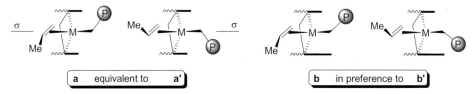

The polymerization on a catalyst with two diastereotopic coordination sites A and B is shown in Figure 9.8. The "normal" reaction course corresponds to the sequence **55** → **56** → **57** → **58** → **55″** → ....[7] Now, however, it must be considered that after a successful insertion step **55** → **56** → **57** the polymer chain can migrate from A to B (**57** → **55′**) (back-skip). Since no olefin is coordinated to B, no insertion takes place ("insertionless migration"). The driving force for this reaction may be the energy difference between **57** and **55′** [172]. The relative rates of olefin coordination/insertion **57** → **58** → **55″** and of "insertionless" migration of the growing polymer chain (**57** → **55′**) determine the "insertion scheme" for the polymerization.

Considering that the selectivity of the coordination sites A and B (*Re* vs. *Si* vs. nonselective) can be markedly different, the cases in Table 9.5 should be distinguished from each other. If chain end stereocontrol is not operative, metallocenes with two nonselective coordination pockets yield atactic polymers (entry 1 in Table 9.5). Entries 2 and 3 correspond to the situation of C_2- and C_s-symmetric catalysts, except that the coordination sites for the olefin are diastereotopic and not homo- or enantiotopic. For entry 4, the insertionless migration of the polymer chain **57** → **55′** occurs so rapidly that the reaction sequence (**55** → **56** → **57** →)$_x$ is achieved and the "normal" reaction sequence (**55** → **56** → **57** → **58** →)$_x$ is fully suppressed. An isotactic polymer is obtained; the stereoselectivity of the coordination

---

7) The growth of the polymer chain **P**$_n$ is indicated by dashes: For **55**, $n = n$; for **55′**, $n = n + 1$; and for **55″**, $n = n + 2$.

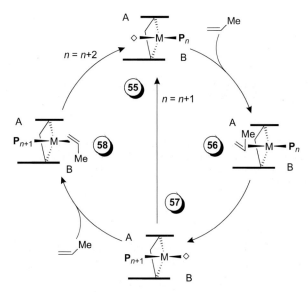

Figure 9.8 Propene polymerization and "insertionless" migration of the polymer chain. P_n indicates the growing polymer chain. In the stylized metallocene complex, one enantio- (A) and one nonselective (B) coordination site is shown, without indicating that one should be limited to this combination.

site B is irrelevant. Insertionless migration of the polymer chain, which does not proceed synchronously with the olefin coordination/insertion, then has no effect on the microstructure of the polymer, if both coordination sites favor a Re coordination (entry 2). By contrast, if the selectivity of the two coordination sites is different, the polymer obtained is only atactic (entry 5). Catalyst systems corresponding to entries 6 and 7 are dealt with below.

Table 9.5 Metallocene catalysts with diastereotopic coordination pockets and polymer structure (adapted from Coates [194]).

Entry	Selectivity of the coordination site[a]		Insertion scheme[b]	Microstructure of the polymer
	A	B		
1	nonselective	nonselective	any[c]	atactic
2	Re	Re	any[c]	isotactic
3	Re	Si	...ABABAB...	syndiotactic
4	Re	any	...AAAAA...	isotactic
5	favoring Re	favoring Si	statistical	atactic
6	Re	nonselective	...ABABAB...	hemiisotactic
7	Re	nonselective	...$(A)_n(B)_m(A)_n(B)_m$...	stereoblock

a) For each entry, Re can be exchanged with Si and vice versa.
b) The insertion scheme specifies in which sequence the coordination sites are included in the migratory insertion.
c) The microstructure of the polymer does not depend on the insertion scheme.

9.4.3.2 Hemitactic Polymers

In hemitactic polymers only every other stereocenter (1, 3, 5, 7, ...) has a precisely defined configuration, while the configurations of the stereocenters in between (2, 4, 6, 8, ...) are random (atactic). As a result, there are two hemitactic polypropylenes, the hemi-isotactic (**59**) and the hemi-syndiotactic polypropylene (**60**).

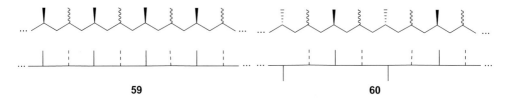

Exercise 9.8
Give the microstructures of the polymers that result from the hemitactic polymers when the configurations of the stereocenters 2, 4, 6, 8, ... are specified to be either iso- or syndiotactic.

A hemi-isotactic polypropylene (*hi*-PP) is produced by zirconocene complex **62** (activated by MAO). The unsubstituted complex **61** is a typical C_s-symmetric pre-catalyst that yields syndiotactic polymer. The methyl substituent in **62** makes the two enantiotopic coordination pockets in **61** diastereotopic, of which one (A) is enantioselective and the other (B) is nonselective. The steric hindrance of the growing polymer chain at B, whether shown in the direction of the fluorenyl ligand (resulting in *Si*-coordinated propene at A; see Formula **62**) or in the direction of the methylcyclopentadienyl ligand (resulting in *Re*-coordinated propene at A; see Formula **62**), is comparable, so no selectivity results. Entry 6 (Table 9.5) applies to this situation [195].

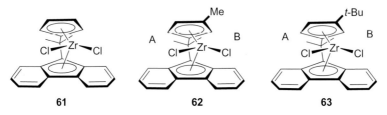

The *tert*-butyl substitution in complex **63** (see structure in Figure 9.7) turns the syndioselectivity of **61** into isoselectivity. Clearly, the growing polymer chain at A points in the direction of the less sterically hindered sector (upwards in formula **63**), so propene at B is coordinated to the *Si* face (Me group downward). The catalyst complex with the polymer chain at B and propene at A is less stable. This case also favors an arrangement where the propene is coordinated to the *Si* face, so an isotactic polymer would result (analogous to entry 2 in Table 9.5). However, there is also the possibility that the olefin always coordinates at position B and the polymer chain after the insertion step undergoes "insertionless" migration to A, since a binding of the bulky polymer chain at B (caused by the *tert*-butyl substituent and the fluorenyl ligand) is energetically too disadvantageous. A mechanism analogous to entry 4 in Table 9.5 corresponding to a reaction cycle ($55 \rightarrow 56 \rightarrow 57 \rightarrow$)$_x$ in Figure 9.8 would occur.

9.4.3.3 Stereoblock Polymers

Stereoblock polymers are composed of blocks of different steric structure, although they are all created from a single monomer. Stereoblock polypropylenes, in which isotactic blocks alternate with atactic, may be obtained when one catalyst center (A in Figure 9.8) is enantioselective and the other (B) is nonselective (Table 9.5, entry 7). Stereoblock polypropylenes are also obtained with nonbridged metallocenes in which a sufficient conformative stability (due to restricted rotation of the cyclopentadienyl substituents around the M–Cp axis) is ensured by suitable substitution at the Cp rings. Thus, for bis(2-arylindenyl)zirconium complexes **64**, three conformational isomers occur: **64a** is (approximately) a *meso* form with nonselective coordination sites; **64b/64b'** (*rac* form) are enantiomeric and each is C_2-symmetric. **64b** and **64b'** produce isotactic sequences with different configured C atoms and, under polymerization conditions, exist in a dynamic equilibrium ("oscillating catalysts"). Recent experiments show that the *meso* form, which would produce an atactic block, does not play a role. The microstructure of the polymers depends markedly on the substitution pattern in the phenyl rings, which substantially influences the isomerization rate **64b** ⇌ **64b'**; it also depends on the cation–anion interactions and the reaction conditions. Thus, for example, stereoblock polymers of the composition **65** are accessible whose microstructure shows a relatively high concentration of isolated *r*-diads that connect stereoblocks of opposite relative configuration. The special microstructure of stereoblock polymers gives access to thermoplastic elastomers (TPEs) [196, 197].

9.4.4
On the Significance of Metallocene Catalysts

Metallocene catalysts made it possible for the first time to steer the polymerization of olefins in such a way that the microstructure of the polymers could be precisely controlled and varied within a wide range. This enabled the targeted synthesis of polymers with "tailor-made" properties. These achievements were made possible by the fact that metallocenes are soluble single-site catalysts with precisely defined structure, and in addition, that the relationships between catalyst structure and polymer architecture are well understood.

In the case of ethene, the polymerization with metallocene catalysts leads to polymers with a narrow molar mass distribution $M_w/M_n = 2$ (for comparison: $M_w/M_n = 5–10$ for Ziegler catalysts) at 0.9–1.2 methyl groups per 1000 C atoms. The molar mass itself depends strongly on the catalyst structure and the reaction conditions. Metallocene catalysts have also proven themselves in copolymerization of ethene with α-olefins (propene, ..., oct-1-ene) to form LLDPE. The majority of the comonomers is statistically distributed in the polymer chain, and high content (up to 30%) of comonomers can be achieved [198].

In the case of propene, all stereoisomeric polymers, including stereoblock polymers, are achievable in high purity by metallocene catalysis. They are distinguished by a narrow molar mass distribution and only contain very low quantities (<0.1%) of low-molecular products (for comparison: 2–4% for Ziegler-Natta catalysts). The copolymerization of ethene with propene in a molar ratio that changes from 1 : 2 to 2 : 1 in the presence of low quantities of a nonconjugated diene (e.g., hexa-1,4-diene) leads to elastomers (EPDM elastomers) with narrow molar mass distribution. Metallocene catalysis allows optically active propene oligomers to be achieved.

Metallocene catalysts are also used for polymerization of other monomers. Examples are the polymerization of styrene to syndiotactic polystyrene (melting point: 275 °C, glass transition temperature: 100 °C), the polymerization of cycloolefins (cyclopentene, cyclobutene, norbornene) without ring-opening to crystalline polymers with high melting points ($\geq$400 °C), and the cyclopolymerization of α,ω-dienes (see Exercise 9.9). Metallocene catalysts have achieved great importance in polymer chemistry from both a scientific and an industrial point of view, and in the future they will become even more important.

Exercise 9.9
- α,ω-dienes are bifunctional monomers that can be cyclopolymerized with metallocene catalysts. Give the course of reaction for the cyclopolymerization of hexa-1,5-diene. Assume an alternation of inter- and intramolecular insertion (primary insertion in each case). Stereoregular polymerization can lead to four different stereoisomers of maximum order. Give the polymer structures.
- Chain transfers (see Section 9.2.1) can be used for *in situ* functionalization of polyolefins. Which polymers are obtained when $HSiR_3$, HBR_2 or HPR_2 are used as chain transfer agents, provided that the chain transfer is the dominant termination reaction? How might the mechanism look for a d^0-metal complex as catalyst?

9.5
Nonmetallocene Catalysts

Extensive investigations on activity and selectivity of metallocenes in olefin polymerization have fostered good understanding of the factors on which selectivity and activity in olefin polymerization depend. They have made it possible to derive structure models that in turn have enabled a directed search for nonmetallocene catalysts. It is safe to assume that the complexes that are catalytically active in olefin polymerization are preferentially coordinatively unsaturated cationic transition metal complexes **66** that can be formed from the precursor complexes **67–69** as shown [199]:

$$L_nM\begin{matrix}R\\X\end{matrix} \quad L_nM\begin{matrix}R\\R\end{matrix} \quad L_nM\begin{matrix}X\\X\end{matrix}$$
$$\quad 67 \qquad\qquad 68 \qquad\qquad 69$$

$$\downarrow \quad L_nM^{\oplus}{-}R\;\square$$
$$\quad\quad 66$$

reagents: $-X^{\ominus}$ (from 67); $-R^{\ominus}$ (from 68); $+R^{\ominus},\, -2X^{\ominus}$ (from 69).

67 → 66. Precursors are alkyl halogeno complexes **67** (X = halide), which react with compounds M′X′ (X′ = weakly coordinating anion such as PF_6^-, BF_4^-, OTf^-, BPh_4^-, [B{3,5-(CF$_3$)$_2$C$_6$H$_3$}$_4$]$^-$, …; M′ = Ag, Tl, alkali metal, …) in a double displacement reaction, with M′X cleaved off, to form [**66**]X′.

68 → 66. Precursors are dialkyl complexes **68** in which, via reaction with [YH]X′, an M–R bond is protolytically cleaved ([YH]$^+$ = [PhNHMe$_2$]$^+$, [R$_2$OH]$^+$, …). Alternatively, Lewis acids such as B(C$_6$F$_5$)$_3$ can cleave off an alkyl ligand R$^-$, whereupon the alkylated Lewis acid then functions as a weakly coordinating counter-ion X′$^-$ for **66**.

69 → 66. Precursors are dihalogeno complexes **69** that are transformed with a cocatalyst, which behaves both as an alkylating agent and a Lewis acid. In the classical Ziegler systems, these could be aluminum alkyls (with the disadvantage that the anions [AlR$_n$X$_{4-n}$]$^-$ still coordinate relatively strongly) or MAO.

The coligands L_n in the catalyst **66** play an important role. Their bulkiness and electronic properties are critical for selectivity and activity, as well as for the stability of the catalyst, thus, for example, preventing undesired redox reactions. Furthermore, they generate stable coordination geometry, which can preferentially be achieved with chelating ligands. This may be essential to ensure *cis* arrangement of the coordinated monomer and growing polymer chain, a critical precondition for a rapid insertion reaction.

9.5.1
Catalyst Systems of Early Transition Metals

They are frequently related to the classical metallocene catalysts of Group 4 (**70**) in that they are d^0-metal complexes. Among these are the so-called "constrained geometry catalysts" (CGC). These are half-sandwich amido complexes of Group 4 (**71**) or half-sandwich phenolato complexes of type **72**. Starting with the Group 4 complexes **70**, a shift to the right in the periodic table and the formal substitution of a Cp ligand by a dianionic ligand lead to neutral Group 5 d^0 precatalysts. Examples are imido complexes of type **73**. It can be expected that, as with metallocenes **70**, cationic 14-*ve* catalyst complexes can be generated from them. For comparison, metallocenes of the rare earth metals **74** are included here that have also proven themselves as precatalysts for olefin polymerization, but the 14-*ve* catalyst complexes that they yield are neutral [200, 201].

70 (M = Zr, Ti)   **71** (M = Zr, Ti)   **72**   **73** (M = V, Ta)   **74**

9.5.2
Catalyst Systems of Late Transition Metals

Table 9.6 presents a selection of precatalysts of late transition metals for the polymerization of ethene. It was found in the 1960s that $[Ni(C_3H_5)(PR_3)]$-$[AlCl_{x+1}Et_{3-x}]$ (**75**) with $PR_3 = PMe_3$ catalyzes the dimerization of ethene, while the significantly bulkier phosphane $P(t\text{-}Bu)_3$ causes polyethylene to be formed, especially when the phosphane is present in excess. Similarly, the SHOP catalyst **76** can be converted into a polymerization catalyst if the strongly binding PPh_3 is removed from the reaction mixture by a phosphane scavenger such as $[Ni(COD)_2]$ or $[Rh(acac)(H_2C=CH_2)_2]$ or if, at the outset, complexes with a more weakly binding ligand L (Ph_3PO, py, ...) are used (**77**). Complexes of type **78** with salicylaldiminato ligands (R, R' = bulky substituents) are similar to those of SHOP-type **76/77**.

Iron(II) and cobalt(II) complexes with tridentate bis(imino)pyridine ligands **80**, after addition of MAO, are sometimes highly active catalyst systems (M. Brookhart, 1998; V. Gibson, 1998). It is possible to control the molar mass of the polymers via the steric bulk of the *ortho*-aryl substituents R. If these *ortho* substituents are smaller, very active oligomerization catalysts (M = Fe) are obtained [202–204].

Similarly, cationic Ni^{II} and Pd^{II} complexes with diimine ligands (**79**) form highly active catalysts with MAO as cocatalyst (M. Brookhart, 1995). In these systems, as well, bulky *ortho*-aryl substituents such as R = *i*-Pr hinder chain transfer and are thus a prerequisite for polymerization. With R = H (M = Ni), oligomers are obtained. The polymers display an interesting microstructure; they are highly branched (up to 100

9.5 Nonmetallocene Catalysts

Table 9.6 Examples of polymerization of ethene by complexes of late transition metals (adapted from Mecking [205]).

Precatalyst	TOF[a] (in h^{-1})	Typical molar mass (in g/mol)[a]		Branching structure of the polyethylene
		M_w	M_w/M_n	
75[b]	very slow	polymers		
76	6·10^3	α-olefin oligomers	Schulz-Flory	linear oligomers
77	2·10^5	10^6 (M_v)	25	linear
78	2·10^5	5·10^5	1.5–3	moderately branched to linear
79 (M = Ni)	4·10^6	>8·10^5	1.5–3	highly branched to linear
80 (M = Fe)	10^7	6·10^5	9	highly linear

a) The values were determined under very different conditions and thus are only comparable to a limited extent.
b) [Ni(C$_3$H$_5$)Br{P(t-Bu)$_3$}]/Al$_2$Cl$_3$Et$_3$ + 3 equiv. P(t-Bu)$_3$ (75′). The analogous PMe$_3$ complex yields ethene dimers.

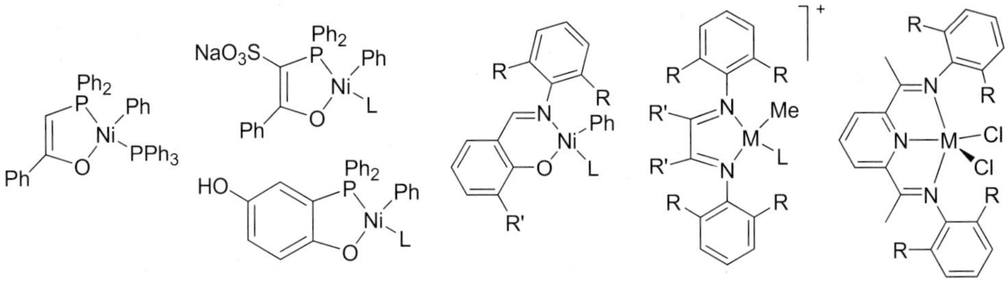

76   77   78   79   80

branches per 1000 CH$_2$ groups) without comonomers having been added. Linear polyethylene is obtained when the insertion of ethene into the growing polymer chain (81 → 82) is followed by further ethene coordination (82 → 81′; 81′ = 81 with lengthened polymer chain). However, if β-H elimination (82 → 83) occurs instead, followed, after rotation of the olefin ligand, by reinsertion with M–C2 bond formation (83 → 84) and ethene coordination (84 → 81′), a methyl-branched linear polyethylene is formed. "Chain running," that is, successive β-H eliminations and reinsertions (84 → 85) that are completed by ethene insertion (85 → 81′), lead to higher branching.

81   82   83   84   85

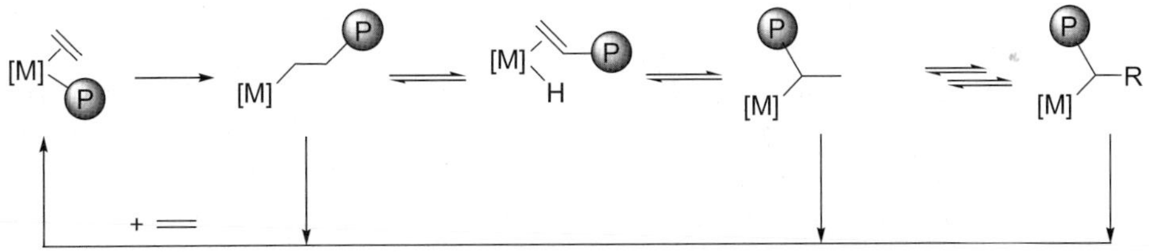

Quantum-chemical calculations, in particular, helped to gain insight into how the steric demands of bulky substituents can raise catalytic activity on the one hand and hinder chain termination on the other. An example is the cationic (diimine)nickel catalyst (**79**, M = Ni, R = *i*-Pr, R' = Me); see Figure 9.9. The aryl substituents are arranged perpendicular to the plane of the complex, so the isopropyl groups hinder the coordination of ligands to the axial, but not to the equatorial, positions. The catalyst complex is a cationic nickel complex [NiP(N⁀N)]⁺ (**86**, Figure 9.10) that has coordinated the growing polymer chain **P** and, with the addition of ethene, enters the resting state [NiP(η^2-H$_2$C=CH$_2$)(N⁀N)]⁺ (**86** → **87**). Ethene insertion leads to chain growth, yielding [Ni(CH$_2$CH$_2$P)(N⁀N)]⁺ (**87** → **88**). Chain termination occurs via β-H transfer from **P** onto the coordinated ethene, forming an ethyl–olefin complex [Ni(CH$_2$CH$_3$)(η^2-H$_2$C=CHP)(N⁀N)]⁺ (**87** → **89**). Chain branching via β-hydride elimination and M–C2 reinsertion is also to be considered (**86** → **90**). These reactions have been calculated (without considering a solvent) for two catalyst models, where the growing polymer chain **P** was modeled by a propyl group (Figure 9.10). In model **I**, the diimine ligand is fully unsubstituted, while model **II** contains the actual ligand.

- *Ethene coordination* (**86** → **87**). The starting complex **86** is stabilized by a β-agostic interaction. Ethene is coordinated to the axial position. Thus, it is plausible that in the unsubstituted model **I**, the interaction energy (**86** + C$_2$H$_4$) is greater and the complex **87** formed is more stable.
- *Chain growth (insertion)* (**87** → **88**). The activation energy for the actual catalyst model **II** is smaller than in model **I**. In the transition state, the coordinated monomer and the growing alkyl chain lie in the plane of the complex. The lower activation energy in the actual model **II** is thus due primarily not to steric interactions with the bulky aryl substituents, but rather to the lower stability of the starting complex **87** in the actual model **II**. The pentyl complex **88** that is formed is stabilized by a β-agostic interaction.

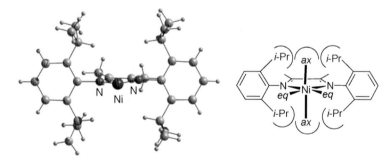

Figure 9.9 Structure of the complex fragment [Ni(ArN=CMe–CMe=NAr)]$^{2+}$ (Ar = 2,5-(*i*-Pr)$_2$C$_6$H$_3$; derived from the complex [Ni(CH$_2$SiMe$_3$)$_2$(ArN=CMe–CMe=NAr)]). This serves as a model for the coordination pocket of ethene and the growing polymer chain in a cationic (diimine)-nickel(II) complex.

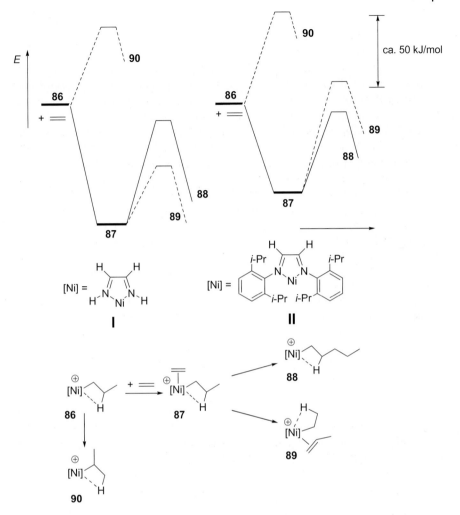

Figure 9.10 Energy profile for ethene coordination (86 → 87), insertion (87 → 88), chain termination (87 → 89) and branching (86 → 90) for the (diimine)nickel-catalyzed ethene polymerization with the generic and the real catalyst model (I and II, respectively). Ground states are indicated by bold dashes and transition states by thin dashes (abbreviated from Michalak and Ziegler [206]).

- *Chain termination* (87 → 89). The activation barrier in the actual catalyst model **II** is about twice as high as in the unsubstituted model complex **I**. The transition state resembles an ethene–hydrido–propyl complex with ethene in one axial position and the α-C atom of the propyl ligand in the other. Thus, the "steric pressure" of the isopropyl substituents in **II** is immediately understandable.

- Chain branching (**86** → **90**). The activation barrier for model **II** is somewhat greater than for **I**. The transition state resembles a hydrido–olefin complex with the hydrido ligand and the α-C atom of the olefin ligand in the plane of the complex. The somewhat larger activation barrier for model **II** can likely be traced back to hindered rotation of the olefin prior to the reinsertion.

Thus, in model **II**, which reflects the actual catalyst structure, the activation barriers are ordered as follows:

chain growth < chain branching < chain termination,

while in model **I** (without steric interactions), they are ordered:

chain termination < chain branching < chain growth.

Compared to the catalysts of the early transition metals, the late transition metals show a higher tolerance toward functional groups. In some cases, this enables the incorporation of polar comonomers and polymerization in polar protic solvents. This is associated with the lower bond polarity of the M–C bonds of late transition metals and their higher kinetic stability [207].

9.5.3
Living Polymerization of Olefins and Block Copolymers

In conventional coordination polymerization of olefins, many polymer chains are produced per catalyst center, as a result of chain termination and chain transfer reactions. In living polymerization, only one chain is formed per catalyst center [208].[8] Polymerization in a living manner allows new well-defined polymer architectures to be created, such as targeted synthesis of block copolymers and end-functionalized polymers. The disadvantage is that at every active metal center, only one polymer chain is formed. Living polymerization is obtained when (in the ideal case) no chain termination or chain transfer occurs. [V(acac)$_3$] (Hacac = acetylacetone) with AlEt$_2$Cl as cocatalyst (anisole can be added as activator) leads to living polymerization of propene at −78 °C. Partially syndiotactic high-molecular polypropylene is obtained. At higher temperatures, the living polymerization behavior declines to an increasing extent. Polymerization of α-olefins catalyzed by (diimine)-nickel complexes/MAO (Table 9.6, complex type **79**) can also be performed in a living manner.

Block copolymers can be prepared via living polymerization by sequential addition of monomers. One example is the synthesis of a block copolymer consisting of blocks of s-PP, EPR (ethene–propene rubber) and s-PP again (**91** → **92**). Termination of living polymerization with iodine leads to iodo-terminated polypropylene (**91** → **93**), to which, for example, tetrahydrofuran can be polymerized cationically (**93** → **94**).

---

[8] Although the formation of C–C bonds is catalyzed, this is no longer, strictly speaking, a catalyst, but rather an initiator. However, we will not make this distinction here.

We will revisit Ziegler's Aufbau reaction. Even at a temperature of 100 °C, at which it runs sufficiently rapidly, β-H transfer and elimination reactions are significant, so no living reaction occurs. However, catalysis of the Aufbau reaction (generalized here as polymerization catalysis by main-group metals M_{mg} including Zn) by a transition or rare earth metal (M_{tr}) now opens the way to living polymerization of ethene. This requires a fast reversible chain transfer according to reaction **a**. The chain growth occurs at the transition metal (reaction **b**). The fast chain transfer ($k_a > k_b$; typically $k_a \approx 100\, k_b$) guarantees that all chains grow. The reaction conditions must be chosen so that [M_{tr}]–P does not undergo termination by route **c**. This must also apply to [M_{mg}]–P, where such reactions are of lower importance in any case. Thus, the growing chains "rest" at the main group metal (or Zn). This is referred to as coordinative chain transfer polymerization.

Examples of catalyst systems are [Li(OEt$_2$)$_2$][SmCp*$_2$Cl$_2$]/Mg(Et)(n-Bu) and a [bis-(imino)pyridine]iron complex of type **80** activated with MAO in the presence of ZnEt$_2$ as chain transfer reagent [209, 210].

Novel olefin block copolymers can be produced via the exchange of polymer chains ("chain shuttling polymerization"). Two different catalysts (**c1** and **c2**) and two olefin monomers (e.g., ethene and oct-1-ene) are used. Each of the two catalysts, during chain growth, forms a copolymer with different microstructure (**95** → **96** and **97** → **98**): assuming that for **c1** the insertion rate of ethene is much greater than that of octene (symbol: —), while for **c2** the reverse is true (symbol: ∿∿). The chain growth is interrupted by a chain transfer (**96** → **97**). This process requires effective reversible chain transfer agents, which can be alkyls of Al, Zn, or Mg. Thus, a uniform block copolymer is created that consists of segments with different microstructure [211].

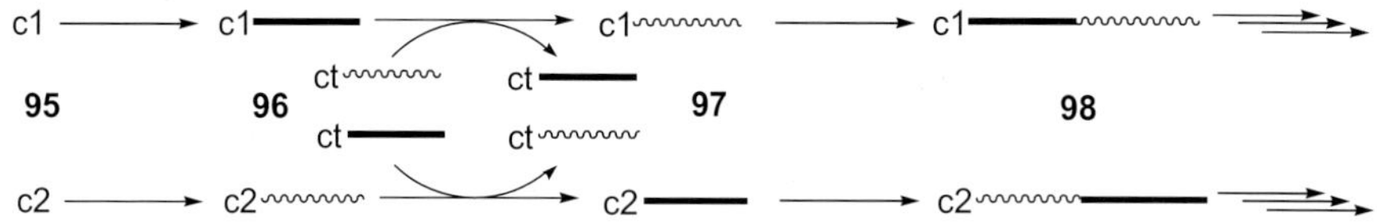

$$c1 \longrightarrow c1 \longrightarrow \underset{ct \longrightarrow}{\overset{ct \sim\sim\sim}{\rightleftarrows}} c1 \sim\sim\sim \longrightarrow c1 \longrightarrow\sim\sim\sim \rightleftarrows$$

$$\underset{95}{} \quad \underset{96}{} \quad \underset{97}{} \quad \underset{98}{}$$

$$c2 \longrightarrow c2\sim\sim\sim \rightleftarrows c2 \longrightarrow c2\sim\sim\sim \rightleftarrows$$

Exercise 9.10

Radical olefin polymerization with living character can be obtained in principle when started by a primary radical R• (e.g., R = alkyl, aryl) in the presence of another radical T• ("reversible spin trap," e.g., Ph$_3$C•) that reacts reversibly with the growing polymer chain ($1 \rightleftharpoons 2$):

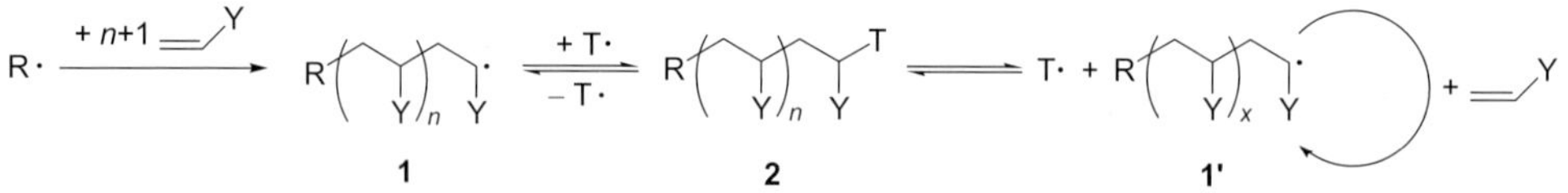

The reversibility of the binding of T• to the polymer chain ensures that the chain can continue to grow when monomer is present ($2 \rightarrow 1'$). If T• is a metal complex, this is called an organometallic radical polymerization (OMRP). Describe the structure of [Mo^{III}Cl$_2$CpL$_2$] (**3**, L$_2$ = (PMe$_3$)$_2$, dppe) and its reaction with alkyl halides. Discuss the reactions ($T = 80\text{–}100\,°C$) of **3** with equimolar quantitites of 1-bromoethylbenzene in the presence of styrene and when **3** is added to a radical polymerization of styrene initiated with AIBN. *Note*: Recall bimolecular oxidative addition reactions (see Section 3.2) and note the bond dissociation enthalpies Mo–X (in kJ/mol) in complexes [Mo^{IV}XCl$_2$CpL$_2$]: X = Me, ca. 100; Br, ca. 130; Cl ca. 180.

9.6
Copolymerization of Olefins and CO

9.6.1
Perfectly Alternating Copolymerization

The alternating copolymerization of ethene and carbon monoxide leads to a polyketone [poly(1-oxotrimethylene)]:

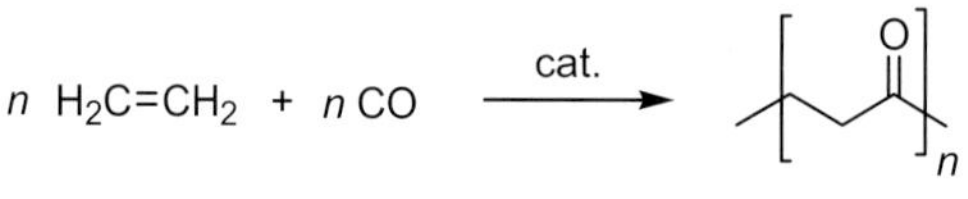

$$n\,H_2C=CH_2 + n\,CO \xrightarrow{cat.} \left[\!\!\begin{array}{c}\\ \end{array}\!\!\right]_n$$

W. Reppe and A. Magin (1948) were the first to conduct a transition-metal-catalyzed copolymerization, with K$_2$[Ni(CN)$_4$] as catalyst. Nowadays, mainly palladium complexes, which can achieve a strictly alternating copolymerization, are in use. Successive carbonyl groups –C(O)–C(O)– are not found at all, while only one tetramethylene unit (–CH$_2$–CH$_2$–CH$_2$–CH$_2$–) resulting from a double ethene insertion error is observed per approximately 10^5–10^6 regular structural elements [212].

Thermodynamically, the homopolymerization of ethene is favored over the alternating copolymerization of ethene and CO. When one considers that in the absence of CO, catalysts of alternating copolymerization generally catalyze the dimerization of ethene to butene or even its polymerization, the (perfectly) alternating CO–ethene copolymerization is a persuasive example of how selectively reactions can be controlled by catalysts.

As precatalysts, palladium(II) complexes [Pd(OAc)$_2$(L⌒L)] with bidentate P⌒P chelating ligands (such as Ph$_2$P(CH$_2$)$_n$PPh$_2$, $n=2$–4) or N⌒N chelating ligands (such as phenanthroline) are used. Reactions are carried out in the presence of Brønsted acids (HX) with non- or weakly coordinating anions X$^-$, such as OTs$^-$, BF$_4^-$, ClO$_4^-$. Sometimes an oxidant such as 1,4-benzoquinone is added to oxidize reduced palladium species to the active divalent form. Typically, the polymerization is carried out in methanol at 80–90 °C and 30–60 bar. The catalysts are highly active (10^4 mol ethene/(mol Pd·h)) and productive ($>10^6$ mol ethene/mol Pd). Thus, from low-cost starting materials (CO/ethene), polyketones representing a class of high-melting innovative thermoplastics can be obtained. Due to the reactive carbonyl groups, they offer many options for functionalization. Furthermore, they contain the highest possible concentration of carbonyl groups in a polymer chain, since carbon monoxide itself cannot be homopolymerized [213, 214].

The catalytically active compound is a square-planar cationic palladium(II) complex that has coordinated a bidentate chelating ligand L$_2$ and the growing polymer chain. This can be either an alkyl- (**99**) or an acylpalladium compound (**101**). In both cases, it has been proven by structural investigations of model complexes and quantum-chemical calculations that with the formation of a five- or six-membered ring, an additional C=O coordination can result.

Starting from **99**, the coordination of CO (**99** → **100**) is followed by an insertion step with formation of an acyl complex (**100** → **101**). Now, ethene is coordinated (**101** → **102**), and its insertion into the palladium–acyl bond completes the cycle (**102** → **99'**). The two insertion reactions proceed in the sense of an alkyl and an acyl migration, respectively.

9 Polymerization of Olefins

The copolymerization proceeds in a strictly alternating manner. A double CO insertion (102 → 103 → 104 instead of 102 → 99') does not occur for thermodynamic reasons. The reaction 103 → 104 is endergonic; this is a result of the very weak C(O)–C(O) bond ($\Delta_d H^\ominus$ in kJ/mol: 307 (MeC(O)–C(O)Me) < 352 (Me–C(O)Me) < 377 (Me–Me)) and the very stable C≡O bond in carbon monoxide (1077 kJ/mol) [215].

A double ethene insertion (99 → 105 → 106 instead of 99 → 100 → 101) is thermodynamically possible, but only occurs to a very minor extent. The (3-oxoalkyl)-palladium complex 99 is additionally stabilized by an intramolecular C=O coordination. The complex can react further with the stronger coordinating CO (99 → 100), but not with the weaker coordinating ethene (99 → 105), so the equilibrium concentration of 105 is very low (25 °C: K ca. 10^4 for 105 + CO ⇌ 100 + C_2H_4; L_2 = dppp). In addition, the insertion reaction that leads to double ethene insertion (ethene into Pd–alkyl: 105 → 106), is slower by about two orders of magnitude than the CO insertion (CO into Pd–alkyl: 100 → 101) that leads to alternating copolymerization [216].

Exercise 9.11

Assume that the Curtin-Hammett principle applies to the reaction scheme above and calculate the ratio of double ethene insertion to alternating copolymerization. Use the data given in the text and the following concentrations in solution for a 1 : 1 CO/ethene mixture: $c_{CO} = 7.3 \cdot 10^{-3}$ mol/l and $c_{C2H4} = 0.11$ mol/l.

9.6 Copolymerization of Olefins and CO

Reactions with the solvent methanol lead to chain termination. Polymers with carbonyl end groups and methoxopalladium complexes (**99′** → **107**) or polymers with ester end groups and hydridopalladium complexes (**101′** → **109**) are thus formed. The reaction **99′** → **107** can be regarded as a protolytic cleavage of an M–C bond. The reaction **101′** → **109** may proceed as a nucleophilic substitution of a Pd⁰ complex fragment [PdL$_2$(MeOH)] by MeOH/MeO⁻ and the subsequent protonation of the Pd⁰ complex.

Neither termination reaction deactivates the catalyst. A new polymer chain can be initiated by insertion of CO (**107** → **108**) or ethene (**109** → **110**). It is clear that a termination with formation of a carbonyl end group enables a reinitiation in which the growing polymer chain displays an ester end group (**99′** → **107** → **108**) and vice versa (**101′** → **109** → **110**).

As precatalysts, compounds [Pd(OAc)$_2$L$_2$] (**111**) are used, which in the reaction with Brønsted acids HX (X = non- or weakly coordinating anion such as OTs⁻, BF$_4$⁻, ClO$_4$⁻) form compounds [PdL$_2$(MeOH)$_2$]X$_2$ (**112**). Deprotonation of the coordinated methanol, which occurs easily due to the high electrophilicity of the doubly positive charged Pd^{II} center, leads to a methoxo complex (**112** → **107**), into which CO inserts, forming a methoxycarbonyl complex (**107** → **108**). (Alternatively, a CO complex can be formed, to which methanol adds, forming the methoxycarbonyl complex **108**.) Insertion of ethene into the Pd–C bond leads to the catalyst complex, which forms a polyketone with at least one ester end group.

In the alternating copolymerization of CO and α-olefins, different stereoisomers can be formed, as with the homopolymerization of α-olefins. The structures of an isotactic (**113**) and a syndiotactic (**114**) polymer are shown below. The stereoselectivity (isotactic vs. syndiotactic vs. atactic) can be controlled via the nature of the catalyst.

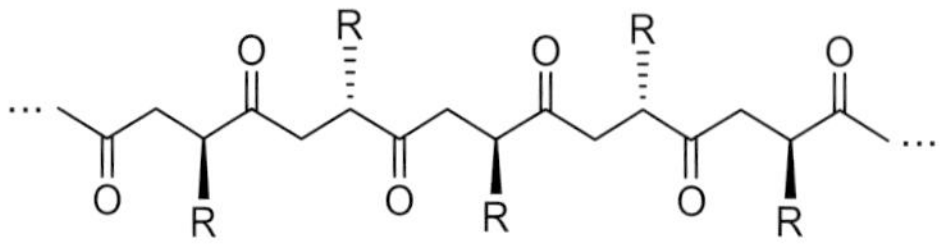

113 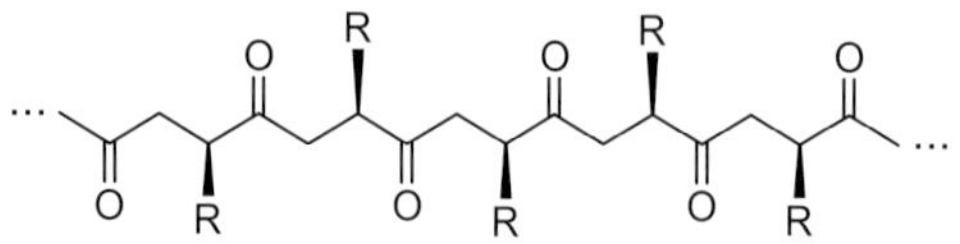   **114**

Exercise 9.12

In the isotactic polymer **113**, the substituents R = alkyl are shown alternately in front and behind. In the syndiotactic polymer **114**, they are all shown in the same direction. For iso- and syndiotactic polypropylene, precisely the opposite applies. Why are the designations nevertheless correct?

9.6.2
Imperfectly Alternating Copolymerization

The copolymerization of ethene with CO, catalyzed by **115** and complexes derived from it, does not proceed in strict alternation. There is an irregular incorporation of between approximately 10 and 30% of the ethene (with **115** itself, up to 15%), so in the polymers **116**, sequences with $x = 2\text{--}4$ can be found in addition to the regular pattern ($x = 1$). Double CO insertion does not occur [217].

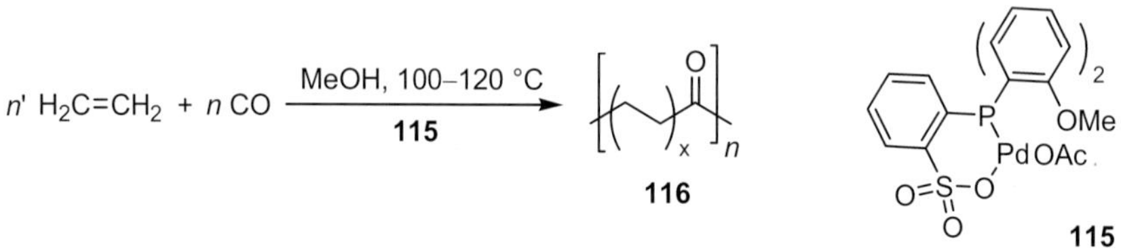

$$n'\ H_2C=CH_2 + n\ CO \xrightarrow[\textbf{115}]{\text{MeOH, 100--120 °C}}$$

116

115

Starting from the alkyl complex **117** ([Pd]–OAc ≡ **115**), pathway **a** is the reaction channel for the alternating copolymerization and pathway **b** for the nonalternating. Quantum-chemical calculations explain why reaction path **b** may be followed [218]: The acyl complex **119** exists in the open-chain form and is not stabilized by a Pd–O bond (**119** versus **119'**). There are two reasons for this: First, a Pd–O coordination is sterically hindered by the *ortho* substituent OMe in **115**, and second, the catalyst complex **119** is a neutral complex, so the electrophilicity of the palladium is in any case lower than in conventional cationic complexes. The missing stabilization from the Pd–O coordination in **119** has the result that the deinsertion of CO proceeds relatively easily (**119** ⇌ **118**), so the reaction channel **b** to double ethene insertion is opened. On the other hand, the ethene addition also proceeds more easily (**119** → **120**), since no Pd–O bond must be broken. The aspect first mentioned predominates, so in this case, the irregular, nonalternating copolymerization competes with the alternating [219].

10
C–C Linkage of Dienes

10.1
Introduction

Buta-1,3-diene **1** is the simplest conjugated diolefin. The terpenes are derived from the methyl-substituted compound 2-methylbuta-1,3-diene (**2**, isoprene). They have the chemical formula $(C_5H_8)_n$ and are to be regarded formally as oligo- or polymerization products of isoprene (Figure 10.1). Terpenes are widespread in nature. They are contained in essential oils and carotenoids and frequently have pronounced biological activity. *cis*-1,4-polyisoprene (natural rubber) is of special significance due to its rubber-elastic properties.

So it is only logical that the catalysis of oligomerization and polymerization reactions of butadiene and isoprene assume a special significance. A significant impetus was the goal of producing synthetic rubber. A milestone, the first industrial synthesis of polybutadiene, occurred in 1937–38, at Buna Factory Schkopau (Saxony-Anhalt, Germany) via anionic polymerization with sodium as initiator (Buna = *bu*tadiene–*na*trium; German: Natrium = sodium). The content of *cis*-1,4 units only reached about 10%, however. A polybutadiene (*cis*-1,4-polybutadiene) structurally analogous to natural rubber (*cis*-1,4-polyisoprene) was first produced with Ziegler-Natta catalysts. With modern catalysts, especially those based on rare earth metals (Nd), a content of >99% of *cis*-1,4 units is achieved. However, organometallic mixed catalysts can catalyze more than just the polymerization of butadiene (G. Natta, 1955–59). In particular, nickel catalysts can also produce cyclooligomers, linear oligomers, and telomers from butadiene (G. Wilke, 1955).

In these catalytic reactions of the 1,3-dienes, allyl complexes appear as intermediates. Allyl ligands[1] can coordinate to metals in various ways giving rise to mechanistic peculiarities. Thus, it is appropriate to precede the description of selected catalytic transformations of dienes with relevant organometallic elementary steps. They correspond, in principle, to those encountered in catalytic reactions of mono-

---

1) Strictly speaking, "allyl" refers to the group $H_2C=CH–CH_2–$. In the broader sense, it also includes substituted compounds such as $RHC=CH–CH_2–$ and $H_2C=CR–CH_2–$. When R = Me, these residues are also called crotyl and methallyl, respectively.

Fundamentals of Organometallic Catalysis. First Edition. Dirk Steinborn
Copyright © 2012 WILEY-VCH Verlag GmbH & Co. KGaA, Weinheim

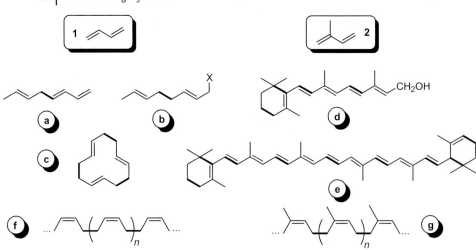

Figure 10.1 Oligomers and polymers derived from butadiene **1** or isoprene **2**. Examples of a butadiene oligomer (**a**), a butadiene telomer (X = OR, NR$_2$, ...) (**b**), a butadiene cyclooligomer (**c**), a terpene alcohol (vitamin A$_1$, **d**), and a carotenoid (α-carotene, **e**), as well as synthetic (**f**) and natural rubber (**g**). The linkage sites for the monomer units are indicated by thick dashes.

olefins (see Section 3), but accommodate the peculiarities of the chemistry of allyl complexes.

10.2
Allyl and Butadiene Complexes

10.2.1
Allyl Complexes

The allyl ligand can be η^1- (σ-) (**3**) or η^3- (π-) bound (**4**) to the metal center, as the structures of the two complexes demonstrate in Figure 10.2. It often makes sense to visualize the reactions of η^3-allyl complexes with the help of the two resonance structures (**4a/b**).

The ease with which the two forms, whose coordination numbers differ by one, interconvert via $\eta^1 - \eta^3 / \eta^3 - \eta^1$ (σ– π/π– σ) rearrangements, depends on the coordination patterns. If equilibrium between the two forms is reached rapidly enough, fluxional molecules result.

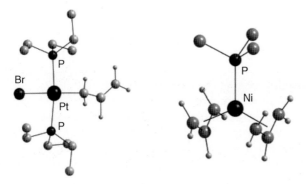

Figure 10.2 Structures of *trans*-[Pt(η^1-C$_3$H$_5$)Br(PEt$_3$)$_2$] (left) and [Ni(η^3-C$_3$H$_5$)$_2$(PMe$_3$)] (right) (H atoms are only shown on the allyl ligands).

- η^1-η^3 (σ–π) *rearrangement.* In η^1-allyl complexes, where the metal is bound to the carbon atom C1 or C3, conversion occurs via an intermediate η^3-allyl complex.

η^1 (M–C1)　　　　　η^3　　　　　η^1 (M–C3)

- *syn/anti isomerization.* In η^3-allyl complexes, the *syn* and *anti* hydrogen atoms[2] change places via an intermediate η^1-allyl complex in which rotation around the C–C and M–C bond is unhindered.

η^3　　　　　η^1　　　　　η^3
(H^1, H^3: *syn*; H$^{1'}$, H$^{3'}$: *anti*)　　　　　　　　　　(H$^{1'}$, H^3: *syn*; H^1, H$^{3'}$: *anti*)

2) In η^3-allyl complexes, the *syn/anti* notation refers to the position of the methylene hydrogen atoms relative to the methine hydrogen atom:

In substituted η^3-allyl complexes, *syn* and *anti* denote the position of a substituent R relative to the methine hydrogen atom.

The activation barrier for the *anti/syn* isomerization is strongly dependent on the structure of complexes. In [Ni(η^3-C$_3$H$_5$)$_2$(PMe$_3$)] (see Figure 10.2), ^{1}H NMR spectroscopic experiments have found the Gibbs free energy of activation to be 40 kJ/mol.

Background: Fluxional Molecules

Molecules are called fluxional when a rearrangement occurs between two (or more) nuclear configurations that are fully equivalent chemically. Every configuration displays a minimum on the energy hypersurface. The energy minima are identical. (By contrast, if the configurations are chemically distinguishable, the rearrangement is considered isomerization.) The definition includes structural rearrangements that occur by breaking and re-forming bonds, but also interconversion of stereochemically nonrigid molecules by such mechanisms as Berry pseudorotation (for example: PF$_5$) [25].

The interchange of identical substituents or ligands in chemically and/or magnetically distinguishable environments is called topomerization, and the indistinguishable species are called topomers. One η^3-allyl ligand permits four topomers (Exercise 10.1).

Example. A fluxional η^3-allyl ligand is easy to identify by ^{1}H NMR spectroscopy [220], since it has four chemically equivalent H atoms (2 H$_a$ ≡ 2 H$_s$) that couple to the central H atom (H$_c$) (spectrum **a**). By contrast, the ^{1}H NMR spectrum of a non-fluxional η^3-allyl complex shows (**b**) three types of chemically non-equivalent protons (2 H$_a$, 2 H$_s$, 1 H$_c$), while an η^1-bound allyl ligand (**c**) has as many as four types of chemically non-equivalent protons. Idealized stick spectra representations (without geminal H–H couplings) for all three types of allyl ligands are shown below (with $^3J_{Hc,Ha} \approx 2\,^3J_{Hc,Hs}$ for **b** and with $^3J_{H1,H3} \approx 2\,^3J_{H1,H2} \approx 2\,^3J_{H1,H4}$ for **c**). The relative intensities are given in parentheses.

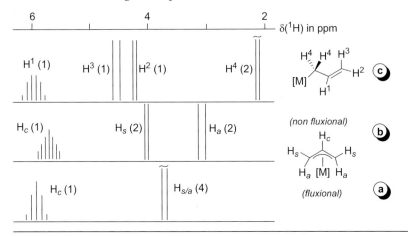

Like unsubstituted allyl complexes, substituted allyl complexes can also undergo *anti/syn* isomerization. In this case, however, these are not fluxional molecules, but rather isomeric:

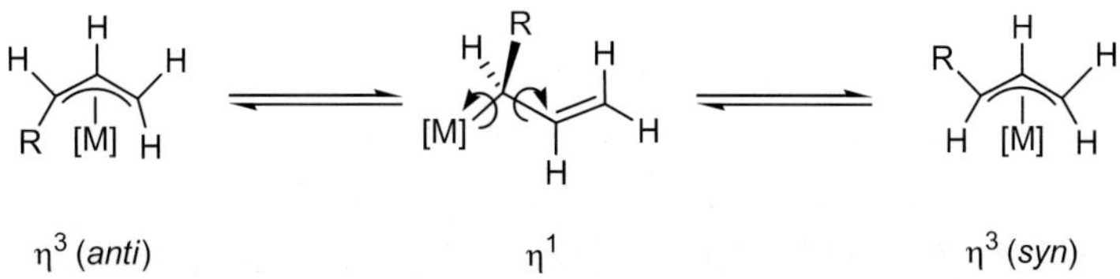

η^3 (anti) ⇌ η^1 ⇌ η^3 (syn)

Exercise 10.1

- Does an *anti/syn* isomerization in the preceding example take place when the η^1-complex formation occurs via the other terminal C atom, hence via the complex [M]–CH$_2$–CH=CHR?
- Give the formulas of all topomers of an η^3-allyl ligand.

10.2.2
Butadiene Complexes

There are two stable conformations of butadiene, the *s-trans* and the *s-cis* conformers:[3]

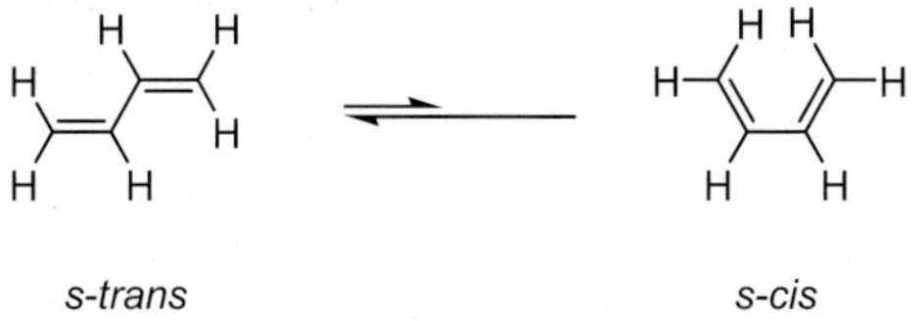

s-trans ⇌ s-cis

The *s-trans* isomer is planar and thermodynamically more stable by about 15 kJ/mol than the *s-cis* isomer (Figure 10.3). Repulsion between the hydrogen atoms on the terminal C atoms causes a minimum structure on the energy hypersurface to be displayed not by the planar *s-cis* conformer, but rather by a form twisted by about 35°. It is designated a nonplanar *s-cis* conformation. However, the distinction is irrelevant to the following, so for simplicity, we only speak of the "*s-cis* conformation."

The butadiene activation in catalytic reactions occurs via formation of π complexes. Both conformers form metal complexes in which one or both double bonds are coordinated (η^2 or η^4 coordination). Thus, four basic types of butadiene complexes result:

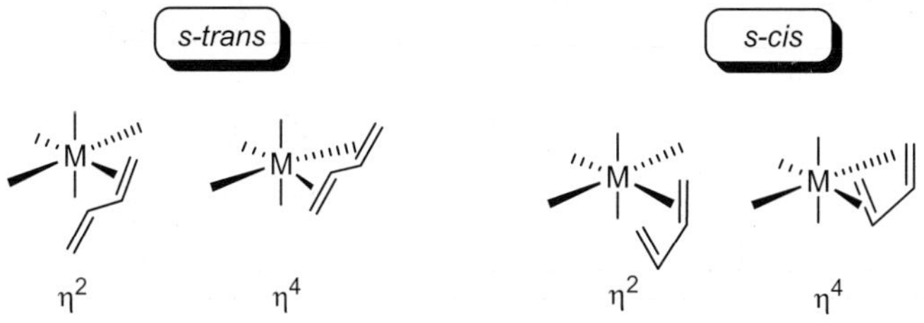

s-trans: η^2, η^4 s-cis: η^2, η^4

---

3) To clarify that the *cis/trans* notation here refers to the central C2–C3 bond – and not to a C=C double bond, as would be customary – an "*s*" (for *single*) is added as a prefix.

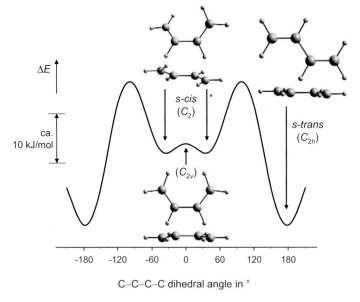

Figure 10.3 Potential curve of butadiene for rotation around the C2–C3 bond and structural models of the two equilibrium structures (the arrow with the star points to the *s-cis* isomer shown) and of the planar *s-cis* transition state.

Examples of structurally characterized butadiene complexes are given in Figure 10.4. As a result, the coordination chemistry of butadiene is versatile. Thus, the (*s-trans*-η^4-butadiene)zirconocene complex **5** is in equilibrium with the *s-cis* complex **6**, which exists as a zirconacyclopentene complex with an additional π-C=C–Zr interaction. **6** undergoes a rapid ring inversion (flipping) (**6a** ⇌ **6b**), and is thus a fluxional molecule [221, 222].

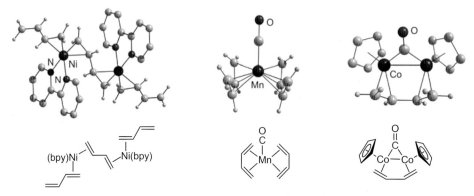

Figure 10.4 Molecular structures of complexes with η^2- and η^4-bound *s-cis*- and *s-trans*-butadiene ligands. To make the structural images easier to understand, only the H atoms for the butadiene ligands are shown.

[structures 5, 6a, 6b with ΔG‡ = 95 kJ/mol and ΔG‡ = 53 kJ/mol]

Such σ-π coordination of butadiene is especially common with diene complexes of early transition metals.

10.2.3
Re/Si and supine/prone Coordination of Allyl and Butadiene Ligands

Substituted η^3-allyl ligands as in **7** (where R could be alkyl) are prochiral, since the addition of a nucleophile can form a chiral C3 atom. Thus, the coordination of the metal to the *Re* face (**7a**) and the *Si* face (**7b**) must be distinguished from each other. The same is true for η^2-coordinated *s-cis-* and *s-trans-*butadiene ligands, which from a coordination chemistry point of view can be regarded as vinyl-substituted ethene ligands. 1,2-addition produces a chiral C2 atom whose configuration depends on whether butadiene coordinates to the *Re* (**8a**) or *Si* face (**8b**). *s-trans-*η^4-Butadiene likewise admits two coordination possibilities (**9a/9b**), while *s-cis-*η^4-butadiene displays *Re/Si* coordination (**10a**).

[structures 7a (Re), 7b (Si), 8a (Re), 8b (Si), 9a (Re/Re), 9b (Si/Si), 10a (Re/Si)]

a) *Re/Si* arrangement for R = primary alkyl.

Furthermore, for η^3-allyl and *s-cis-*η^4-butadiene complexes, a distinction with respect to a reference ligand L can be made between a *"supine"* (**11a/12a**) and *"prone"* orientation (**11b/12b**),[4] according to whether the π-bound ligand lies on its "back" or "stomach" if the reference ligand L is oriented upwards [223].

[structures 11a (supine), 11b (prone), 12a (supine), 12b (prone)]

4) The conventional terms *exo (supine)* and *endo (prone)* for the position of L relative to the π ligand are only suitable to a limited extent.

10.3
Organometallic Elementary Steps of Allyl Ligands

10.3.1
Oxidative Coupling and Reductive Cleavage

Bis(butadiene)metal complexes **13** can react, via oxidative coupling (oxidative addition with C–C bond linkage), to form C_8H_{12} complexes with an $\eta^1-\eta^3$- (**14**) or $\eta^3-\eta^3$-allyl structure (**15**). Starting from a bis(η^2-butadiene) complex, the linkage occurs so as to form a C–C bond between the terminal C atoms of the uncoordinated double bonds (C4/C4′). The reverse reaction is called reductive cleavage (of a C–C bond) or reductive decoupling.

13   **14** ($\eta^1-\eta^3$)   **15** ($\eta^3-\eta^3$)

Quantum-chemical calculations offer a more accurate look at the oxidative coupling of bis(butadiene)nickel(0) complexes with PH_3 as model ligand. The starting complexes are the bis(η^2)-complexes [Ni(η^2-C_4H_6)$_2$(PH_3)] (**16**) and not the corresponding η^2,η^4- or bis(η^4)-complexes. Due to the *s-cis/s-trans* isomerism and the possibility that both (prochiral!) butadiene ligands can be coordinated to the same (*Re/Re* or *Si/Si*) or different faces (*Re/Si*), six different starting complexes **16** result. Each of these complexes now leads to another η^1,η^3-octadienediyl complex [Ni(η^3,η^1-C_8H_{12})(PH_3)] (**17**). Figure 10.5 shows the reaction profile for the most energetically favorable path (**16b → 17b**), which starts from the η^2-(s-

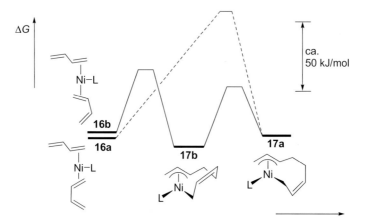

Figure 10.5 Reaction profile for the oxidative coupling of butadiene starting from [Ni(η^2-C_4H_6)$_2$L] (**16**, L = PH_3) (shortened from Tobisch and Ziegler [224]).

cis),η^2-(s-trans) complex in which the two butadiene ligands are coordinated to different faces (Re/Si). For comparison, another reaction path is shown. It starts from the same complex, but it displays Re/Re butadiene coordination (**16a** → **17a**), which presents an activation barrier that is almost twice as high. This is an impressive demonstration that stereoelectronic effects can govern the course of such reactions. The two complexes **17a** and **17b** interconvert via a ring inversion. Coupling of two η^2-s-trans- or two η^2-s-cis-bound butadiene ligands has a higher activation barrier than reaction **16b** → **17b**.

10.3.2
Butadiene Insertion and β-Hydrogen Elimination

For the insertion of butadiene into an M–H or M–C bond, a 1,2- or 1,4-addition to η^1-allyl (**18a, 19**) or but-3-enyl complexes (**18b**) and the formation of η^3-allyl complexes with a *syn* (**20**) or *anti* structure (**21**) must be considered. For coordinatively unsaturated transition metal compounds, η^3-allyl complexes are preferentially formed due to their particular stability. In 1,2-additions, butadiene behaves like a terminal olefin H$_2$C=CHR' (R' = vinyl) and thus is prochiral.

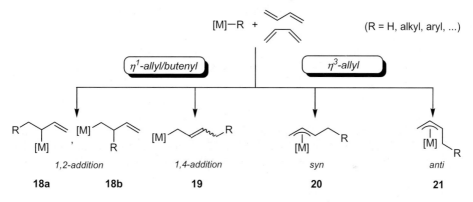

For R = H, the reverse reactions, which are β-hydrogen eliminations, also occur. They are encountered more frequently with allylpalladium compounds than with allylnickel complexes.

10.3.3
Allyl Insertion

The insertion of butadiene into the M–C bond of an allyl complex is called allyl insertion. Two principal mechanisms have been substantiated:

- σ-*Allyl insertion mechanism* (**22** → **23**). Starting from a complex with an η^1-bound allyl ligand and an η^2- or η^4-coordinated butadiene ligand (**22**), the insertion of butadiene into the M–C bond of the allyl ligand occurs with C1–C1' linkage and generation of an η^3-allyl structure using the C2'–C4' atoms of the butadiene.

- π-*Allyl insertion mechanism* (**24** → **23**). The mechanism is analogous, but starts from a complex with an η^3-bound allyl ligand (**24**) and an η^2- or η^4-coordinated butadiene ligand.

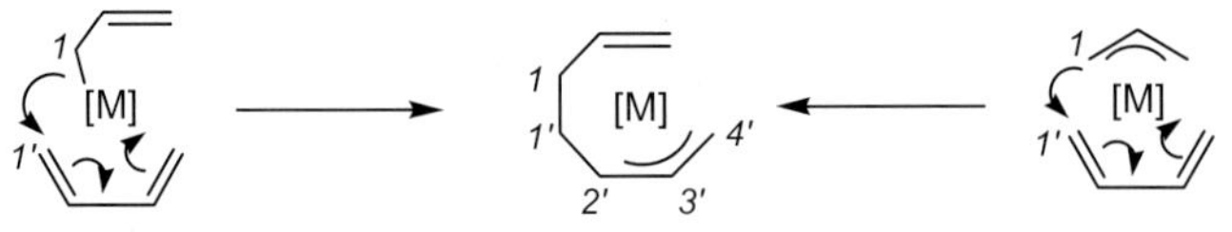

22   σ-allyl insertion   **23**   π-allyl insertion   **24**

Note: Allyl insertion reactions may be easier to comprehend if they are (*formally!*) separated into partial steps that correspond to those for simple olefins. These are the "transformation" of π- into σ-allyl complexes (**24′** → **22′**) and vice versa (**25′** → **23′**), as well as the insertion of butadiene in the sense of a 1,2-addition of the diene (compare with the previous reaction) into the σ-M-C bond so as to form an M-C2′ bond (**22′** → **25′**).

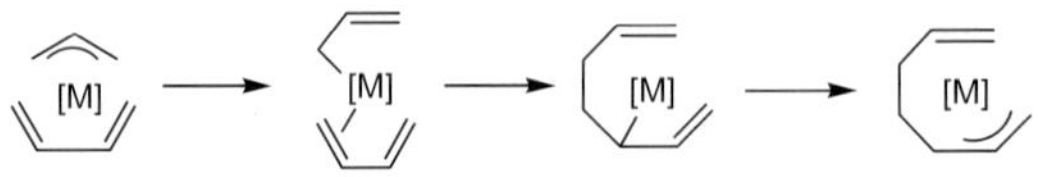

24′   **22′**   **25′**   **23′**

However, one must remember that this does not describe the actual reaction course of either a σ- (**22′** → **25′** → **23′**) or π-allyl insertion (**24′** → **22′** → **25′** → **23′**)!

10.3.4
Oxidative Addition and Reductive Elimination

Oxidative additions of allyl compounds $XCH_2\text{-}CH=CHR$ ($X = Cl, Br, I, OAc, CN, \ldots$; $R = $ alkyl, aryl, $\ldots$) to low-valent metal complexes and the corresponding reductive eliminations as reverse reactions have the peculiarity that η^1- and η^3-allylmetal complexes can be involved:

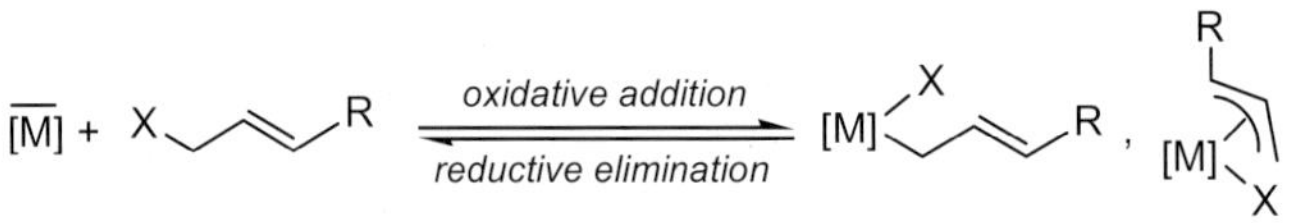

If bis(allyl) complexes undergo reductive elimination, hexa-1,5-diene (diallyl) is formed, which can coordinate to the low-valent metal fragment:

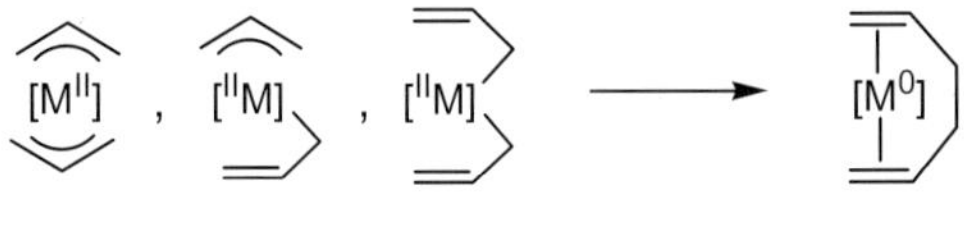

The starting point can be either π- or σ-bound allyl groups. Thus, bis(η³-allyl)-palladium **26** is converted by diphosphanes into the bis(η¹-allyl) complex **27**, which, above −30 °C, undergoes reductive elimination, with formation of **28**.

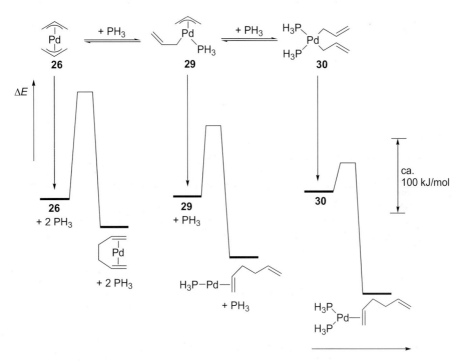

In accord with this result, quantum-chemical calculations show that in the equilibrium of [Pd(η³-C₃H₅)₂] (**26**), [Pd(η³-C₃H₅)(η¹-C₃H₅)(PH₃)] (**29**) and [Pd(η¹-C₃H₅)₂(PH₃)₂] (**30**), complex **26** is indeed the main component (Figure 10.6), but the activation barrier for the reductive elimination of hexadiene decreases in the order **26** > **29** ≫ **30**.

Figure 10.6 PH₃-induced reductive elimination of diallyl from bis(allyl)palladium(II) complexes. In the presence of PH₃, the complexes **26**, **29** and **30** are in dynamic equilibrium. Since the PH₃ coordination to **26** (**26** → **29**) and to **29** (**29** → **30**) is only weakly endothermic (ΔE = 2 or 10 kJ/mol), in the presence of PH₃ the reductive elimination of diallyl from the bis(η¹-allyl)palladium complex **30** is favored, although at equilibrium **30** exists only at low concentration (shortened from Méndez and Echavarren [225]).

10.3.5
anti/cis and syn/trans Correlations

The stereochemistry of insertion and reductive elimination reactions with participation of allyl ligands is determined by the following correlations:

- (a) η^3-Allyl ligands with *anti* structure yield, after reductive elimination, a *cis* olefin, and those with *syn* structure yield a *trans* olefin.

anti ⟶ cis   syn ⟶ trans

- (b) The insertion of *s-cis*-butadiene yields an *anti*-allyl structure, and the insertion of *s-trans*-butadiene yields a complex with *syn*-allyl structure. Note that according to (a), a *trans* double bond is formed from the initial *syn*-allyl group.

s-cis ⟶ anti   s-trans ⟶ syn

The theoretical background for the *anti/cis* and *syn/trans* correlations is the principle of least structural variation: In the *anti*- and *syn*-allyl structure, the *cis*- or *trans*-olefin structure, respectively, is already "preformed," just as *s-cis*- and *s-trans*-butadiene are in a direct structural relationship with an *anti*- or *syn*-allyl structure, respectively.

10.4
Oligomerization and Telomerization of Butadiene

Nickel(0) complexes have shown themselves to be particularly well-suited to the catalysis of oligomerization reactions of butadiene. Figure 10.7 gives an overview [226, 227]. Cyclo-cooligomerization of butadiene with olefins and alkynes is not discussed in this context.

10.4.1
Cyclotrimerization of Butadiene

10.4.1.1 Mechanism
In the nickel-complex-catalyzed cyclotrimerization of butadiene, butadienenickel(0) complexes **31/32** function as actual catalysts. They are obtained by reduction of divalent nickel compounds (preferentially [Ni(acac)$_2$]) with AlEt$_2$(OEt) in the

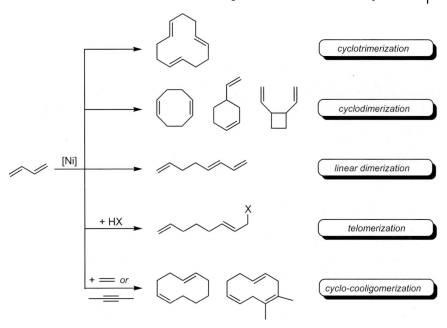

Figure 10.7 Overview of nickel-catalyzed oligomerization reactions of butadiene.

presence of butadiene (**a**), by reductive elimination of diallyl from [Ni(η^3-C$_3$H$_5$)$_2$] in the presence of butadiene (**b**) or from nickel(0) complexes such as [Ni(COD)$_2$] (COD = cycloocta-1,5-diene) or [Ni(CDT)] (CDT = cyclododeca-1,5,9-triene) by ligand displacement with butadiene (**c**). Nickel in nickel(0) complexes with ligands that are readily displaceable (especially by butadiene) is also called "naked" nickel (e.g., nickel in [Ni(COD)$_2$]).

In butadiene, there is an equilibrium between different butadiene complexes [Ni(C$_4$H$_6$)$_x$] (x = 2, 3); the differences in energy are low. Calculations show that for bidentate coordination, the η^4-s-cis form is favored, and for monodentate coordination, the η^2-s-trans form is favored. For the tetrahedral 18-ve complexes (x = 2) the one with two η^4-s-cis-butadiene ligands (**31**) is the stablest, and for the trigonal-planar 16-ve complexes (x = 3), the stablest is the one with three η^2-s-trans-butadiene ligands (**32**) [228].

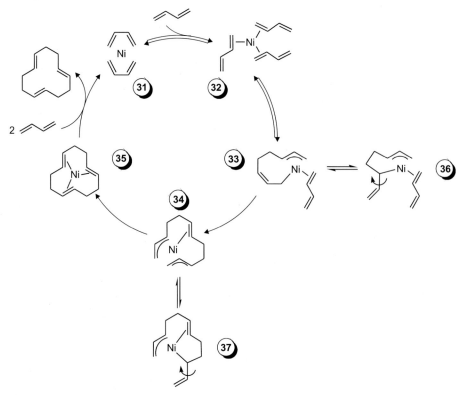

Figure 10.8 Principal reaction scheme for the nickel-catalyzed cyclotrimerization of butadiene (without explicit consideration of the stereochemistry).

In the cyclotrimerization of butadiene, three of the four isomeric cyclododeca-1,5,9-trienes are obtained:

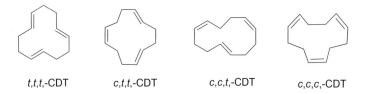

t,t,t,-CDT   c,t,t,-CDT   c,c,t,-CDT   c,c,c,-CDT

The main product – with a selectivity >85% ($T = 0$–40 °C in liquid butadiene) – is the *all-trans* isomer (*t,t,t*-CDT), with low quantities of *c,t,t*-CDT and *c,c,t*-CDT. Formation of the *all-cis* isomer (*c,c,c*-CDT) is not observed. Cyclodimers and higher oligomers appear as side products. The principal mechanism of cyclotrimerization is shown in Figure 10.8. The following individual reaction steps can be named:

31 → 32: *Ligand attachment/cleavage*. The catalyst complex exists in different forms, with an equilibrium rapidly established.

32 → 33: *Oxidative coupling/reductive cleavage.* In the oxidative addition with C–C bond formation, the terminal C atoms of the uncoordinated double bonds are linked, forming an η^3,η^1-octadienediylnickel(II) complex (16 ve) with an additional butadiene ligand. The reaction is reversible.

33 → 34: *Allyl insertion.* Insertion of butadiene into the η^3-allyl group of **33** leads to an η^3,η^3-dodecatrienediylnickel(II) complex, which is an 18-ve complex if the internal double bond is also coordinated. The bis(η^3-*anti*) isomer has been isolated and fully characterized via NMR spectroscopy.

34 → 35: *Reductive elimination.* Reductive elimination with formation of a C–C bond between the two terminal C atoms yields a (cyclododecatriene)nickel(0) complex (16 ve).

35 → 31: *Ligand substitution.* Via substitution of the CDT ligand by butadiene, the catalyst complex **31** is re-formed.

33 ⇌ 36; 34 ⇌ 37: *Allyl isomerization.* The allyl intermediates **33** and **34** undergo *syn/anti* isomerization, which occurs via complexes with $\eta^1(C^3)$-bound allyl ligands **36** and **37**, respectively.

10.4.1.2 *cis/trans* Selectivity – A Closer Look

For understanding the *cis/trans* selectivity of the double bonds in CDT, it becomes necessary to take a closer look that considers *anti/cis* and *syn/trans* correlations (see Section 10.3.5):

33a → 34b → 35b: Complex **33a** (labeled as **33** in Figure 10.8) is the η^3-*syn*,$\eta^1(C^1)$, Δ-*cis* isomer[5] with an *s-trans*-butadiene ligand. The *s-trans*-butadiene is inserted into the η^3-*syn*-allyl group, forming a new η^3-*syn*-allyl group and a Δ-*trans* double bond.

[Scheme showing equilibria between **33a** and **33b**, leading down to **34a** bis(η^3-*syn*), **34b** η^3-*anti*/η^3-*syn*, and **34c** bis(η^3-*anti*), with anti-syn isomerizations, and further down to **35a** [Ni(*t,t,t*,-CDT)], **35b** [Ni(*c,t,t*,-CDT)], and **35c** [Ni(*c,c,t*,-CDT)].]

---

5) The terms Δ-*cis* and Δ-*trans* refer to the configuration of the double bond.

This is accompanied by conversion of the $\eta^1(C^1),\Delta$-cis-allyl group into an η^3-anti-allyl group, resulting in a dodecatrienediylnickel complex (η^3-anti/η^3-syn,Δ-trans isomer) (**34b**). In the reductive elimination, a *cis* or *trans* double bond is formed from the *anti*- or *syn*-allyl structure, respectively, generating the [Ni(*c,t,t*-CDT)] complex **35b**.

33b → **34c** → **35c**: The octadienediyl complex **33b**, which contains an *s-cis*-butadiene ligand, undergoes butadiene insertion into the η^3-*syn*-allyl group to yield the bis(η^3-*anti*) complex **34c**, which, after reductive elimination, forms [Ni(*c,c,t*-CDT)] (**35c**).

34a → **35a**: The main isomer (*t,t,t*-CDT) is formed via **34a** and **35a** (labeled 34 and 35, respectively, in Figure 10.8). However, complexes **33** offer no direct access to the bis(η^3-*syn*) complex **34a**, which consequently must be formed via *anti/syn* isomerization from the η^3-*anti*/η^3-*syn* complex **34b**.

Quantum-chemical calculations on the DFT level provide a detailed look at the reaction course in which all plausible isomers are thoroughly considered. The simplified reaction profile is shown in Figure 10.9.

Oxidative coupling (32 → 33). The more thermodynamically stable tris(η^2-*s*-*trans*-butadiene) complex **32a** is first converted into a bis(η^2-*s*-*trans*-butadiene)–(η^2-*s*-*cis*-butadiene) complex **32b**, since the oxidative coupling (**32** → **33**) between an η^2-*s*-*trans*- and an η^2-*s*-*cis*-butadiene ligand has a significantly lower activation barrier than that between two η^2-*s*-*trans*- or two η^2-*s*-*cis*-butadiene ligands. The oxidative coupling (**32b** → **33a**) is approximately thermoneutral, hence reversible.

Butadiene insertion (33 → 34). This is strongly exergonic and favors the formation of **34b**, but also **34c**. All three isomeric dodecatrienediyl complexes **34** are of approximately equal thermodynamic stability. The activation barriers for the *anti/syn* isomerization (**34a** ⇌ **34b** ⇌ **34c**) lie between 55 and 85 kJ/mol. However, they are more than 25 kJ/mol lower than for the subsequent reductive eliminations. Thus, the *syn*- and *anti*-allyl complexes exist in dynamic equilibrium.

Reductive elimination (34 → 35). The dodecatrienediyl complexes **34** represent the "thermodynamic sink," considering their heavily exergonic formation from **33** and the high activation barrier for the reductive elimination to form **35**. Thus, the reductive elimination **34** → **35** is rate-determining and there exist preestablished dynamic equilibria between the dodecatrienediyl complexes **34**. It follows that the Curtin-Hammett principle (see Background, Section 4.3.3) is critical here. The transition state **34a** → **TS** → **35a** displays the lowest Gibbs free energy. As a result, the formation of [Ni(*t,t,t*-CDT)] (**35a**) predominates over that of [Ni(*c,c,t*-CDT)] (**35c**) and [Ni(*c,t,t*-CDT)] (**35b**). This can essentially be traced back to the fact that the transition state **34a** → **TS** → **35a** is stabilized by an additional η^2 coordination of butadiene, which, for steric reasons, is not the case for the other two reactions forming **35b** and **35c**.

Exercise 10.2
- Write a hypothetical reaction path for the nickel-catalyzed formation of *c,c,c*-CDT and give reasons why it is not formed.

10.4 Oligomerization and Telomerization of Butadiene

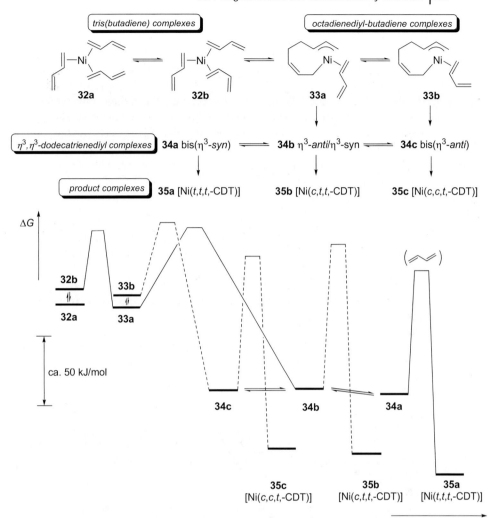

Figure 10.9 Reaction profile for the nickel-catalyzed cyclotrimerization of butadiene. The isomeric complexes that are in dynamic equilibria (**32a/32b**, **33a/33b**, **34a–34c**) are connected by equilibrium arrows without identifying the activation barriers. The favored reaction pathway is drawn with a solid line (adapted from Tobisch [228]).

- Quantum-chemical calculations show that, based on the standard Gibbs free energies of formation, both the *all-trans* isomer of CDT and its nickel(0) complex are the most stable thermodynamically. Compared to the *all-cis* isomer, the following applies: [Ni(t,t,t,-CDT)] (**35a**)/[Ni(c,c,c,-CDT)] (**35d**), $\Delta\Delta G = -28$ kJ/mol; t,t,t,-CDT (**38a**)/c,c,c,-CDT (**38d**), $\Delta\Delta G = -48$ kJ/mol. Which complex is more stable, and which reaction between **35a** and **38d** do you expect?

10.4.1.3 Industrial Synthesis of CDT

Titanium-containing catalyst systems react with butadiene, forming c,t,t-CDT (**39**). Thus, the Ziegler system $TiCl_4/Al_2Cl_3Et_3$ is suited to industrial synthesis of **39** [229]. As opposed to ethene polymerization, the system remains homogeneous and delivers, at practically complete conversion (30–75 °C) with >90% yield, the cyclotrimer, which is hydrogenated, oxidized to form the ketone, and finally converted to the oxime (**39** → **40**). Beckmann rearrangement yields a lactam (**40** → **41**), which then is further processed, producing polyamide 12 (**41** → **42**, Nylon 12, Vestamid®).

39 (c,t,t,-CDT)   **40**   **41**   **42**

10.4.2 Cyclodimerization of Butadiene

10.4.2.1 Mechanism

In the presence of a P-donor L, the cyclotrimerization previously described is redirected into a cyclodimerization of butadiene. The principal cyclodimers formed are 4-vinylcyclohexene (VCH) and cis,cis-cycloocta-1,5-diene (COD), but the cyclodimer cis-1,2-divinylcyclobutane (DVCB) is also generated. The selectivity is controlled by the steric and electronic properties of L. The cyclotrimer/cyclodimer ratio is essentially determined by the steric properties of L, while the COD/VCH ratio mainly depends on the electronic properties of L. A simplified reaction scheme is shown in Figure 10.10.

The catalytically active complex **43**, a ligand-containing bis(butadiene)nickel(0) complex, is obtained by addition of a P-donor to the cyclotrimerization catalyst. Through oxidative coupling and allyl isomerization, ligand-containing η^3,η^1- (**44**) and η^3,η^3-octadienediylnickel(II) complexes (**45a/45b**) are formed. Complexes **44** and **45** appear in numerous isomers (η^3-syn/anti; Δ-cis/trans) that exist in equilibrium with each other.

The position of the equilibria depends on the reaction conditions and can be controlled by the electronic and steric properties of L. Figure 10.10 shows the immediate precursor complexes that undergo reductive elimination under ring closure, yielding the nickel(0) complexes with the cyclodimers as ligands (**44/45** → **46**). Substitution by butadiene releases the cyclodimers and re-forms the catalyst complex (**46** → **43**). The reductive C–C eliminations with formation of COD and VCH are irreversible. Due to the highly strained four-membered ring, the reductive elimination forming DVCB is reversible, so it is only obtained under kinetic control.

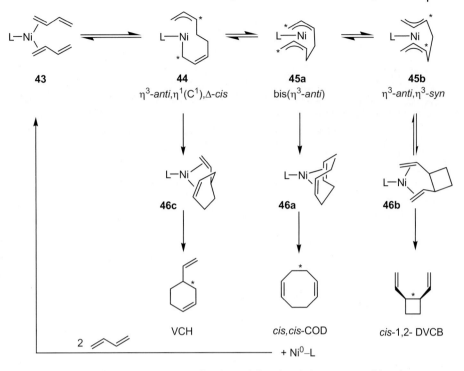

Figure 10.10 Simplified reaction scheme for the nickel-catalyzed dimerization of butadiene (L = phosphane or phosphite). The C atoms that are linked together in reductive elimination, as well as the C–C bonds formed, are indicated by a star (*).

Central intermediates in the catalytic cycle are ligand-containing octadienediylnickel(II) complexes **44/45**. In the reaction of [Ni(CDT)(PCy$_3$)] with isoprene, this intermediate stage was also able to be structurally characterized (Figure 10.11). Quantum-chemical calculations with the model ligand L = PH$_3$ show that the most stable isomer is the bis(η^3-syn)-octadienediyl complex (**a**) (Figure 10.12). Other bis(η^3)- and η^3,η^1-complexes are only insignificantly less stable. By contrast, all bis(η^1)-complexes are less stable by more than 100 kJ/mol

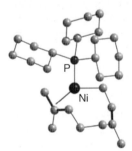

Figure 10.11 Structure of [Ni(η^3,η^1-Me$_2$C$_8$H$_{10}$)(PCy$_3$)] (without hydrogen atoms; the double bond and the bonds of the allyl group are marked in bold).

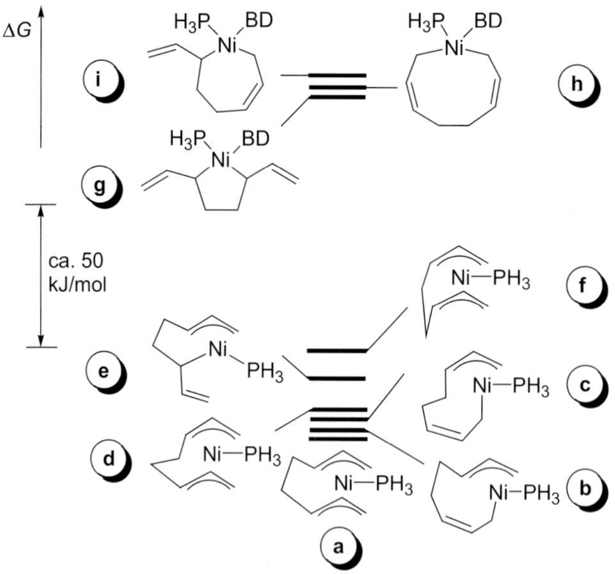

Figure 10.12 Stability of phosphane(octadienediyl)nickel(II) complexes: The bis(η^1)-complexes g–i require an additional butadiene molecule (BD) for stabilization (after Tobisch and Ziegler [224]).
(a) bis(η^3-syn).
(b) η^3-syn,η^1(C^1),Δ-cis.
(c) η^3-anti,η^1(C^1),Δ-cis.
(d) η^3-anti,η^3-syn.
(e) η^3-syn,η^1(C^3).
(f) bis(η^3-anti).
(g) bis(η^1(C^3)) + BD.
(h) bis(η^1(C^1),Δ-cis) + BD.
(i) η^1(C^1),η^1(C^3),Δ-cis + BD.

(ΔG), so thermodynamic considerations eliminate them as possible intermediates in the catalytic cycle.

Exercise 10.3
Which isomer (according to Figure 10.12) exists in the structurally characterized compound from Figure 10.11? Give the bis(isoprene) complex from which the complex in Figure 10.11 could have been created by oxidative coupling.

10.4.2.2 Selectivity Control

The product distribution for the nickel-catalyzed cyclooligomerization of butadiene is compiled for a series of representative phosphane/phosphite ligands L in Table 10.1. This makes clear that on the one hand, even ligand-containing catalysts can form substantial quantities of CDT and on the other, "naked" nickel leads to the formation of cyclodimers as side products.

The dependence of the selectivity of the nickel-catalyzed cyclooligomerization of butadiene on the steric and electronic properties of the phosphane/phosphite ligand

10.4 Oligomerization and Telomerization of Butadiene

Table 10.1 Ligand control[a] of the cyclooligomerization of butadiene ($[Ni(COD)_2] : L : C_4H_6 = 1 : 1 : 170$; $T = 60\,°C$, $t = 48\,h$) (after Heimbach and Schenkluhn [230, 231]).

L	θ (in °)	ν (in cm^{-1})	COD + VCH (in %)[b]	CDT (in %)[b]
P(t-Bu)(i-Pr)$_2$	167	2058	45.8 (1.9)	49.6
P(i-Pr)$_3$	160	2059	68.8 (1.7)	23.6
PEt$_3$	132	2062	64.5 (1.6)	29.0
PPh$_3$	145	2069	85.0 (3.0)	14.8
P(OMe)$_3$	107	2080	38.0 (1.3)	59.8
P(OPh)$_3$	128	2085	87.4 (10.7)	12.2
P(O-o-Tol)$_3$	141	2084	97.6 (11.7)	1.4
—[c]			14.3 (0.6)	81.7

a) L is characterized by Tolman's steric (θ) and electronic parameters (ν).
b) The ratio of COD and VCH is given in parentheses. The sum of the values differs from 100% due to unknown and open-chain butadiene oligomers.
c) For comparison, without the addition of a P-ligand: Ni(acac)$_2$/AlEt$_2$(OEt), $T = 20\,°C$.

L is well understood. Beginning in the 1960s, this opened for the first time the possibility of targeted control of the selectivity of an organometallic-catalyzed reaction via the nature of the ligand L ("ligand tuning"; "ligand-property control").

Background: Steric and Electronic Effects of Phosphorus Ligands

Within a catalyst complex, a targeted variation in electronic and steric properties of ligands not directly involved in the reaction ("spectator ligands"; "control ligands") makes it possible to control activity, selectivity and/or productivity of the catalyst. To characterize ligands accordingly, C. A. Tolman introduced a quantitative measure for the frequently used (monodentate) phosphorus ligands [232]:

Steric ligand parameter. The steric demand is described by a cone angle θ (**I**). The apex of the cone is 2.28 Å from the P atom, and the perimeter just touches the van der Waals spheres of the substituents of the phosphorus ligands in their most compact conformation.

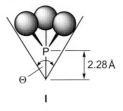

Electronic ligand parameter. The electronic parameter ν of a ligand L is the wave number ν in cm^{-1} of the CO stretching frequency (A$_1$ symmetry) in [Ni(CO)$_3$L] complexes in CH$_2$Cl$_2$ solution. Although the parameter ν only sums up the effect of σ-donor and π-acceptor strength, it can be assumed that the smaller the magnitude of ν, the greater the donor strength.

The plot shows the relationship between steric and electronic parameters for selected ligands (values from [232]).

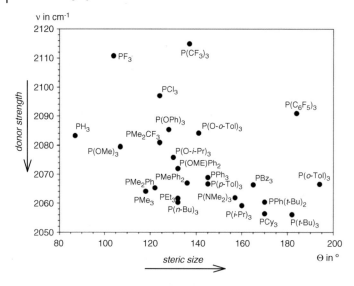

The Tolman electronic ligand parameter has been validated by DFT calculations that have also included other ligand types (especially NHC ligands). However, these calculations also show that donor properties of ligands cannot always be adequately represented by a single parameter [233]. More advanced concepts for a separate analysis of σ-donor and π-acceptor effects, and for a more detailed description of steric effects of ligands, have been developed, respectively, by W. P. Giering [234] and A. J. Poë [235].

An equilibrium between the primarily formed octadienediyl complexes with the ligand L (**44a**) or with butadiene (**33a**) can be taken as the basis for further explanation (Figure 10.13). This is probably the decisive crossover between the reaction channels to the C_8 and C_{12} cyclooligomers. It can be assumed that the ligand substitution (L vs. butadiene) does not display a notable kinetic barrier, so the two complexes are connected through a rapidly established equilibrium. The activation barrier for the further reaction to form either CDT or COD/VCH is substantially larger, so a typical Curtin-Hammett situation exists (see Background, Section 4.3.3): The selectivity is determined by the difference between the Gibbs free energies of the transition states of the reactions that lead to CDT or COD/VCH, and not by the position of the equilibrium **33a** $\rightleftharpoons$ **44a** (Figure 10.13, a).

Ligand-property control must be discussed separately for pure σ-donor ligands and for those that also possess π-acceptor properties.

σ-*Donor ligands* ($L = PR_3$). Three factors determine the ligand effect: *i*) A high σ-donor strength of L stabilizes **44a**. *ii*) Steric bulk on L facilitates ligand cleavage, and thus shifts the equilibrium **33a**/**44a** in favor of **33a**. *iii*) Bulky σ-donor ligands L lower the activation barrier for reductive elimination to form COD/VCH. Thus, high steric strain from the ligands L favors the formation of both CDT (see *ii*) and COD/VCH

10.4 Oligomerization and Telomerization of Butadiene

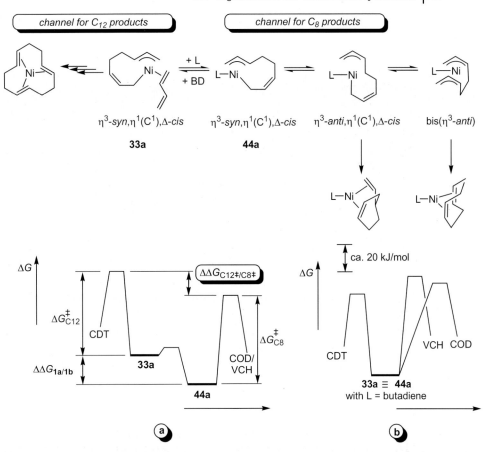

Figure 10.13 C_8/C_{12} selectivity for the nickel-catalyzed cyclooligomerization of butadiene. (a) Qualitative reaction profile. (b) Reaction profile (simplified) of cyclooligomerization with "naked" nickel (L = butadiene) (after Tobisch [228]).

(see *iii*). The factor *ii*) seems to dominate, so it is logical that the CDT proportion for L = P(*t*-Bu)(*i*-Pr)$_2$ is by far the largest (Table 10.1).

π-Acceptor ligands (L = P(OR)$_3$). π-Acceptor ligands L stabilize the ligand complex **44a** only to a moderate extent (relative to **33a**); phosphites affect the position of the equilibrium **33a/44a** less than phosphanes do. However, bulky π-acceptor ligands lower the activation barrier for the C_8 reaction channel, which particularly favors the formation of COD.

The formation of low quantities of cyclodimers during synthesis of CDT with "naked" nickel still remains to be explained (see Table 10.1). Butadiene is a weakly coordinating ligand, so the starting complexes for the C_8 (**44a**, L = butadiene) and C_{12} reaction channels (**33a**) are identical. This simplifies the reaction scheme (Figure 10.13, b), since the reaction channel to form CDT and the one to form the cyclodimers start from the same compound. Quantum-chemical calculations show the lowest activation barrier for the formation of CDT, in accordance with the experiment.

10.4.3
Linear Oligomerization and Telomerization of Butadiene

In the presence of active hydrogen compounds (H–X = H–OR, H–OC(O)R, H–OH, H–NR$_2$, ...), the nickel(0)-catalyzed cyclodimerization of butadiene changes to linear dimerization to form octatrienes or to telomerization to form functionalized octadienes. The mechanism is similar to that for cyclodimerization. The starting point is a bis(butadiene)nickel(0)-ligand complex **43** that – as in cyclodimerization – can react to form $\eta^3,\eta^1(C^1)$- and $\eta^3,\eta^1(C^3)$-octadienediyl-nickel(II) complexes **44b** and **44c** via oxidative coupling and allyl isomerization (*cis/trans*, *syn/anti* isomerism is ignored throughout this discussion). This is followed by a protolytic cleavage of the σ-Ni–C bond (**44b** → **47a**, **44c** → **47b**), forming η^3-allyl complexes in which anionic ligands X as well as L are coordinated to nickel.

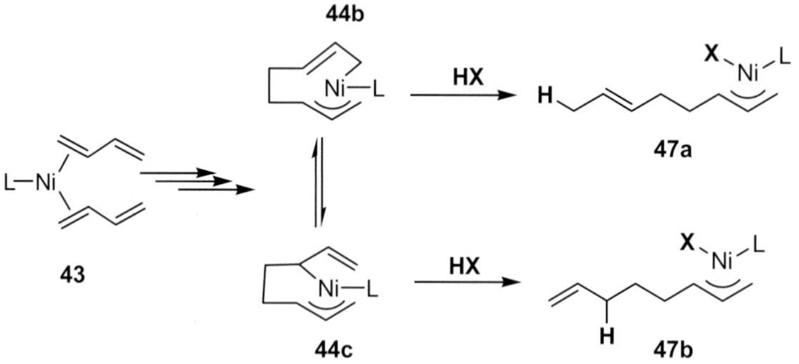

In order to make the reaction easier to understand, HX is marked in bold in all products. The further reaction for the isomer **47a** is shown as an example:

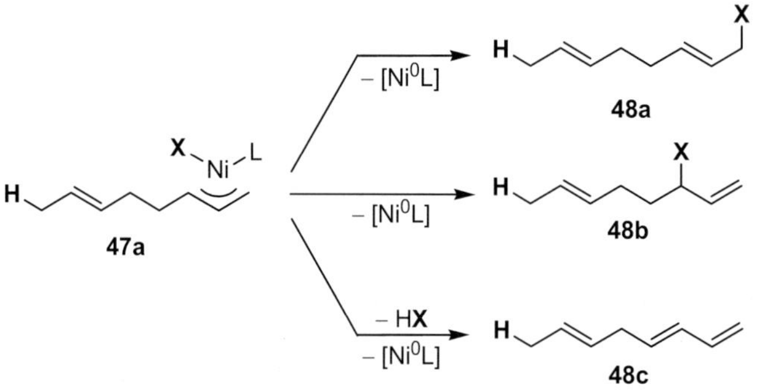

The assumed reaction mechanism presumes reductive elimination with formation of a C–X bond. The catalyst complex [Ni^{0}L] is thus regenerated, and octadienes with terminal (**47a** → **48a**) or internal (**47a** → **48b**) functional groups are formed. However, via β-hydride elimination from **47a**, an octatriene can be formed (**47a** → **48c**), whereby formally an [NiH(X)L] complex is created that decomposes into HX and the catalyst complex [Ni⁰L].

The formation of the octatriene **48c** represents a linear dimerization of butadiene. Catalytic quantities of HX are sufficient; HX is a cocatalyst. The formation of the X-functionalized octadienes (**48a/48b**) represents a telomerization of butadiene. Here, HX is a reactant and thus stoichiometric quantities are necessary. When water is used as telogen (H–X = H–OH), hydroxyoctadienes are obtained. This reaction is called "hydrodimerization" of butadiene.

Background: Telomerization

Telomerization is a special form of oligomerization or polymerization in which a molecule A–B (the telogen) reacts with n molecules of a monomer (mon) to form telomers according to the following equation:

$$A\text{–}B + n\,\text{mon} \longrightarrow A\text{–}(\text{mon})_n\text{–}B$$

Generally, telomers have a relatively low molar mass and thus find industrial application as plasticizers, adhesive components, or surfactants. Frequently, telomerization is carried out as radical solution polymerization with the telogen A–B as solvent (carbon tetrachloride, mercaptans, alcohols, hydrocarbons). Hence, the radically initiated telomerization of ethene in carbon tetrachloride leads to chlorofunctionalized oligomers ($n = 4\text{–}6$).

$$n\ H_2C=CH_2 \xrightarrow{+\,\cdot CCl_3} Cl_3C{-}[CH_2{-}CH_2]_{n-1}{-}CH_2{-}\dot{C}H_2 \xrightarrow[-\,\cdot CCl_3]{+\,CCl_4} Cl_3C{-}[H_2C{-}CH_2]_{n-1}{-}CH_2{-}CH_2{-}Cl$$

Thus, telomerization assumes a position between the addition of A–B to olefins/dienes and their polymerization [178, 236].

Exercise 10.4

Which products are formed from the η^3-octadienylnickel(II) complex **47b**? Sketch a formula scheme like that offered previously for isomer **47a**. In your answer, ignore cis/trans isomers.

The product spectrum obtained is more varied than previously indicated. For one thing, cis/trans double-bond isomers and other positional isomers can be formed. The product distribution can be controlled within certain limits and depends markedly on HX, the ligand L, and the reaction conditions. As side products, cyclooligomers, polymers, and (if octatrienes are involved in the reaction) higher linear oligomers can be formed. In addition, functionalized butenes can be formed. They are likely created by hydridonickel(II) complexes **49**, which appear as intermediates in the reaction. They can also be made via partial protonation of nickel(0) complexes. Insertion of butadiene into the Ni–H bond produces a crotylnickel complex **50**, which cleaves off X-functionalized butenes in a reductive C–X elimination.

$$\text{"LNi}\begin{smallmatrix}H\\X\end{smallmatrix}\text{"} \quad \xrightarrow{\diagup\!\!\!\diagdown} \quad \text{X}\diagdown\text{Ni}\diagup\text{L} \quad \xrightarrow{-[\text{Ni}^0\text{L}]} \quad \diagup\!\!\!\diagdown\text{X} \; , \; \diagdown\!\!\!\!\!\diagup_X$$

49   **50**

Starting from **50**, insertion of butadiene opens an additional path to formation of C$_8$ compounds **47a**.

Exercise 10.5

In the reaction of butadiene with [Ni{P(OEt)$_3$}$_4$]/CF$_3$COOH in ROH (R = Et, i-Pr, ...), the products formed include a novel cyclodimer, namely 1-methylene-2-vinylcyclopentane (**51**). The reaction sequence 43 → 44c → 47b → ... (L = P(OEt)$_3$; X = CF$_3$COO$^-$) has been discussed as a reaction path. Complete the reaction.

51

Linear dimerization and telomerization of butadiene are also catalyzed by palladium. In general, these reactions proceed more selectively than the nickel-catalyzed counterparts and no cyclooligomers such as COD or CDT are obtained. This is due to the larger Pd atoms and the fact that hydride eliminations proceed more readily. Thus, ligand-free palladium(0) catalysts catalyze the linear trimerization of butadiene, forming various isomers of dodecatetraene **54** [237]. In the reaction of [Pd(dba)$_2$] (**52**, dba = dibenzylideneacetone, PhCH=CH–CO–CH=CHPh) with butadiene, an intermediate **53** analogous to that for nickel has been identified, but rather than reacting further to form CDT, hydrogen migration occurs (formally, a β-hydride elimination and a reductive C–H elimination) to form **54** [238].

[Pd(dba)$_2$] $\xrightarrow[-2\text{ dba}]{3\;\diagup\!\!\!\diagdown}$ Pd(...) $\xrightarrow{-\text{"Pd"}}$ (...)

52   **53**   **54**

Palladium(0) catalysts containing ligands are often generated from [Pd(PR$_3$)$_4$] or Pd(OAc)$_2$/PR$_3$, where in the last case, *in situ*[6] reduction to Pd^0 occurs. The linear dimerization to form octa-1,3,7-triene (**55**) is catalyzed with a Pd/L ratio (L = PR$_3$) of 1/1 in aprotic solvents. In the presence of HX, telomerization proceeds, forming functionalized octadienes (**56**). Functionalization at the terminal positions is favored. PdL$_2$ complexes catalyze the formation of functionalized butenes (**57**) or cocyclization of butadiene with heteroolefins (aldehydes, ketones, imines, ...) to form six-membered heterocycles (**58**) [239].

---

[6] Pd(OAc)$_2$ is very readily reduced, for example, by CO, alcohols, tertiary amines and olefins. It reacts with butadiene to form Pd^0 and AcOCH$_2$CH=CHCH$_2$OAc.

10.4 Oligomerization and Telomerization of Butadiene

Palladium-catalyzed telomerization reactions such as the hydrodimerization of butadiene have achieved industrial significance [240]. This is carried out in water/sulfolane ((CH_2)$_4SO_2$) in a CO_2 atmosphere. Pd(OAc)$_2$ acts as precatalyst in the presence of a large excess of the phosphonium salt [PPh$_2$(C$_6$H$_4$-*m*-SO$_3$Li)-(CH$_2$CH=CHR)](HCO$_3$) and triethylamine. The primary product, with a selectivity of >90%, is octa-2,7-diene-1-ol (**59**). Extraction with hexane enables separation of product without destroying the thermally sensitive catalyst [241].

Hydrogenation of **59** forms octan-1-ol (**60**), an important starting material for the synthesis of plasticizers for PVC. A heterogeneously catalyzed isomerization of **59** and subsequent hydroformylation leads to a dialdehyde (**59** → **61** → **62**), the starting material for synthesis of important 1,ω-diamines, -alcohols and -carboxylic acids (**62** → **63**). In addition, 1-methoxyoctadiene **64** is produced industrially by palladium-catalyzed telomerization of butadiene with methanol. Subsequent hydrogenation and methanol cleavage on heterogeneous catalysts leads to oct-1-ene (**65**), which is required as comonomer for the production of LLDPE. Recycling of methanol leads to the synthesis of oct-1-ene from inexpensive butadiene and hydrogen.

10.5
Polymerization of Butadiene

10.5.1
Mechanism

Buta-1,3-diene can be polymerized with C1–C4 or C1–C2 linkage to form 1,4- or 1,2-polybutadiene, respectively. In the 1,2-polymerization, as in polymerization of a terminal olefin, a stereocenter is generated. If the polymerization proceeds regio- and stereoselectively, either cis- (**66a**) or trans-1,4-polybutadiene (**66b**), or iso- (**67a**) or syndiotactic 1,2-polybutadiene (**67b**) is formed. If the polymerization proceeds regioselectively with C1–C2 linkage but not stereoselectively, it leads to atactic 1,2-polybutadiene.

66a   **66b**   **67a**   **67b**

Exercise 10.6
Write the formulas of all regio- and stereoregular polymers of isoprene.

In the mid-1950s, G. Natta became the first to prepare all four types of stereoregular polybutadienes with a high degree of steric purity (>98%) using organometallic mixed catalysts (Ziegler-Natta catalysts) and to demonstrate regio- and stereoregularity of the diene polymerization. cis-1,4-Polybutadiene has similar properties to natural rubber (cis-1,4-polyisoprene), and both are important in the tire industry.

In the catalyst complex **68**, it is essential for catalysis that the monomer (butadiene) and the growing polymer chain (via a terminal allyl group) be coordinated. The growth of the polymer chain occurs by π-allyl insertions (**68** → **69**). Butadiene coordination completes the cycle (**69** → **68′**).

allyl insertion   butadiene coordination

68   **69**   **68′**

The π-allyl insertion proceeds in such a way that, in principle, the growing polymer chain is linked to the terminal C atom of the coordinated butadiene (C_{1BD}). The remaining three C atoms of the butadiene (C_{2BD}–C_{4BD}) form a new η^3-allyl ligand. The regioselectivity (1,4 vs. 1,2 bond linking) of the polymerization is now deter-

10.5 Polymerization of Butadiene

mined by which C atom of the polymer chain (C_{1P} or C_{3P}) bonds to the C1 atom of the coordinated butadiene (C_{1BD}): C_{1P}–C_{1BD} linkage yields a 1,4-C_4 unit (**68** → **69**) and C_{3P}–C_{1BD} linkage yields a 1,2-C_4 unit (**68** → **70**):

[Scheme showing structures 68, 69, 70 with 1,4-polymerization and 1,2-polymerization pathways]

The stereoselectivity in the 1,4-polymerization (*cis*- vs. *trans*-1,4-polybutadiene) is determined by the structure of the allyl group in the growing polymer chain. The *syn/trans* and *anti/cis* correlation (see Section 10.3.5) apply (Figure 10.14):

- From a *syn*-allyl structure, a *trans*-1,4-C_4 unit forms (**71a/68a** → **69a**), and from an *anti*-allyl structure, a *cis*-1,4-C_4 unit forms (**71b/68b** → **69b**).
- Allyl insertion of butadiene in the *s-trans* form leads to a *syn*-η^3-allyl group (**68a** → **69a**) and in the *s-cis* form to an *anti*-η^3-allyl group (**68b** → **69b**).

If *syn*- and *anti*-η^3-allyl groups could not interconvert, an insertion of *s-trans*-butadiene would have to lead to *trans*-1,4-polybutadiene (**71a** → **68a** → **69a**), and an insertion of *s-cis*-butadiene to *cis*-1,4-polybutadiene (**71b** → **68b** → **69b**). This is not the case, however, since *syn*- and *anti*-η^3-allyl groups can undergo a *syn/anti* or *anti/syn* isomerization (**71a** ⇌ **71b**, **68a** ⇌ **68b**). If this occurs before the next insertion step (**68a** → **69a** or **68b** → **69b**), the other stereoisomer will be generated. Hence, the stereoselectivity is decided by the structure of the allyl group not at the time it is formed, but rather at the time at which it reacts with the coordinated butadiene.

As for polymerization of monoolefins, in the case of diene polymerization, hydrogen-transfer reactions can lead to chain termination without deactivation of the catalyst. Thus, starting from **69a/69b** (in the following scheme, the reaction is shown with **69b**) hydrogen transfer from the C4 atom of the growing polymer chain **P** onto the metal leads to a hydridometal complex to which the polymer molecule with a diene end group is coordinated (**69b** → **72**; the relevant CH_n groups are henceforth explicitly given in the formulas). Substitution of the coordinated polymer by butadiene (**72** → **73**) as well as insertion of butadiene into the M–H bond and further butadiene coordination (**73** → **68b′**) regenerate the catalyst complex. Starting from **68b** (the **68a** case is similar), the hydrogen transfer can also occur directly onto the coordinated monomer (**68b** → **74**). Substitution of the polymer, which is coordinated to the metal through a terminal diene group, by butadiene leads to the formation of catalyst complex **68b′**. In both cases, chain termination occurs, but the catalyst is not deactivated.

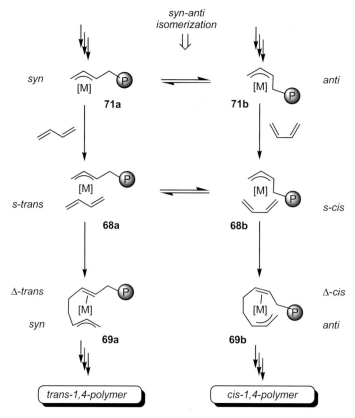

Figure 10.14 General reaction scheme for *cis/trans* selectivity in butadiene polymerization. The *syn/anti* isomerization **68a** ⇌ **68b** need not include a *trans/cis* isomerization of the coordinated butadiene (as shown in the scheme). Furthermore, the insertion of an *s-cis*-coordinated butadiene into a *syn*-allyl group, or of an *s-trans*-coordinated butadiene into an *anti*-allyl group, is also possible (though not shown in the scheme).

As in the polymerization of simple olefins, the polymer chain can be transferred onto the aluminum alkyl in the case of Ziegler systems ($[M(\eta^3\text{-}CH_2\text{---}CH\text{---}CHP)]$ +

$AlR_3 \rightarrow$ [M]–R + R_2Al–CH_2–CH=CHP) without deactivating the catalyst. By contrast, a homolytic bond cleavage ([M(η^3-CH_2═CH═CHP)] $\rightarrow$ [M]· + ·CH_2CH=CHP) leads not only to chain termination, but also to catalyst deactivation.

Exercise 10.7

In addition to hydrogen, ethene and α-olefins are used to control the molar mass. Give a possible mechanism.

10.5.2
Butadiene Polymerization Catalyzed by Allylnickel(II) Complexes

Partial protolysis of the C_{12}-diallylnickel(II) complex **75** by acids HX (X = weakly coordinating anion) such as tetrafluoroboric acid produces dodecatrienylnickel(II) complexes **76** (Figure 10.15), which are highly active single-component catalysts of the 1,4-polymerization of butadiene. Complexes of type **76** can thus be seen as models for the catalyst complex if the methyl group at the end (see arrow in Figure 10.15) is replaced by the growing polymer chain. The coordination of the two double bonds leads to a 16-ve complex that gains additional stability through the weak cation–anion interactions. If a still weaker coordinating anion such as tetrakis-[3,5-(trifluoromethyl)phenyl]borate is used (**77** → **78**), no direct cation–anion interactions can be observed in the catalyst complex **78**, but there is probably an additional weak contact between a double bond in the growing polymer chain and the nickel (Figure 10.15).

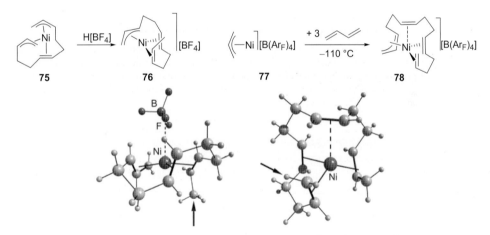

Figure 10.15 Formation of [Ni($C_{12}H_{19}$)][BF_4] (**76**, $C_{12}H_{19}$ = η^3-syn,η^2(Δ-trans),η^2(Δ-cis)-dodecatrienyl) and [Ni($C_{15}H_{23}$)][B(Ar$_F$)$_4$] (**78**, $C_{15}H_{23}$ = η^3-anti,(Δ-cis),η^2(Δ-cis),η^2-pentadecatetraenyl, Ar$_F$ = 3,5-(CF_3)$_2C_6H_3$) and structures of **76** (left) and the cation of **78** (right). Weak interactions via the "axial" fifth coordination site of the nickel are suggested by the dashed lines. The allyl group that remains on the nickel(II) after protolysis of **75** → **76** is subject to anti/syn isomerization. In the catalyst complexes for allylnickel(II)-catalyzed butadiene polymerization, the methyl group or an H atom from the terminal double bond (see arrows in the structure models) is replaced by the growing polymer chain.

Table 10.2 Catalytic selectivity and activity of allylnickel(II) complexes in stereoselective butadiene polymerization (after R. Taube, G. Sylvester in [M6], 1st ed., VCH, Weinheim 1996, p. 280).

Complex	L or X/X′	TOF[a]	1,4-cis[b]	1,4-trans[b]
(a)	X′ = Cl	0.1 (65)	92	6
	X′ = Br	2.4 (65)	46	53
	X′ = I	30 (65)	—	95
(b)	L = PPh$_3$	10	3	90
	L = SbPh$_3$	10 000	85	11
	L = P(OPh)$_3$	200	4	96
(c)	L = PPh$_3$	650 (50)	59	36
	L = PCy$_3$	90 (50)	52	39
	L = P(O-o-Tol)$_3$	5400	90	8
(d)	X = PF$_6$	12 000	91	8
	X = BF$_4$[c]	7500	75	13
	X = CF$_3$SO$_3$	10	17	80

a) In mol butadiene/(mol Ni · h); reaction temperature in parentheses (in °C) when it differs from 25 °C.
b) Microstructure of the polymers in %. Differences from 100 % correspond to the fraction of 1,2-polymers.
c) Complex **76** in Figure 10.15.

Allylnickel(II) complexes were the first single-component catalysts for butadiene polymerization (Table 10.2), including neutral dinuclear allylnickel(II) complexes [{Ni(C$_3$H$_5$)(μ-X′)}$_2$] (**a**: X′ = halide, carboxylate), cationic allylbis(ligand)nickel complexes [Ni(C$_3$H$_5$)L$_2$][PF$_6$] (**b**: L = PR$_3$, P(OR)$_3$, ...), (C$_8$-allyl)mono(ligand)nickel complexes [Ni(C$_8$H$_{13}$)L]X (**c**: X = weakly coordinating anion) and the previously mentioned "ligand-free" C$_{12}$-allylnickel(II) complexes [Ni(C$_{12}$H$_{19}$)]X (**d**). All these complexes catalyze the formation of 1,4-polybutadiene with very low quantities of 1,2-polybutadiene.

The reaction model for the allylnickel-catalyzed 1,4-polymerization of butadiene, formulated and experimentally verified by R. Taube (University of Halle, Germany) at the beginning of the 1990s, is displayed in Figure 10.16. The following individual reaction steps are worthy of discussion:

- *Catalyst formation (ligand cleavage/butadiene coordination).* The allylbis(ligand) complexes **79** (R = H) are the precatalysts from which the catalyst complexes are generated (**79** → **80**) by cleavage of L and coordination of butadiene.[7] Additional cleavage of L forms the ligand-free complexes (**80** → **81**), which likewise are catalytically active. The coordination of a double bond from the growing polymer chain contributes to the stability of the complexes **81**. Starting from **80**,

7) Complexes **80** and **81** are shown in the figure with a growing polymer chain **P**, hence after the cycle has completed at least once.

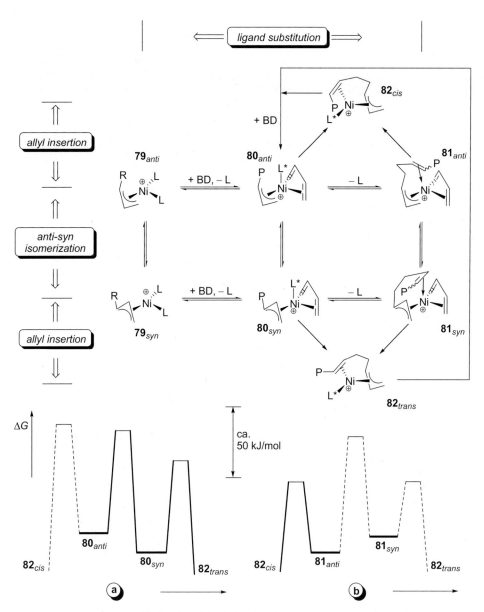

Figure 10.16 Reaction model for the allylnickel-catalyzed 1,4-polymerization of butadiene (BD = butadiene, L = ligand, L* = ligand or double bond from the growing polymer chain, **P** = growing polymer chain, R = H or **P**). (a) Reaction profile for the cationic monoligand complex [Ni(C$_3$H$_4$R)(C$_4$H$_6$)L]$^+$ (**2**, L = P(OMe)$_3$). (b) Reaction profile for the cationic ligand-free complex [Ni(C$_3$H$_4$R)(C$_4$H$_6$)]$^+$ (**3**). In each case, the energetically favored reaction pathway is drawn with solid lines (abbreviated from Tobisch [242]).

substitution of butadiene by L leads to catalytically inactive bis(ligand) complexes, but now with the growing polymer chain as ligand (**80** → **79**, R = **P**). All these ligand substitution reactions are associated with negligible activation barriers, as is generally the case for spin-paired d^8 nickel complexes.
- *anti/syn isomerization.* The complexes **79–81** are subject to *anti/syn* isomerization ($79_{syn} \rightleftharpoons 79_{anti}$; $80_{syn} \rightleftharpoons 80_{anti}$; $81_{syn} \rightleftharpoons 81_{anti}$). In complexes **79** and **80**, the *syn* form is thermodynamically more stable. In complexes **81** with the η^3,η^2-chelating polymer chain, the *anti* form is thermodynamically more stable.
- *Chain growth (allyl insertion).* It has been shown by experiments and calculations that in the insertion step (**80/81** → **82**), an *anti*-allyl structure is primarily formed, regardless of whether *cis-* or *trans-*1,4-polybutadiene is generated. This clarifies that in **80/81**, butadiene is bound in the *s-cis* conformation. It also follows that *trans-*1,4-polymerization must include an *anti/syn* isomerization.
- *Catalyst re-formation (butadiene coordination).* In **82**, at least one double bond in the growing polymer chain is coordinated to nickel; it can be replaced by butadiene in an exothermic process (**82** → **80**).

The reaction model predicts two typical reaction channels, one for *trans-* and one for *cis-*1,4-polymerization:

- *trans-1,4-polymerization* (Figure 10.16, **a**). For a monoligand complex (L* = ligand), the following reaction sequence occurs: 80_{anti} → 80_{syn} → 82_{trans} → ... The favoring of the *trans* polymerization can be traced to the fact that the *syn* form of **80** is both more thermodynamically stable and kinetically more reactive than the *anti* form. The *anti/syn* isomerization is the rate-determining step.
- *cis-1,4-polymerization* (Figure 10.16, **b**). For a ligand-free complex, the following reaction sequence occurs: 81_{anti} → 82_{cis} → $80_{anti} \equiv 81_{anti}$[8] → ... The transition states for the insertion 81_{anti} → TS → 82_{cis} and 81_{syn} → TS → 82_{trans} are very similar and their Gibbs free energies are almost identical. It follows that the thermodynamically more stable ligand-free complex 81_{anti} displays the lower reactivity. Only the fact that the activation barrier for the isomerization $81_{anti} \rightleftharpoons 81_{syn}$ is relatively high makes the reaction channel leading to *cis-*1,4-polybutadiene operative. The allyl insertion is the rate-determining step.

For the *cis/trans* regulation, it must still be considered that the catalyst complexes **80** and **81** are formed via the preequilibria **79** ⇌ **80** ⇌ **81**. The concentration of **80** and **81** is decided by the position of these equilibria, which thus determine the catalytic activity and the *cis/trans* selectivity as well. The position of these equilibria now depends substantially on the steric and electronic properties of the ligands L, so their effect on the activity and selectivity of catalysis is understandable.

In a similar way, anion coordination of varying strength, not considered here, can influence the course of catalysis.

---

[8] These are ligand-free complexes. Thus, 80_{anti} (L* = double bond in the growing polymer chain!) is identical with 81_{anti}.

Exercise 10.8

Quantum-chemical calculations verify that complex **82′** is the resting state for the polymerization of butadiene with **77** (Figure 10.15) as catalyst, as long as 1,4-*cis*-polymerization occurs exclusively. A complex of type **83** has been confirmed experimentally as the resting state. Explain, and make sure that you understand the reactivity of **83** with additional butadiene.

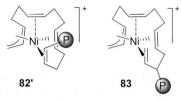

10.5.3
Synthesis and Properties of Polybutadienes and Polyisoprenes

Rubber is a term for a non-cross-linked, but cross-linkable (vulcanizable) polymer with rubber-elastic properties at room temperature. In 1909, the first usable synthetic rubber were obtained at the Farbenfabrik Bayer (Leverkusen, Germany) by thermal polymerization of dienes (isoprene, butadiene, 2,3-dimethylbutadiene) in a moderate temperature range (50–150 °C). Shortly thereafter, it was found that low quantities of sodium (1–3%) substantially accelerated the polymerization of dienes. On this basis, in 1912, BASF (Germany) developed an industrial process for synthesis of poly-2,3-dimethylbutadiene ("methyl rubber B"). In 1937–1938, the continuous polymerization of butadiene using sodium (Buna = *bu*tadiene-*natrium*; German: Natrium = sodium) began on a large scale in the Buna Factory Schkopau (Saxony-Anhalt, Germany), where the rubber types were known as "number Bunas" (Buna 32, Buna 85, . . .) in the market. A significant step forward in the quality of the rubber and the execution of the process was achieved by the development of radical polymerization of butadiene in aqueous solutions (emulsion polymerization) and the emulsion copolymerization of butadiene with styrene ("Buna S") and acrylonitrile ("Buna N").[9] All these products display very low 1,4-*cis* fractions, typically 10–20% ("*low-cis*-polybutadiene"). The synthesis of *high-cis*-polybutadienes (>95% 1,4-*cis*-content) first became possible with transition metal catalysts in the mid-1950s.

In 2009, worldwide production of rubber amounted to $22 \cdot 10^6$ metric tons, of which 44% was natural and 56% synthetic. About 1/4 of the synthetic rubber was polybutadiene rubber, but more than 40% was styrene–butadiene rubber (SBR). About 60% of the synthetic rubber was used for production of tires. In addition to tire production, 1/3 of the produced polybutadiene rubber is used in modification of plastics (especially of polystyrene: "rubber-modified polystyrene"; "high-impact polystyrene").

---

9) So-called "letter Bunas," known in the United States as GR-S (GR = government rubber) and GR-A, respectively.

- *cis*-1,4-Polybutadiene
  Anionic polymerization initiated with lithium alkyls (preferably lithium butyl) is closely related to the sodium-initiated butadiene polymerization. The starting reaction is the formation of a lithium allyl by nucleophilic attack of BuLi on butadiene. Insertion of butadiene leads to chain growth with re-formation of an allyllithium group at the end of the chain. As allyllithium compounds are very stable, it is possible to maintain "living polymerization" reaction conditions under which chain termination (with and without deactivation of the catalyst) or other side reactions practically do not occur. This permits the synthesis of chain-end functionalized (end-capped) polymers, such as polymers with terminal –COOH or –CH$_2$CH$_2$OH groups formed by reaction with CO$_2$ or ethylene oxide. The 1,4-*cis* content of the polymer is relatively low (<40%); see Table 10.3. Much of the Li-BR (BR = butadiene rubber) is used in the production of rubber-modified polystyrene.

  Catalyst systems containing titanium (TiI$_4$/AlR$_3$; R = Et, *i*-Bu) were the first which were patented (1955, Phillips Petroleum, USA) for the production of polybutadiene with a high 1,4-*cis* content (>90%). Soon thereafter, cobalt- and nickel-based Ziegler-Natta catalysts were introduced and by the beginning of the 1960s, these catalyst systems were already being used to produce *high-cis*-polybutadiene rubber on an industrial scale. In the 1980s, systems containing neodymium were first used industrially [243]. Table 10.3 offers an overview of the various processes and the microstructure of the polymers and their properties.

- *trans*-1,4- and 1,2-Polybutadienes
  The heterogeneous Ziegler system VCl$_3$/AlEt$_3$ enables polymerization of butadiene into high-molecular, very pure (>99%) *trans*-1,4-polybutadiene (melting point 145 °C). Titanium-containing systems such as α-TiCl$_3$/AlEt$_3$ are less stereoselective. Cobalt-containing systems (e.g., Co(acac)$_3$/AlR$_3$/CS$_2$) in benzene as solvent yield syndiotactic 1,2-polybutadienes with a purity of >99%, with a high melting point (200–216 °C) and up to 80% crystallinity. Chromium-containing systems such as Cr(acac)$_3$ or [Cr(CO)$_5$(py)] in combination with aluminum alkyls produce, according to reaction conditions, mixtures of syndiotactic and isotactic 1,2-polybutadiene, from which the isotactic can be separated in crystalline form.

- Polyisoprenes
  The initiation of the polymerization of isoprene with alkyllithium (in hydrocarbons or without solvent) leads to a product with high *cis* content (96% *cis*-1,4-, 4% 3,4-polyisoprene). With titanium (TiCl$_4$/AlEt$_3$) and neodymium catalysts, *cis*-1,4-polyisoprene is also obtained (>95%). Similar to its behavior with butadiene, α-TiCl$_3$/AlEt$_3$ produces *trans*-1,4-polyisoprene.

  cis- and *trans*-1,4-polyisoprene correspond in their structure to natural rubber and gutta-percha/balata, respectively. Synthetic *cis*-1,4-polyisoprene has a market share of about 7% among synthetic rubber types. The synthesis of *trans*-1,4-polyisoprene is of minor industrial importance only. Like gutta-percha itself, it is used in the manufacture of golf balls, but also finds applications in medicine.

Table 10.3 Catalysts for the industrial production of polybutadiene rubber and typical properties of different rubber types (BR = butadiene rubber)[a] (compiled according to H.-D. Brandt et al. in [M14] ("Rubber, 3. Synthetic") and R. Taube, G. Sylvester in [M6], p. 285).

	Li-BR	Ti-BR	Co-BR	Ni-BR	Nd-BR
Catalyst system[b]	LiBu	$TiCl_4 + I_2 + Al(i\text{-}Bu)_3$ (1 : 1.5 : 8)	$Co(O_2CR)_2 + H_2O + AlEt_2Cl$ (1 : 10 : 200)	$Ni(O_2CR)_2 + AlEt_3 + BF_3 \cdot OEt_2$ (1 : 8 : 7.5)	$Nd(O_2CR)_3 + Al_2Et_3Cl_3 + Al(i\text{-}Bu)_2H$ (1 : 1 : 8)
Solvent	hexane/cyclohexane	aromatics	benzene/cyclohexane	aromatics	aliphatics/cycloaliphatics
Productivity[c]		4–10	40–160	30–90	7–15
Microstructure of the Polymers (in %)					
1,4-*cis*	36–38	93	97	97	98
1,4-*trans*	52–53	3	1	2	1
1,2 (vinyl groups)	10–12	4	2	1	1
Properties of the Polymers					
T_g (in °C)[d]	–93	–103	–106	–107	–109
Molar mass distribution	narrow	moderate	moderate	broad	very broad
Linearity	unbranched	low branching	variable	(low) branching	highly linear

a) As a result of the variability in catalyst systems and polymerization conditions, the properties vary within certain limits, so the data can only be used as a guide.
b) R indicates a longer alkyl chain, such as those that appear in fatty acids; this provides sufficient solubility of the catalyst system in the specified solvents.
c) In kg polybutadiene/g metal.
d) The glass transition temperature T_g depends on the vinyl group content and is thus almost identical for all systems with a high *cis* content.

Exercise 10.9

The monomer for biosynthesis of natural rubber is 3-methylbut-3-enyl diphosphate (**1**; IPP – isopentenyl pyrophosphate). The mechanism of polymerization is not fully clear; it may be a living cationic polymerization. If this is the case, then the growing polymer chain displays an allylic end group >C=CH–CH$_2$X (X = diphosphate), which is ionized by an enzyme (in combination with divalent metal cations) (**2** → **3**). Reaction with the monomer **1** produces a tertiary carbocation (**3** → **4**). Cleavage of HX regenerates the allylic chain end (**4** → **2'**). Why does the enzyme not react with the monomer **1**? Which chain initiation is to be considered for the biosynthesis when **1** is the only substrate available?

11
C–C Coupling Reactions

11.1
Palladium-Catalyzed Cross-Coupling Reactions

11.1.1
Introduction

Back in 1855, A. Wurtz found that alkyl halides R–X in the presence of sodium formed C–C bonds, producing longer-chain alkanes:

$$2\ R\text{–}X\ +\ 2\ Na\ \longrightarrow\ R\text{–}R\ +\ 2\ NaX$$

However, Wurtz reactions are accompanied by noticeable side reactions (elimination, rearrangement). About 50 years later, analogous coupling was found for organomagnesium compounds (V. Grignard, Nobel Prize in Chemistry, 1912). The reaction of R′–X with Mg to form Grignard compounds comprises an "Umpolung"[1] of the reactivity of R′–X such that the carbanionoid C atom in the Grignard reagent reacts with alkyl halides R–X with C–C bond linkage to form alkanes:

$$R'\text{–}X\ \xrightarrow{+\ Mg}\ R'\text{–}Mg\text{–}X\ \xrightarrow[-\ MgX_2]{+\ R\text{–}X}\ R\text{–}R'$$

Such reactions, in which two different organyl residues R and R′ are linked in a targeted way, are called "C–C cross-couplings." These cross-couplings are only formed at an acceptable rate when the halide in R–X is activated, as is particularly the case in allyl and benzyl halides. However, cross-coupling reactions can be catalyzed effectively with phosphane–nickel(0) complexes (M. Kumada, 1972):

$$R\text{–}X\ +\ R'\text{–}Mg\text{–}X\ \xrightarrow{[Ni^0]}\ R\text{–}R'\ +\ MgX_2$$

In the following years, palladium has proven to be the element of choice for catalysis of cross-couplings. Today there is a wide palette of palladium-catalyzed cross-coupling reactions in which a compound R–X (**1**; X = Hal, OTs, OTf, . . .) which contains an electrophilic C atom, reacts with an organoelement (metal or semi-metal)

---

[1] The German term "Umpolung" is used to describe a chemical modification of a functional group by which its commonly accepted donor/acceptor reactivity pattern is reversed.

Fundamentals of Organometallic Catalysis. First Edition. Dirk Steinborn
Copyright © 2012 WILEY-VCH Verlag GmbH & Co. KGaA, Weinheim

Table 11.1 Important palladium-catalyzed C–C cross-coupling reactions [244].

M	Year	Author[a]	[M]–R′	Comments
Mg	1972	Kumada-Tamao	XMg–R′	Ni-catalyzed (Pd-catalyzed[b])
Li	1975	Murahashi	Li–R′	
Cu	1975	Sonogashira	Cu–C≡CR′	in situ from HC≡CR′ + CuI + base
Zn[c]	1976–77	Negishi	R′Zn–R′; XZn–R′	analogous: AlR′$_3$[d] and [ZrCl(R′)Cp$_2$][d]
B	1979	Suzuki-Miyaura	[(base)(OH)$_2$B–R′]$^-$	in situ from BR′(OH)$_2$ + anionic base
Sn	1979	Stille	R″$_3$Sn–R′	R″ = alkyl
Si	1988	Hiyama	[F$_{n+1}$Me$_{3-n}$Si–R′]$^-$	in situ from SiR′Me$_{3-n}$F$_n$ + F$^-$

a) Name reactions also frequently appear with the second author omitted.
b) Murahashi (1975).
c) Also Al and Zr [245].
d) In the presence of ZnCl$_2$ or ZnBr$_2$, where necessary.

compound [M] – R′ (**2**), which has a nucleophilic C atom, to form coupling products **3**. The most important are collected in Table 11.1 [246].

$$\text{R–X} + [M]\text{–R}' \xrightarrow{[Pd]} \text{R–R}' + [M]\text{–X}$$

1   **2**   **3**

The Nobel Prize in Chemistry 2010 was awarded to Ei-ichi Negishi (Purdue University, West Lafayette, Indiana), Akira Suzuki (Hokkaido University, Sapporo, Japan) and Richard F. Heck (University of Delaware, Newark) for the development of palladium-catalyzed cross-coupling, one of the most sophisticated tolls available in organic synthesis today.

11.1.2
Mechanism of Cross-Coupling Reactions

The principal mechanism of cross-coupling is shown in Figure 11.1. The following individual reaction steps can be identified:

> **4 → 5**: *Ligand cleavage/attachment.* Starting from [PdL$_4$] (L = phosphane), ligand cleavage forms the actual catalyst, such as a bis(phosphane) complex (n = 2, Pd^0, 14 ve). If a Pd^{II} compound is used as precatalyst, reduction occurs first.
>
> **5 → 6**: *Oxidative addition/reductive elimination.* Oxidative addition of R–X to the Pd^0 complex [PdL$_2$] first yields an organyl palladium(II) complex *cis*-[PdR(X)L$_2$], which isomerizes to form the *trans* complex. The reaction can be reversible.
>
> **6 → 7**: *Metathesis-like (double displacement) reaction; in the broader sense: trans-metallation.*[2] Reaction with [M]–R′ (M = main group element) generates a

---

2) Transmetallations are in the stricter sense reactions between organometallics and metals that proceed with organyl group transfer: M + M′R$_n$ → M′ + MR$_n$. In the broader sense, transmetallations denote reactions in which an organyl group of a metal is transferred to another, which include metal–metal exchange reactions ([M]–R + [M′]–R′ → [M]–R′ + [M′]–R) and metathesis-like (double displacement) reactions ([M]–X + [M′]–R → [M]–R + [M′]–X). Particularly in the context of cross-couplings, the latter are classified as transmetallations.

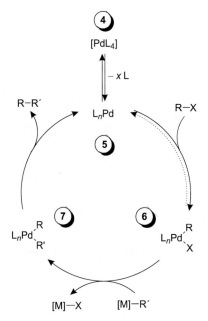

Figure 11.1 Basic mechanism of palladium-catalyzed cross-coupling (L = phosphane, $n = 1$ or 2).

diorganylpalladium(II) complex. In general, the transmetallation or the oxidative addition is the rate-determining step.

7 → 5: *Reductive elimination.* In a rapid, irreversible reaction, the product R–R′ is cleaved off and the catalyst complex is regenerated. The reductive elimination is substantially accelerated when it proceeds from a T-shaped intermediate [PdR-(R′)L] [247].

The precatalysts used are palladium(0) complexes ([PdL$_4$], [Pd(dba)$_2$]/[Pd$_2$(dba)$_3$] + x L; dba = dibenzylideneacetone, PhCH=CH–CO–CH=CHPh) or palladium(II) complexes ([PdCl$_2$L$_2$], Li$_2$[PdCl$_4$] + x L, Pd(OAc)$_2$ + x L, ...), preferably with L = PPh$_3$. The palladium(II) complexes are first reduced to palladium(0) complexes. This can occur, for example, by organylation of the divalent palladium and subsequent reductive elimination. Futhermore, nucleophiles Nu$^-$ can replace a phosphane ligand intra- (**8 → 10**) or intermolecularly (**9 → 10**), forming, in addition to [Pd^0], a phosphonium salt, which converts into a phosphane oxide through, for example, reaction with water [248].

$$[Pd^{II}] \overset{Nu}{\underset{PR_3}{\curvearrowright}} \quad \overset{Nu|^\ominus}{[Pd^{II}]-PR_3}$$
$$\mathbf{8} \quad\quad\quad \mathbf{9}$$

$$[Pd^0] + Nu-\overset{\oplus}{P}R_3 \xrightarrow{+ H_2O, - H^\oplus} NuH + OPR_3$$
$$\mathbf{10}$$

The precursor complex can determine the course of cross-coupling. If [PdL$_4$] (4) is the starting point, the equilibrium position ([PdL$_4$] ⇌ [PdL$_2$] + 2 L) causes the concentration of the catalytically active species [PdL$_2$] (5) to be very low. On the other hand, if L is present in excess, it stabilizes the complex and hinders the formation of inactive palladium clusters or metallic palladium. If Pd^{II} complexes are the starting point, anions X'$^-$ such as Cl$^-$ or OAc$^-$ are present. Kinetic experiments and quantum-chemical calculations show that reduction of Pd^{II} complexes then leads to anionic palladium(0) complexes of type [PdX'L$_2$]$^-$ (5'), which undergo oxidative addition of R–X more easily than the neutral complexes [PdL$_2$] (5) [249, 250].

Cross-coupling has a broad synthetic potential and the residues R and R' can be varied to a wide extent (Figure 11.2). The greatest restriction is probably that alkyl compounds R–X can only be used to a limited extent, since for residues R with β-H atoms, the intermediates [PdR(X)L$_n$] (6) are subject to rapid β-H elimination. In contrast, alkylmetal compounds [M]–R' can be used as transmetallation agents, since the reductive elimination of R–R' from the intermediates 7 can compete successfully with β-H elimination.

In the oxidative addition of R–X, the reactivity of the leaving group X plays a role in which the ordering I > OTf > Br ≫ Cl generally applies. For the less reactive electrophiles with sp^2- and sp-hybridized C atoms, better leaving groups (I, OTf, Br) are required, while for the more reactive benzyl and allyl compounds, chlorides can be used.

Cross-coupling is generally stereospecific. sp^2-hybridized C atoms in R and R' react while retaining their configuration, while for electrophiles R–X with sp^3-hybridized C atoms, both retention and inversion are observed.

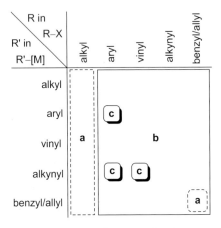

Figure 11.2 Extent of variation of R/R' in cross-coupling. (a) Generally unfavorable. (b) Favorable, but may be demanding. (c) Generally favorable and not highly demanding (adapted from E.-i. Negishi [251], p. 215).

11.1.3
Selected Types of Cross-Coupling

11.1.3.1 Cross-Coupling with Organolithium, Organomagnesium, and Organozinc Reagents

Organolithium, -magnesium, and -zinc compounds are so reactive that the transmetallation step in Murahashi, Kumada, and Negishi coupling requires no activation and can occur under milder reaction conditions than Suzuki and Stille coupling.

Coupling is limited to functionalities in R/R' that are tolerated by R'Li, R'MgX or ZnR'_2/ZnR'X compounds, respectively. Organozinc compounds offer advantages since they display very high reactivity but also tolerate a broad range of functional groups (–COR, –CO_2R, –CN, –Hal, –C≡CH, ...). As long as the organolithium or organomagnesium compound is easily available, synthesis of the corresponding organozinc compound (*in situ*) requires only a simple metathesis-like (double displacement) reaction with zinc halides (2 R'Li + ZnX_2 → ZnR'_2 + 2 LiX).

Cross-coupling with Grignard reagents R'MgX (but not with R'Li) have the advantage that they can be carried out with nickel catalysts (Kumada coupling). However, when disruptive β-H elimination occurs, the palladium-catalyzed reactions with bidentate diphosphane ligands have the advantage.

Either palladium(0) (e.g., [PdL_n], L = phosphane) or palladium(II) or nickel(II) complexes (e.g., [MX_2L_2]; M = Pd, Ni) are employed as precatalysts. Pd^{II} and Ni^{II} precatalysts are organylated first by the transmetallation reagent, and the actual Pd^0 or Ni^0 catalyst is then formed via reductive elimination.

Cross-couplings with Grignard reagents can also be catalyzed with iron. Even as early as 1971, it was shown by J. K. Kochi that in the presence of iron salts, alkyl Grignard compounds react with alkenyl halides under C_{sp^3}–C_{sp^2} bond formation. Since that time, the range of substrates amenable to such transformations has been expanded substantially and nowadays aryl chlorides and – even more exciting – alkyl halides can be used as electrophilic coupling partners. In contrast to the palladium-catalyzed reactions, low-valent iron compounds that are also known as "inorganic Grignard reagents," such as the cluster [$\{Fe^{-II}(MgX)_2\}_n$] (**11**), formed by reaction of RMgX with iron(II) chloride, play a role, at least as precatalysts. The significance of highly reduced iron complexes for such catalysis is also highlighted by the fact that the iron(−II) complex [Li(tmeda)]$_2$-[Fe(H_2C=CH_2)$_4$] is a very efficient precatalyst. However, it has not been conclusively established whether the catalytic cycle itself is based on a change in oxidation state Fe^{-II}/Fe^0, or whether Fe^0/Fe^{II} or Fe^I/Fe^{III} redox systems play a role (in which case iron(−II) complexes are "only" precatalysts) [252–254].

Iron also catalyzes the Grignard formation reaction itself, where the Fe^{-II} cluster **11** probably plays a decisive role. This has led to direct coupling of two electrophiles (Ar–Br and R–Br; R = alkyl), without requiring the extra step of a separate synthesis of a Grignard compound. It is assumed that first **11**, produced *in situ*, catalyzes formation of the Grignard compound (**12** → **13**) and then the same catalyst catalyzes the cross-coupling (13 → 14), so this is an iron-catalyzed domino reaction[3] [255].

$$\text{Ar–Br} + \text{R–Br} + \text{Mg} \xrightarrow[\text{THF (0} \rightarrow \text{20 °C)}]{\text{FeCl}_3 \text{ (5 mol\%)/TMEDA (1.2 equiv.)}} \left\{ \begin{array}{c} \text{Ar–Mg–Br} + \text{R–Br} \\ \text{or} \\ \text{Ar–Br} + \text{R–Mg–Br} \end{array} \right\} \xrightarrow{- \text{MgBr}_2} \text{Ar–R}$$

12   **13**   **14**

11.1.3.2 Suzuki Coupling

In Suzuki coupling, organyl groups R′ (alkyl, alkenyl, aryl, alkynyl, …) of organoboronic acids (**15a**) or boronic acid derivatives (**15b**, **15c**) are transferred onto palladium. Due to the relatively high stability of the B–C bonds, an anionic base (for example NaOH, NaOMe, NaOAc, Na_2CO_3, [N(n-Bu)$_4$]F) must be added to raise the carbanion activity so that tetracoordinated boronate anions (**15d**, R″ = H, alkyl, aryl, …; X = OH, OMe, F, …) will ultimately work as organylating agents. From the outset, tetracoordinated boron compounds (organotrifluoroborates **15e**) can also be used [256–258].

15a   **15b**   **15c**   **15d**   **15e**

Aryl bromides, iodides, and triflates are the preferred electrophilic coupling partners. Solvents such as THF, dioxane, EtOH, and benzene and aqueous solutions of the base are frequently used, while [Pd(PPh$_3$)$_4$] is commonly the precatalyst [259].

One advantage of Suzuki coupling is the easy availability of organoboronic acids, for example, via hydroboration, and their high tolerance for substituents such as –OH, –OR, –NR$_2$, –CHO, –C(O)R, –C(O)OR, –C≡N, –NO$_2$ (R = alkyl, aryl, …). For the synthesis of unsymmetric biaryls (including heterobiaryls), Suzuki coupling is the method of choice. Thus, on the order of about 100 metric tons/year of 2-cyano-4′-methylbiphenyl **16** is manufactured via Suzuki coupling as a pharmaceutical intermediate product. The product is isolated by phase separation, so the catalyst can be recycled.

---

3) A domino reaction, also known as a tandem reaction, consists of a sequence of two or more bond-forming transformations (very often C–C bonds) that follow one another without (in a strict sense) the reaction conditions having to be changed or new reagents or catalysts having to be added. The subsequent reaction in the cascade is a consequence of the functionality formed in the preceding step. In principle, each reaction step could also be carried out separately. In a broader sense, domino/tandem processes can also include the delayed addition of a reactant or catalyst.

11.1 Palladium-Catalyzed Cross-Coupling Reactions

[Reaction scheme: 4-NC-C6H4-B(OH)2 + Cl-C6H4-CN → biaryl product **16**, using [Pd]/P(m-NaOSO$_2$C$_6$H$_4$)$_3$, glycol/base (120 °C)]

Exercise 11.1

Difunctional (or multifunctional) components enable the construction of complex molecules by successive Suzuki coupling steps. To obtain these molecules selectively, and thus to achieve an orthogonal functionalization,[4] one strategy, among others, is to use difunctional building blocks with a temporarily inactive ("protected"/ "masked") boronic acid derivative such as R–B(dan) (**2**). What is the source of this decrease in reactivity? Build up compound **3** by four successive Suzuki cross-couplings (the bonds to be formed are marked in bold).

[Scheme: R–B(OH)$_2$ (**1**) + 1,8-diaminonaphthalene (H$_2$dan), toluene (111 °C) / H$^+$ in THF/H$_2$O (25 °C) → R–B(dan) (**2**); and target compound **3** (quaterphenyl with Me, OMe, OMe, B(dan) substituents, positions labeled 1, 2, 3, 4)]

11.1.3.3 Hiyama Coupling

Hiyama C–C coupling is the basis of the following palladium-catalyzed reaction (R = alkenyl, aryl, allyl, …; X = Cl, Br, I, OTf, …):

[Scheme: –Si–R' (**17**) + F$^\ominus$ → pentacoordinate fluorosilicate **18**; R–X + **18** → R–R' + –Si–F, with [Pd], –X$^\ominus$]

Corresponding to the general mechanism for cross-coupling, in a transmetallation reaction (see reaction **6 → 7** in Figure 11.1), organyl groups R' (alkyl, alkenyl, aryl, alkynyl, …) are transferred from organosilicon compounds **17** onto palladium. The high stability and low reactivity of the Si–C bond necessitates activation, as for B–C

---

4) The concept of orthogonality was introduced at the end of the 1970s into peptide chemistry and indicates the property of protecting groups or linkers that permit the removal, modification, or cleavage of one such structure without affecting others. In the broader sense, a molecule that can react at multiple reactive sites is said to have *orthogonal functionalization* when only one reaction occurs, selectively. For orthogonal cross-couplings, the reactivity difference between the electrophilic coupling partners (C–I > C–Br >> C–Cl), for example, could be put to advantage, or in the given example, a masking/demasking strategy could be used with the boron reagent.

bonds in Suzuki coupling. To facilitate the activation of the Si–C bond, the exceptionally high stability of Si–F bonds ($\Delta_d H = 565$ kJ/mol) is used by adding at least stoichiometric quantities of fluorides (such as [N(n-Bu)$_4$]F, KF) or fluoride donors such as [(Et$_2$N)$_3$S][SiF$_2$Me$_3$] to the reaction mixture. These react with the organosilanes **17** to form the actual organylating agents, namely pentacoordinated fluoro(organyl)silicates **18**. The corresponding methyl- or fluoro(methyl)silicon compounds Me$_{3-n}$F$_n$Si–R' ($n = 0$–3) are frequently employed as substrates **17**. For alkylation (R' = alkyl), alkylsilicon trifluorides F$_3$Si–R' are preferred and fluorides are added in excess in order to trap the resulting SiF$_4$ in the form of [SiF$_5$]$^-$/[SiF$_6$]$^{2-}$.

There is also a fluoride-free variant of Hiyama coupling, which is illustrated using the example of an aryl–alkenyl coupling: Alkali metal silanolates (produced from the corresponding silanols and a base such as KOSiMe$_3$, Cs$_2$CO$_3$, NaOR, KH, ...) react with the intermediates **6** (Figure 11.1) with cleavage of the alkali metal salt M'X to form an arylpalladium silanolate **6'**. Intramolecular transmetallation (**6'** → **7**) results releasing the polysiloxane (Me$_2$SiO)$_n$. Reductive elimination of ArCH=CHR' and oxidative addition of ArX as in Figure 11.1 (**7** → **5** → **6**) complete the reaction cycle. Kinetic experiments show that, unexpectedly, no intermediate product with pentacoordinated silicon appears [260].

11.1.3.4 Stille Coupling

The tendency to transfer the organyl group R' from Sn onto Pd grows with increasing electronegativity of R' in the order:

$C(sp^3)$ $< C(sp^2)$ $< C(sp)$
alkyl ≪ benzyl/allyl < aryl < alkenyl < alkynyl

Thus, from trialkyltin organyls R"$_3$Sn–R' (R" = alkyl), only R' is transferred onto palladium, as long as this is not itself an alkyl group. Functional groups are tolerated to a large extent. The advantage of Stille coupling over Suzuki coupling is that base-sensitive groups are tolerated, but the disadvantage is that organotin compounds have a higher toxicity [261].

In addition to palladium(0) complexes ([Pd(PPh$_3$)$_4$], [Pd$_2$(dba)$_3$]/PPh$_3$), palladium(II) complexes such as Pd(OAc)$_2$/PPh$_3$ or [PdCl$_2$(PPh$_3$)$_2$], which first are reduced to Pd^0 complexes, are used as precatalysts. DMF is the preferred solvent.

Transmetallation can be understood as electrophilic substitution (S$_E$2) as exemplified here by the reaction R"$_3$Sn–R' + [PdX(R)L$_2$] (R' = alkyl). Investigations to elucidate the mechanism show that a cyclic transition state **19** can form, as long as X is a ligand (such as a halide) capable of forming Pd–X–Sn bridges. Hence, L and then

R"₃SnX can be cleaved so that [PdR(R')L], which readily undergoes reductive elimination of R–R', is obtained directly. The configuration of the α-C atom of R' is retained. With a nonbridging ligand X (e.g., OTf) that can be easily replaced, the palladium complex exists as a cationic complex [PdRL₂(s)]⁺, especially in strongly solvating solvents s. In these cases, the formation of an open transition state **20** may be favored in the reaction with R"₃Sn–R'. The transmetallation then proceeds with inversion of the configuration at the α-C_{sp^3} atom of R' [248, 262].

Exercise 11.2

The addition of olefins can exert a substantial influence on the activity and selectivity of transition-metal-catalyzed cross-couplings [263]. An example is the reaction of 1-bromoethylbenzene with tetramethyltin catalyzed by [PdEt₂(bpy)], which produces styrene (**2**) as main product, while addition of fumaronitrile (L) or use of [Pd(bpy)L] as catalyst yields isopropylbenzene (**1**) as main product. Upon which reaction course the formation of **1** and **2** is based? Interpret the result.

[Pd]	1 (in %)	2 (in %)
[PdEt₂(bpy)]	2	34
[PdEt₂(bpy)] + L (1:7)	77	12
[Pd(bpy)L]	34	8

11.1.3.5 Sonogashira Coupling

In Sonogashira coupling, terminal alkynes **21** react with aryl or vinyl compounds R–X (**22**, R = aryl, vinyl; X = Cl, Br, I, OTf) with stoichiometric quantities of a base (e.g., NEt₃, NHEt₂, piperidine) in the presence of a palladium catalyst such as [Pd(PPh₃)₄] and catalytic quantities of CuI to form **23**.

R'≡H + R–X + NR"₃ —[Pd]/CuI→ R'≡R + [NHR"₃]X
   21   **22**   **23**

The overall equation shows that the alkyne hydrogen atom is replaced by R. Thus, Sonogashira coupling is the alkyne variant of a Heck reaction, in which a vinylic H atom is replaced by R (see Section 11.2). It has proven to be an elegant synthetic method for formation of C_{sp^2}–C_{sp} bonds and thus of enynes and enediynes, which are significant in the chemistry of natural products.

Terminal alkynes react with excess bases such as trialkylamines or piperidine, in the presence of copper(I) iodide to form alkynylcopper reagents that transfer the alkynyl group onto the palladium. This regenerates the copper(I) halide, so the reaction is catalytic not only with respect to palladium but copper as well [264, 265].

Exercise 11.3

There is a copper-free variant of the Sonogashira reaction, for which the following mechanism is proposed: After oxidative addition of Ar–I to a Pd^0 complex, the terminal alkyne Ar′–C≡C–H is coordinated (**24** → **25**). If phenylacetylenes with electron-withdrawing substituents are used, this is followed by a base-assisted deprotonation of the alkyne (**25** → **26**) as rate-determining step, while for phenylacetylenes with electron-donating substituents, there is evidence for the reaction path **25** → **25′** → **26′**. Finally, the product Ar–C≡C–Ar′ is cleaved off by reductive elimination (**26/26′** → …). Give reasons for the change in mechanism.

11.1.3.6 Ligand Effects

The classical catalysts for cross-coupling use phosphanes, particularly PPh_3, as coligand L. A significant increase in activity and stability of catalysts is achieved by targeted variation in steric *and* electronic properties of L.

If instead of the relatively weakly basic PPh_3, a strongly basic *and* sterically demanding phosphane such as $P(t\text{-}Bu)_3$ or $PPh(t\text{-}Bu)_2$ is used (compare to the Tolman parameters of PPh_3 with those of $P(t\text{-}Bu)_3/PPh(t\text{-}Bu)_2$, Section 10.4.2.2), the formation of monophosphane–palladium(0) intermediates (see Figure 11.1, $n = 1$) is favored. Coordination of solvent molecules and/or agostic interactions can contribute to further stabilization. On the whole, all reaction steps of cross-coupling with PdL complexes (12 ve!) seem to proceed more easily than with PdL_2 complexes (14 ve) [266, 267].

An example is the Suzuki coupling of chloroaromatics **27** to form biaryls **28**, which is not possible with PPh_3 (yield 5%), whereas with $PBu(Ad)_2$ (Ad = adamantyl) yields > 90% (TON = 17400) are obtained. Even electron-rich chloroaromatics (R = 4-MeO, 2,6-Me$_2$) react with high turnover numbers (TON > 10^4).

L	TON
PPh$_3$	50
P(t-Bu)$_3$	8200
PBu(Ad)$_2$	17400

N-Heterocyclic carbenes (see Background, Section 7.1.3), derived from imidazole (**29**, R = t-Bu, i-Pr), represent an additional class of strongly basic (nucleophilic) and bulky ligands that, when coordinated to palladium, form highly active catalyst systems for cross-coupling. Monocarbene–palladium(0) complexes (12 ve) are probably the species that are catalytically active. Surprisingly, however, very productive catalyst systems with bidentate and even multidentate P-, N-, and carbene coligands can be formed. These ligands may hinder the aggregation of Pd^0 to form catalytically inactive clusters [268, 269].

11.1.3.7 Alkyl–Alkyl Coupling

In cross-coupling reactions, C_{sp^3}–C_{sp^3} bond formation between two alkyl residues is generally not possible, since alkyl derivatives R–X cannot be used as electrophiles (Figure 11.2). The reason is that the oxidative addition of R–X to [M'^{0}] (M' = Pd, Ni) (**30** → **31**) runs very slowly and furthermore, if R has a β-H atom, the alkylmetal complex undergoes a rapid β-hydride elimination (**31** → **33**). Thus, the transmetallation (**31** → **32**), which is usually slower, cannot compete, so the coupling product (**32** → **34**) is not formed [270–272].

Nowadays, it has become possible in many cases to use primary alkyl halides as electrophilic coupling partners in cross-coupling, as is increasingly also true for secondary alkyl halides, where steric hindering further impedes oxidative addition [273].

A nickel-catalyzed coupling reaction of alkyl iodides or bromides R–X with dialkylzinc compounds and alkylzinc iodides ZnR'$_2$/R'ZnI succeeds in the presence of π acceptors such as 4-fluoro- or 3-trifluoromethylstyrene. An advantage is that these coupling reactions show a high tolerance for functional groups [274].

$$R-X + R'-Zn-R' / R'-Zn-I \xrightarrow[\pi \text{ acceptor}]{[Ni(acac)_2] ([NBu_4]I)} R-R'$$

It seems to be crucial for success that transmetallation runs rapidly and that the added π acceptor accelerates the reductive elimination and/or blocks the vacant coordination sites necessary for a β-H elimination from [Ni]–R.

A further example of alkyl–alkyl coupling is Suzuki coupling of alkyl derivatives R–X (X = Cl, Br, OTs) with R'–(9-BBN) (R' = alkyl, 9-BBN = 9-borabicyclo[3.3.1]-nonane) in the presence of bases such as $K_3PO_4 \cdot H_2O$, $CsOH \cdot H_2O$ or NaOH [275].

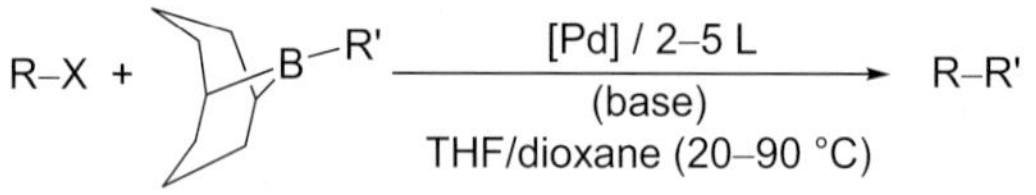

$$R-X + \text{(9-BBN-R')} \xrightarrow[\text{THF/dioxane (20-90 °C)}]{[Pd] / 2-5 \text{ L} \atop \text{(base)}} R-R'$$

$Pd(OAc)_2$ or $[Pd_2(dba)_3]$ is employed as the [Pd]/L precatalyst in the presence of an excess (!) of a strongly basic but bulky phosphane L such as PCy_3 or $PMe(t\text{-Bu})_2$. Apparently, the excess of L ensures the formation of bis(phosphane) complexes that are sufficiently resistant to undesired β-H elimination. However, cleavage of L seems to occur sufficiently easily to permit the desired reaction cycle to proceed. A careful selection of the base, the phosphane ligand, and the reaction conditions (specially attuned to the substituent X in R–X) for each individual case is crucial for success.

Exercise 11.4
Alkyl–alkyl coupling according to the following equation (R, R' = alkyl) is effectively catalyzed by $NiCl_2$ in the presence of butadiene or isoprene:

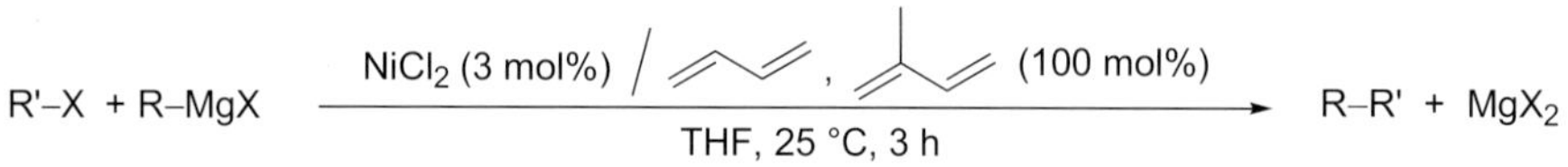

$$R'-X + R-MgX \xrightarrow[\text{THF, 25 °C, 3 h}]{NiCl_2 \text{ (3 mol\%)} / \text{butadiene, isoprene (100 mol\%)}} R-R' + MgX_2$$

The findings that β-H elimination only plays a minor role and that the oxidative addition of R'–X (R' = alkyl!) runs relatively rapidly point to a completely different mechanism. Present a proposal that first considers the formation of Ni^0 (how?) and then its reaction with butadiene.

11.1.3.8 Enantioselective Cross-Coupling

Atropisomeric o,o'-substituted biaryls are axially chiral. They are of interest as chiral ligands and as intermediates in the synthesis of natural products. Thus, as shown in the following reaction scheme, Suzuki coupling in the presence of the chiral P,N-ligand **35** produces the biaryl (**36**, R = $P(O)(OMe)_2$; ee = 86%), which, through reaction with PhMgBr and subsequent reduction, produces the chiral *P*-ligand **36** (R = PPh$_2$) in high enantiomeric purity [276].

Similarly, but via catalysis by nickel, axially chiral biaryls can be obtained by Kumada coupling from aryl Grignard compounds and 1-bromonaphthalenes. Another concept is implemented in the Kumada coupling of **38**. In **38**, the two triflate substituents are enantiotopic. The palladium complex of **37** is chiral and can distinguish between the two enantiotopic positions in **38**. Kumada coupling with PhMgBr yields the axially chiral biphenyl **39a** (*ee* = 93%) with the achiral diphenylated product **39b** as side product.

Despite the previously discussed principal difficulties in using alkyl halides as electrophilic coupling partners, even secondary alkyl halides, which as reactants are particularly suited to the formation of stereocenters, have been employed recently. It has thus been possible to achieve nickel-catalyzed enantioselective Negishi, Hiyama, and Suzuki cross-coupling using chiral bidentate and tridentate *N*-coligands of types **40** and **41**. Examples are alkyl–alkyl Suzuki coupling reactions starting from *racemic* secondary alkyl halides and R″–(9-BBN) derivatives (R′, R″ = alkyl) that have yielded alkyl compounds **42** with *ee* values up to 94% [277].

11.1.3.9 Carbonylative Cross-Coupling

In the presence of carbon monoxide, cross-coupling occurs as "carbonylative coupling," since CO insertions into Pd–C bonds proceed quickly. The reaction then yields not hydrocarbons R–R′, but rather unsymmetrically substituted ketones R–C(O)–R′. The mechanism in Figure 11.1 must be modified as follows: As it indicates, oxidative addition of R–X to L_nPd (5 → 6) occurs first. However, this is followed not by transmetallation (6 → 7), but rather CO insertion (6 → 6′). Transmetallation (6′ → 7′) occurs next and finally, in a reductive elimination reaction, the ketone is cleaved off and the catalyst is regenerated (7′ → 5).

$$L_nPd \xrightarrow{RX} L_nPd\begin{smallmatrix}R\\X\end{smallmatrix} \xrightleftharpoons{CO} L_nPd\begin{smallmatrix}C(O)R\\X\end{smallmatrix} \xrightarrow{[M]-R'\ \ [M]-X} L_nPd\begin{smallmatrix}C(O)R\\R'\end{smallmatrix} \longrightarrow R-C(O)-R' + L_nPd$$

(5)   (6)   (6′)   (7′)   (5)

Carbonylative coupling substantially extends the synthetic potential of cross-coupling. Acid chlorides, for instance, can only be used in some cases as electrophilic coupling components for the direct synthesis of ketones due to their high reactivity. In a Suzuki coupling reaction, this would not be possible, as aqueous bases are present. Carbonylative cross-coupling is often the method of choice for synthesizing unsymmetrically substituted ketones. Carbonylative Stille coupling reactions are important for ketone synthesis with a variety of sensitive substituents [278].

Exercise 11.5

Propose (a) a synthesis of **2** (starting from **1**) and (b) a synthesis of the ketone **3** (X = OH, COOH, NH_2, …) by Stille coupling.

1   **2**   **3**

11.2 The Heck Reaction

In Heck reactions (R. F. Heck, 1972; Nobel Prize in Chemistry 2010), a C–C bond is formed under palladium catalysis, whereby the vinylic hydrogen atom is replaced by an organyl group R (aryl, vinyl, benzyl, allyl) while retaining the double bond:

$$\text{C=C(H)} + R-X + B \xrightarrow{[Pd]} \text{C=C(R)} + (BH)X$$

The reaction requires stoichiometric quantities of a base B. Over the course of the last three decades, Heck reactions have established themselves as one of the most important methods for synthesizing styrene derivatives (R = aryl), 1,3-dienes (R = vinyl) and allylbenzenes (R = benzyl).

Typically, Heck reactions are conducted in polar aprotic solvents (MeCN, DMF, DMSO; if necessary, some water may be added) at temperatures of 50–150 °C. Preferred substrates R–X are bromides, iodides and triflates. Chlorides are rarely used, since the oxidative addition proceeds with more difficulty or not at all. The most frequently used bases are secondary or tertiary amines, carbonates, or hydrogen carbonates [279].

11.2.1
Mechanism of Heck Reactions

Although all details have not yet been clarified, when phosphane–palladium complexes are used as catalysts, the reaction mechanism is very probably that in Figure 11.3. The following individual reactions occur:

43 → 44 → 45: As with cross-coupling, the Heck reaction begins with the formation of the catalytically active palladium(0) complex **44** and the subsequent oxidative addition of RX to form **45**.

45 → 46: Olefin coordination/insertion. Olefin coordination and subsequent insertion into the σ-Pd–C bond form an alkylpalladium(II) complex.

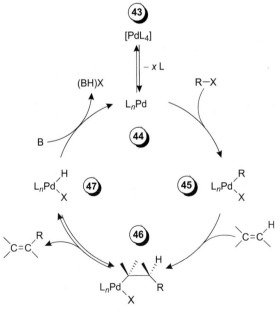

Figure 11.3 Mechanism of the Heck reaction with phosphane–palladium complexes as catalysts (L = phosphane; $n = 1, 2$).

46 → 47: *β-H elimination/insertion.* β-Hydrogen elimination leads to formation of the product and a hydridopalladium(II) complex. This reaction step is reversible. Starting from terminal olefins R'CH=CH$_2$, either the more thermodynamically stable *trans* olefin (E)-R'CH=CHR or CH$_2$=CRR' is obtained. The reactions **45 → 46** and **46 → 47** proceed in a stereochemically defined manner as *syn* addition and *syn* elimination, respectively. Hence, the *syn* elimination must be preceded by a rotation around the C–C bond such that a β-C–H bond is synperiplanar to the Pd–C bond.

47 → 44: *Reductive elimination.* The cleavage of HX with stoichiometric quantities of a base B leads to regeneration of the catalyst complex **44**.

In a fundamental distinction from the previously mentioned cross-coupling reactions, the Heck reaction causes C–C bond formation via olefin insertion into a Pd–C bond (**45 → 46**) and not via reductive C–C elimination. Regarding RX, Heck reactions are subject to the same restrictions as cross-couplings: The residue R must not contain β-H atoms, since decomposition via β-H elimination of [PdR(X)L$_2$] (**45**) would proceed more easily than olefin insertion.

11.2.2
Mechanism – A Closer Look

Anion influence (X$^-$) on the course of reaction is not considered in the (simplified) mechanism above. However, it substantially affects the activity, regioselectivity, and stereoselectivity of the reaction. As with cross-couplings, in the presence of chloride or acetate anions X$^-$, the complexes formed are not neutral PdL$_2$ complexes, but rather anionic complexes [PdXL$_2$]$^-$ that act as precursors for the Heck cycle, leading to easier oxidative addition [280].

Depending on the ligand L and the anionic ligand X$^-$, two different reaction paths for olefin insertion have been demonstrated (s = solvent; dashed arc indicates that L$_2$ can also be a chelating ligand) [281, 282]:

- *Nonpolar route via neutral complexes.* Solvent-assisted substitution of a neutral ligand L by the olefin (**48 → 49a**).
- *Polar route via cationic complexes.* Solvent-assisted substitution of an anionic ligand X$^-$ by the olefin (**48 → 49b**).

In particular, the mechanism can be controlled via the nature of the anion X$^-$: Anions that are easily cleaved off such as triflate, or halides in the presence of AgX' or

11.2 The Heck Reaction

TlX′ (X′ = weakly coordinating anion) favor the polar route. Competitive experiments have shown that the neutral complexes in the nonpolar route favor the coordination of electron-deficient olefins (poor σ donors, good π acceptors) whereas the cationic complexes in the polar route favor the coordination of electron-rich olefins (good σ donors, poor π acceptors). The regioselectivity for an insertion starting from **49a** is dominated by steric factors, whereas for **49b**, electronic factors are important. For the polar route, this leads increasingly to the formation of branched products, as illustrated by the following examples [281, 283]:

	=Y b)	=Ph	=OH	⟩=OH	=⌒OH	=n-Bu
nonpolar pathway a)	100	100	100	90	80	80
polar pathway a)	100	60	100	95	90	80–85

a) The arrows point to the favored site for substitution; figures in percent (differences to 100% correspond to the other regioisomer). b) Y = COOR, CONH$_2$, CN.

11.2.3
Ligand Effects

Palladium acetate reacts with P(o-Tol)$_3$ or P(Mes)$_3$ with metallation of an *ortho*-methyl substituent to form dinuclear palladacycles **50** (R = o-Tol, Mes; R′ = H, Me).

50 **51**

Type **50** complexes are air and moisture stable and, in the solid state, unusually thermally stable (**50**, R′ = H; $T_{dec.}$ = 250 °C). They have shown themselves to be extraordinarily active precatalysts for the Heck reaction, even allowing Heck coupling of chloroaromatics with electron-withdrawing substituents (in the presence of bromides such as [N(n-Bu)$_4$]Br as promoter). It seems certain that they do not participate directly in the catalytic process. They serve as a resting state for the actual catalyst, a monophosphane–palladium(0) complex **51**, which they release in a very slow reaction. Due to its high catalytic activity, **51** then reacts in a very rapid oxidative addition reaction with R–X. Accordingly, complexes [PdAr(Br){P(o-Tol)$_3$}$_2$] show an activity and selectivity similar to the palladacycles **50**. The catalytically active compound **51**, generated from **50**, belongs to the group of PdL complexes with strongly σ-basic phosphane ligands of high bulkiness. The previous discussion on the extraordinary catalytic activity of these complexes in cross-coupling reactions can also be applied to Heck reactions. Strongly basic (nucleophilic) N-heterocyclic carbenes, derived from imidazole (see Background, Section 7.1.3), also form very stable monocarbene– and dicarbene–palladium(II) complexes (**52**, **53**), which are precatalysts of very high activity [269, 284, 285].

From the mechanism of the Heck reaction it can be seen that no strongly σ-bonding ligand is required for any of the catalytically relevant reaction steps. Catalyst systems without such ligands are called "ligand-free." Thus, for example, aryl iodides react with cycloalkenes or methylacrylates in a Heck reaction under Jeffery-Larock conditions, that is, Pd(OAc)$_2$ is used as precatalyst and NaHCO$_3$ or KOAc as base (DMF, 25–50 °C) in the presence of quaternary ammonium salts such as [N(n-Bu)$_4$]Cl. These function, among other roles, as phase-transfer catalysts (solid-liquid PTC). In aqueous solvent or in water itself, liquid-liquid phase-transfer catalysis can also occur [279].

Pd^0 nanoparticles,[5] which are used directly or can form through decomposition reactions of a catalyst system, thus also catalyze the Heck reaction. However, they are not catalytically active themselves, but rather serve only as a reservoir from which either Pd atoms or, after oxidative addition of RX, soluble Pd^{II} species are cleaved off. There are indications that at higher temperatures (>120 °C), soluble palladium(0) colloids formed by decomposition are crucial for catalysis, largely independent of the precursor complex [286, 287].

A "ligand-free" catalyst system is the basis of the industrial synthesis of cinnamic acid derivatives such as p-methoxycinnamic acid **54**. The 2-ethylhexyl ester (**54'**, R = CH$_2$CH(Et)n-Bu) of **54** ("Octinoxate") is a component of sunscreens

---

[5] Pd^0 nanoparticles or colloids denote here a soluble agglomeration of palladium atoms in the nanometer range (typically 1.5–7.0 nm). They are generally stabilized by salts or ligands that hinder formation of insoluble larger aggregates ("palladium black") that have little or no catalytic activity. Halides are often used as salts, which also make the formation of anionic species such as [Pd^{0}X]$^-$ or [Pd^{II}RX$_2$]$^-$ possible.

11.2.4
Enantioselective Heck Reactions

In general, no stereocenter is formed in the Heck reaction. With monosubstituted olefins **55**, the intermediate **56/56'** does contain a stereogenic C atom, but it is lost again by β-H elimination (**56'** → **57**). Since reaction **55** → **56** is a *syn* addition and **56'** → **57** is a *syn* elimination, the latter must be preceded by a rotation about the C–C bond (**56** → **56'**) so that the Pd–C and C–H bonds are synperiplanar.

In the Heck reaction, the temporarily generated stereocenter is only retained when disubstituted olefins **58** (R' ≠ H) are used as substrates. After olefin insertion (**58** → **59**), β-H elimination of the hydrogen atom H' can occur (**59** → **60**; Heck reaction with double-bond isomerization). With R" = H, a tertiary stereogenic C atom is formed; with R' ≠ R" ≠ H it is quaternary [276].

If a tertiary stereogenic C atom (R" = H) is to be formed, the β-H elimination, which leads to the "normal" Heck product (in which R" = H is replaced by R), must be suppressed. This is especially the case for intramolecular Heck reactions of cycloalkenes, since rotation around the C–C bond (**56** → **56'**) is then impossible. Thus, fused ring systems can be favorably synthesized in enantioselective Heck reactions, as demonstrated by the synthesis of decalins (**61** → **62**). With (R)-BINAP (for formula, see Figure 4.5 in Section 4.3.1.) as chiral P,P-ligand, *ee* values of up to 93% have been achieved.

The generation of a quaternary stereogenic C atom by an intramolecular Heck reaction is demonstrated by the example of the reaction **63** → **64**. With (R)-BINAP as chiral P,P-coligand, regioselective asymmetric cyclization occurs, forming **64** with *ee* = 95%. **64** is an intermediate product in the synthesis of the tricyclic diterpene **65** ((−)-abietic acid, a resin acid, main component of colophonium).

63 →[Pd(OAc)₂/(R)-BINAP, (K₂CO₃), toluene, 50 °C]→ **64** (ee 95%) ⇌ **65**

Exercise 11.6
- Complete the reaction cycle **61** → **62** by giving the formulas of the intermediates. Explain why *cis*-decalins are formed and why a double-bond shift occurs.
- In principle, there are four possible ways the conjugated double-bond system of **63** can insert into the Pd–C$_{Ar}$ bond. Discuss them and give reasons why **64** is formed exclusively.

There are many coligands other than BINAP (see example above) that can be used for chiral induction. Among them are 2-(phosphinophenyl)oxazolines of type **66** (PHOX ligands), which give high catalytic activity and high *ee* values in intermolecular Heck arylation and alkenylation of, for example, dihydrofurans. It is evident that in both intra- and intermolecular asymmetric Heck reactions, coordination of the coligand in a chelating fashion allows for better chiral induction than in a monodentate, so one should endeavor to achieve the polar mechanism for the insertion reaction.

11.3
Palladium-Catalyzed Allylic Alkylation

11.3.1
Principles and Mechanism

The substitution of X (X = OAc or OCO₂R, or, less commonly, Cl, Br, OH, OPh, SO₂Ph, CN, …) in allyl derivatives **67** by nucleophiles Nu⁻ is catalyzed effectively by palladium (Tsuji–Trost reaction). Allylic substitutions by stabilized (soft) carbanions such as malonic acid derivatives **69** (Y, Y' = electron-withdrawing substituents: COOR, COR, CN, …) and by hard carbanions **70** such as unstabilized alkyl and aryl anions are of particular value in synthesis.[6]

R⌒⌒X + Nu⁻ →[Pd] R⌒⌒Nu + X⁻   Nu⁻ = ⁻CH(Y)(Y'), R'⁻

67 → **68**   **69**   **70**

---

[6] We restrict ourselves to C-nucleophiles. Many heteroatomic nucleophiles (amines, phenols, …) belong to the soft nucleophiles; H⁻ is a hard nucleophile. The generic term for all these reactions is "allylic substitution"; reactions with C-nucleophiles are often called "allylic alkylation."

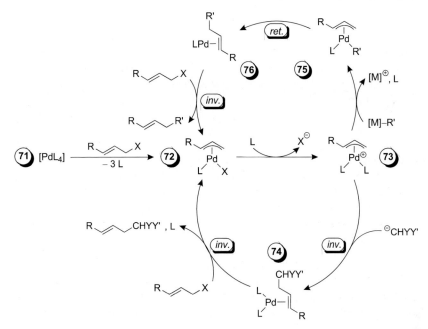

Figure 11.4 Mechanism of palladium-catalyzed allylic alkylation (X = Br, Cl, OAc, …; L = phosphane, …; R, R′ = H, alkyl, aryl). ret./inv.: Reaction proceeds with retention (ret.) or inversion (inv.) of the configuration at the C1 atom of the allyl group.

The mechanism of palladium-catalyzed allylic alkylation is shown in Figure 11.4. Pd^0 complexes or Pd^{II} complexes that must first be reduced are used as precatalysts. Furthermore, (η^3-allyl)palladium(II) complexes of type **72** (Figure 11.4) may also be used as catalysts. The following individual reactions occur:

71 → 72: *Oxidative addition.* Starting from a Pd^0 complex as precatalyst, oxidative addition of the allyl derivative generates a neutral (η^3-allyl)palladium(II) complex.

72 → 73: *Ligand substitution.* Substitution of the anionic ligand X^- by L produces a cationic (η^3-allyl)palladium(II) complex. While the neutral complexes **72** are relatively inert against nucleophiles, the cationic complexes **73** – due to the higher electrophilicity of the palladium – are very reactive against nucleophiles.

73 → 74: *Nucleophilic addition.* Soft C-nucleophiles react in an intermolecular addition reaction with C–C bond formation, forming an olefinpalladium(0) complex.[7] The addition is to one of the two terminal C atoms at the π-allyl

---

7) It may be easier to understand the reaction if one of the two π-allyl ligand resonance structures **4a/b**, (see Section 10.2.1) is taken as the formal (!) starting point. Then the reaction represents a nucleophilic substitution at the C1 atom of the allyl group with a PdL_2(olefin) unit (with acceptance of the electron pair, hence, reduction of Pd^{II} to Pd^0!) as leaving group:

group; for unsymmetrically substituted allyl groups, it is generally to the one that is less substituted. However, it is possible to control the regioselectivity to some extent via the choice of the substrate and the reaction conditions [288].

73 → 75 → 76: *Transmetallation, reductive elimination.* Hard nucleophiles such as Zn, B, or Sn alkyls/aryls attack the palladium directly and react, causing transmetallation. Reductive elimination with C–C bond formation generates the product, which is intermediately coordinated to Pd^0.

74/76 → 72: *Ligand cleavage, oxidative addition.* Cleavage of the product and oxidative addition of the allyl compound regenerates the catalyst complex. The oxidative addition proceeds in a stereochemically defined manner, with inversion.

Exercise 11.7

In the reaction of (η^3-allyl)palladium complexes in the presence of strong σ-donor ligands such as tmeda with less stabilized nucleophiles, an attack on the C2 atom of the allyl ligand is also observed. One example is the following reaction, which proceeds with cyclopropanation:

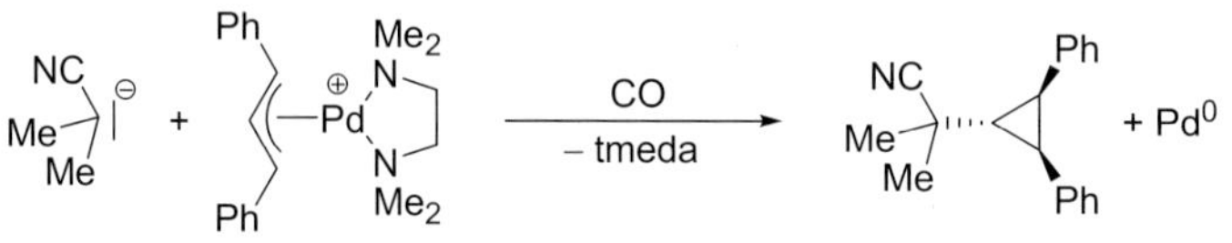

In the absence of CO, an intermediate product has been isolated and structurally characterized. Give a formula for it.

The stereochemically uniform course of oxidative addition (74/76 → 72) is due to the fact that the palladium attacks the allyl system on the face opposite substituent X. This always causes the allyl ligand to coordinate with the same prostereogenic face (either *Re* or *Si*) to Pd ("enantiofacial differentiation/selection"). The oxidative addition 77 → 78 is an example. (R serves as a "marker" for distinguishing one face of the cyclohexenyl ligand from the other.)

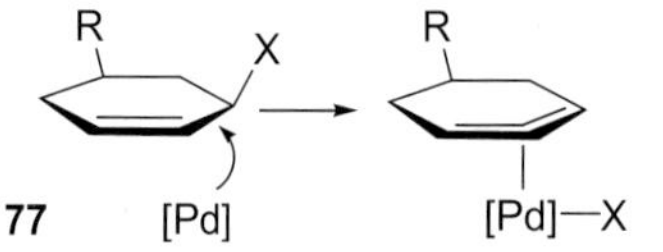

The stereochemically uniform course of reaction is only ensured if no racemization ("stereoscrambling") occurs, which, in the case of π-allyl systems, would be associated with a switch in *Re/Si* coordination. This can happen intramolecularly by π-σ-π-allyl rearrangement (see Section 10.2.1)[8] or intermolecularly by Pd–Pd exchange (79 ⇌ 80).

$$\text{79 (Si face)} \rightleftharpoons \text{coordination at the ...} \rightleftharpoons \text{80 (Re face)}$$

The reaction path 73 → 74 vs. 73 → 75 → 76 (Figure 11.4) determines the stereochemistry of the overall reaction. The intermolecular addition 73 → 74 is a *trans* addition that proceeds with inversion of the configuration at the C1 atom of the allyl group, whereas in transmetallation/reductive elimination (73 → 75 → 76), the configuration of the C1 atom is retained. Since the oxidative addition of the allyl compound (74/76 → 72) proceeds with inversion, soft nucleophiles cause overall retention and hard nucleophiles cause inversion of the configuration, provided there is no stereoscrambling.[9]

11.3.2
Chirality Transfer in Asymmetric Allylation

The stereochemically uniform reaction course of allylic substitution reactions is the basis for stereo- and enantioselective synthesis [289, 290]. Three cases of allylic substitutions by soft nucleophiles, including both substrate- and ligand-controlled stereoselective syntheses, are discussed below:

- (a) *Allylic substitution with unsymmetrically 1,3-substituted substrates:* These are substrate-controlled stereospecific reactions, and the chiral information of an allyl substrate is generally fully transferred to the product. If the starting point is an enantiomerically enriched substrate, the product displays the same *ee* value as the substrate.

  Unsymmetrically 1,3-disubstituted allyl compounds **81** yield an unsymmetric π-allyl complex **82** with two different terminal C atoms. Assuming that the addition of the nucleophile (which generally occurs at the sterically more accessible C atom) takes place regioselectively to C1, **83** is obtained from **81**.

$$\mathbf{81} \xrightarrow[-X^{\ominus}]{PdL_2} \mathbf{82}\ (\oplus PdL_2) \xrightarrow{Nu^{\ominus}} \mathbf{83}$$

---

8) π-σ-π-Allyl rearrangements associated with a change of the *Re/Si* coordination are ruled out for cyclic allyl systems such as **78**, so these systems are particularly suitable for studying stereochemical aspects.

9) This is most easily understood for cyclic allyl systems **77**: If the overall reaction occurs with retention, the entering substituent Nu⁻ finds itself on the same side of the ring system as the leaving substituent X⁻. For inversion, the opposite is true.

If the other enantiomer of **81** or the racemate is used as starting material, the other enantiomer of **83** or the racemate, respectively, is obtained. The chiral information of the allyl substrate is transferred completely to the product. This assumes that no stereoscrambling (see above) occurs, which would lead to partial or complete loss of the chiral information.

- (b) *Allylic substitution with symmetrically 1,3-substituted substrates:* The chiral information of the allyl substrate is lost entirely. Even when the initial substrate is enantiomerically pure, the racemate is obtained.

Symmetrically 1,3-disubstituted allyl compounds **81a/81b** yield a symmetric π-allyl complex **82'**. (σ indicates a symmetry plane perpendicular to the plane of the paper.) The two terminal C atoms of the allyl group are symmetry-equivalent (*meso* complex) and react with the same probability. **83a** and **83b** (obtained from the reaction at C1 or C3, respectively, of the π-allyl intermediate) are created in the same amount, regardless of whether **81a** or **81b** or the racemate **81a/81b** is used as the substrate. It has lost its chiral information.

The intermediate **82'** is an ionic intermediate that need not be fully symmetric due to cation–anion interaction. As a result, for example, more **83a** may be formed from **81a**, and more **83b** may be formed from **81b**. The catalyst has demonstrated a "memory" as to whether the substrate was one enantiomer (**81a**) or the other (**81b**), although this information should have been lost ("memory effect").

- (c) *Enantioselective allylic alkylation.* With chiral ligands L$\overset{*}{\frown}$L, allylic alkylation can be enantioselectively designed. Examples are a C_2-symmetric bis-(phosphinobenzamide) ligand with a rigid chiral scaffold (**84**) and C_1-symmetric 2-(phosphinophenyl)oxazoline ligands (**85**, PHOX; R = *i*-Pr, *t*-Bu, ...; Ar = Ph, biphenyl-2-yl, ...) [290, 291].

Enantioselective reactions of symmetrically 1,3-disubstituted allyl derivatives are of special interest, since – without a chiral ligand on Pd – the chiral information in the substrate is lost (see b). However, in a chiral coordination pocket, the two C atoms (C1 and C3) are no longer equivalent and the ligand L$\overset{*}{\frown}$L controls the stereochemistry. Enantioselective alkylation requires regioselective attack on one of the two C atoms. If the addition occurs only at C1 of the π-allyl intermediate, for instance, only **83a** is formed, regardless of whether **81a**, **81b**, or the racemate **81a/81b** was the starting point.

Memory effects can play a large role in asymmetric allylation reactions.

Exercise 11.8

- Conventional methods (hydrogenation, hydrolysis of the protecting group, and esterification of the OH groups) convert **1** to Famciclovir (**2**), an antiviral chemotherapeutic agent. Starting from an allyl derivative, design a synthesis plan for **1**.
- "Hard" metal enolates of type **6** (M = Li, MgX; R ≠ R') are suitable prochiral (!) nucleophiles for Tsuji-Trost allylation in which a stereogenic center can be formed not only in the allylic position, but also in the homoallylic position [292]. Design a diastereo- and enantioselective synthesis of **7** starting from cyclohexanone.

12
Hydrocyanation, Hydrosilylation, and Hydroamination of Olefins

12.1
Introduction

The addition of compounds with an element–hydrogen bond H–X such as H–CN, H–SiR$_3$ or H–NR$_2$ (R = alkyl, aryl, H) to olefins leads to functionalized alkanes.

$$\text{>C=C<} \;+\; \text{H–X} \;\xrightarrow{\text{cat.}}\; \text{H–C–C–X}$$

In general, such reactions require catalysis, since in the synchronous addition in the transition state, the overlap integrals of the two HOMO–LUMO interactions are nearly zero (Figure 12.1).

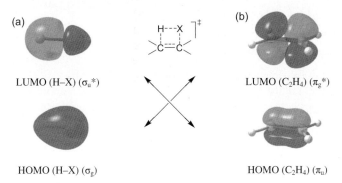

Figure 12.1 Transition state for the synchronous addition of H–X to olefins and the orbitals that determine the reactivity of H–X (a) and ethene (b). The double arrows point to the two possible HOMO–LUMO interactions.

Fundamentals of Organometallic Catalysis. First Edition. Dirk Steinborn
Copyright © 2012 WILEY-VCH Verlag GmbH & Co. KGaA, Weinheim

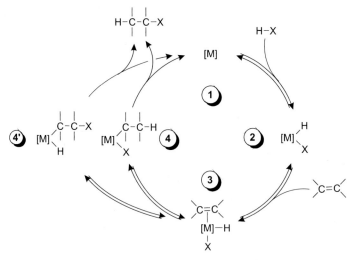

Figure 12.2 Possible mechanism for the addition of H–X (H–CN, H–SiR$_3$, H–NR$_2$) to olefins catalyzed by metal complexes.

Here, we will discuss metal-catalyzed addition of hydrogen cyanide, silanes and amines to C–C multiple bonds (preferably olefins). There are different types of mechanisms. In many cases, the following reactions are involved (Figure 12.2):

1 → 2: *Oxidative addition/reductive elimination.* Activation of H–X by oxidative addition, which is generally reversible.

2 → 3: *Substrate activation.* Activation of the olefin by π-complex formation. The reaction is reversible.

3 → 4: *Insertion.* Insertion of the olefin into the M–H bond leads to an alkylmetal complex that still has the anionic ligand X$^-$ coordinated to it. This reaction can be reversible.

4 → 1: *Reductive elimination.* Reductive C–X elimination cleaves off the product, usually in an irreversible reaction step.

Alternatively, the olefin coordination can occur before the oxidative addition of H–X. Furthermore, the olefin can insert into the M–X bond (**3 → 4′**) rather than into the M–H bond (**3 → 4**) such that the product is released through reductive C–H elimination (**4′ → 1**).

12.2
Hydrocyanation

12.2.1
Principles and Mechanism

The hydrocyanation of olefins leads to nitriles **5/6** according to the following equation.

12.2 Hydrocyanation

```
                        [Co₂(CO)₈]        NC  H
                       ───────────→    R─C─C─H    5
  R       H                             H   H
   \     /
    C=C       + HCN
   /     \
  H       H                             H   CN
                       [Ni{P(OR')₃}₄]   R─C─C─H    6
                       ───────────→     H   H
```

The regioselectivity of the HCN addition (Markovnikov vs. anti-Markovnikov addition to **5** or **6**, respectively) depends on the catalyst. Thus, [Co₂(CO)₈] as precatalyst favors the formation of branched nitriles **5**, while with [Ni{P(OR')₃}₄], terminal nitriles **6** are obtained. The mechanism corresponds in its general outline to the cycle in Figure 12.2 (H–X = H–CN). Lewis acids A such as AlCl₃, ZnCl₂ and BPh₃ are promoters. The way they work is not yet fully understood. They may facilitate the cleavage of phosphite ligands from the precatalyst and thus facilitate the oxidative addition of HCN. Alternatively, coordination to the cyano ligand [Ni]–CN ··· A in **7** may facilitate reductive elimination (**7** → **6**).

```
                                 ────→ [Ni] + NC–CH₂–CH₂R
        CH₂–CH₂R                               6
 [Ni]
        CN
         7                               CN
                        + HCN      [Ni]       + H₃C–CH₂R
                        ────→           CN
                                         8
```

Nickel catalysts are deactivated if, instead of a reductive elimination of the alkyl nitrile (**7** → **6**) in a second reaction with HCN, a protolysis of the σ-Ni–C bond occurs, resulting in the formation of catalytically inactive dicyanonickel(II) complexes. This reaction requires a vacant coordination site on the nickel and is thus suppressed by the presence of phosphite as a competitive donor.

12.2.1.1 Mechanism – A Closer Look

Kinetic measurements and NMR spectroscopic investigations have given a deeper insight into the mechanism (Figure 12.3) of the hydrocyanation of ethene with [Ni(η²-C₂H₄)L₂] (**10**, L = P(O-o-Tol)₃) as catalyst.

Starting from **10**, oxidative addition of HCN and cleavage of L form a cyano(ethene)-hydridonickel(II) complex (**10** → **11** → **12**). Then the ethene is coordinated and inserts into the Ni–H bond (**12** → **13**). After addition of L (**13** → **14**), propionitrile cleaves off via reductive elimination (**14** → **10**). As the reaction profile (Figure 12.3) shows, this step is rate-determining and irreversible.

The formation of the catalyst **10** by ligand substitution starting from the tetrakis-(phosphite)nickel(0) complex **9** is well known. Reaction of **13** with HCN leads to protolysis of the σ-Ni–C bond, which results in a deactivation of the catalyst with formation of a dicyanonickel(II) complex and ethane (**13** → **15**) [293].

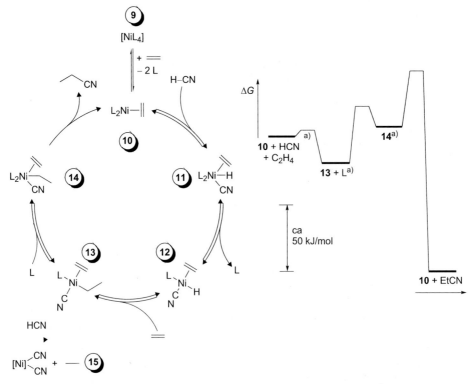

Figure 12.3 Mechanism of the hydrocyanation of ethene with [Ni(η^2-C$_2$H$_4$)L$_2$] (**10**, L = P(O-o-Tol)$_3$) as catalyst and the reaction profile at −40 °C (after McKinney and Roe [294]).
a) Values estimated.

12.2.2
The DuPont Adiponitrile Process

The double hydrocyanation of butadiene leads to adiponitrile. Most of the adiponitrile produced worldwide is manufactured by a process with phosphite–nickel(0) catalysts developed by DuPont in 1972. Adiponitrile is an important intermediate product in the synthesis of nylon-6,6.

The DuPont adiponitrile process includes three steps [295]:

- *Hydrocyanation of butadiene (synthesis of unsaturated mononitriles).* In the presence of tetrakis(phosphite)nickel(0) complexes NiL$_4$ (**16**), butadiene reacts with HCN to form pent-3-enenitrile (**19a**) and 2-methylbut-3-enenitrile (**19b**) in a ratio of approximately 2:1. The mechanism is analogous to the hydrocyanation of monoolefins except that the insertion of butadiene into the Ni–H bond generates an allyl intermediate (**17** → **18**). Reductive elimination with C1–CN bond formation yields the linear product (**18** → **19a**), and with C3–CN bond formation, the branched product (**18** → **19b**). In this special case – namely due to the formation of allyl cyanides **19a/19b** – the reductive elimination is reversible.

$$NiL_4 \;(\mathbf{16}) \xrightleftharpoons[]{+\,HCN,\,-L} L_3Ni(H)(CN) \;(\mathbf{17}) \xrightleftharpoons[-L]{+\,\text{butadiene}} L_2Ni(H)(CN) \;(\mathbf{18}) \xrightarrow{-NiL_2} \mathbf{19a} / \mathbf{19b}$$

The introduction of bis(arylphosphonite) chelating ligands in place of monodentate phosphite ligands has led to catalysts with higher activity and selectivity [296].

- *Isomerization of pentenenitriles.* The isomerization of **19a** to form pent-4-enenitrile (**19c**) is accomplished with cationic hydridonickel complexes **20**, which are obtained from the reaction of [NiH(CN)L$_3$] (**17**) with Lewis acids A such as ZnCl$_2$ or BPh$_3$. The isomerization of pent-3-ene- to form pent-4-enenitrile (**19a** → **19c**) is kinetically controlled. The pent-2-enenitrile (**19d**), which is more thermodynamically stable due to the conjugation of the π bonds, is only formed slowly. Due to the reversibility of the reductive elimination of pent-3-enenitrile and 2-methylbut-3-enenitrile (**18** ⇌ **19a/19b**), isomerization of the branched nitrile to the linear nitrile occurs in addition to the double-bond isomerization, so over 90% linear pentenenitriles can be obtained.

$$L_3Ni(H)(CN) \;(\mathbf{17}) \xrightleftharpoons[]{A} L_3Ni(H)(CN\text{---}A) \xrightleftharpoons[]{} [L_3Ni\text{–}H][A(CN)] \;(\mathbf{20})$$

$$\mathbf{19a} \xrightleftharpoons[]{[L_3Ni\text{–}H][A(CN)] \;(\mathbf{20})} \mathbf{19c} \;/\; \mathbf{19d}$$

- *Hydrocyanation of pentenenitriles (synthesis of hexanedinitrile [adiponitrile]).* The third step involves the hydrocyanation of the mixture of pent-3-ene- and pent-4-enenitrile (**19a/19c**), which leads to adiponitrile (**21a**) as main product with 2-methylglutaronitrile (**21b**) and ethylsuccinonitrile (**21c**) as side products. The mechanism corresponds to that of the hydrocyanation of monoolefins. The selectivity with respect to adiponitrile depends heavily on the Lewis acid promoter and can amount to > 90%.

The very slow formation of the most thermodynamically stable linear pentenenitrile **19d** (equilibrium composition **19d** : **19a** : **19c** approximately 78 : 20 : 2 at 50 °C) and the very rapid establishment of equilibrium between **19a** and **19c** are crucial for the selectivity with respect to adiponitrile **21a**. This "kinetically controlled isomerization" can probably be traced to the fact that **19a** favors coordination to nickel via the nitrile group (and not via the double bond). However, double-bond isomerization to **19d** would lead, after insertion of the double bond into the Ni–H bond, to a [Ni]–CH(Et)–CH$_2$CN intermediate that does not permit further nitrile coordination [297].

Although the inner olefin **19a** is significantly more stable thermodynamically than the terminal olefin **19c** (equilibrium composition **19a** : **19c** approximately 10 : 1) and adiponitrile (**21a**) is only obtained from **19c**, a selectivity with respect to **21a** of > 90% can be achieved. This is primarily a result of the fact that bulky ligands L and sterically demanding Lewis acids A impede the formation of the cyano(isoalkyl)nickel(II) complexes that appear in the formation of **21b** and **21c** as intermediates.

As a result, with bulky Lewis acid promoters such as BPh$_3$, a significantly higher selectivity with respect to adiponitrile is achieved (96%; L = P(O-p-Tol)$_3$, 50 °C) than with ZnCl$_2$ (82%) and AlCl$_3$ (50%).

12.2.3
Outlook

12.2.3.1 Enantioselective Hydrocyanation

Hydrocyanation of vinyl aromatics, exemplified here by the unsubstituted styrene **22**, generally proceeds as a Markovnikov addition, so the formation of branched nitriles **23** is favored. This can be traced to the special stability of the benzylnickel intermediate **24a** with an allyl-type bond. No such stabilization occurs for the 2-phenylethylnickel compound NC–[Ni]–CH$_2$CH$_2$Ph (**24b**). This is the corresponding intermediate for an anti-Markovnikov addition, which would lead to linear products.

Enantioselective synthesis of branched nitriles **23** is possible [298]. If GLUP-type carbohydrate-based diphosphinites such as **25** are employed as coligands, *ee* value up to 90% can be achieved; as an example, see the hydrocyanation of **26**. Hydrolysis of **27** (–CN → –COOH) yields Naproxen, which belongs to the 2-arylpropionic acid class of analgesics.

Mechanistic studies have shown that olefin coordination is followed by the oxidative addition of HCN (**28** → **29**). However, the reverse sequence has also been demonstrated. Insertion of the olefin into the Ni–H bond yields a (π-allyl)nickel(II) complex (**29** → **30**), from which the product is cleaved off (**30** → **31**) with reductive C–C elimination. Coordination of the styrene derivative and oxidative addition of HCN (or vice versa) to the Ni(0) complex regenerates **28/29** [299].

It has been experimentally shown that the enantioselectivity is determined not by the coordination of the olefin to the *Re* or *Si* face to form (*Re*)-**28** or (*Si*)-**28**, respectively, but rather that the activation barrier for the styrene insertion (**29** → **30**) and/or the irreversible reductive elimination forming (*S*)-**31** is lower than that for forming (*R*)-**31**. The *ee* values turn out to be heavily dependent on the electronic properties of the coligand **25** and of the substrate [300].

12.2.3.2 Hydrocyanation of Alkynes

Nickel phosphite catalysts also catalyze the hydrocyanation of alkynes, which leads to α,β-unsaturated nitriles **32a/32b**, which have a high synthetic potential in Michael addition reactions and as dienophiles in cycloaddition reactions.

12 Hydrocyanation, Hydrosilylation, and Hydroamination of Olefins

Generally, *syn* addition occurs. The regioselectivity is determined by steric and electronic factors. For terminal alkynes ($R^1 = H$), the branched product **32a** is predominantly formed via Markovnikov addition, unless the substituent R^2 is very bulky [301].

Exercise 12.1
Explain the stereoselectivity of the reaction.

The hydrocyanation of acetylene with copper(I) salts as catalyst (**a**) was the most important process for manufacture of acrylonitrile until the 1960s. According to Reppe, it can be seen as vinylation of hydrocyanic acid.

Today, acrylonitrile is predominantly produced via ammoxidation of propene at around 450 °C in a heterogeneously catalyzed reaction (**b**; SOHIO process: Standard Oil of Ohio) [302].

12.2.3.3 Hydrocyanation of Polar C=X Bonds

The addition of HCN to C=O double bonds in aldehydes to form cyanohydrins is an equilibrium reaction that can be catalyzed by acids or bases. Enantiomerically pure cyanohydrins are important building blocks for the synthesis of α-hydroxycarboxylic acids and β-amino alcohols. Sharpless-type catalysts (see Section 13.2.3.1) are employed; trimethylsilylcyanide has proven itself as a cyanation agent ("silylcyanation"). For benzaldehydes (R = aryl), *ee* values between 90 and 96% are achieved.

Furthermore, chiral monometallic, yet bifunctional, catalysts, whose structure is shown schematically in **a**, are used. The complexes **33** and **34** are examples for which *ee* values > 90% have been achieved.

In silylcyanation and hydrocyanation, the carbonyl compound is activated by coordination of the C=O group to the Lewis acidic metal atom (M = Al, Ti), *and* the cyano compound (HCN, Me$_3$SiCN) by interaction with the Lewis basic catalyst center, that is, the oxygen atom of the phosphane oxide group (D = O=PPh$_2$–). In the transition state, the two substrate molecules are bound to the monometallic but bifunctional catalyst molecule. A structural precondition for high catalytic efficiency is that no deactivation occurs by intramolecular donor–acceptor interaction (M ← D). In addition, the catalyst must bind to the two reaction partners such that they assume a suitable spatial orientation toward each other [303, 304].

Similarly, aldimines and ketimines are employed as substrates (RR'C=NR" + HCN (or Me$_3$SiCN) → RR'C*(NHR")–CN) which, in a Strecker-type synthesis, gain access to chiral α-amino acids [305, 306].

Exercise 12.2

Aldimines react enantioselectively with HCN in the presence of organocatalysts (i.e., metal-free catalysts) such as axially chiral BINOL phosphates **1** to form aminonitriles, which are precursors of amino acids. Formulate the reaction equation and classify the catalysis according to the information in Table 2.1 (see Section 2.1). Give a possible mechanism that also explains the enantioselectivity of the reaction. Which solvents should be suitable?

12.3
Hydrosilylation

12.3.1
Principles and Mechanism

The hydrosilylation of olefins produces alkylsilanes:

$$\text{>C=C<} + \text{H–SiR}_3 \xrightarrow{\text{cat.}} \text{H–C–C–SiR}_3$$

Exercise 12.3

Estimate from the mean bond dissociation enthalpies (C–C 348, C=C 612, C–H 412, Si–C 311, Si–H 318 kJ/mol) the enthalpy of the reaction. Determine whether an addition in the form of a radical chain reaction is possible. What regioselectivity do you expect for a radical hydrosilylation of terminal olefins?

Most addition reactions of silanes to olefins are based not on homolytic, but on heterolytic Si–H bond cleavage. They can be catalyzed by Lewis acids such as AlCl$_3$ and especially by transition metal complexes [307]. This was first shown by J. L. Speier in 1957.

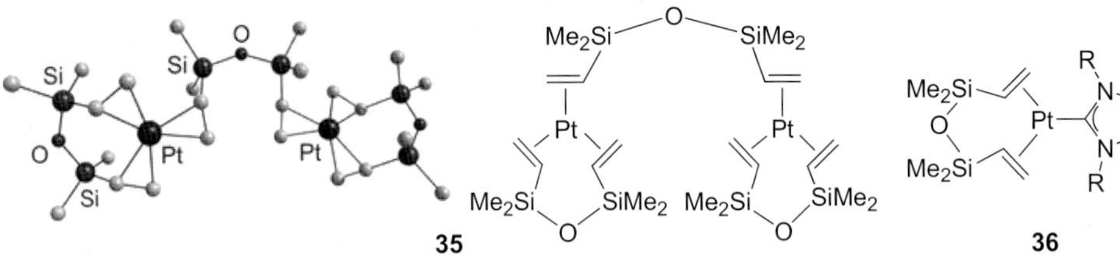

Figure 12.4 Dinuclear (**35**) and mononuclear (**36**) (divinyldisiloxane)platinum complexes as precatalysts for hydrosilylation.

For transition-metal-catalyzed hydrosilylation of olefins, platinum compounds are predominantly used. Speier catalyst (hexachloroplatinic acid in isopropanol) and Karstedt catalyst (hexachloroplatinic acid treated with a vinylsiloxane such as (H_2C=CH)Me_2Si–O–$SiMe_2$(CH=CH_2)) have proven themselves as precatalysts. In both cases, Pt^0 complexes are catalytically active. In Karstedt solutions, a binuclear platinum(0) complex **35** has been identified (Figure 12.4) as well as the formation of platinum colloids, on which the catalysis can take place. The substitution of the μ-siloxane ligand in **35** by N-heterocyclic carbene ligands leads to mononuclear complexes **36** (Figure 12.4), which are superior to the Karstedt catalyst with regard to chemo- and regioselectivity [308].

For substituted olefins, Si–H additions generally follow the anti-Markovnikov rule, so products with terminal silyl groups are formed. They proceed uniformly, from a stereochemical point of view, as *syn* additions. Terminal double bonds are generally easier to hydrosilylate than inner double bonds. In addition to platinum, many other transition metals catalyze hydrosilylation of olefins, especially metals of Groups 8–10 (Rh, Co, Fe, Ir, Ru, …). In the case of rhodium, complexes of the Wilkinson type [RhX(PR_3)$_3$] and [RhX(CO)(PR_3)$_2$] are used as precatalysts.

The mechanism of platinum-catalyzed hydrosilylation (A. J. Chalk, J. F. Harrod; 1965) corresponds in its basic outlines to that shown in Figure 12.2 (H–X = H–SiR_3; [M] = [Pt]) with the reaction sequence (Figure 12.5):

37 → 38: Oxidative addition of HSiR_3 to a hydrido(silyl)platinum(II) complex.
38 → 39: Olefin coordination.
39 → 40: Olefin insertion into the Pt–H bond, forming an alkyl(silyl)platinum(II) complex.
40 → 37: Reductive C–Si elimination with regeneration of the catalyst complex.

Later, a modification was proposed in which the olefin inserts into the Pt–Si bond (39′ → 40′) and then the reaction cycle is completed via reductive C–H elimination (40′ → 37).

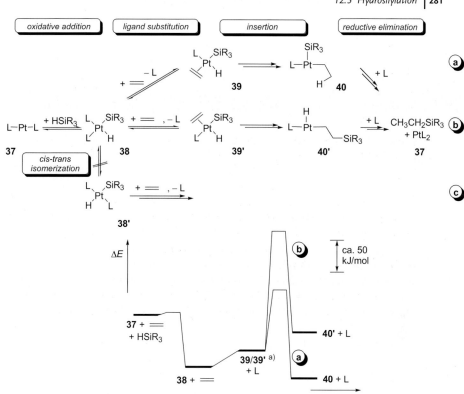

Figure 12.5 Mechanism of addition of HSiR$_3$ (R=H, Me, Cl) to ethene, catalyzed by PtL$_2$ (L=PH$_3$). The energies given refer to R=H, but the same applies to R=Me, Cl (shortened from Sakaki [309]).
a) The energy difference between **39** and **39'** is marginal.

However, quantum-chemical calculations on the addition of H–SiR$_3$ to ethene with [Pt(PH$_3$)$_2$] as model catalyst[1] show that the mechanism originally proposed (reaction channel **a**, Figure 12.5) exhibits a significantly lower activation barrier than the modified one (reaction channel **b**). This result should not be generalized, since the opposite is true for rhodium complexes, for example [310].

The oxidative addition of HSiR$_3$ to the Pt^0 complex is associated with a very low activation barrier (**37** → **38**), as is true for the subsequent ligand substitution (**38** → **39/39'**), which proceeds according to the addition-elimination mechanism. In the Chalk-Harrod mechanism (Figure 12.5, **a**), the insertion of ethene into the Pt–H bond and the isomerization of the primarily formed insertion product proceed

---

1) Phosphane complexes such as [Pt(η^2-H$_2$C=CH$_2$)(PPh$_3$)$_2$] are also catalytically active, but less than the Speier and Karstedt catalysts.

in such a way that the ethyl ligand is coordinated *trans* to the PH$_3$ ligand (**39** → **40**). This reaction is rate-determining.

Alternatively, ethene can insert into the Pt–Si bond (reaction channel **b**: modified Chalk-Harrod mechanism; **39'** → **40'**). The reaction profile (Figure 12.5) shows that this activation barrier is significantly greater, so the Chalk-Harrod mechanism is favored. A *cis/trans* isomerization of **38** must also be considered (**38** → **38'**). The *trans* complex **38'** opens the reaction channel **c**, which enables the insertion of ethene into the Pt–H or Pt–Si bond analogously to the channels for the *cis* complex **38**. However, calculations show that the *cis/trans* isomerization (**38** → **38'**) is kinetically inhibited, so reaction channel **a**, which is to say the normal Chalk-Harrod mechanism, remains the favored reaction [309].

Bis(silyl)(cyclooctadiene)platinum(II) complexes **41**, which have been fully characterized spectroscopically and structurally, have shown themselves to be catalytically active in the hydrosilylation of olefins. It has been shown that these reactions probably are based on a catalytic cycle with a change in oxidation state Pt^{II} ⇌ Pt^{IV}. This flexibility in oxidation states for catalytic behavior – Pt^0/Pt^{II} and Pt^{II}/Pt^{IV} – could also be one of the reasons for the multitude of platinum compounds catalytically active in hydrosilylation reactions [308].

When using other Group 8–10 metal catalysts instead of platinum catalysts for the hydrosilylation of olefins, dehydrogenative silylation reactions are frequently observed as a side reaction. They lead to vinylsilanes and hydrogen (**a**) and/or to vinylsilanes and alkanes (**b**) [311].

As a starting point for the formation of the vinylsilanes, hydrido(2-silylalkyl)-metal complexes **42** are to be considered, as they appear in the modified Chalk-Harrod mechanism as intermediates (see complexes **4'/40'** in Figures 12.2 and 12.5). Reductive C–H elimination leads to the normal hydrosilylation products (**42** → **43**) with regeneration of the catalyst complex [M]. By contrast, β-H elimination yields, with cleavage of a vinylsilane, a dihydrido complex that reacts to form H$_2$ and [M], with reductive elimination (**42** → **44** → **43**). Insertion of the olefin into the M–H bond of **42** leads to an alkyl–silylalkyl complex **45**, which can, by β-H and reductive C–H elimination, cleave off a vinylsilane and an alkane, respectively, also regenerating [M] (**45** → **46** → **43**) [311, 312].

[Scheme showing mechanism with compounds 42–46 and 43]

$$[M] \xleftarrow{-R'CH_2-CH_2SiR_3} [M]\underset{H}{\overset{CH-CH_2SiR_3}{<}}\underset{R'}{} \xrightarrow{-R'HC=CHSiR_3} [M]\underset{H}{\overset{H}{<}} \xrightarrow{-H_2} [M]$$

43   42   44   43

$\big| + R'HC=CH_2$

$$[M]\underset{CH-CH_3}{\overset{CH-CH_2SiR_3}{<}} \xrightarrow{-R'HC=CHSiR_3} [M]\underset{R'}{\overset{H}{<}}CH-CH_3 \xrightarrow{-R'CH_2-CH_3} [M]$$

45   46   43

Exercise 12.4

Silylenes can be considered homologues of carbenes and form metal complexes, like carbenes do; see **1** (shown without stabilizing ether molecule) for an example. In reactions with olefins such as *n*-hex-1-ene, complexes of this type have been found to display a new type of Si–C bond-forming reaction (**1** → **2**).

[Structures: Complex 1: Cp*Ru(P(i-Pr)_3)(H)(H)(Si(H)Ph) cation with [B(C_6F_5)_4]⁻ counterion; n-hex-1-ene; Complex 2: Cp*Ru(P(i-Pr)_3)(H)(H)(Si(n-Hex)Ph) cation with [B(C_6F_5)_4]⁻ counterion]

1   **2**

Complex **1** even catalyzes the hydrosilylation of olefins with primary silanes. Experimental findings such as high selectivity (only primary silanes react, forming exclusively secondary silanes), tolerance of steric hindrance around the C=C bonds, and exclusive formation of anti-Markovnikov products point to a novel mechanism. Formulate the mechanism, keeping in mind that silylene complexes can be formed from silyl complexes by an α-H shift from the silicon atom to the metal atom. Find analogies to the hydroboration of olefins and consider why the cationic character of complex **1** is significant.

12.3.2
Significance of Hydrosilylation and Outlook

12.3.2.1 Applications

Hydrosilylation has achieved industrial significance, especially in the synthesis of organosilanes and in polymer chemistry for cross-linking silicones and for the synthesis of graft copolymers. Furthermore, it is also possible to prepare polymers such as **47** and **48** by "hydrosilylation polymerization" from monomers that contain Si–H and Si–CH=CH$_2$ groups [313].

Dendrimers with Si atoms as branching junctures, such as **49** (represented as a graph), are built up via controlled stepwise growth by alternating hydrosilylation and Grignard coupling reactions. Thus, for instance, tetraallylsilane **50** can react with H–SiCl$_2$Me, catalyzed by platinum, to form **51**. Subsequent reaction with a vinyl Grignard compound yields the first generation of the dendrimer **52** with eight vinyl groups in the periphery. Repeated hydrosilylation and Grignard coupling lead to the following generations. Species **49** is a schematic representation of the third generation with 32 terminal vinyl groups (Vi) [313].

12.3.2.2 Enantioselective Hydrosilylation

A precondition for enantioselective hydrosilylation reactions of terminal olefins is a Markovnikov addition. Palladium complexes formed from [{PdCl(η^3-C$_3$H$_5$)}$_2$] and chiral monophosphane ligands (X-MOPs) **53** (Ar = Ph; X = H, MeO, OR, Ar, COOR, ...) ensure that this will occur. MOPs are atropisomeric 1,1′-dinaphthyl ligands, which, in contrast to the BINAP ligand, only contain one P-ligator atom. The hydrosilylation of olefins (R = alkyl) with HSiCl$_3$ is surprisingly regio- (**54a** : **54b** about 9 : 1) and enantioselective (about 95% *ee*). In the hydrosilylation of styrene, high *ee* values with H-MOP (**53**, X = H, Ar = 3,5-(CF$_3$)$_2$C$_6$H$_3$) have been achieved [314–316].

From EtOH and the chiral alkyl(chloro)silanes **54a**, alkyltriethoxysilanes are obtained, which then are oxidized with H_2O_2 in the presence of KF to form the chiral alcohols **55**.

- *Activity.* In contrast to the high activity of the complexes with X-MOP ligands, bis(phosphane)palladium complexes have proved to be catalytically inactive. The reason is probably that the catalysis of hydrosilylation occurs at three coordination sites. Only monophosphane ligands L* allow the formation of stable square-planar palladium(II) complexes of the type $[Pd(SiCl_3)H(\eta^2\text{-}H_2C=CHR)L^*]$ (**56**).
- *Enantioselectivity.* From the reactants **A**, via the diastereomeric complexes (*Si*)-**56**/(*Re*)-**56**, the diastereomeric alkyl complexes (*S*)-**57**/(*R*)-**57** are formed by insertion and finally, the two enantiomers (*S*)-**54a**/(*R*)-**54a** by reductive Si–C elimination. For styrenes, it has been shown that the high enantioselectivity is due less to high enantiofacial differentiation during olefin coordination ((*Re*)-**56** vs. (*Si*)-**56**), and more to a rapid β-H elimination of the alkylpalladium intermediates ((*S*)-**57** → (*Si*)-**56** and (*R*)-**57** → (*Re*)-**56**) paired with a very selective reductive elimination ((*S*)-**57** → (*S*)-**54a**, but not (*R*)-**57** → (*R*)-**54a**). Thus, due to the much lower reactivity of (*R*)-**57** toward reductive elimination, it is in a dynamic equilibrium with (*S*)-**57** and ultimately reacts to form (*S*)-**54a**.

Rhodium complexes, among others, have also proven effective for the catalysis of enantioselective hydrosilylation of polar double bonds (ketones, imines) to form optically active secondary alcohols and amines [317–319].

12.3.2.3 Hydrosilylation of Alkynes

Alkynes can be hydrosilylated via transition metal catalysis to form vinylsilanes. For terminal alkynes, three different reaction products can be formed, corresponding to an anti-Markovnikov addition (**58**) and a Markovnikov addition (**59**).

In general, the achievable regio- and stereoselectivity is not particularly high. With platinum catalysts, the formation of vinylsilanes with a terminal silyl group *trans* to R (*E*)-**58** is favored. Such hydrosilylation reactions (**60** → **61**) can be combined with palladium-catalyzed cross-couplings (**61** → **62**) in an intermolecular domino reaction so that, for example, 1,2-disubstituted (*E*)-alkenes can be synthesized in high yield and with high stereoselectivity (dvds = ($H_2C=CH$)$Me_2Si-O-SiMe_2$($CH=CH_2$)) [320].

$$R\!-\!\!\equiv\!\!-\ +\ HSiMe_2-O-SiMe_2H\ \xrightarrow[\text{THF, r.t.}]{[Pt(dvds)\{P(t\text{-}Bu)_3\}]}\ (R\!-\!\!\!\diagup\!\!\!\diagdown\!\!-SiMe_2)_2O\ \xrightarrow[\text{THF, r.t.}]{R'\text{-}I\,/\,(NBu_4)F,\ [Pd(dba)_2]}\ R\!-\!\!\!\diagup\!\!\!\diagdown\!\!-R'$$

60 → **61** → **62**

Exercise 12.5
Diethynylmethylsilane is a monomer of AB_2 type. A (Si–H) and B (Si–C≡CH) are two functional groups that react with each other but not with themselves. Platinum-catalyzed hydrosilylation leads to a hyperbranched polycarbosilane. Formulate the reaction. Give the possible structures of the Si centers in the polymer.

12.3.2.4 σ Complexes of Silanes
Similarly to dihydrogen, hydrosilanes **63** (i.e., silanes containing a Si–H bond) can form σ complexes with transition metals (**64**).

$$[M] + H-SiR_3 \longrightarrow [M]\!\!\leftarrow\!\!\begin{array}{c}H\\+\\SiR_3\end{array} \longrightarrow [M]\!\!\diagdown\!\!\begin{array}{c}H\\SiR_3\end{array}$$

63 → **64** → **66**

IIa ← → IIb

$$\left[[M]\!\!\leftarrow\!\!\begin{array}{c}H\\+\\SiR_3\end{array} \longleftrightarrow [M]\!\!\diagdown\!\!\begin{array}{c}H\\SiR_3\end{array} \right]$$

64a ↔ **64b**

As silanes are only weak σ donors, not only σ donation, but also π back-donation is important for the stability of the η^2-Si–H bond to the metal. **IIa** shows the transfer of electron density from the bonding σ-Si–H orbital into a vacant metal orbital of σ-symmetry (σ donation); here the d_{z^2} valence orbital is shown. **IIb** shows the transfer of electron density from an occupied *d* orbital of the metal into the σ*-Si–H orbital (π back-donation). Within the framework of the VB model, each of these binding components is depicted by a resonance formula **64a/64b**.[2]

Concerted oxidative addition reactions of hydrosilanes **63** to form hydrido–silyl complexes **66** proceed via a σ-Si–H complex **64** as intermediate (**63** → **64**). If this is to

---

[2] It should be noted that **64b** is a resonance formula for describing the bond in **64** and hence must not be identified with the hydrido–silyl complex **66**.

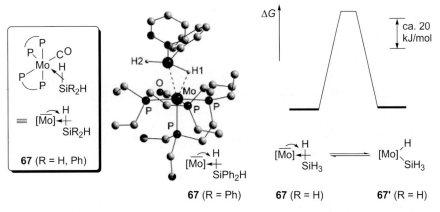

Figure 12.6 Molecular structure of [Mo(η²-SiH₂R₂)(CO)(Et₂PCH₂CH₂PEt₂)₂] (**67**, R = Ph; Si–H1 1.66(6), Si–H2 1.54(6) Å) and reaction profile (as determined by NMR spectroscopy) for the equilibrium (60 °C in toluene) between the silane complex **67** (R = H) and the hydrido–silyl complex **67'** (R = H). The oxidative addition is almost thermoneutral (after Vincent, Kubas, Lledós [321]).

be stable, the extent of π back-donation must be well balanced. If it is too low, the complex formation is too weak, while if it is too high, oxidative addition will result (**64 → 66**) [322]. In accord with the binding model for η²-Si–H complexes, the π back-donation weakens the Si–H bond by electron transfer into the σ*-Si–H orbital. This leads to bond lengthening of 0.1–0.4 Å (**63** vs. **64**) [323].

In classical cis hydrido–silyl complexes **66**, Si···H distances greater than 2.5 Å can be found. Si···H distances between 1.9 and 2.5 Å (**65**) point to (increasingly weaker) attractive Si···H interactions. The boundaries between σ-Si–H complexes and hydrido–silyl complexes are not well defined [324, 325]. Figure 12.6 shows, as an example, the molecular structure of an η²-silane complex and the energetics of its oxidative addition to the hydrido–silyl complex.

12.4 Hydroamination

12.4.1 Principles

The synthesis of alkylamines by direct hydroamination of olefins according to the following scheme is an attractive alternative to the two-step process with alcohols as intermediate. Addition of amines to olefins is thermodynamically allowed, thus, for example, addition of NH_3, $EtNH_2$ and Et_2NH to ethene is exergonic by -15 to -33 kJ/mol.

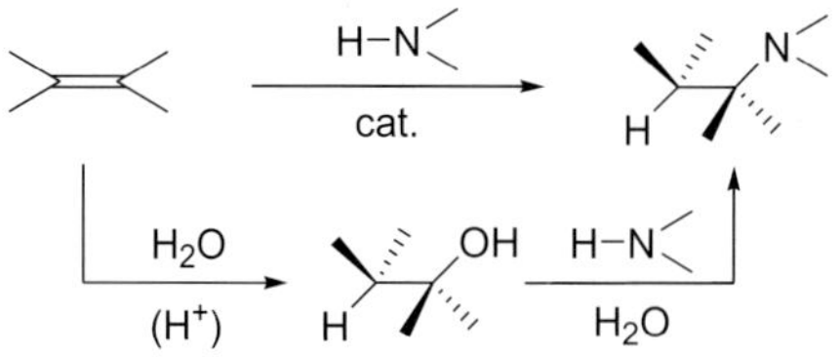

Exercise 12.6
Explain why in alkylamines the α-C–H bonds, but not the N–H bonds have low stability ($\Delta_d H^\ominus$ for $MeNH_2$: C–H 393, N–H 425 kJ/mol). From the mean bond dissociation enthalpies (C–C 348, C=C 612, C–H 412, C–N 305 kJ/mol), predict the course a radical-initiated addition of $MeNH_2$ to ethene will take.

So far it has been possible to show that catalysis of the hydroamination of olefins can be carried out in many ways. Although significant progress has been made in the last 10 years, it has not yet been enough to develop industrially usable catalysts for non-activated olefins. Further clarification of the mechanism of this catalysis and insights into relationships between catalyst structure and its effect on the catalysis are therefore required [326–328].

The crucial step in the catalysis of hydroamination of olefins is the C–N bond formation. After activating the amine (by deprotonation, or by formation of a metal amide [M]–NR_2 either via oxidative N–H addition or via protolysis of an M–C bond with HNR_2) and/or activating the olefin (by coordination or by insertion into an M–H bond), the following reactions are to be considered for the formation of C–N bonds:

- (a) Addition of R_2N^- to an olefin:

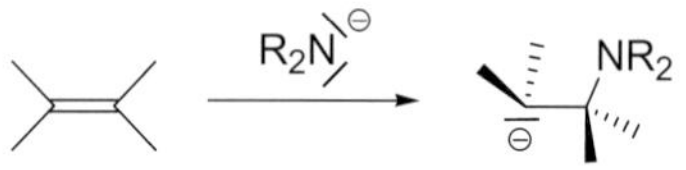

- (b) Insertion of an olefin into an M–N bond:

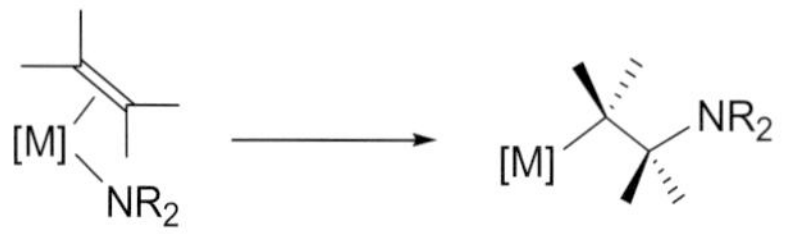

- (c) Reductive C–N elimination:

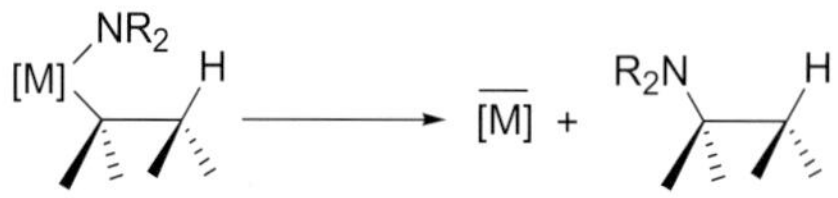

- (d) (Intermolecular) addition of an amine to a coordinated olefin:

$$[M]-\!\!\parallel \;\xrightleftharpoons[]{H-NR_2}\; [M]\text{—}\!\!\!\!\!\!\text{—}\!\!\!\!\!\!\!\!\!\!\!\text{—}^{NHR_2}\; \xrightleftharpoons[]{-H^+}\; [M]\text{—}\!\!\!\!\!\!\text{—}\!\!\!\!\!\!\!\!\!\!\!\text{—}^{NR_2}$$

This multiplicity of mechanisms makes it easy to understand that there is a broad selection of catalyst systems for the hydroamination of olefins, which extends from alkali metal amides to lanthanoid complexes and complexes of the late transition metals.

12.4.2
Catalyst Types

12.4.2.1 Alkali Metal Amides as Catalysts

As far back as the 1950s, alkali metals M and alkali metal hydrides MH were used as precatalysts. They react with the amine HNR_2 (R = alkyl, aryl, H) to form alkali metal amides MNR_2. NR_2^- adds nucleophilically to the olefin (reaction a, Section 12.4.1), forming an alkali metal alkyl (**68** → **69**). Protolysis of the M–C bond by the amine releases the product **70** with regeneration of MNR_2.

$$\text{C=C} \;\xrightarrow{+ MNR_2}\; M\text{—C—C—}NR_2 \;\xrightarrow[- MNR_2]{+ HNR_2}\; H\text{—C—C—}NR_2$$

    68   **69**   **70**

Addition to nonactivated olefins requires comparatively harsh reaction conditions (M = Na, K: 100–200 °C; up to 100 bar pressure), while lithium amides react under somewhat milder conditions. Furthermore, the reactions are not very selective [329].

12.4.2.2 Platinum Group Metals as Catalysts

The first transition-metal-based homogeneous catalyst system for hydroamination of a nonactivated olefin (ethene) with secondary amines, $RhCl_3 \cdot 3H_2O$ (180–200 °C; 5–14 MPa), was described at the beginning of the 1970s. $[RhCl(\eta^2\text{-}C_2H_4)(HNC_5H_{10})_2]$ (**71**) was identified as the catalytically active species for the reaction of ethene with piperidine to form N-ethylpiperidine (**74**, $NR_2 = NC_5H_{10}$). $[IrCl(\eta^2\text{-}C_2H_4)_2(PEt_3)_2]$ (in the presence of $ZnCl_2$ as cocatalyst) catalyzes the hydroamination of norbornene with aniline to form exo-2-(N-phenylamino)norbornane (**79**), where the 14-ve species $[IrCl(PEt_3)_2]$ (**75**) is the actual catalyst.

$$[Rh^I]\text{—}\!\!\parallel \;\xrightarrow{HNR_2}\; [Rh^I]^{\ominus}\text{—}\!\overset{\oplus}{NHR_2}\text{H} \;\longrightarrow\; [Rh^{III}]\text{—}\!\overset{NR_2}{H} \;\longrightarrow\; [Rh^I] + H\text{—}\!NR_2$$

   71   **72**   **73**   **74**

The two reactions are based on different mechanisms. In the rhodium-catalyzed reaction the formation of the C–N bond seems to proceed by intermolecular addition of the amine to the activated olefin according to reaction **d** (see Section 12.4.1) (71 → 72; [Rh'] = RhCl(HNC$_5$H$_{10}$)$_2$). The β-ammonioethyl complex 72 can, via H transfer, be converted into a (β-aminoethyl)hydridorhodium(III) complex, which then cleaves off the alkylated amine in a reductive C–H elimination (72 → 73 → 74). However, a direct protolysis of the M–C bond should also be considered, such that no hydrido complex appears as intermediate (72 → 74). By contrast, the iridium-catalyzed reaction is based on an oxidative N–H addition (75 → 76; [Ir'] = IrCl(PEt$_3$)$_2$). This is followed, according to reaction **b** (see Section 12.4.1), by insertion of the olefin into the Ir–N bond (76 → 77 → 78) and reductive C–H elimination (supported by the Lewis acid cocatalyst) (78 → 79) [326, 328].

In the ionic liquid [P(n-Bu)$_4$]Br (melting point 100–103 °C), PtBr$_2$ catalyzes, without any further addition of ligands at 150 °C, the N–H addition of anilines to ethene, norbornene and hex-1-ene. The mechanism corresponds to the rhodium-catalyzed reaction above. In the presence of a proton source such as CF$_3$SO$_3$H, which promotes the cleavage of the Pt–C bond, TON > 200 is achieved [330].

Mechanistic investigations on intramolecular hydroamination of alkenyl-amines 80 to form pyrrolidine derivatives 81 with the palladium complex 82 as catalyst give evidence of an amine addition after coordination of the olefin to the highly electrophilic dicationic palladium(II) center (80 → 83 → 84) according to reaction **d** (see Section 12.4.1). Complex 84 is the resting state; the protolysis of the Pd–C bond (84 → 81) is the turnover-limiting step.

12.4 Hydroamination

The proton released by the amine addition 83 → 84 is in a protonation-deprotonation equilibrium with the amine (2 80 + H$^+$ ⇌ 85). In the presence of stronger bases such as tertiary amines and pyridines, no catalysis occurs. Then the reaction stops with the complexes 84, but can be reactivated again by addition of acids such as HOTf or H[BF$_4$]. This gives insight into how the Brønsted acidity/basicity of all components affects the reaction. This might be the basis for further optimization of hydroamination reactions when other substrates are used [331].

The intermolecular Markovnikov hydroamination of vinyl aromatics with arylamines is effectively catalyzed by palladium complexes [Pd(OTf)$_2$(P⌢P)] (86)/HOTf (P⌢P = Xantphos, DPPF, ...) [332].

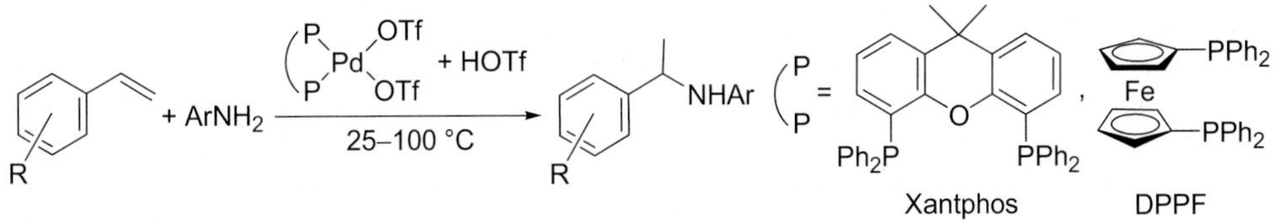

The reaction mechanism is shown in Figure 12.7. The precatalyst 86 is reduced by the arylamine to a diphosphane–palladium(0) complex (86 → 87) to which styrene is coordinated (87 → 88). Protonation yields a palladium(II) complex with an η^3-bound benzyl ligand, which has also been structurally characterized (88 → 89). (Alternatively, HOTf could be oxidatively added to 87 and then styrene inserted into the Pd–H bond of [PdH(OTf)(P⌢P)].)

Stereochemical investigations prove that the arylamine attacks the benzyl carbon atom intermolecularly, thus completing the cycle with cleavage of PhCHMeNHAr/HOTf (89 → 90 → 87). This reaction step is analogous to the Tsuji-Trost reaction (see Section 11.3.1). With chiral diphosphane ligands such as (R)-BINAP, asymmetric hydroamination can be implemented [333].

12.4.2.3 Gold Complexes as Catalysts

Cationic gold(I) complexes are strong Lewis acids with a relatively low capability for π back-donation, so the electrophilicity of an olefin is increased by coordination and thus activated for an attack by nucleophiles (reaction d, Section 12.4.1). Reaction with amines leads to a 2-ammonioethyl complex. Cleavage of the proton and protolysis of the Au–C bond generate the product and the catalyst complex:

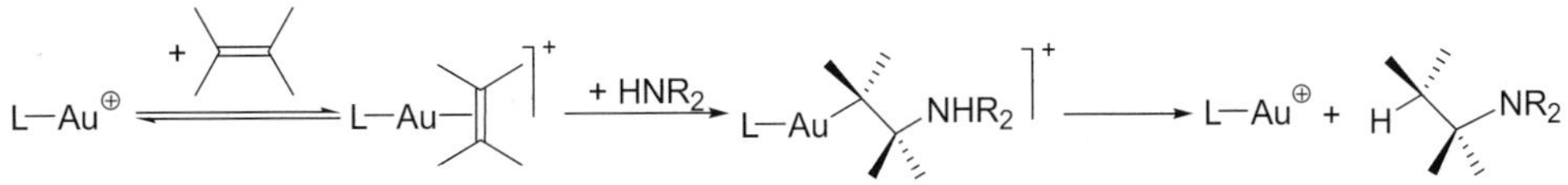

The ligand L is usually a phosphane or phosphite, but NHC ligands also come into use. For nonactivated olefins, intramolecular hydroaminations of alkenylamines to form N-heterocycles (e.g., of H$_2$C=CH–CH$_2$–CPh$_2$–CH$_2$–NHTs to the corresponding pyrrolidine derivative, catalyzed by [Au(OTf)(PPh$_3$)]) proceed more readily than

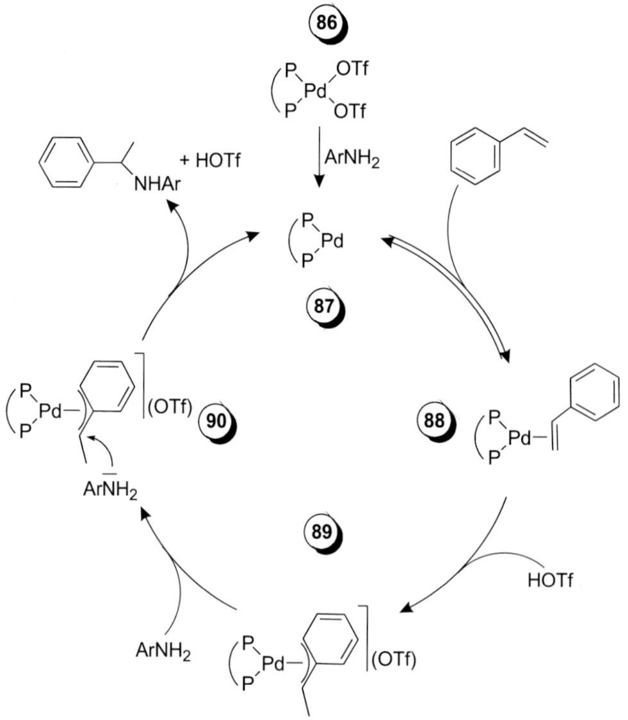

Figure 12.7 Mechanism for the palladium-catalyzed hydroamination of vinyl aromatics with arylamines.

intermolecular ones. Intermolecular reactions are implemented with high yields with sulfonamides as *N*-nucleophiles (e.g., the reaction of cyclohexene with *p*-toluenesulfonamide, catalyzed by [Au(OTf)(PPh$_3$)]) [334, 335].

12.4.2.4 Lanthanoid Complexes as Catalysts

Lanthanoid compounds, especially metallocene complexes of type [LnR(Cp*)$_2$] (**91**, Ln = La, Nd, Sm, Y, ...; Cp* = η^5-C$_5$Me$_5$; R = H, Me, CH(SiMe$_3$)$_2$, N(SiMe$_3$)$_2$, ...), catalyze, with cyclization, intramolecular hydroamination of aminoolefins **92** (*n* = 1–3), forming 5- to 7-membered azaheterocycles **93**. In many cases, conversions > 95% are achieved.

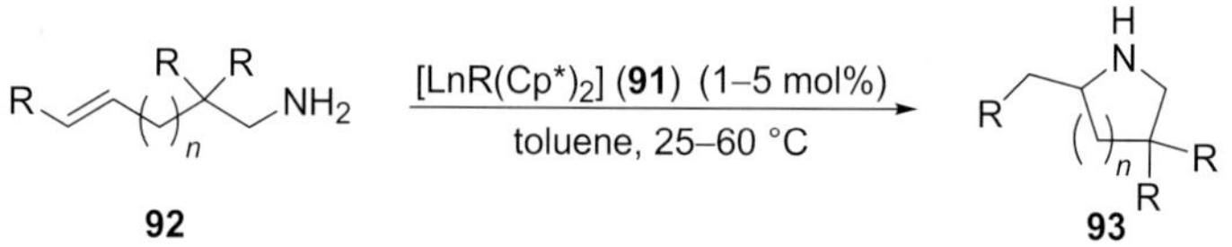

The mechanism is shown in Figure 12.8. Protolysis of the Ln–C bond with the substrate forms a lanthanoid amide as the actual catalytically active species

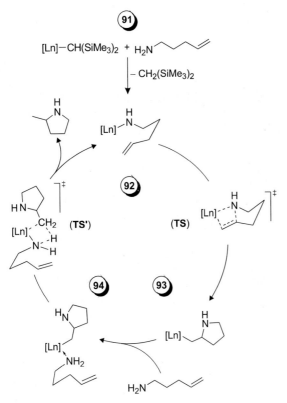

Figure 12.8 Simplified reaction mechanism of the lanthanoid-catalyzed cyclohydroamination (hydroamination/cyclization) of aminoolefins using as an example the reaction of $H_2C=CH\text{-}(CH_2)_3NH_2$ ([Ln] = Ln(Cp′)$_2$; Cp′ = Cp, Cp*). Further amine molecules (not drawn here) can be coordinated to the lanthanoid center (after Hong and Marks [336] and Hunt [337]).

(**91** → **92**). Insertion of the double bond into the Ln–N bond (see reaction b, Section 12.4.1), probably via a cyclic transition state, yields a β-aminoalkyl compound (**92** → **TS** → **93**). N coordination of the substrate (**93** → **94**) and cleavage of the product by protolysis of the Ln–C bond as a σ-bond metathesis via a four-membered cyclic transition state complete the catalytic cycle (**94** → **TS′** → **92**). The key steps are thus protolytic cleavages of Ln–C bonds by N–H functionalities (**91** → **92**, **94** → **92**) and an olefin insertion into an Ln–N bond (**92** → **93**).

Lanthanoid catalysts are also able to catalyze intermolecular hydroamination of alkenes, though these proceed slower by 2–3 orders of magnitude. With chiral lanthanoid precatalysts, asymmetric hydroamination of aminoolefins can be achieved [338–340].

With the previously described lanthanoid catalysts, aminoalkynes **95** ($n = 1-3$) can undergo a cyclohydroamination reaction. Enamines **96** are obtained as products, which, with R′ = H, tautomerize to form imines **97**.

In addition to the lanthanoid catalysts, titanium and zirconium complexes (particularly amido and imido complexes) have also proven themselves as d^0 precatalysts for intermolecular hydroaminations of alkynes [341, 342].

Exercise 12.7
Numerous d^0 complexes of Group 4 such as [Ti(NMe$_2$)$_4$], [ZrCp$_2$(NHAr)$_2$] and [Ti(Cp′)$_2$Me$_2$] (Cp′ = Cp, Cp*) have shown themselves to be precatalysts for the addition of primary amines to alkynes. In accord with the restriction to primary amines, imido complexes [M^{IV}]=NR have been proven to be the catalytically active species. Propose a mechanism for the catalysis.

13
Oxidation of Olefins and Alkanes

13.1
The Wacker Process

13.1.1
Introduction

As early as 1894, F. C. Phillips found that in aqueous solution, palladium(II) chloride oxidizes ethene to form acetaldehyde (**a**). In the process, the divalent palladium is reduced to metallic palladium. This is thus a stoichiometric reaction: Acetaldehyde and palladium are formed in equimolar quantities. It was not until between 1956 and 1959 that Wacker Chemie (Consortium für elektrochemische Industrie, Munich, Germany [343, 344]) succeeded in carrying out a catalytic reaction in which the reoxidation of the metallic palladium formed in stoichiometric quantities was performed with copper(II) salts (**b**). The Cu^I formed in this process is then oxidized with oxygen to form Cu^{II} (**c**). Thus, the Wacker process is formally based on the oxidation of ethene by oxygen (**d**):

$$Pd^{2+} + H_2C=CH_2 + H_2O \longrightarrow Me\text{-}CHO + Pd^0 + 2\,H^+ \quad (a)$$

$$Pd^0 + 2\,Cu^{2+} \longrightarrow Pd^{2+} + 2\,Cu^+ \quad (b)$$

$$2\,Cu^+ + 1/2\,O_2 + 2\,H^+ \longrightarrow 2\,Cu^{2+} + H_2O \quad (c)$$

$$H_2C=CH_2 + 1/2\,O_2 \xrightarrow[(H_2O)]{Pd^{2+}/Cu^{2+}} Me\text{-}CHO \quad (d)$$

Fundamentals of Organometallic Catalysis. First Edition. Dirk Steinborn
Copyright © 2012 WILEY-VCH Verlag GmbH & Co. KGaA, Weinheim

The overall equation (d), however, does not allow one to conclude that the source of the aldehyde oxygen atom is molecular oxygen (O_2): The aldehyde oxygen atom comes from the solvent (water), which becomes particularly clear when the previously formulated equations are represented as coupled reaction cycles:

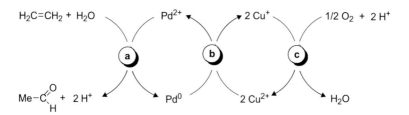

There are many metal-catalyzed oxidation reactions of organic substrates in which the direct reoxidation of the metal by O_2 or H_2O_2, two particularly suitable oxidants (from both an economic *and* an ecological perspective), does not succeed. In these cases, so-called electron transfer mediators are used, such as the redox pair Cu^I/Cu^{II} employed in the Wacker process. This principle is also found often in nature, as in the aerobic respiratory chain, in which enzyme complexes with a great number of redox cofactors such as cytochromes and ubiquinones assume this role [345].

Ethene oxidation (equation **d**) is strongly exergonic ($\Delta G^\ominus = -197$ kJ/mol). The Wacker process uses acidic aqueous solution of palladium chloride and copper chloride at 100–130 °C and a pressure of 4–10 bar. The process can be carried out in one or two stages. In the single-stage process, the formation of acetaldehyde and the reoxidation of the catalyst with oxygen take place in the same reactor. In the two-stage process, these steps are separated and air can be used to reoxidize the palladium. In both variants, the yield of acetaldehyde amounts to about 95%. As side products, chlorinated aldehydes and other products are formed, some of which are highly toxic and require much expenditure to be removed from the wastewater.

The coal-chemistry-based synthesis of acetaldehyde, starting from CaO/C, via calcium carbide and acetylene, is now of very low industrial importance.

$$CaO \xrightarrow[\text{electric arc}]{C} CaC_2 \xrightarrow{H_2O} HC\equiv CH \xrightarrow[\text{cat.}]{H_2O} MeCHO$$

Until 1990, acetaldehyde was produced in great quantities on this basis in the Buna Factory Schkopau (Saxony-Anhalt, Germany). The heterogeneously catalyzed oxidation of alcohol with oxygen (or air) in the gas phase also has low industrial significance.

In 2009, about $0.9 \cdot 10^6$ metric tons of acetaldehyde were produced worldwide by the Wacker process. The importance of acetaldehyde as raw material, however, has decreased in the last years, since it has been replaced as precursor for a number of large-volume chemical products. Thus, acetic acid is now scarcely produced by oxidation of acetaldehyde; it is mostly produced by methanol carbonylation. Furthermore, instead of producing C_4 aldehydes via an aldol reaction from acetaldehyde, propene can be easily hydroformylated.

13.1.2
Mechanism of Ethene Oxidation

The heart of the Wacker process is the palladium(II)-mediated oxidation of ethene to form acetaldehyde. Some details of the mechanism have not yet been clarified. A possible reaction course is shown in Figure 13.1, starting from a tetrachloropalladate(II) complex (**1**), which is predominant in a hydrochloric acid solution of $PdCl_2$. The following individual reaction steps can be named [346]:

1 → 2: *Ligand substitution reaction.* Substitution of a chloro ligand by ethene yields an (η^2-ethene)palladate(II) complex **2** which is the palladium analogue to the anion of Zeise's salt.

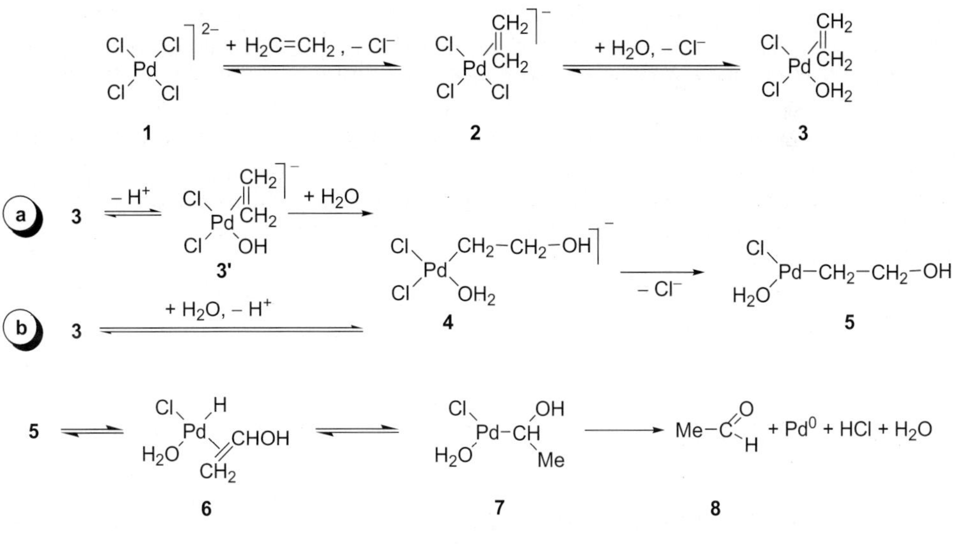

Figure 13.1 Mechanism of the Wacker process.

2 → 3: *Ligand substitution reaction.* Substitution of a chloro ligand by water leads to a neutral (η^2-ethene)palladium(II) complex **3**.

Now there are two possible reaction paths:

(a) **3 → 4:** *Deprotonation/insertion/ligand attachment.* Deprotonation leads to a cis ethene–hydroxo complex **3'**; insertion of ethene into the Pd–OH bond and subsequent attachment of water forms the (2-hydroxyethyl)palladate(II) complex **4**.

(b) **3 → 4:** *Intermolecular addition of a nucleophile/heterolytic fragmentation.* Intermolecular addition of water to the coordinated ethene and subsequent deprotonation yields the (2-hydroxyethyl)palladate(II) complex **4**. The reverse reaction is a heterolytic fragmentation.

4 → 5: *Ligand cleavage.* Cleavage of Cl⁻ leads to the coordinatively unsaturated (2-hydroxyethyl)palladium(II) complex **5**.

5 → 6 → 7: *Isomerization.* Initially, β-H elimination forms a hydrido–η^2-vinyl alcohol complex and subsequent reinsertion of the π-bound vinyl alcohol into the Pd–H bond yields the (1-hydroxyethyl)palladium complex **7**.

7 → 8: *Deprotonation/ligand cleavage.* Deprotonation of the 1-hydroxyethyl ligand and heterolytic cleavage of the Pd–C bond leads to formation of acetaldehyde. Calculations indicate that the reaction passes through a transition state **TS₇→₈** in such a way that, synchronously with the cleavage of Cl, the hydrogen from the OH group is transferred onto the Cl. *Formally,* this reaction can also be seen as β-hydride elimination with subsequent decomposition of the aldehyde(hydrido)palladium(II) complex [PdCl(H)(MeCHO)(H₂O)]. However, for the transition state that corresponds to this β-H elimination, an energy has been calculated which is significantly higher than the one for the transition state previously mentioned.

With experiments in D₂O, in which no deuterium was incorporated into the formed acetaldehyde, the reaction course **5 → 8** was confirmed. This excludes the possibility that the π-vinyl alcohol complex **6** cleaves off vinyl alcohol that subsequently tautomerizes into acetaldehyde.

Exercise 13.1

- Kinetic experiments indicate that the ligand substitution reaction **2 → 3** is a rather complex reaction and that first the chloro ligand *trans* to the ethene ligand is replaced by water. Why is this reaction favored over direct substitution of a *cis* chloro ligand?
- Explain why aqua complexes such as **3** can be deprotonated easily (**3 → 3'**).
- Explain why the aquadichloro(η^2-ethene)palladium complex **3** is better able to add water intermolecularly (**3 → 4**) than the (η^2-ethene)palladate complexes **2** and **3'** are.

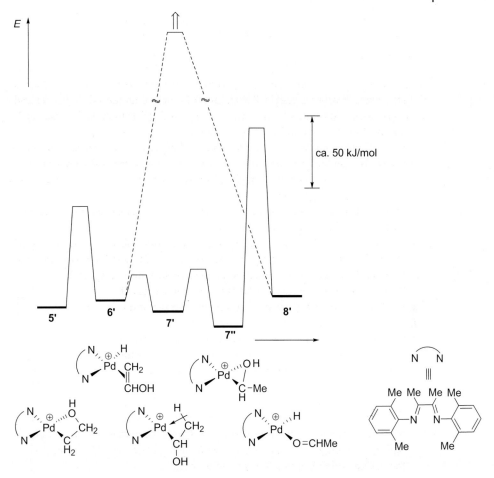

Figure 13.2 Profile of the potential energy for the formation of acetaldehyde from [Pd(CH$_2$CH$_2$OH-κC,κO)(N⌒N)]$^+$ (QM/MM calculations in the gas phase; shortened from DeKock and Ziegler [347]).

Quantum-chemical calculations on the cationic metal complex entity [Pd(N⌒N)]$^{2+}$ (Figure 13.2)[1]) show that all intermediate complexes **5′–8′** have comparable energies. These are a 2-hydroxyethyl complex (**5′**), a hydrido(π-vinyl alcohol)palladium complex (**6′**), two isomeric 1-hydroxyethyl complexes (**7′**: κC-coordinated ligand with an additional β-agostic C–H··· Pd interaction; **7″**: κC,κO-coordinated ligand) and an aldehyde(hydrido)palladium complex (**8′**). The activation

1) Brookhart's catalyst for ethene polymerization, [PdMe(H$_2$C=CH$_2$)(N⌒N)]$^+$, decomposes in water in the presence of ethene, probably in a Wacker-like reaction forming palladium and acetaldehyde. We start our discussion with complex **5′**, which could be formed from Brookhart's catalyst after protolytic cleavage of the methyl ligand by H$_2$O in reactions analogous to 3 → 4 → 5 in Figure 13.1.

barriers for the insertion of the vinyl alcohol into the Pd–H bond (**6'** ⇌ **7'**) and for the isomerization of the two 1-hydroxyethyl complexes (**7'** ⇌ **7''**) are low. The rate-determining step is the β-hydride elimination from the OH group (**7''** ⇌ **8'**). The direct path from **6'–8'** – the migration of the OH hydrogen atom to the unsubstituted olefin C atom in a concerted reaction (tautomerization of a vinyl alcohol complex to form an acetaldehyde complex) – has almost twice as high an activation barrier and thus must be ruled out as a possible reaction path. The formation of Pd^0 and acetaldehyde can be initiated by H^+ abstraction from the O–H group in **7''** or from the Pd–H moiety in **8'**. Hence, the deprotonation of complex **7''** is associated with a lower activation barrier (not shown in Figure 13.2) than reaction **7''** ⇌ **8'** so that **8'** most likely does not occur as intermediate.

The decisive reaction step in the Wacker process is the formation of the C–O bond, which either proceeds intramolecularly as an insertion (Figure 13.1, **a**) or intermolecularly as an addition of a nucleophile to coordinated ethene (Figure 13.1, **b**). Although it is in principle possible to distinguish between the two reactions, it has not yet been possible to definitely clarify which reaction is rate-determining: the insertion of ethene into the Pd–OH bond (**3'** → **4**) or the cleavage of chloride from the (2-hydroxyethyl)palladium complex (**4** → **5**), which is preceded by rapid establishment of equilibrium **3** ⇌ **4** [348]. Apparently, which reaction occurs depends heavily on the reaction conditions and concentration ratios.

With the Wacker catalyst at very high concentrations of chloride ions,[2] 2-chloroethanol (ethylene chlorohydrin) or ethylene oxide is formed as reaction product according to the following scheme:

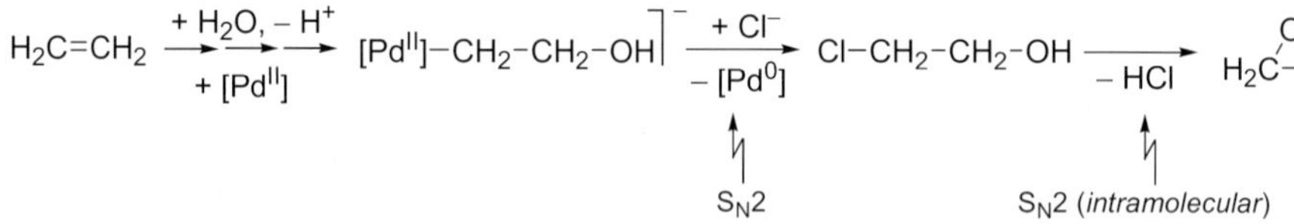

$$H_2C=CH_2 \xrightarrow[+ [Pd^{II}]]{+ H_2O, -H^+} [Pd^{II}]-CH_2-CH_2-OH]^- \xrightarrow[- [Pd^0]]{+ Cl^-} Cl-CH_2-CH_2-OH \xrightarrow{- HCl} \underset{H_2C-CH_2}{\overset{O}{\triangle}}$$

$$S_N2 \qquad\qquad S_N2\ (intramolecular)$$

The stereochemistry of the formation of 2-chloroethanol (S_N2 reaction with inversion of configuration) and of ethylene oxide (intramolecular S_N2 reaction with inversion of configuration) is known. Since the first reaction step is either an intramolecular *syn* addition or an intermolecular *anti* addition, so by selecting an appropriate olefinic reaction partner, the stereochemical course of this reaction step can be deduced. When (*E*)-1,2-dideuteroethene (**9**) is used, *threo*-1,2-dideutero-2-chloroethanol (**10**) or *cis*-1,2-dideuteroethylene oxide (**11**) is obtained as reaction product [349].

---

2) At a concentration as low as 2.5 mol/l for Cl^- (e.g., added as LiCl) and 3 mol/l for $CuCl_2$, the formation of 2-chloroethanol is a significant side reaction.

13.1 The Wacker Process

[Scheme: compound 9 → 10 (threo) → 11, with reagents +H₂O, −H⁺, [Pd^II]; +Cl⁻; −HCl]

The reaction courses are compared below for an intermolecular *anti* addition (a) and an intramolecular *syn* addition (*cis* insertion) (b).[3)]

[Scheme (a): intermolecular *anti* addition via erythro intermediate → threo → epoxide]

[Scheme (b): intramolecular *syn* addition via threo intermediate → erythro → epoxide]

The formation of *threo*-1,2-dideutero-2-chloroethanol or *cis*-1,2-dideuteroethylene oxide demonstrates an intermolecular *trans* addition of water to the ethenepalladium complex. However, due to the different reaction conditions, particularly the high Cl⁻ concentration, no conclusion can be drawn about the mechanism of the Wacker reaction under industrial conditions. Kinetic studies of the acetaldehyde formation in systems with a low Cl⁻ concentration under conditions approximating those of the industrial process suggest *cis* insertion of ethene into the Pd–OH bond [348, 350].

A different reaction mechanism at high and low Cl⁻ concentrations has also been found for palladium-catalyzed addition reactions of MeO⁻ to chiral allyl alcohols **12**. At low concentrations of Cl⁻, the allyl alcohol is oxidized to form the corresponding carbonyl compound (**12** → **13** → **14**). This reaction course corresponds to the Wacker reaction; compare with the reactions **5** → ... → **8** in Figure 13.1. At high concentrations of Cl⁻, a (nonoxidative) double-bond shift occurs (**12** → **13** → **15**).

---

3) In compounds with two neighboring asymmetric C atoms and the general constitution C(Xab)–C(Yab), stereoisomers are called "*erythro*" when the substituents a and b are on the same side in the Fischer projection (see below). Then, in the Newman projection, they can be simultaneously superimposed pairwise (a ↔ a; b ↔ b), which can be seen if C(Xab) is rotated by 180°. If that is not the case, the stereoisomer is called the "*threo*" form. The nomenclature has its origin in the carbohydrate chemistry of the tetroses erythrose and threose.

[Fischer and Newman projections of erythro and threo forms; D-erythrose (2R,3R) and L-threose (2R,3S)]

The double bond in the chiral allyl alcohol **12** is prochiral and as a result, an additional stereocenter is generated in the σ-organopalladium(II) intermediate **13**, whose configuration allows the reaction course (*syn* vs. *anti* addition) to be determined. The stereochemistry of the products obtained demonstrates (see Exercise 13.2) that the reaction to form "Wacker product" **12** → **13** → **14** proceeds via insertion of the allylic double bond into the Pd–OMe bond (*syn* addition), while at high chloride concentrations **12** → **13** → **15**, intermolecular *anti* addition takes place. A similar finding has been obtained for the addition of OH⁻ and Ph⁻ (instead of MeO⁻) [346, 351].

The reason for the change in mechanism (*syn* vs. *anti*) and in the reaction products (Wacker product vs. double-bond isomerization) is probably the following [351]:

- Only at low chloride concentrations can hydroxo- and aqua(olefin)palladium(II) complexes be formed. That is a precondition for the Wacker reaction via *cis* insertion (3' → 4 in Figure 13.1) and the β-H elimination leading to the Wacker products (5 → 6 in Figure 13.1) to run sufficiently rapidly.
- At high chloride concentrations, the trichloro(olefin)palladate(II) complexes are found. That causes an intermolecular *anti* addition. Since the [Pd(CH$_2$CH$_2$OH)-Cl$_3$]$^{2-}$ complex that is formed contains no easily displaceable aqua ligands, it is relatively stable against β-H elimination (5 → 6 in Figure 13.1), which opens the reaction channel to the Wacker products. Instead, in the presence of copper(II) chloride, the formation of chlorohydrin becomes increasingly important. In the absence of copper(II) chloride, in the case of allyl alcohols, heterolytic fragmentation leading to double-bond isomerization (**13** → **15**) becomes more significant.

Exercise 13.2

In the reaction to form **14** or **15** starting from (*R*)-**12**, an olefin palladium(II) complex **12'** is first formed. Write down the four reaction products formed from **12'** during *syn* and *anti* addition of MeOH at low ($c_{Cl^-} = 0.1$ M) and high ($c_{Cl^-} > 2$ M) chloride ion concentrations.

13.1.3
Oxypalladation of Olefins

13.1.3.1 Types of Oxypalladation
Initiated by the Wacker process, further palladium-catalyzed oxidative functionalizations of olefins have been developed. If an O-nucleophile is involved, the reaction is called oxypalladation [352]. The principal reaction course corresponds to the Wacker reaction and is shown schematically below.

It encompasses olefin activation by complex formation with Pd^{II} (16 → 17), *syn* or *anti* addition of the nucleophile XOH, which is deprotonated in the process (17 → 18), and β-H elimination, which leads to cleavage of the products (18 → 19) with formation of Pd^0. Finally, oxygen oxidizes Pd^0 to Pd^{II} (19 → 16) in the presence of a copper catalyst.

- **XOH = H_2O**
  With ethene as substrate and a Pd/Cu catalyst, this is the prototypical Wacker reaction. At high concentrations of chloride and $CuCl_2$, 2-chloroethanol can be produced in a targeted manner. In comparison with ethene, the Wacker reaction runs significantly more slowly with terminal olefins and generally leads to methyl ketones. Inner olefins react even more slowly, so chemoselective oxidation of diolefins to form methyl ketones with retention of the inner double bond is possible [353]. Interestingly, $PdCl_2$ in N,N-dimethylacetamide catalyzes the oxidation of internal olefins to ketones (e.g., (E)-oct-4-ene + H_2O → octan-4-one) in the absence of copper species. Under the reaction conditions (80 °C, 3 bar O_2), Pd^0 is reoxidized directly by oxygen. Investigations on the course of the catalysis reveal that, indeed, a copper compound promotes the reoxidation of Pd^0 but inhibits the oxidation of internal olefins [354].
- **XOH = ROH**
  Unsaturated alcohols **20** undergo an intramolecular oxidative cyclization that, due to the favoring of H over H′ in the β-H elimination reaction in the intermediate

complex **21**, leads to allyl ethers **22**. This is used to synthesize allyl-substituted tetrahydrofurans and -pyrans.

$$\text{20} \xrightarrow{+ [Pd^{II}], - H^+} \text{21} \xrightarrow{- Pd^0, - H^+} \text{22}$$

The regioselectivity is dependent to some extent on the anionic ligand on the palladium. With $PdCl_2$, the formation of six-membered rings seems to predominate; with $Pd(OAc)_2$, five-membered rings are apparently favored.

- **XOH = AcOH**

The palladium-catalyzed addition of acetic acid to ethene yields vinyl acetate. Since the reaction generates water, some of the vinyl acetate is hydrolyzed to form acetaldehyde and acetic acid. This process has even been run industrially, but cannot compete with a heterogeneously catalyzed gas phase reaction (palladium and alkali metal salts on an oxidic support):

$$H_2C=CH_2 + MeCOOH + 1/2\, O_2 \xrightarrow[140\,°C,\ 0.5-1.2\,MPa]{cat.} Me-C(=O)OCH=CH_2 + H_2O$$

The reaction of butadiene with acetic acid leads to 1,4-diacetoxy-but-2-ene (**23**), which subsequently reacts with hydrogen and water to form butane-1,4-diol (**24**). This process is operated industrially (with a heterogeneous Pd/C catalyst) and is an alternative to the acetylene-based synthesis of butynediol ($HC\equiv CH + 2\, HCHO \rightarrow HOCH_2C\equiv CCH_2OH$) and its hydrogenation to form **24**.

$$\text{butadiene} \xrightarrow[(PdCl_2/CuCl_2)]{HOAc/O_2} \text{23 (AcO-CH}_2\text{-CH=CH-CH}_2\text{-OAc)} \xrightarrow[2)\ H_2O]{1)\ H_2} \text{24 (HO-CH}_2\text{-CH}_2\text{-CH}_2\text{-CH}_2\text{-OH)}$$

Exercise 13.3

N-Heterocycles, as in the example, can be synthesized via aza-Wacker cyclization. Give a possible mechanism.

$$\text{alkenyl-NHTs} \xrightarrow[O_2,\ (NaOAc)\ (DMSO,\ 25\,°C)]{[Pd(OAc)_2]} \text{N-Ts pyrrolidine with vinyl} \quad (>90\%)$$

13.1.3.2 Enantioselective Oxypalladation

If the oxidation of ethene is carried out with [PdCl$_3$(py)]$^-$ (instead of [PdCl$_4$]$^{2-}$), 2-chloroethanol (ethylene chlorohydrin) is obtained, even at low concentrations of chloride ions. That opens the possibility to enantioselective synthesis of chlorohydrins **25/25'** from terminal olefins if palladium complexes with chiral ligands as precatalysts are used. These can be neutral mononuclear complexes **26** with chiral diphosphane ligands. For solubility reasons, sulfonated BINAP ligands (**27**, Ar = C$_6$H$_4$-m-SO$_3$Na, Ph), among others, were used. Higher ee values, some > 90%, were achieved with cationic binuclear palladium complexes **28** (s = solvent; R, R' = Me, Ph, CF$_3$) containing chiral μ-L⌒L ligands such as BINAP, DIOP or DACH [276, 355].

13.1.3.3 Palladium Oxidase Catalysis

The Wacker reaction and the other reactions previously described are Pd^{II}-catalyzed aerobic oxidations. The catalysis can be separated into two partial reactions, the oxidation of the substrate by Pd^{II} and the copper-catalyzed reoxidation of Pd^0 by dioxygen. This is a reaction principle that is encountered with oxidases. Oxidases are metalloenzymes that catalyze aerobic oxidation where dioxygen functions both as electron and proton acceptor without an oxygen atom being transferred to the substrate.[4] The oxidase withdraws electrons from the substrate (**29 → 30**; SuH$_2$ = reduced substrate), transferring them to O$_2$ in the course of reoxidation of the enzyme, generating H$_2$O or H$_2$O$_2$ (**30 → 29**). Oxidases may contain metals such as copper, iron, and/or molybdenum.

---

4) Oxidase enzymes thus oxidize a substrate without transferring an oxygen atom from O$_2$. This is an important difference from oxygenases, which transfer either one or both oxygen atoms from O$_2$ (mono- or dioxygenases) to the substrate.

To rightfully call the palladium-catalyzed reactions discussed above "palladium oxidase catalysis," the reoxidation of Pd^0 must not be copper-catalyzed, but rather ought to be accomplished directly by dioxygen. That is possible in principle, as the formation of (η^2-peroxo)palladium(II) complexes by reaction of Pd^0 complexes with O_2 demonstrates (**31** → **32**). However, mechanisms can also be imagined that have no Pd^0 intermediates at all; instead the O_2 can be activated by insertion into a Pd–H bond (**33** → **34**) [356, 357].

$$[Pd^0] \xrightarrow{O_2} [Pd^{II}]\begin{pmatrix}O\\O\end{pmatrix} \rightleftharpoons \qquad [Pd^{II}]-H \xrightarrow{O_2} [Pd^{II}]-OOH \rightleftharpoons$$

  31   **32**   **33**   **34**

13.2
Epoxidation of Olefins

13.2.1
Introduction

The favored reactions of dioxygen with hydrocarbons are radical oxidation reactions that lead to complete combustion, forming CO_2 and H_2O, as long as no special precautions are encountered. These reactions are strongly exergonic, but encumbered with a relatively high activation barrier. The essential reason for this is the triplet structure of O_2 in the ground state ($^3O_2$), so the reaction with singlet molecules would require a spin inversion, which has low probability at temperatures that are not particularly high ("spin conservation rule"). This does hinder selective oxidation of organic compounds with molecular oxygen, but ensures their stability in the presence of O_2 and thus makes possible the life forms found on earth.

The oxidation of olefins with dioxygen to form epoxides is exergonic, but its complete combustion to form CO_2 and H_2O is also exergonic, as shown in the following reaction sequence for ethene:

$$\xrightarrow{\Delta G^\ominus = -80 \text{ kJ/mol}}_{+\ 1/2\ O_2} \triangle\!\!\!O \xrightarrow{\Delta G^\ominus = -1251 \text{ kJ/mol}}_{+\ 5/2\ O_2} 2\ CO_2 + 2\ H_2O$$

The addition of an oxygen atom ("oxene") to an olefin to form an epoxide (**37**) is formally analogous to that of a carbene or nitrene to form a cyclopropane (**35**) or aziridine (**36**), respectively.

$$=\ +\ \overset{R}{\underset{R}{>}}C: \longrightarrow \overset{R\ \ R}{\triangle} \quad =\ +\ R-\ddot{N} \longrightarrow \overset{R}{\underset{}{\triangle}}\!\!N \quad =\ +\ \ddot{O}: \longrightarrow \triangle\!\!\!O$$

       35   **36**   **37**

Oxo- (**a**) and peroxometal complexes (**b/b'**: R = H, alkyl, ...) can transfer oxygen to olefins. Reactions (**a**) can proceed in a concerted way, so the two C–O bonds form simultaneously (**a₁**). The high electrophilicity of the oxo oxygen atom required for this reaction can be achieved by a high oxidation state of the metal, and additionally, as the case may be, by a positive charge on the complex. If radical intermediates occur during the reaction (a_2), the two C–O bonds are formed in succession. Peroxo complexes in reactions (b/b′) react like "oxenoids."[5)]

$$[M]=O \;+\; {=\!=} \quad\longrightarrow\quad \overline{[M]} \;+\; \triangle\!\!\!\!O \qquad \left([M]{=}O{\cdots}\!\!\!\!\bigtimes \;;\; [M]\text{–}\overset{\bullet}{O}{\cdots}\!\!\!\!\bigtimes \right) \qquad (a)$$

$$\begin{array}{c}\;\;\overset{R}{\underset{O}{\bigcirc}}\\ [M]{<}_{O}^{O}\end{array} \;\left([M]\!\!\diagdown_{O}\!\!\diagup^{O\text{–}R} \right) \;+\; {=\!=} \quad\longrightarrow\quad [M]\text{–}O\text{–}R \;+\; \triangle\!\!\!\!O \qquad (b)$$

$$[M]{<}_{O}^{O}\!\!\overset{O}{} \;+\; {=\!=} \quad\longrightarrow\quad [M]{=}O \;+\; \triangle\!\!\!\!O \qquad (b')$$

While the oxidation number of M remains unchanged by epoxidation with peroxometal complexes (b/b′), it is lowered by two by epoxidation with oxometal complexes (a). Thus, the only metal complexes which can catalyze reactions according to path a are those that can easily undergo a change of two in their oxidation number.

The transfer of an oxygen atom from an oxygen donor X′O to an oxygen acceptor X proceeds according to equation (a). With H_2O_2 as reference donor, the reaction Gibbs free energy $\Delta_r G$ from equation (b) is a measure of the thermodynamic oxygen transfer potential (TOP) of the XO/X pair.

$$X + X'O \;\longrightarrow\; XO + X' \quad (a) \qquad\qquad X + H_2O_2 \;\longrightarrow\; XO + H_2O \quad (b)$$

Table 13.1 lists quantum-chemically calculated values of $\Delta_r G$ for several oxygen atom donors XO and acceptors X, respectively.

Reactions a proceed readily (from a thermodynamic perspective; possible kinetic inhibition of the reactions is not considered!) when $\Delta_r G(XO/X) < \Delta_r G(X'O/X')$. The arrow in Table 13.1 thus shows the direction of the oxygen transfer, where the pairs XO/X are arranged according to the values $\Delta_r G$ in the gas phase. The pair further up in the table is reduced (X′O → X′ + O) and the pair further down in the table is oxidized (X + O → XO).

13.2.2
Epoxidation of Ethene and Propene

13.2.2.1 O₂ and ROOH as Oxygen Transfer Agents

So far, the large-scale industrial production of epoxides by catalytic oxidation of olefins with dioxygen has only been achieved for ethene. In a heterogeneously catalyzed reaction on a silver catalyst (Ag on Al_2O_3) ethene is oxidized to ethylene oxide at 200–300 °C (1–3 bar) with air or oxygen, with selectivity of 80–90%. In this way, about $1.5 \cdot 10^6$ metric tons/year (2000) of ethylene oxide are produced worldwide.

[5)] In analogy to carbenoids [M]–CR₂X, which react like carbenes, "oxenoids" are metal complexes [M]–O–X (X = leaving group) that can transfer an oxene ([M]–O–X → [M]–X + "O"). A typical property of oxenoids is the electrophilic character of the oxygen atom.

13 Oxidation of Olefins and Alkanes

Table 13.1 Quantum-chemically (DFT) calculated thermodynamic oxygen transfer potentials (TOP), for the coupled pairs XO/O in the gas phase and in aqueous solution ($\Delta_r G$ in kJ/mol at 298 K) (after Deubel [358]).

XO	X	$\Delta_r G$ (gas phase)	$\Delta_r G$ (water)
Me$_2$COO[a)]	Me$_2$CO	42	49
[ReO(O$_2$)$_2$Me]	[ReO$_2$(O$_2$)Me]	21	
PhIO	PhI	18	−8
[ReO(O$_2$)$_2$Me(H$_2$O)]	[ReO$_2$(O$_2$)Me(H$_2$O)]	5	8
H$_2$O$_2$	H$_2$O	0	0
MeC(O)OOH	MeC(O)OH	−8	−3
MeOOH	MeOH	−28	−26
HSO$_5^-$	HSO$_4^-$	−30	−10
[MoO(O$_2$)$_2$(OPH$_3$)]	[MoO$_2$(O$_2$)(OPH$_3$)]	−44	−47
[MoO$_2$(O$_2$)(OPH$_3$)]	[MoO$_3$(OPH$_3$)]	−53	−49
Me$_3$NO	Me$_3$N	−72	−100
ClO$_4^-$	ClO$_3^-$	−99	−73
C$_5$H$_5$NO[b)]	C$_5$H$_5$N (py)	−114	−116
[OsO$_3$(OCH$_2$CH$_2$O)][c)]	[OsO$_2$(OCH$_2$CH$_2$O)]	−149	−144
Me$_2$SO (DMSO)	Me$_2$S	−184	−198
C$_2$H$_4$O[d)]	H$_2$C=CH$_2$	−192	−205
MeOH	CH$_4$	−224	−249
Me$_2$SO$_2$	Me$_2$SO (DMSO)	−273	−284
Me$_3$PO	Me$_3$P	−382	−407

a) Me–O–O–Me (dioxirane structure) b) pyridine N-oxide c) osmate ester d) ethylene oxide

To synthesize propylene oxide industrially, as an alternative to using the chlorohydrin process (propene → propylene chlorohydrin → propylene oxide), propene can be catalytically oxidized with hydroperoxides (**38/39** → **40/41**), which in turn are obtained by oxidation of hydrocarbons by oxygen (**42** → **39**). The production of propylene oxide is thus coupled to coproduction of alcohols (**41**) (Halcon-ARCO process).

$$\text{CH}_2{=}\text{CHCH}_3 + \text{ROOH} \xrightarrow{\text{cat.}} \text{propylene oxide} + \text{ROH}$$

38   **39**   **40**   **41**

                ↑
         R–H + O$_2$
              42

The organic hydroperoxides, *tert*-butyl hydroperoxide (**39a**, R = *t*-Bu) and 1-phenylethyl hydroperoxide (**39b**, R = CH(Ph)Me), have become prevalent in industry. **39a** is obtained from isobutane and oxygen (120–140 °C, 25–35 bar). The *tert*-butanol **41a** produced is dehydrated to isobutene which is treated with methanol to give methyl *tert*-butyl ether (MTBE), which is used as a fuel additive. **39b** is obtained from ethylbenzene and air (130–145 °C, 2 bar); the 1-hydroxyethylbenzene **41b** generated in the process is dehydrated to form styrene. The theoretical mass ratio for C_3H_6O : MTBE is about 1 : 1.5 and for C_3H_6O : styrene it is about 1 : 1.8; due to the process technology, about 3 times the quantity of MTBE or 2.5 times the quantity of styrene, respectively, are produced. However, the coproduction of both alcohols, although they can be profitably processed further, comes with a disadvantage inherent to the process: The quantities of MTBE or styrene that are inevitably produced do not always correspond to market demand. Furthermore, they make additional demands on the propylene oxide producer regarding infrastructure and distribution.

The reaction of propene with *t*-BuOOH is carried out in toluene (100–120 °C, 40 bar) and the reaction with PhCH(OOH)Me is performed without solvent (100 °C, 35 bar). As precatalysts, molybdenum salts such as molybdenum naphthenates are used. Under the reaction conditions, oxidation occurs to form peroxo or hydroperoxo complexes of the hexavalent molybdenum, which are the actual catalytically active compounds. As an alternative to these homogeneous catalytic systems (ARCO), a heterogeneous titanium catalyst (Ti^{IV} on SiO_2) is used (Shell) for the ethylbenzene-based process.

About $5 \cdot 10^6$ metric tons of propylene oxide a year are produced worldwide, of which about half is produced by the chlorohydrin process and 1/4 each by oxidation with hydroperoxides coupled with production of MTBE and styrene. Propylene oxide is primarily (about 2/3) used for the synthesis of polyethers with terminal hydroxy groups, which are further processed into polyurethanes. The hydrolysis of propylene oxide yields propane-1,2-diol, which finds applications as starting material for the synthesis of polyester resins and in the food, pharmaceutical, and cosmetics industries.

13.2.2.2 Mechanism

Investigations on model complexes have been performed to obtain insight into the mechanism of the molybdenum-catalyzed oxygen transfer of hydroperoxides to olefins. From stoichiometric reactions of olefins with peroxomolybdenum(VI) complexes $[MoO(O_2)_2L]$ (**43**, L = HMPA, ...), two contradictory mechanisms have been proposed [359]:

- *Stepwise mechanism (Mimoun, 1970).* After complex formation of the olefin (**43** → **44**), it inserts into an Mo–O bond ([2 + 2] cycloaddition; **44** → **45**). The epoxide cleaves from the dioxamolybdacyclopentane complex by cycloreversion (**45** → **46**).
- *Concerted mechanism (Sharpless, 1972).* Nucleophilic attack of the olefin on one of the two peroxo oxygen atoms (**43** → **44′**) yields the epoxide (**44′** → **45′** → **46**) via a transition state **45′** (originally formulated with broken O–O bond).

DFT calculations have confirmed the Sharpless mechanism (Figure 13.3). Starting with [MoO(O$_2$)$_2$L] (**43**; L = OPH$_3$) and H$_2$C=CH$_2$, an exothermic reaction yields [MoO$_2$(O$_2$)L] and C$_2$H$_4$O directly (**43** → **45′**(TS) → **46**). In the gas phase, the addition of ethene to **43** is weakly endothermic (**43** → **44**). The associated activation barrier is low, so an equilibrium between **43** + H$_2$C=CH$_2$ and **44** is to be assumed. However, no path from **44** to **45** has been found, but one directly from **43** + H$_2$C=CH$_2$ to **45** exists. However, the activation barrier is significantly higher than that for the reaction of **43** to **44** and than that of the concerted reaction directly from **43** to **46**. Furthermore, it has been shown that the cycloreversion of **45** would not generate the epoxide, but rather the "wrong" product, namely acetaldehyde (**45** → **47**).

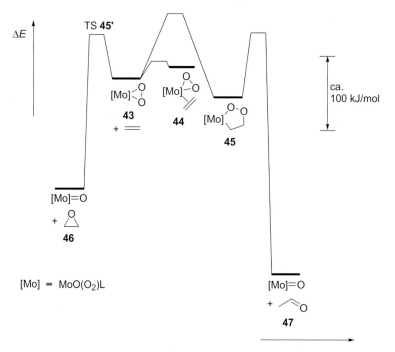

Figure 13.3 Oxygen transfer from [MoO(O$_2$)$_2$L] (**43**, L = OPH$_3$) to ethene: stepwise vs. concerted mechanism. In each case, only the reaction path that is energetically most favorable is shown (after Deubel and Frenking [360]).

13.2 Epoxidation of Olefins

The decisive orbital interaction in the transition state TS **45'** is an electron transfer from the π-C=C orbital of the olefin (nucleophile) into the antibonding σ*-O–O orbital of the peroxo ligand (electrophile) (**I**). In accord with this, a higher electrophilicity of the peroxo ligand leads to a higher reactivity. Accordingly, the peroxo ligand is activated by protonation (**48/48'**), which can occur intramolecularly (by H transfer from a neighboring aqua ligand) or intermolecularly. The hydroperoxo complex **48/48'** is attacked at the α-O atom by the olefin.

In the molybdenum-catalyzed synthesis of propylene oxide from propene using alkyl hydroperoxides as oxygen transfer agents, an analogous mechanism can be assumed: The epoxide formation occurs through nucleophilic attack by propene on an alkylperoxo complex of hexavalent molybdenum (**49 → 50 → 51**, R = t-Bu, CH-(Ph)Me). Protolytic cleavage of the Mo–OR bond by the alkyl hydroperoxide completes the catalytic cycle (**51 → 49**) [361, 362].

No intermediates with M–C bonds are involved in the formation of an epoxide. The olefin is *not* activated by complex formation. Other metals with low oxidation potential and high Lewis acidity in their highest oxidation states, such as W, V, and Ti, are also catalytically active.

A single electron transfer (SET) from a metal to the alkyl hydroperoxide leads to O–O bond cleavage (RO–OH + e^- → RO• + $^-$OH) and causes formation of side products. Metals such as Co, Mn and Fe, which have such a tendency, are not suitable catalysts.

13.2.2.3 H$_2$O$_2$ as Oxygen Transfer Agent

Hydrogen peroxide is an inexpensive and environmentally friendly oxidant [363]. Its use in the epoxidation of propene is desirable, since only water is "coproduced" (see Section 13.2.2.1, equation **38/39 → 40/41**, R = H) and the previously mentioned disadvantages associated with the coproduction of MTBE or styrene (R = t-Bu, CH-(Ph)Me) disappear.

Heterogeneously catalytic. With a Ti^{IV}-substituted silicalite (an aluminum-free zeolite of MFI structure, similar to that of ZSM-5) as catalyst, which in contrast to the Shell catalyst (Ti^{IV} on SiO$_2$) has a hydrophobic (inner) surface, the oxidation of propene with H$_2$O$_2$ was successful (Evonik-Uhde). A large-scale industrial plant for this hydrogen peroxide to propylene oxide (HPPO) process (capacity: 100 000 metric tons/year) has been built in Ulsan (Korea).

Homogeneously catalytic. An example of an effective homogeneous catalyst for epoxidation of alkenes with H_2O_2 (**52** → **53**) is methyltrioxorhenium(VII) (**54**) [364]. Complex **54** reacts with H_2O_2 to form a diperoxo complex (**54** → **55**), which is the actual epoxidation catalyst at high H_2O_2 concentrations. As with the Mo^{VI} catalysts, the oxygen transfer to the olefin occurs by electrophilic attack of a peroxo oxygen atom from **55** in a concerted reaction.

$$R\text{=} \xrightarrow[\text{54}]{+H_2O_2,\ -H_2O} \underset{R}{\triangle} \quad\quad \underset{\text{54}}{\overset{Me}{\underset{O}{\overset{|}{\underset{\|}{Re}}}}\underset{O}{\overset{O}{\diagdown}}} \xrightarrow[-H_2O]{+2\ H_2O_2} \underset{\text{55}}{\overset{O}{\underset{Me}{\overset{\|}{\underset{OH_2}{Re}}}}}$$

52   **53**   **54**   **55**

As side products, diols appear whose formation is suppressed by addition of Lewis bases such as pyridine or pyrazole, so selectivity with respect to epoxide formation of >95% can be achieved [365].

Reaction-controlled phase-transfer catalysis. A heteropolytungstate with an N-alkylpyridinium cation, $(C_5H_5NC_{16}H_{33})_3[PW_4O_{16}]$ (**56**), is employed as precatalyst. The anion in **56** is derived from the tetratungstate $[W_4O_{16}]^{8-}$. In the tetratungstate structure, four edge-sharing WO_6 octahedra are assembled so as to yield a heterocubane structure in which the vertices of a cube are alternatingly occupied by W and μ_3-O atoms. Thus, the cube consists of two interpenetrating W_4 and O_4 tetrahedra. In **56**, a P atom is found in the center of the cube, so **56** can also be written as $(C_5H_5NC_{16}H_{33})_3[PO_4(WO_3)_4]$, emphasizing that each W atom binds three terminal oxo ligands.

A water–toluene–tributylphosphate mixture does not dissolve **56** by itself, but under the influence of H_2O_2 (52%), it does. This forms the catalytically active complex $(C_5H_5NC_{16}H_{33})_3[PO_4\{WO_2(O_2)\}_4]$ (**56** → **57**), in which one oxo ligand is replaced by a peroxo ligand at each W atom. The peroxo complex **57** oxidizes propene to propylene oxide (**57** → **56**; 65 °C, yield and selectivity >90%). When the H_2O_2 in the reaction mixture has been used up, **56** precipitates; it can be filtered off and used again [366, 367].

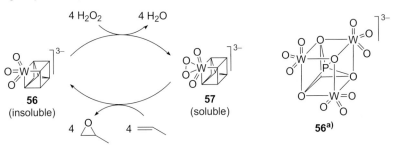

a) Structure of the anion of **56**; one WO_3 group is not drawn.

This reaction principle is called reaction-controlled phase-transfer catalysis. It assumes that the precatalyst is insoluble in the reaction mixture (here **56**), but goes into a soluble catalytically active form (here **57**) under the influence of a reactant. When a reactant has been used up (here H_2O_2), the catalyst goes back into the insoluble form and can be separated from the product by filtration. This novel procedure can accomplish the sometimes difficult separation of the catalyst from the product in a homogeneously catalyzed process.

13.2.3
Enantioselective Oxidation of Olefins

13.2.3.1 Epoxidation of Allyl Alcohols

The catalyst of choice for asymmetric epoxidation of allyl alcohols with *tert*-butylhydroperoxide is tetrakis(isopropoxo)titanium in the presence of dialkyl tartrates (DAT) as chiral ligands (Sharpless, 1980; Nobel Prize in Chemistry 2001) [368]. Epoxidation with (S,S)-DAT yields (2R)-epoxides (**58** → **59a**). With the naturally occurring (R,R)-DAT enantiomer, (2S)-epoxides are obtained (**58** → **59b**). This corresponds to an attack of the peroxo oxygen atom on the *Si* and *Re* face of the prochiral allyl alcohol, respectively.

a) The R/S and Re/Si categorization refers to the middle C atom of the allyl system and applies when R = H, Me, among others.

For a catalytic reaction, a strictly water-free reaction medium, achieved by the presence of a molecular sieve, is required. First, a bis(isopropoxo)tartratotitanium(IV) complex **60** is formed, which reacts with the allyl alcohol and *t*-BuOOH to form **61**. Now, the oxygen transfer (**61** → **62**) occurs by nucleophilic attack of the allylic double bond on the peroxo ligand. Reaction with the substrates leads to cleavage of the product and regeneration of the catalyst complex (**62** → **61**).

The catalyst complex **61** is dinuclear. Its probable structure is sketched in **61′** (X = COOR). There are two (distorted) octahedrally coordinated titanium atoms with one tridentate and one bidentate tartrato ligand. In addition, the *t*-BuOO ligand and the allyl alcoholate are coordinated in a meridional arrangement to one titanium atom.

The Sharpless epoxidation of allyl alcohols finds broad application in organic synthetic chemistry, including industrial-scale synthesis of enantiomerically pure (R)- and (S)-glycidol (2,3-epoxypropan-1-ol), which is required as a starting material for the synthesis of various pharmaceuticals. Nonfunctionalized olefins cannot be asymmetrically epoxidized with Sharpless catalysts.

13.2.3.2 Epoxidation of Nonactivated Olefins

Salen complexes[6] of trivalent manganese [Mn(salen)X] (**63′**) (X: weakly coordinating anion such as PF_6^- or coordinating anion such as Cl^-, which then is bound as an axial ligand), can epoxidize nonfunctionalized olefins. Oxidants [O] such as NaOCl, PhIO and pyridine-N-oxide convert **63** into an oxo(salen)manganese(V) complex **64**, which reacts with olefins, with epoxidation and regeneration of **63**.

If a chiral 1,2-diamine, $H_2NCHR'–CHR'NH_2$, is used in ligand synthesis, instead of ethylenediamine, a chiral salen complex **65** is obtained. Substitution of the H atoms *ortho* to the phenolate oxygen atoms by bulky substituents R leads to enantioselective epoxidation catalysts (E. N. Jacobsen, T. Katsuki, 1990).

The epoxidation of cyclic and acyclic disubstituted (Z)-olefins with a conjugated double-bond system, $R^1HC=CHR^2$ (**66**, R^1 = aryl, alkenyl, alkynyl; R^2 = alkyl), is not stereoselective. It yields a mixture of *cis* (**68a**) and *trans* epoxide (**68b**), which points to a radical mechanism (see Section 13.2.1, route **a** via **a₂**). The C–O bonds are formed sequentially,

---

6) The condensation of salicylaldehyde with ethylenediamine in the molar ratio 2 : 1 generates N,N′-bis-(salicylidene)ethylenediamine (H_2salen), which forms N,N′-bis(salicylidene)ethylenediaminato(2−) complexes ("salen complexes") with deprotonation of the two hydroxy groups.

so **67** occurs as intermediate. If the second C–O bond is formed immediately, the *cis* epoxide (**67** → **68a**) is formed. If rotation around the C–C bond occurs before the bond is formed, the *trans* epoxide (**67** → **68b**) is generated [369, 370].

$$R^1\text{—}R^2 \xrightarrow{\ ^+[Mn^V]=O\ } \left\{ \begin{array}{c} [Mn^{IV}] \\ | \\ O \\ R^1 \cdot\ \cdot R^2 \end{array} \right\} \xrightarrow{-[Mn^{III}]^{\oplus}} \begin{array}{c} \xrightarrow{\text{recombination}} R^1\triangle R^2 \ \ \mathbf{68a} \\ \\ \xrightarrow[\text{recombination}]{\text{rotation}} R^1\triangle^{\prime\prime}R^2 \ \ \mathbf{68b} \end{array}$$

66   **67**

The direction from which the olefin approaches the Mn=O bond determines the enantioselectivity. If the approach occurs in the direction of the arrow (a/a′), the direction (a) is favored, since the complexes **65** are not planar (see the front view **65′**, from which the C_2-symmetry of the salen ligand becomes obvious). Due to the bulky substituents R on the salen ligand, the olefin will now approach so as to show its sterically less hindered side toward R (see drawing of **65**). Hence, there is an enantiofacial differentiation. For (Z)-olefins **66**, enantioselectivity of >90% *ee* can be achieved, while the *ee* values for the corresponding (E)-olefins are significantly lower.

On the other hand, (Z)-olefins with isolated double bonds, such as *t*-BuHC=CHEt, react stereoselectively to form the *cis* epoxide, which indicates a concerted mechanism (see Section 13.2.1, route **a** via **a₁**). However, the reactions only run slowly, and in most cases enantioselectivities of only 0–60% are achieved. Asymmetric epoxidation of olefins has also been implemented with chiral salen and porphyrin complexes of other metals (Fe, Ru) [371, 372]. With titanium complexes that have bound chiral reduced salen-type ligands, asymmetric epoxidation of terminal non-functionalized olefins succeeds with hydrogen peroxide, a particularly attractive oxidant from an economic and ecological standpoint [373, 374].

13.2.4
Monooxygenases

In living organisms, the transfer of an oxygen atom from O_2 to organic substrates Su is catalyzed by monooxygenases, to form, in addition to water, the oxidized substrate SuO [375]. Nicotinamide adenine dinucleotide (NADH) or nicotinamide adenine dinucleotide phosphate (NADPH) serves in many cases as a biogenic reductant. Monooxygenases are capable, in particular, of oxidizing C–H bonds to form alcohols and double bonds to form epoxides. They play an important role in the oxidative degradation of substances endogenous and exogenous to the body.

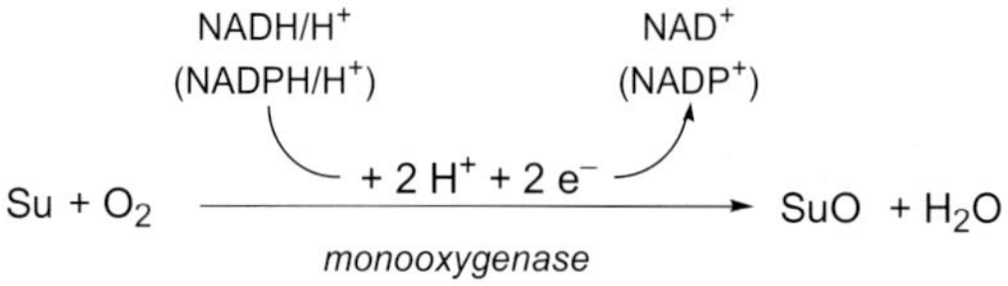

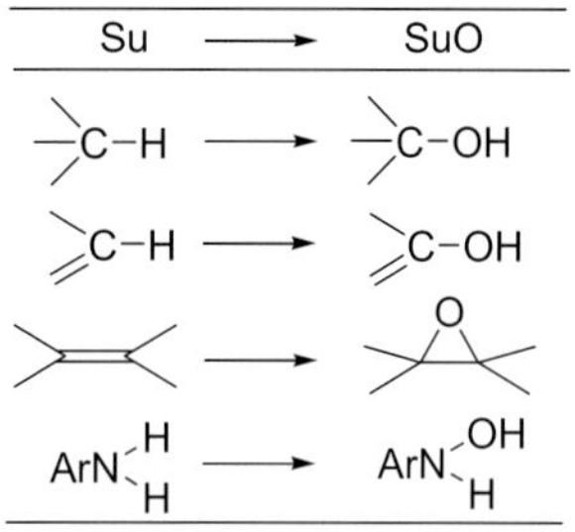

13 Oxidation of Olefins and Alkanes

$$Su + O_2 \xrightarrow[\text{monooxygenase}]{\substack{\text{NADH/H}^+ \\ \text{(NADPH/H}^+\text{)}} \xrightarrow{\text{NAD}^+} \text{(NADP}^+\text{)} \\ +2\,H^+ + 2\,e^-} SuO + H_2O$$

$$Su \longrightarrow SuO$$

$$\begin{aligned} &\text{>C-H} \longrightarrow \text{>C-OH} \\ &\text{>C-H} \longrightarrow \text{>C-OH} \\ &\text{>C=C<} \longrightarrow \text{epoxide} \\ &\text{ArN(H)(H)} \longrightarrow \text{ArN(OH)(H)} \end{aligned}$$

Examples of monooxygenases are the cytochrome P-450 enzymes, which belong to the large class of hemoproteins. The protein can be composed of about 400 amino acids. The heme group is bound via an axial Fe–S bond to a cysteinate of the protein. In the resting state of the enzyme, a low-spin iron(III) complex is present that has coordinated water as sixth ligand to the other axial binding site (**69**; in the schematic structure on the right, the porphyrin ligand is symbolized by a thick dash, while the protein is indicated by the two arcs).

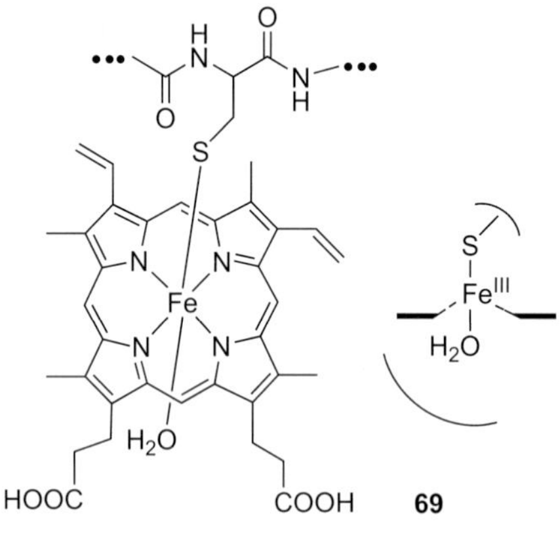

69

Cleavage of the aqua ligand and bonding of the substrate (Su) to the protein near the sixth coordination site leads to a high-spin Fe^{III} complex (**69** → **70**), which is reduced by a biogenic reductant to an Fe^{II} complex (high-spin) (**70** → **71**); see Figure 13.4. Now, dioxygen is coordinated (**71** → **72**) and another reduction occurs (**72** → **73**). Complex **72** is probably an Fe^{II} complex (low-spin) with a dioxygen ligand, and **73** is probably an Fe^{III} complex with a peroxo ligand (O_2^{2-}). Double protonation leads via a hydroperoxoiron(III) complex as intermediate (**73** → **74**) to cleavage of the terminal oxygen atom as water (**74** → **75**). To gain insight into the electronic structure of **75**, the following resonance formulas must initially be considered (HCys = cysteine, H$_2$por = porphyrin):

13.2 Epoxidation of Olefins

(Cys⁻)(por²⁻)Fe^V=O   (**75a**),   (Cys⁻)(por²⁻)Fe^IV–O•   (**75b**),
(Cys⁻)(por⁻•)Fe^IV=O   (**75c**),   (Cys•)(por²⁻)Fe^IV=O   (**75d**).

Spectroscopic and structural investigations, as well as quantum-chemical calculations, show that an Fe^IV complex is present whose ground state within the framework of VB theory is best described as resonance **75c ↔ 75d** [376]. Intermediate **75** transfers the oxygen to the substrate and is thus the actual oxidizing agent. Radical intermediates appear in the epoxidation of olefins. The first step could be an electron transfer from the alkene to **75**. The reactive intermediate **75** can also be generated by direct oxidation of **70** with external oxygen donors such as PhIO, IO₄⁻ or RC(O)OOH (**70** → **75**). Experiments on the dependence of the reactivity of oxo(porphyrin)iron(IV) model complexes on the electron density in the porphyrin ring (adjusted via selection of appropriate peripheral *meso* substituents), and also on the nature of the axial ligand, which in the model complexes replaces the Cys⁻ ligand, have led to a deeper under-

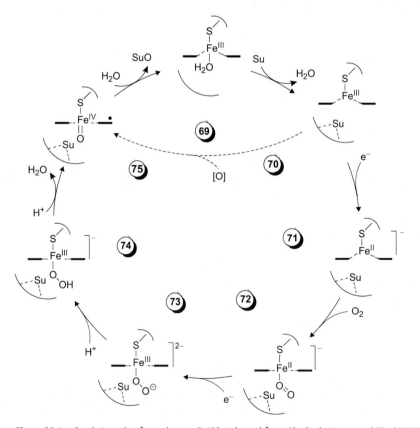

Figure 13.4 Catalytic cycle of cytochrome P-450 (adapted from Shaik, de Visser and Thiel [377]).

standing of the function and mode of operation of cytochrome P-450 enzymes [378].

Exercise 13.4
Complex 75 is an iron(IV) complex with a terminal oxo ligand. What do you know about the stability of electron-rich transition metal complexes with terminal oxo ligands?

Background: Oxidation States of Metals in Complexes
See "Background: Oxidation states of metals in olefin and alkyne complexes" (Section 3.2). To assign the oxidation number of M in metal complexes, it can be assumed that the ligands are generally more electronegative than M. Thus, the electrons in M–L bonds are counted to the ligands. The following cases are to be analyzed:

- L_xMO complexes (**1**). In complexes **1**, the electronic structure can be expressed by the resonance structures **1a–1d**. Normally, an appropriate description is given by **1a/1b** with an oxide ion as ligand. If M is in a very high oxidation state, then the other two resonance structures **1c/1d** become increasingly important with a radical anion $O^{\cdot -}$ or an oxene O as ligand. Consequently, the oxidation number of M decreases by one or two.

$$L_xM{-}\overset{\oplus}{\underset{\ominus}{O}}| \quad \longleftrightarrow \quad L_xM{=}\overline{O} \quad \longleftrightarrow \quad L_x\overset{\cdot}{M}{-}\overline{O}\cdot \quad \longleftrightarrow \quad L_x\overline{\overline{M}}{-}\overline{\overline{O}}$$

	1a	1b	1c	1d
ON(M)/ligand:	$+n/O^{2-}$	$+n/O^{2-}$	$+(n-1)/O^{\cdot -}$	$+(n-2)/O$

- $L_xM(O_2)$ complexes (**2**). For complexes **2**, the resonance structures **2a–2e**, which represent a dioxygen complex **2a** (ligand: O_2), a hyperoxo complex (superoxido complex) **2b/2c** (ligand: $O_2^{\cdot -}$) and a peroxo complex **2d/2e** (ligand: O_2^{2-}), respectively, should be drawn. In accord, the oxidation number of M increases by one or two.

$$L_xM{-}\overset{O}{\underset{O}{|}} \quad \longleftrightarrow \quad L_x\overset{\cdot}{M}{-}\overset{O\cdot}{\underset{O^{\ominus}}{|}} \quad \longleftrightarrow \quad L_x\overset{\cdot}{M}{-}\overset{O^{\ominus}}{\underset{O\cdot}{|}} \quad \longleftrightarrow \quad L_xM{=}\overset{O^{\oplus}}{\underset{O^{\ominus}}{|}} \quad \longleftrightarrow \quad L_xM{-}\overset{O}{\underset{O}{|}}{}^{\ominus}$$

	2a	2b	2c	2d	2e
ON(M)/ligand:	$+n/O_2$	$+(n+1)/O_2^{\cdot -}$	$+(n+1)/O_2^{\cdot -}$	$+(n+2)/O_2^{2-}$	$+(n+2)/O_2^{2-}$

- *Electron-variable complexes.* See example in Section 3.8.

It follows that an appropriate assignment of the oxidation number of M in a metal complex can only be made when the electronic structure is known. Corresponding findings in this area can be obtained from magnetic measurements and from spectroscopic (e.g., ESR and Mössbauer spectroscopy) and structural investigations, as well as from quantum-chemical calculations [24].

13.3
C–H Functionalization of Alkanes

13.3.1
Introduction

Alkanes are the least reactive hydrocarbons. The lower homologues, particularly methane, are main components of natural gas, which exists in abundance as a raw material. The selective creation of value-added products by direct utilization of methane as a starting material through C–H functionalization

$$R–H \longrightarrow R–X$$

(X: functionality such as OR', NR'$_2$, Cl, …; R = alkyl; R' = H, alkyl, …), is of fundamental scientific and industrial interest. Effective execution of a catalytic reaction is one of the great challenges of organometallic catalysis.

Such reactions are based on cleaving very stable C–H bonds. This is called C–H activation, which can only be achieved with difficulty. The C–H bonds are almost nonpolar (methane: pK_a about 48) and are among the strongest single bonds to be found. The homolytic bond dissociation enthalpies $\Delta_d H^\ominus$ of C_{sp^3}–H bonds lie between 400 and 440 kJ/mol. The stability ordering $H_3C–H > RH_2C–H > R_2HC–H > R_3C–H$ leads one to expect that radical reactions favor the formation of branched products, which are frequently undesired. Furthermore, regioselectivity becomes a problem in higher alkanes, due to the multitude of similar C–H bonds. Moreover, the introduction of a substituent X in most cases weakens the C–H bonds on the same C atom (example: $\Delta_d H^\ominus$ in CH$_3$X in kJ/mol: 439, X = H; 419, X = Cl; 402, X = OH; 393, X = NH$_2$), so multiple functionalizations may dominate.

Metal-catalyzed radical oxidative functionalizations of hydrocarbons with oxygen, such as the oxidation of p-xylene to form terephthalic acid (Co/Mn), of cyclohexane to cyclohexanone (Co), or of butane to acetic acid (see Section 6.1), have great industrial significance. The redox-active metal ions or complexes (Co^{III}/Co^{II}, Mn^{III}/Mn^{II}) are involved via electron transfer in the formation of radicals R• from RH and in reactions with oxidation intermediates such as ROO•/ROOH. C–H activation in conjunction with formation of organometallic compounds does not occur in these cases.

13.3.2
C–H Activation of Alkanes

13.3.2.1 Cyclometallation and Orthometallation
C–H activation is an inherent component of C–H functionalization reactions. Reactions that proceed stoichiometrically with intramolecular C–H activation are easier to implement than those based on an intermolecular one. C–H activation can lead to hydrido–organyl complexes via oxidative addition of the C–H bond, which is often preceded by an agostic C–H⋯M interaction (**76 → 77**). However,

13 Oxidation of Olefins and Alkanes

metallacycles can also be formed (**76'** → **77'**) with cleavage of an anionic ligand X⁻ (Cl⁻, AcO⁻, ...) in the form of HX.

C–H activation **76** → **77**/**76'** → **77'** in which Y is any coordinatively bound atom is called cyclometallation. If YR$_n$ is a neutral donor group (NR$_2$, PR$_2$, ...), complexes of type **77**/**77'** are referred to as organometallic inner complexes. Cyclometallation reactions based on activation of an aromatic *ortho*-C–H bond proceed particularly easily; these are called orthometallation reactions. An example is the orthopalladation of benzylamines (**78** → **79**), which occurs even at room temperature.

In conventional palladium-catalyzed C–C cross-coupling reactions (see Figure 11.1 in Section 11.1.2), organohalides RX and similar reagents are used, where, starting from a Pd⁰ complex, an organyl palladium(II) complex [Pd](R)X is formed via oxidative addition. In principle, complexes of this type can also be generated (**80** → **81**) from Pd(OAc)$_2$ by C–H activation (especially of aryl C–H bonds), so that the usage of human- and ecotoxic organohalides can be avoided in cross-coupling reactions. As with cross-coupling, the C–C coupling product can then be released via transmetallation and reductive elimination (**81** → **82** → **83**). This generates Pd⁰, which then must be reoxidized (**83** → **80**). Despite all the progress that has been made, such processes do not display yet the same notably high level of efficiency and executability offered by conventional cross-coupling [379, 380].

Exercise 13.5
- Formulate the reaction of 3-methylbenzoic acid with potassium trifluorophenylborate with Pd(OAc)$_2$ as catalyst in the presence of air or oxygen. Explain the selectivity and give a possible mechanism.

- In special cases, C–H activation (especially orthometallation) can be the basis for a palladium-catalyzed biaryl synthesis which proceeds as cross dehydrogenative coupling. The reaction of an acetanilide **1** with o-xylene **2** to form biaryl **3** serves as an example:

$$\text{1} + \text{2} \xrightarrow[\text{O}_2 \text{ (1 bar), EtCOOH} \atop (120\,°\text{C, 7 h})]{\text{Pd(OAc)}_2/\text{Cu(OTf)}_2} \text{3}$$

This is a biaryl synthesis in which, unlike conventional cross-couplings, neither organohalide nor organometallic compounds are used as substrates. Formulate a possible mechanism that contains an orthometallation of **1** and a C–H activation of **2** via proton transfer to a carboxylato ligand.

Cyclometallations with activation of C_{sp^3}–H bonds are more difficult to achieve than with C_{sp^2}–H bonds and were not demonstrated until 1965, by J. Chatt: The reduction of transition metal halides with sodium naphthalide in the presence of $Me_2PCH_2CH_2PMe_2$ (dmpe) leads to dmpe complexes of zero-valent metals (M = V, Cr, Mo, Fe, Co, …). In contrast, in the reaction of [RuCl$_2$(dmpe)$_2$] with intermolecular C_{sp^2}–H activation, a hydrido(naphthyl)ruthenium(II) complex (**84**) is formed. At 150 °C, complex **84** undergoes reductive elimination of naphthalene, and with activation of a C_{sp^3}–H bond, a dinuclear hydrido complex is obtained (**84** → **85**), which was originally described as mononuclear.

$$[\text{RuCl}_2(\text{dmpe})_2] \xrightarrow[-2\,\text{NaCl},\,-\text{C}_{10}\text{H}_8]{+\,2\,\text{Na[C}_{10}\text{H}_8]} \textbf{84} \xrightarrow[-\text{C}_{10}\text{H}_8]{T} \tfrac{1}{2}\,\textbf{85}$$

13.3.2.2 Intermolecular C–H Activation of Alkanes

With regard to catalytic functionalization, the most important C–H activation reactions of alkanes are [381–383]:

- *Oxidative addition* (**86** → **87**) yielding alkyl(hydrido)metal complexes with an increase of two in the oxidation number of M. A σ-alkane complex **86′** as intermediate ensures substrate activation.

$$[\text{M}] + \text{R–H} \longrightarrow [\text{M}]\cdots\underset{\text{H}}{\overset{\text{R}}{|}} \longrightarrow [\text{M}]\underset{\text{H}}{\overset{\text{R}}{\diagup}}$$

    86   **86′**   **87**

Oxidative C–H addition is typical of late transition metals in low oxidation states. The immediate precursor complex (**86** or **88** in the following example) is usually generated *in situ*, for instance, by thermal or photochemically induced reductive elimination, as is shown in the following by an example (R–H = cyclohexane, neopentane, …):

Further examples of complex fragments to which oxidative addition of alkanes has been achieved are d^{10} Pt^0L_2 and T-shaped d^8 $Pt^{II}L_3$ fragments [384]. A well-studied example is the reaction of the cationic complex $[Pt(CH_3)(s)(N\frown N)]^+$ (**89**, $N\frown N = ArN=CMe–CMe=NAr$ with Ar = 3,5-di-*tert*-butylphenyl; anion: $[B(C_6F_5)_3(OCH_2CF_3)]^-$) in CF_3CH_2OH (s) as solvent with cyclohexane. The oxidative addition of cyclohexane (**89** → **90**) and the reductive elimination of methane (**90** → **90'**) are preceded in each case by the formation of a σ complex. The final β-H elimination then leads to a cationic cyclohexene–hydrido complex **91** ($[Pt] = Pt(N\frown N)$) [385].

- *Electrophilic substitution* (**92/92'** → **94**) of C–H bonds is substitution of H^+ by a metal complex: $[M]^+ + R–H \rightarrow [M]–R + H^+$. The oxidation number of M does not change. The alkane can be activated by σ-complex formation (**93/93'**). The proton is either accepted by an external base X^- (**93** → **94**) or directly transferred to a ligand adjacent to R–H (**93'** → **93''** → **94**). The latter reaction mentioned should thus be classified as σ-bond metathesis.

Such electrophilic substitutions of C–H bonds are typical of cationic, strongly electrophilic metal complexes with central atoms in normal and high oxidation states. They are found with typical transition metals (Pd^{II}, Pt^{II}, Pt^{IV}, ...), but also with Group 12 and main-group metals (Hg^{II}, Tl^{III}) (see Sections 13.3.3.1–13.3.3.3).

- σ-*Bond metathesis* with C–H activation of alkanes has been demonstrated for d^0 complexes of early transition metals and was discussed in Section 7.4. in connection with alkane metathesis.
- *C–H activation by carbene complexes.* An example of a C–H activation which occurs as 1,2-addition of a C–H bond of $SiMe_4$ across a Mo=C double bond is the reaction of the bis(neopentyl) complex **95** via the neopentylidene intermediate **95'** to form the mixed trimethylsilylmethyl–neopentyl complex **96**, which runs quantitatively at room temperature.

Metal-bound carbenes (X = CRR′) can insert into C–H bonds,

$$[M]{=}X + \!\!\!\overset{\big|}{\underset{\big|}{C}}{-}H \longrightarrow [M] + \!\!\!\overset{\big|}{\underset{\big|}{C}}{-}X{-}H ,$$

whereby the metal atom is not thought to interact directly with the alkane C–H bond. Starting from diazo compounds as carbene precursors these reactions can be performed catalytically. Analogous reactions of nitrenes (X = NR) are also described [386, 387].

- *C–H Activation by metalloradicals.* Dinuclear porphyrinatorhodium(II) complexes **97** ([Rh] = Rh(por); H_2por = tetraxylylporphyrin, tetramesitylporphyrin) have a very weak Rh–Rh bond[7] and exist in equilibrium with the monomeric Rh^{II} complexes (d^7). They react in a reversible reaction with methane to form a mixture (1 : 1) of the methyl- and hydridorhodium(III) complex **99a/99b**. The reaction very likely proceeds via a linear four-center transition state **98**.

13.3.3
C–H Functionalization

If $\chi(X) > \chi(C)$ (χ = electronegativity), functionalization of an alkane R–H (**100**) → R–X (**101**) corresponds to oxidation of the alkane. Reactions that proceed according to

---
7) $\Delta_d H$ is about 70 kJ/mol for the octaethylporphyrinato complex [Rh]–[Rh].

the general equation

$$\text{R-H} + \text{X}^- \xrightarrow{\text{cat.}} \text{R-X} + \text{H}^+ + 2\,e^{\ominus}$$
$$\quad\; 100 \qquad\qquad\qquad\qquad 101$$

include as partial reactions a C–H activation (cleavage of a C–H bond) and a functionalization (formation of a C–X bond). Furthermore, the electron balance makes it clear that a redox reaction is a necessary component of the process: An oxidant is required in stoichiometric quantity.

The homogeneously metal-catalyzed reactions to be discussed here include chlorination of alkanes ($X^- = Cl^-$) and oxidation of alkanes to form alcohols ($X^- =$ OH^-) or alkyl esters ($X^- = HO_3SO^-$) [388].

13.3.3.1 The Shilov Catalyst System

At the beginning of the 1970s, A. E. Shilov reported platinum(II)-catalyzed alkane functionalization (**100** → **101**) in an aqueous acidic medium, but with hexachloroplatinic acid as primary oxidant [389].

$$\text{R-H} + \text{Cl}^- \xrightarrow[\quad [PtCl_4]^{2-} \quad]{[PtCl_6]^{2-} \quad [PtCl_4]^{2-} + 2\,Cl^-} \text{R-Cl} + \text{H}^+$$
$$\quad 100 \qquad\qquad\qquad 120\,°\text{C} \qquad\qquad 101$$

According to today's state of knowledge, substitution with Pt^{II} as electrophile (**102** → **103**) should be assumed from a mechanistic point of view, probably preceded by a ligand substitution (formation of $[PtCl_3(s)]^-$ and possibly also of $[PtCl_2(s)_2]$; s = solvent). Possibly a σ-alkane complex is primarily formed, from which a proton is cleaved off. Splitting off H^+ is substantially favored by its solvation in the polar protic solvent. Oxidation of the alkylplatinum(II) complex with hexachloroplatinate yields an alkylplatinum(IV) complex (**103** → **104**) via electron transfer (and *not* via alkyl group exchange!). Cleavage of chloride and nucleophilic attack of Cl^- on the alkyl ligand lead to the formation of RCl and the regeneration of $[PtCl_4]^{2-}$ (**104** → **105** → **102**). Other mechanistic alternatives cannot be ruled out.

In contrast to radical chlorination, for Pt-catalyzed reactions, the selectivity follows the order C–H (primary) > C–H (secondary) > C–H (tertiary). Alcohols can be obtained in a similar way if H_2O is used as a nucleophile instead of Cl^- [390].

13.3.3.2 The Catalytica System – Hg^{II} as Catalyst

R. A. Periana at Catalytica Inc. (California, USA) invented a selective mercury-catalyzed system for oxidation of methane to form methanol. The starting point was a stoichiometric, almost quantitative reaction of methane with mercury(II) triflate to form methyltriflate with reduction of Hg^{II} to form Hg^{I} (106 → 107) [391].

$$CH_4 + 2\ Hg(CF_3SO_3)_2 \xrightarrow{180\ °C} CF_3SO_3CH_3 + Hg_2(CF_3SO_3)_2 + CF_3SO_3H$$
$$\text{106} \hspace{6cm} \text{107}$$

In anhydrous sulfuric acid, the corresponding methyl ester is formed (106 → 108) and sulfuric acid assumes the role as oxidant, so the reaction is catalytic with respect to Hg^{II}.

$$CH_4 + 2\ H_2SO_4 \xrightarrow[180\ °C]{[Hg^{II}]} CH_3OSO_3H + 2\ H_2O + SO_2$$
$$\text{106} \hspace{6cm} \text{108}$$

Hydrolysis of 108 (108 + H$_2$O → MeOH + H$_2$SO$_4$) and reoxidation of SO$_2$ (SO$_2$ + 1/2 O$_2$ → SO$_3$) result in an overall reaction (106 + 1/2 O$_2$ → MeOH) that demonstrates the oxidation of methane with oxygen to form methanol. This illustrates that, in principle, hydrocarbons can be selectively oxidized in a homogeneously catalyzed process.

The mechanism of oxidation of methane (106 → 108) is in some respects analogous to the Shilov process. Electrophilic substitution of methane leads to formation of a methylmercury(II) compound (C–H activation: 109 → 110). The intermediate 110 has been identified by NMR spectroscopy. Then, with reduction of Hg^{II} and H$_2$SO$_4$, the methyl ligand in 110 is cleaved off as methyl hydrogen sulfate (alkane functionalization: 110 → 111). This esterification of methanol is crucial because thus it is protected from overoxidation. Reoxidation of Hg^{I} by sulfuric acid completes the catalytic cycle (111 → 109).

$$Hg(OSO_3H)_2 \xrightarrow[-\ H_2SO_4]{+\ CH_4} HO_3SO-Hg-CH_3 \xrightarrow[-\ CH_3OSO_3H,\ -\ H_2O,\ -\ 1/2\ SO_2]{+\ 3/2\ H_2SO_4} 1/2\ Hg_2(OSO_3H)_2$$
$$\text{109} \hspace{4cm} \text{110} \hspace{5cm} \text{111}$$

$$\xleftarrow[-\ H_2O,\ -\ 1/2\ SO_2]{+\ 3/2\ H_2SO_4}$$

The use of strong acids as solvents has at least two advantages: The conjugated bases (here: hydrogen sulfate) are only weakly coordinating, so the central atom (here: Hg^{II}) is highly electrophilic. Furthermore, esterification reliably protects the alcohol from further oxidation.

13.3.3.3 The Catalytica System – Pt^{II} as Catalyst

A similar process with a (2,2'-bipyrimidine)platinum(II) complex **112** (X=Cl) as precatalyst has been developed. With cleavage of HCl, the bis(hydrogen sulfato) complex **112** (X=OSO$_3$H) is formed first.[8] Cleavage of a hydrogen sulfate anion leads to a highly electrophilic Pt^{II} cation (14 ve), which is probably the actual catalyst [392].

$$CH_4 + 2\,H_2SO_4 \xrightarrow[180-220\,°C]{112} CH_3OSO_3H + 2\,H_2O + SO_2$$

106 → **108**

112

This catalyst system shows extraordinarily high activity, selectivity, and stability (yield of **108** > 70% with a selectivity > 90% and $TON > 300$) for methane activation. However, water and methanol act as inhibitors, so the reaction must be carried out in fuming sulfuric acid.

The high activity and stability is probably due to protonation of at least one of the nonligated N atoms of the bipyrimidine ligand whereby the ligand remains strongly bound to Pt^{II}. This protonation leads to a substantial increase in the electrophilicity of the platinum atom on the one hand, and it lowers the electron density in the aromatic system, thus hindering electrophilic substitution reactions and oxidative degradation on the other. In contrast to the platinum-containing system, PdSO$_4$/H$_2$SO$_4$ catalyzes the oxidation of methane (180 °C, $TON = 18$ in 7 h) to form acetic acid as main product [389, 390].

13.3.3.4 Cytochrome P-450

The active complex of cytochrome P-450 enzymes (**75** in Figure 13.4) is capable of transferring an oxygen atom to many organic compounds, so it catalyzes, for example, hydroxylation of aliphatic C–H bonds:

$$\text{>C-H} \xrightarrow{[O]} \text{>C-OH}$$

75 → **69** (with H$_2$O, ROH)

Complex **75** (Figure 13.4) is an Fe^{IV} complex with an oxidized ligand system [(Cys$^-$)(por$^{\bullet}$)Fe^{IV}=O] ↔ [(Cys$^{\bullet}$)(por^{2-})Fe^{IV}=O], and complex **69** (Figure 13.4) is an Fe^{III} complex with an unoxidized ligand system [(Cys$^-$)(por^{2-})Fe^{III}–OH$_2$]. Thus, the reaction **75** → **69** includes an oxygen transfer to the substrate (R–H) and a formal two-electron transfer from the substrate to the heme-Fe complex. One

---

8) This complex is surprisingly stable. In 20% oleum at 200 °C over the course of 50 h, no decomposition was observed.

possible mechanism is an H abstraction from R–H by **75** (**75** → **75′**), followed by a binding of the alkyl radical to the oxygen atom (**75′** → **75″**) and substitution of the alcohol ligand by water (**75″** → **69**). The reactions **75** → **75′** and **75′** → **75″** each include the transfer of one electron from the substrate to the iron complex [376, 377].

$$[Fe]=O + H-R \longrightarrow [Fe]-OH \cdots \cdot R \longrightarrow [Fe]-O\genfrac{}{}{0pt}{}{R}{H} \xrightarrow[-ROH]{+H_2O} [Fe]-O\genfrac{}{}{0pt}{}{H}{H}$$

75   **75′**   **75″**   **69**

14
Nitrogen Fixation

14.1
Fundamentals

Nitrogen is an essential element for all living beings. It is a component of amino acids and nucleobases, from which proteins and nucleic acids are built. Without those two fundamental classes of biopolymers, life on earth in its existing form would not be possible. In many cases, too low a content of nitrogen in the soil limits the growth of crops. Thus, nitrogen fertilization has assumed a tremendous significance in ensuring production of food. More than 99% of the "mobile" nitrogen (N in the atmosphere, hydrosphere, and biomass) is found as dinitrogen in the atmosphere, a form in which it is bioavailable only to a very limited extent. Hence, the importance of nitrogen fixation, the reduction of N_2 to form ammonia, the most important bioavailable inorganic nitrogen compound, is evident. Biological nitrogen fixation, the only biological process in which N_2 is converted into a form in which it can be used by organisms, is limited to microorganisms. It proceeds according to the following schematic equation, with enzyme (nitrogenase) catalysis, under physiological conditions (room temperature, normal pressure):

$$N_2 + 8\,H^+ + 8\,e^- \xrightarrow[\text{nitrogenase}]{\text{1 bar, 20 °C}} 2\,NH_3 + H_2$$

The necessary energy is provided by adenosine triphosphate (ATP), and dihydrogen is coproduced. Abiogenic nitrogen fixation has been performed for almost 100 years according to the Haber-Bosch process:

$$N_2 + 3\,H_2 \xrightarrow[\text{Fe catalyst}]{\text{150–250 bar, 400–500 °C}} 2\,NH_3$$

BASF put the first plant into operation in Ludwigshafen-Oppau (Germany) in 1913. Soon after that (1916), the Merseburg ammonia factory (which later became the Leuna factories) was established. For the development of ammonia synthesis from its elements, Fritz Haber (KWI für Physikalische Chemie und Elektrochemie, today the Fritz-Haber-Institut der MPG) was awarded the 1918 Nobel Prize in Chemistry. Synthesis of ammonia from its elements supplanted the industrial production of

Fundamentals of Organometallic Catalysis. First Edition. Dirk Steinborn
Copyright © 2012 WILEY-VCH Verlag GmbH & Co. KGaA, Weinheim

ammonia by hydrolysis of calcium cyanamide (nitrolime) with superheated water vapor:

$$CaO \xrightarrow[\text{electric arc}]{+3\,C,\,-CO} CaC_2 \xrightarrow[1000\,°C]{+N_2} CaCN_2 + C \xrightarrow[-C]{+3\,H_2O} CaCO_3 + 2\,NH_3$$

Calcium cyanamide has been available from calcium carbide and nitrogen via the Frank-Caro process since 1898 and is used today directly as nitrogen fertilizer that also possesses herbicidal and fungicidal properties.

The production of ammonia, which after sulfuric acid is the most widely produced inorganic bulk chemical, amounted to about $124 \cdot 10^6$ metric tons in 2006. It consumes about 2% of the natural gas and about 1% of the energy produced worldwide. About 85% of the ammonia produced is further processed into fertilizers. Biological nitrogen fixation by nitrogenases and anthropogenic nitrogen fixation occur on the same order of magnitude. Other natural activation of N_2, such as oxidation by atmospheric processes to form NO_x, have a lower significance (<10%).

Dinitrogen is extraordinarily chemically inert. The only element with which it reacts at room temperature – albeit only slowly – is lithium, to form Li_3N. A large number of other elements, among them many transition metals, also react with N_2, breaking the N–N bond to form nitrides, but only at higher temperatures.

Exercise 14.1

Crystalline Ti reacts with N_2 only above 1000 °C, to form TiN (**1**). On the other hand, it has been shown in matrix isolation experiments that at 10 K, dimeric Ti_2 (but not atomic Ti) and N_2 react in an exothermic reaction ($\Delta H^\ominus = -396$ kJ/mol), without a significant activation barrier, to form Ti_2N_2 (**2**).

$$2\,Ti_{(s)} + N_2 \xrightarrow{>1000\,°C} 2\,TiN_{(s)} \qquad Ti_2 + N_2 \xrightarrow{10\,K} Ti\begin{smallmatrix}N\\ \\N\end{smallmatrix}Ti$$

                     1   **2**

2 is D_{2h}-symmetric. The N $\cdots$ N distance (2.47 Å) demonstrates that the N–N bond has been fully broken. Give the thermodynamics of the reactions that lead to **1** and **2** in a diagram and calculate the enthalpy of the reaction of 1 mol of **2** to form 2 mole of **1**. Consider what causes the activation barrier for the formation of **1** from its elements (standard enthalpy of formation of monoatomic $Ti_{(g)}$ from the elemental form $\Delta_{at}H^\ominus = 473$ kJ/mol; standard enthalpy of formation of TiN $\Delta_f H^\ominus = -338$ kJ/mol; standard bond dissociation enthalpy of Ti_2 $\Delta_d H^\ominus = 114$ kJ/mol).

The high thermodynamic stability of N_2 is demonstrated by the strength of its triple bond (Table 14.1), which is among the strongest bonds known. However, this cannot be the only reason for its inertness, since the isoelectronic CO is known to be highly reactive, although the CO bond is stronger. The oxidation of CO to form CO_2 is exergonic ($\Delta G^\ominus = -257$ kJ/mol), while that of N_2 to form 2 NO_2 is endergonic ($\Delta G^\ominus = 103$ kJ/mol). The comparison in Table 14.1 makes it evident that N_2 is more

Table 14.1 Bond dissociation enthalpies and electronic parameters for N_2 and CO.

	N_2	CO
Bond dissociation enthalpy $\Delta_d H^\ominus$ in kJ/mol[a]	945	1077
Ionization energy E_i in eV (kJ/mol)[a]	15.58 (1503)	14.10 (1361)
First electronic excitation energy τ in eV (kJ/mol)[a]	7.8 (753)	6.3 (608)
Orbital energy ε(HOMO) in eV (kJ/mol)[b]	−11.95 (−1153)	−10.53 (−1016)
Orbital energy ε(LUMO) in eV (kJ/mol)[b]	−1.13 (−109)	−1.20 (−115)

a) Experimental values.
b) Calculated values [393].

difficult to oxidize and electronically excite than CO. The HOMO and LUMO energies show that N_2 is both a weaker electron-pair donor and acceptor, than CO. However, these same characteristics are important for complex formation with transition metals to activate N_2, and the formation of dinitrogen complexes is critical for a homogeneously catalytic process.

Hence, the first isolated dinitrogen complex, $[Ru(NH_3)_5(N_2)]X_2$ (X = Br, I, BF_4), was not obtained until 1965 by A. D. Allen and C. V. Senoff via reaction of ruthenium(III) chloride with hydrazine in aqueous solution. Important synthetic methods for N_2 complexes are:

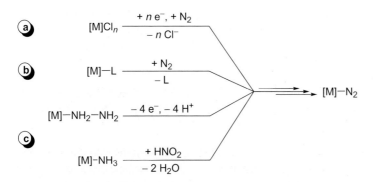

- (a) *Addition of N_2*. Coordinatively unsaturated complexes of low-valent transition metals, obtained by *in situ* reduction, can react with N_2 to form dinitrogen complexes. Thus, for example, complexes of type $[Mo(N_2)_2(PR_3)_4]$ (R = alkyl, aryl) can be achieved by reaction of $[MoCl_2(PR_3)_4]$ or $[MoCl_4(PR_3)_2] + 2\ PR_3$ with sodium or sodium amalgam under N_2. $[\{Zr(Cp^*)_2(\eta^1-N_2)\}_2(\mu-\eta^1:\eta^1-N_2)]$ is obtained from $[Zr(Cp^*)_2Cl_2]$ and N_2 with Na/Hg as reductant.
- (b) *Ligand substitution reactions*. Weakly bound ligands can be substituted by N_2. The reaction of $[Ru(NH_3)_5(H_2O)]^{2+}$ to form $[Ru(NH_3)_5(N_2)]^{2+}$ serves as an example. The bis(dihydrogen) complex $[RuH_2(\eta^2-H_2)_2(PCy_3)_2]$ reacts with N_2 to form the bis(dinitrogen) complex $[RuH_2(\eta^2-N_2)_2(PCy_3)_2]$. The reaction is reversible in an atmosphere of hydrogen.

- (c) *Ligand transformation reactions.* Hydrazine ligands can be oxidized to form N_2 ligands, preserving the N–N bond. In special cases, ammine ligands can also be oxidized to N_2 ligands, forming a new N–N bond. Hence, the oxidation of [MnCp-(CO)$_2$(N$_2$H$_4$)] with H_2O_2 in the presence of copper(II) salts generates [MnCp-(CO)$_2$(η^1-N$_2$)]. In a complex reaction which, overall, amounts to the diazotization of an ammine ligand, [Os(NH$_3$)$_5$(N$_2$)]$^{2+}$ reacts with nitrous acid to form [Os-(NH$_3$)$_4$(N$_2$)$_2$]$^{2+}$.

For almost all transition metals, dinitrogen complexes are now known in which the N_2 ligand displays different coordination modes. The most important are mononuclear complexes with N_2 ligands bound end-on (**1**) and dinuclear complexes with a μ-N_2 ligand, which can be bound end-on (**2**) or side-on (**3**, **4**). The M_2N_2 unit can be planar (**3**) or folded ("roof-like") with or without an additional M–M bond (**4**).[1)]

[M]—N—N	[M]—N—N—[M]	[M]⟨N·N⟩[M]	[M]⟨N·N⟩[M] (folded)
1 (η^1)	**2** (μ-η^1:η^1)	**3** (μ-η^2:η^2)	**4**

In complexes with N_2 coordinated end-on (**1/2**), dinitrogen functions as an *n*-EPD via the nonbonding electron pairs, whereas in those with N_2 coordinated side-on (**3/4**), it functions as a π-EPD, via the π-electron pairs. In both cases, efficient back-donation into the π* orbitals of N_2 is essential for a stable M–N_2 bond to form. The lengthening of the N–N bond induced by the coordination can be seen as a measure of the degree of N_2 activation. Metal complex formation of type **1** is generally associated with a low degree of activation of the N–N bond (typical N–N bond lengths: 1.10–1.15 Å; compare to 1.0975 Å in N_2). In contrast, in complexes of types **2–4**, the N–N bonds can be significantly longer, especially in those of type **3**, where N–N bonds longer than 1.50 Å have been observed. This points to substantial activation of the N_2 ligand. Based on the bond lengths in the diimine (diazene) *trans*-HN=NH (1.252 Å) and in hydrazine $H_2N–NH_2$ (1.449 Å), a notably lengthened N–N bond in N_2 complexes reveals a (formal) reduction of the ligand to form N_2^{2-} or N_2^{4-}, respectively. In the framework of valence bond theory, complexes of type **2** can be represented by the three resonance structures **2a–2c**, where those with a very long N–N bond are best represented by **2c**.

$$[\overline{M}]-N\equiv N-[\overline{M}] \longleftrightarrow \begin{array}{c}[M]=N=N=[\overline{M}]\\ {[\overline{M}]=N=N=[M]}\end{array} \longleftrightarrow [M]\equiv N-N\equiv[M]$$

	2a	**2b**	**2c**
ligand:	$|N\equiv N|$	$\langle N=N\rangle^{2\ominus}$	$|\overline{N}-\overline{N}|^{4\ominus}$
ON(M):	n	n+1	n+2

---

1) The formulas show only the topology (i.e., the connectivity) of the respective molecules; they only specify which atoms are bound to each other.

This must be taken into account when the oxidation number is assigned to the metal (an increase of one or two, respectively, compared to the neutral N_2 ligand). To evaluate this issue, however, other aspects such as the M–N bond length, the N–N stretching frequency, the magnetism, and the coordination geometry of the complex must be included (compare to the Background on oxidation states in Section 3.2. and 13.2.4). Examples of the dinitrogen complexes are shown in Figure 14.1.

Exercise 14.2

Using a qualitative molecular orbital diagram, describe the π-bond system in a linear complex $[(ML_5)_2(\mu\text{-}\eta^1\text{:}\eta^1\text{-}N_2)]$. It should be based on approximately fourfold symmetry. Consider how activation of the N–N bond depends on the d electron configuration of the two central atoms.

$L_5M\text{—}N\equiv N\text{—}ML_5$

Hint: When you position the coordinate system in the specified manner, you need only to consider the d_{xy}, d_{xz}, and d_{yz} orbitals of the two ML_5 units (why?). For the N_2 ligand, the two π orbitals that extend into the yz and xz planes and the corresponding π^* orbitals (or the p_x/p_y orbitals) must be considered.

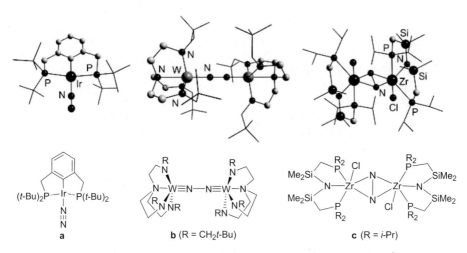

Figure 14.1 Structures of dinitrogen complexes with terminal (**a**: N–N 1.107(2)–1.109(2) Å) and bridging $\mu\text{-}\eta^1\text{:}\eta^1\text{-}$ (**b**: N–N 1.39(2) Å) or $\mu\text{-}\eta^2\text{:}\eta^2\text{-}$ligands with a planar Zr_2N_2 unit (**c**: N–N 1.548(7) Å), respectively. Hydrogen atoms are not shown. Peripheral alkyl groups of the coligands are indicated only in the wire model. In **b** and **c** there are exceptionally long N–N bonds, so they should be regarded as hydrazido(4–) complexes of W^V and Zr^{IV}, respectively.

14.2
Heterogeneously Catalyzed Nitrogen Fixation

14.2.1
Principles

Ammonia synthesis from its elements is, under standard conditions, exothermic and also weakly exergonic:

$$\tfrac{1}{2}\,N_2 + \tfrac{3}{2}\,H_2 \longrightarrow NH_3 \qquad \Delta_f H^{\ominus} = -46 \text{ kJ/mol}, \; \Delta_f G^{\ominus} = -16 \text{ kJ/mol}$$

It follows from these facts, as well as from Le Chatelier's principle, that low temperatures and high pressures result in high equilibrium concentrations of ammonia (Figure 14.2). Under the physiological conditions under which biological nitrogen fixation occurs, the equilibrium lies completely on the ammonia side. However, the catalysts used in Haber-Bosch synthesis require temperatures between 400 and 500 °C, so at normal pressure, the equilibrium concentration of NH_3 amounts to <1 mol%. At pressures of 150–250 bar typically employed, conversions of about 15–20 mol% are achieved. This makes circulation of the reaction gases necessary.

The steel contact tubes used in the first technical trials did not withstand the hydrogen under the reaction conditions; they quickly became unusable and burst due to decarburization and ensuing embrittlement. Carl Bosch of BASF constructed a reactor whose inner lining consisted of soft (i.e., largely carbon-free) iron, which remained unaffected by hydrogen, but was not impervious to diffusion and pressure. Thus, it was covered with a steel pipe provided with small holes through which H_2 could escape to prevent decarburization. For his contribution to the invention and development of chemical high pressure methods, Bosch (together with Friedrich Bergius) was awarded the Nobel Prize in Chemistry in 1931 [394–396].

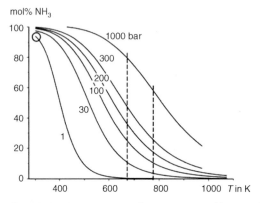

Figure 14.2 Concentrations of ammonia in equilibrium with N_2 and H_2 ($p(N_2)/p(H_2) = 1/3$) as a function of pressure and temperature (numerical values – some extrapolated – from [397]). The p/T pair marked with a circle corresponds to the conditions at which nitrogenases operate. The typical temperature range in which Haber-Bosch synthesis is performed is marked by dashed lines.

The N_2/H_2 mixture used for ammonia synthesis is predominantly produced from natural gas (methane), air, and water. Hydrogen is obtained by steam reforming of methane using nickel catalysts (NiO on Al_2O_3):

$$CH_4 + H_2O \longrightarrow 3 H_2 + CO \qquad \Delta H^{\circ} = 206 \text{ kJ/mol}$$

Of the hydrogen produced, 2/3 comes from the methane, 1/3 from the water. In parallel to this reaction, the water-gas shift reaction (see Section 6.5.1) takes place:

$$CO + H_2O \rightleftharpoons H_2 + CO_2$$

First, a portion of the methane is allowed to react in a primary reformer at 700–900 °C and 20–40 bar. Then, in the following secondary reformer, using the same nickel catalysts as in the primary reformer, part of the gas mixture is burned under the controlled introduction of air. The reaction is exothermic and the temperature climbs to about 1200 °C. The endothermic steam reforming reaction cools the gas mixture to about 1000 °C. Under these conditions, a residual methane content of <0.5% is achieved. The quantity of air introduced into the secondary reformer is measured so that the ratio of hydrogen and nitrogen required for ammonia synthesis is achieved. CO, CO_2, and sulfur compounds are poisons for the Haber-Bosch catalyst and must be completely removed.

14.2.2
Mechanism of Catalysis

Substantial contributions to the detailed clarification of the mechanism were made by G. Ertl, who in 2007 was honored with the Nobel Prize in Chemistry. The rate-determining step in ammonia synthesis is the dissociative chemisorption of dinitrogen on the surface of the catalyst (iron) (a). This includes adsorption of N_2, cleavage of the N–N bond, and the bonding of the two N atoms to the iron surface. H_2 is likewise dissociatively chemisorbed onto the surface (b). The hydrogenation occurs through successive reaction of each surface-bound nitrogen atom with three surface-bound hydrogen atoms (c). The reaction completes with desorption of NH_3 (d) [396].

The energy profile diagram for the formation of ammonia on iron is shown in Figure 14.3. The energetically highest state is the transition state for the reaction

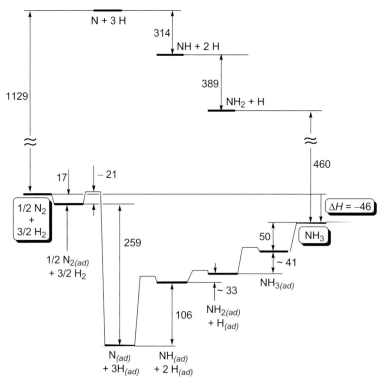

Figure 14.3 Energy profile diagram for the formation of ammonia on iron (bottom part) and for the uncatalyzed homogeneous reaction in the gas phase (top part). Energies are given in kJ/mol (after Ertl [398]).[2]

$1/2\,N_{2(ad)} + 3/2\,H_2 \rightarrow N_{ad} + 3\,H_{ad}$. The energetically most stable state is the one in which four atoms are bound to the surface ($N_{ad} + 3\,H_{ad}$). From there, the reaction goes energetically "uphill," but the overall reaction is exothermic at −46 kJ/mol. The energy profile of the uncatalyzed homogeneous reaction is completely different. If four atoms (N + 3 H) are present, they represent the highest energetic state, and from there, the reaction proceeds energetically "downhill." This reaction cannot be performed under chemically feasible conditions. Even at 1000 K (1 bar), the degree of dissociation of N_2 and H_2 only amounts to about 10^{-22} or 10^{-9}, respectively, so the uncatalyzed gas phase reaction is not a feasible method.

The schematic profile of the potential energy V during dissociative chemisorption of a diatomic molecule A_2 is shown in Figure 14.4. First, A_2 is bound relatively weakly to the surface ($A_{2(ad)}$) and then two A atoms are bound to the surface after overcoming an activation barrier with cleavage of the A–A bond. The potential-energy profile

2) With the kind permission of the author and Springer Science and Business Media, adapted from G. Ertl in Catalytic Ammonia Synthesis (J.R. Jennings, ed.), Plenum Press, New York 1991, Figure 3.13 (p. 128) and Figure 3.3 (p. 114).

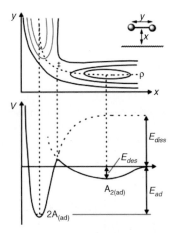

Figure 14.4 Schematic potential energy (V) diagram and contour plot of the energetics for dissociative chemisorption of a diatomic molecule A_2 approaching a surface. For simplicity, η^2 coordination is assumed for the adsorption, so only two coordinates, namely the distance $A \cdots A$ (y) and the distance between A_2 and the surface (x) are sufficient for description (E_{diss} = bond dissociation energy of A_2, E_{ad} = adsorption energy of 2 A, E_{des} = desorption energy of A_2) (after Ertl [398]).[2)]

corresponds to two intersecting curves. One describes the process $A_{2(g)} \rightarrow A_{2(ad)}$ and the other $2\,A_{(g)} \rightarrow 2\,A_{(ad)}$.

In experiments on the N_2 chemisorption on the (111) surface of crystalline α-iron without promoters, two surface N_2 complexes have been identified:

- γ-N_2: η^1 coordination (end-on coordination), which only leads to low N–N activation ($\nu_{N\equiv N} = 2100\,\text{cm}^{-1}$; compare to N_2 (g): $\nu_{N\equiv N} = 2194\,\text{cm}^{-1}$ for $(^{15}N)_2$)
- α-N_2: η^2 coordination (side-on coordination), associated with stronger bond activation ($\nu_{N\equiv N} = 1490\,\text{cm}^{-1}$ for $(^{15}N)_2$).

Through theoretical and kinetic studies, other surface complexes have been identified.

Iron occurs with three modifications, which interconvert enantiotropically (i.e., reversibly):

$$\alpha\text{-Fe} \xrightleftharpoons{906\,°C} \gamma\text{-Fe} \xrightleftharpoons{1401\,°C} \delta\text{-Fe} \xrightleftharpoons{1535\,°C} \text{Fe}_{(l)}$$

α- and δ-iron have a cubic body-centered structure, while γ-iron is cubic face-centered (i.e., cubic closest-packed). The (111) surface of α-iron is of special importance for NH_3 synthesis. Experiments on single crystals of iron have demonstrated the following order of activity for ammonia synthesis:

$$(111) > (211) > (100) \approx (210) > (110)$$

However, the activity of the less active (100) and (110) surfaces (but not of the (111) surface) can be substantially increased by pretreatment with NH_3 at elevated temperatures. This reveals restructuring of the surface, probably by formation of surface nitrides.

14 Nitrogen Fixation

Exercise 14.3
Draw the cubic body-centered structure of α-iron and give the positions of the previously mentioned crystal faces. Which is the most closely packed face? Give the degree of coverage of a (111) and a (110) surface by the Fe atoms of the uppermost layer.

The (111) surface of α-iron is very rough and contains "hills" and "valleys" (Figure 14.5). It is an "open" face giving access not only to the iron atoms that lie directly on the surface (henceforth called the *n*-th layer), but also to those in the underlying two layers ($n-1$ and $n-2$). In contrast, in the most closely packed (110) surface, only the surface atoms (layer *n*) are fully accessible (Figure 14.5).

In the "valleys" of the (111) surface, highly coordinated iron atoms ($CN = 7$) are accessible (Figure 14.5). These C7 sites have proven themselves in model studies to be particularly active in ammonia synthesis. These Fe atoms display a high electron density, making them especially capable of π back-donation onto chemisorbed N_2, which facilitates N–N bond cleavage.

14.2.3
The Industrial Catalyst

Originally, Haber worked with osmium and uranium as catalysts, since they first proved more active than iron. In further experiments, Alwin Mittasch (BASF)

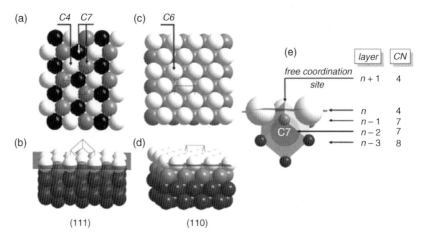

Figure 14.5 Structure of the (111) surface of α-iron viewed from above (a) and from the side (b). For comparison, the structure of the most closely packed (110) surface is shown (c, d). The iron atoms from the top layer (*n*-th layer) are white, while those in the underlying layer $n-1$ are light gray and all the lower layers ($n-2, n-3, \ldots$) are shown as black. The coordination number of the iron atoms is given by C*x* ($x = 4, 6, 7$). The coordination pattern of a C7 site in the $n-2$ layer of a (111) surface is shown in (e). For clarity, the next few neighbors of the iron atom with $CN = 7$ are shown with a smaller radius, except for those directly on the surface.

found in 1909 that the activity of iron can be substantially increased by adding compounds of alkali and alkaline earth metals. The precatalyst conventionally used in industry contains magnetite (Fe_3O_4) as its main component, with aluminum oxide, potassium oxide, and calcium oxide additives. Reduction with H_2 or N_2/H_2 leads to so-called "ammonia iron." The aluminum oxide (about 2%) is a structural promoter that hinders sintering to a material with low surface area. By contrast, K_2O (<1%) functions as an electronic promoter, so it exerts a direct influence on at least one elementary step in the reaction and thus raises the specific activity of the catalyst. The surface becomes especially rich in potassium. It transfers electric charge onto the iron surface atoms, which thus become capable of higher π back-donation into the π^* orbital of the chemisorbed N_2, which facilitates the N–N bond cleavage. This effect is partially reduced by the coadsorbed O. An additional activity-enhancing effect comes from the fact that the adsorption energy of NH_3 is lower on the (more electron-rich) potassium-modified iron surface than on pure iron.

Furthermore, the industrial catalyst is a polycrystalline material with a complex structure and a high specific surface area. A catalyst grain consists of aggregated nanoparticles with oxidic spacers as stabilizers. The nanoparticles contain an iron core, surrounded by small, single-crystalline iron platelets. Under the conditions of the synthesis, the iron is partially nitrided.

It is very likely that in addition to the (111) surfaces, the active centers in the polycrystalline industrial catalyst are step edges. In particular, surfaces of metal crystals consist of larger flat "terraces" and monoatomic steps, as shown in Figure 14.6 for a close-packed surface of α-iron. Such step edges display electronic and structural peculiarities; for example, they have accessible atoms with the coordination number $CN=7$, while the surface atoms of the flat terraces have the coordination number $CN=6$ (Figure 14.5). In addition, the step edges seem to have an optimal geometry for the dissociating N_2 and the stabilization of the transition state of the dissociative chemisorption. After cleavage of the N–N bond on a step edge, one N atom remains on the lower and one on the upper terrace that form the step. The activated N atoms can move across the terraces (i.e., they are subject to surface diffusion), which act only as a reservoir for them. Calculations show that the activation barriers for the dissociative chemisorption of N_2 on a step edge of a (110) surface and on an "open" (111) surface of α-iron are very

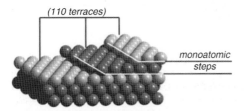

Figure 14.6 Step-shaped construction of a close-packed (110) surface of α-iron (schematic). For clarity, the close-packed layers of iron atoms are shaded differently.

similar. Since this is the rate-determining step in NH$_3$ synthesis, surface irregularities of the polycrystalline iron are critical for the activity of the industrial catalyst [394, 396].

14.2.4
Ruthenium Catalysts

High catalyst activity requires, on the one hand, an interaction of N$_2$ with the metal surface that is sufficiently strong to cleave an N–N bond. On the other hand, this interaction must not be too strong, or the adsorbate will not be sufficiently reactive. Hence, it will not react further, and the product will not desorb [399]. According to Sabatier's principle, this leads to a typical "volcano plot" when the activity of different metals is plotted against the adsorption energy of nitrogen (Figure 14.7). On the left, molybdenum and other d-electron-deficient metals cleave N$_2$ readily, but bind the N atoms too strongly, so the resulting catalytic activity is low. In contrast, the metals on the right, which are richer in d electrons, are not capable of dissociative chemisorption of N$_2$. This relationship between the formation rate of NH$_3$ and the binding energy of N is a consequence of a (linear) Brønsted–Evans–Polanyi relation. The optimum catalytic activity can be found with the metals of Group 8 of the periodic system.

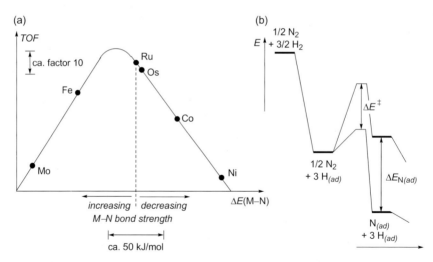

Figure 14.7 Calculated turnover frequencies (TOF) for NH$_3$ synthesis (400 °C, 50 bar) as a function of the adsorption energy of nitrogen for different metals, considering the effect of promoters (a). Energy profile diagram for the dissociative chemisorption of N$_2$ (b) on one metal with high M–N bond energy and another with low M–N bond energy ($\Delta E^\ddagger$ – difference between the activation energies of N–N bond cleavage; $\Delta E_{N(ad)}$ – difference of the adsorption energies of nitrogen) (shortened from Jacobsen [400]).

Ruthenium-based catalysts have been used industrially since the mid-1990s, under the acronym "KAAP" (*Kellogg* [Brown & Root] *Advanced Ammonia Process*; KBR Inc., Houston, Texas). Carbon-supported ruthenium catalysts with alkali metal and alkaline earth metal promotion are in use. The active centers are the edges of monoatomic steps on Ru(0001) terraces. A very active catalyst system was discovered with a barium-promoted ruthenium catalyst on an oxidic support (MgO). At 300–350 °C, it is almost 10 times more active than the industrial iron catalysts. To achieve the same ammonia yield with the conventional iron catalysts, higher temperatures are required and pressures must be doubled, which is what essentially causes the high energy requirements of the Haber-Bosch synthesis [401].

The synthesis of ammonia is an excellent example of a heterogeneously catalyzed reaction, understood in detail on the atomic and molecular levels, so the elementary steps have been correlated quantitatively with the macrokinetics. The investigations on well-defined single-crystal surfaces ("single crystal approach") have been instrumental in this regard. Single crystals as model catalysts possess a value for heterogeneous catalysis comparable with that of studies on (defined) model complexes for homogeneous catalysis.

Background: Sabatier's Principle and the Brønsted–Evans–Polanyi Relation

The elementary steps of a heterogeneously catalyzed reaction typically include chemisorption, dissociation, and activation of the substrates, as well as surface diffusion processes, recombination reactions, and desorption of the products. Over the course of the reaction, bonds are formed and cleaved, processes in which surface atoms of the catalyst are involved. *Sabatier's principle* (P. Sabatier, Nobel Prize 1912) states that the best catalyst forms intermediates that are neither too strongly nor too weakly adsorbed onto the surface of the catalyst. This is evident in a "volcano plot" in which, for different catalysts, the reaction rates of a reaction (ln r) are plotted against a quantity that is a measure for the reactant–catalyst surface interaction, such as the heat of adsorption $\Delta_{ad}H$. In the range of weak interactions (area **1** in the diagram), the catalysts are not sufficient to activate the substrate. The activation of the reactants (e.g., the dissociative adsorption) is rate-determining. The surface coverage is low. The decrease in catalytic activity with strong interaction (here, area **3**) can be traced to increasing coverage of the surface. The desorption, and hence the regeneration of the surface, is rate-determining. In the range of maximum reaction rate (area **2**), there is an intermediate amount of surface coverage.

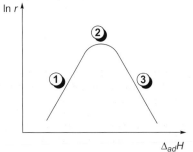

Brønsted–Evans–Polanyi relations are empirical. With a heterogeneously catalyzed reaction for different catalysts, they establish a linear relation between a non-equilibrium property (e.g., activation energy) and an equilibrium property (e.g., reaction energy or adsorption energy). A proportionality constant near 1 is an indication of a late transition state, and one near 0 indicates an early transition state. The structure of the transition state is thus "product-like" or "reactant-like" [402–404, T10].

14.3
Enzyme-Catalyzed Nitrogen Fixation

Nitrogenases are metalloenzymes that can reduce N_2 to NH_3. Based on the transition metals they contain, three types are distinguished, namely Fe/Mo-, Fe/V-, and Fe-containing nitrogenases. The overwhelming majority of nitrogenases are Mo-dependent.[3] They consist of two units [405]:

- *Fe protein.* The Fe protein is called dinitrogenase reductase. It is a homodimer and has a Fe_4S_4 cluster bound between its two subunits. It also has two binding sites for MgATP. The task of the Fe protein is to reduce the MoFe protein. It thus provides the electrons needed to reduce the substrate. The Fe_4S_4 cluster has a heterocubane structure as presented in Figure 6.9 (see Section 6.5.2). Instead of the terminal benzylthiolato ligands given there, each Fe atom is bound to cysteines from the protein, with deprotonation of the SH group.
- *MoFe protein.* The MoFe protein is the actual dinitrogenase. It is a dimer of an αβ-protein dimer. Each αβ-dimer contains, in the α-unit, an FeMo cofactor (M cluster, $MoFe_7S_9N$ cluster), and between the α- and β-unit, an Fe_8S_7 cluster, known as the P cluster. The structure of the P cluster is shown in Figure 14.8. It consists of two Fe_4S_4 clusters with heterocubane structure, which share a "sulfur vertex." This sulfur atom is surrounded by six iron atoms (μ_6-S). The iron atoms have the coordination number $CN = 4$. Free valences are saturated by a total of six S-bound cysteines (4 × terminal, 2 × μ-S), of which three each belong to the α- and the β-protein.

The nitrogenase-catalyzed reduction of N_2 involves the sequential transfer of electrons from the Fe protein to the MoFe protein and then to the substrate (N_2). As a result, two cycles can be distinguished, the Fe protein cycle and the MoFe protein cycle, whose interaction is shown schematically in Figure 14.9 [406].

---

3) Root nodule bacteria, the most important nitrogen-fixing microorganisms, only synthesize the "conventional" (Mo/Fe) nitrogenase. Their nitrogenase activity occurs in a close symbiosis (host–guest relationship) with leguminous plants. This is reflected, for example, in the synthesis of leghemoglobin, where the protein component comes from the plants and the heme moiety from the bacteria. This leghemoglobin furnishes the aerobic nodule bacteria with the necessary oxygen, but it also binds O_2 and thus protects the very oxygen-sensitive nitrogenase complex from oxidative destruction. The symbiosis is such effective that it not only provides the plant with nitrogen, but also releases NH_3 into the soil.

14.3 Enzyme-Catalyzed Nitrogen Fixation

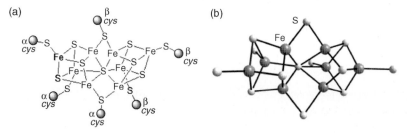

Figure 14.8 (a) Structure of the P cluster [Fe$_8$S$_7$(Cys)$_6$] in the reduced (native) state of a MoFe protein from a bacterial nitrogenase. Of the cysteine molecules in the structural model (b), only the sulfur atoms are shown. Binding sites on the protein (α- or β-Cys) are shown in (a) by circles shaded gray.

14.3.1
The Fe Protein Cycle

In the Fe protein cycle, the following reaction steps occur (Figure 14.9):

5 → 6: The iron protein in its oxidized form with two bound MgADP moieties is reduced by a biogenic reductant such as ferredoxin or flavodoxin, and ADP is replaced by ATP.

6/7 → 8: The reduced Fe protein associates with one of the oxidized forms of the MoFe protein forming a protein–protein complex. Since the MoFe protein contains two FeMo cofactors, these are 2 : 1 (Fe protein : MoFe protein) complexes.

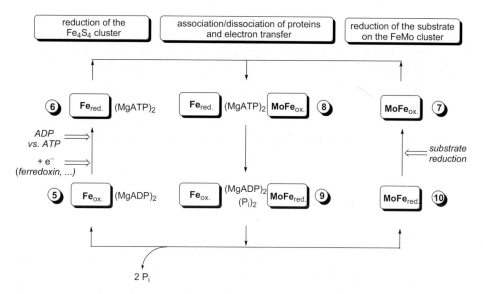

Figure 14.9 The interconnecting Fe protein and MoFe protein cycles of nitrogenase catalysis (Fe and MoFe indicate the Fe and MoFe proteins, respectively; P_i = inorganic phosphate HPO$_4^{2-}$) (adapted from W. E. Newton in [M9], Vol. 1, p. 300).

8 → 9: The electron transfer from the Fe$_4$S$_4$ cluster to the FeMo cluster occurs via the P cluster:

$$[Fe_4S_4] \xrightarrow{e^-} [Fe_8S_7] \xrightarrow{e^-} [MoFe_7S_9N]$$
$$\qquad\qquad\quad (P\ cluster) \qquad\quad (FeMo\ cluster)$$

In the protein complex, Fe$_4$S$_4$ and Fe$_8$S$_7$ clusters, as well as Fe$_8$S$_7$ and MoFe$_7$S$_9$N clusters, are each separated by about 14 Å, a distance that the electron must bridge. The process is accompanied by hydrolysis of ATP to form ADP and HPO$_4^{2-}$ (P$_i$).

9 → 5/10: The protein–protein complex dissociates, forming oxidized Fe protein with bound MgADP and the MoFe protein in one of its reduced forms. This dissociation step is rate-limiting and prevents any flow of electrons from the FeMo cluster back onto the Fe$_4$S$_4$ cluster (9 → 8). This is essential for multiple reduction of the FeMo cluster.

The reduction of the FeMo cluster proceeds in steps. Each time, one electron is transferred, so the Fe protein cycle (5 → 6 → 8 → 9 → 5) has to run eight times to reduce one dinitrogen molecule and two protons. Thus, the following overall equation is the result of the nitrogenase-catalyzed reduction of dinitrogen:

$$N_2 + 8\,H^+ + 8\,e^- + 16\,ATP \longrightarrow 2\,NH_3 + H_2 + 16\,ADP + 16\,P_i$$

14.3.2
The MoFe Protein Cycle

The structure of the FeMo cofactor, as obtained from an X-ray crystal structure analysis of a bacterial nitrogenase complex, is shown in Figure 14.10. Two heterocubanes M$_4$S$_4$ with a missing sulfur vertex (composition: [MoFe$_3$S$_3$] and [Fe$_4$S$_3$]) are linked via three Fe–S–Fe bridges. In the interior of the cavity formed can be found an interstitial N atom linked to six iron atoms (μ_6-N) in a trigonal-prismatic coordination.[4] The cluster core has the overall composition [MoFe$_7$S$_9$N].[5] It consists of two heterocubanes [Fe$_4$S$_3$N] and [MoFe$_3$S$_3$N] sharing a nitrogen, so in the middle of the cluster, three puckered Fe$_4$S$_4$ eight-membered rings ("crown form") are formed. The octahedal coordination of Mo is completed with a bidentate bound homocitrate ligand and a histidine that belongs to the protein matrix. The terminal Fe atom of the [Fe$_4$S$_3$] building block, which is not bound to the interstitial nitrogen atom, is bound to a cysteine from the protein.

---

[4] The X-ray crystal structure analysis supported the proposition that the cluster contains either a N, O, or C as an interstitial atom. However, the precision was not sufficient to differentiate between the three elements. Experimental studies and quantum-chemical calculations indicate that the atom might be N [407, 408].

[5] In the resting state, an overall charge of 0 for the cluster core of the FeMo cofactor has been determined. It can be helpful, for the sake of understanding, to arbitrarily (!) assign the atoms these *formal* (!) oxidation numbers: $[(Mo^{IV})(Fe^{III})_3(Fe^{II})_4(S^{-II})_9(N^{-III})]^0$.

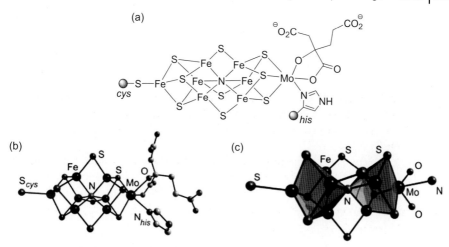

Figure 14.10 Structure of the FeMo cofactor [MoFe$_7$S$_9$N(Cys)(His)(Hcit)] (Hcit = homocitrate) in a bacterial nitrogenase. Binding sites on the protein are indicated in (a) by circles shaded gray. Of the amino acids in the structural model (b), only the sulfur atom (Cys) and the imidazole ring (His) are shown. The polyhedral diagram (c) elucidates the cluster core [407].

The mechanism of the reduction of N_2 on the FeMo cofactor is not yet clear in many regards. A schematic representation of the eight consecutive electron/proton transfers can be found in Figure 14.11. Starting from E_0, three or four electrons and the corresponding number of protons are first transferred to the FeMo cluster ($E_0 \rightarrow \ldots \rightarrow E_3H_3/E_4H_4$). Now coordination of N_2 and liberation of coproduced H_2 ($E_3H_3 \rightarrow E_3N_2H$ and $E_4H_4 \rightarrow E_3N_2H_2$, respectively) occurs. The intermediates during further reduction and protonation can be seen in Figure 14.11. In which reaction step the N–N bond cleavage and release of the first ammonia molecule occur is the subject of discussion. It may occur after three protons and the corresponding number of reduction equivalents have been transferred to the N_2 complex. This corresponds to the "distal reaction path," in which first the terminal nitrogen atom is protonated three times, and then the first NH_3 molecule is cleaved off. This is followed by the stepwise protonation and reduction of the nitrido complex formed, and finally, in the last reaction step, the second NH_3 molecule is cleaved off and the initial state (E_0) is regenerated. In the "alternating reaction path" the protonation proceeds alternately on the two N atoms of the N_2 complex, so the first NH_3 molecule does not cleave off until the penultimate reaction step [409].

The coproduction of H_2 ($E_3H_3 + N_2 \rightarrow E_3N_2H + H_2$ or $E_4H_4 + N_2 \rightarrow E_3N_2H_2 + H_2$) is intrinsically associated with the reduction of N_2. This causes the loss of at least two reduction equivalents (25%!). A larger quantity of H_2 can be formed if dihydrogen is liberated from the intermediates E_xH_m ($x = m = 2$–4). This nonobligatory (unproductive) generation of H_2 is indicated with dashes in Figure 14.11. In many cases, nitrogen-fixing organisms possess hydrogenases that gain energy by oxidizing H_2 to H^+, and thus part of the energy lost is won back.

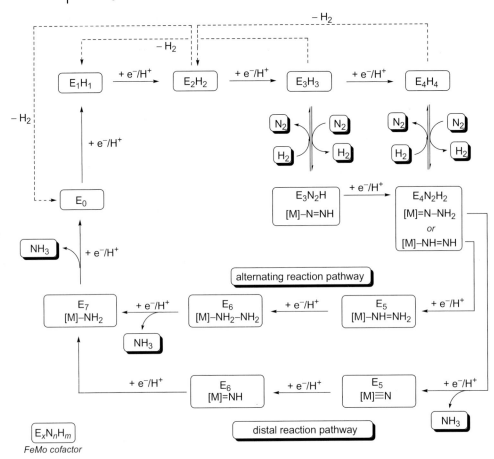

Figure 14.11 Presumed intermediates in the reduction of dinitrogen to form ammonia and hydrogen (modified Thorneley-Lowe cycle). A nonobligatory production of H_2 is shown with dashes. The FeMo cofactor is shown schematically for the first reaction steps by $E_xN_nH_m$, where x gives the number of electrons transferred. For the later reaction steps, the presumed structures of the intermediates are given separately for the distal and the alternating path, where η^1 coordination is assumed for the sake of simplicity (adapted from Barney and Seefeldt [410] and from Dance [411] and Hoffman [409]).

The sites of the protonations and the coordination of N_2 at the FeMo cofactor are not known for certain. There are indications that the entry site for the protons is one of the μ_3-S atoms. Before N_2 is coordinated, either three or four protons and electrons are transferred to the FeMo cofactor, respectively. In addition to the sulfido ligands, the iron atoms from the puckered Fe_4S_4 faces of the triply or quadruply reduced complex are a possible site for the protonations, whereby hydrido and/or dihydrogen iron complexes would be formed. The dinitrogen is probably coordinated to an Fe atom (Fe^2/Fe^6 in **11**) from an Fe_4S_4 eight-membered ring. In principle, η^1- (end-on) or η^2- (side-on) coordination is possible, either in endo or exo position (**11a** or **11b**, respectively; the same is true for coordination to Fe^2). The endo η^1-N_2 coordination to Fe^6 may be the favored one. In the schematic formulas **11a/b**, the protonation is indicated symbolically by –S(H), although this is neither a requirement, nor a restriction [411].

11a, Fe^6-N_2 (endo)   **11b**, Fe^6-N_2 (exo)   **11c**, μ-η^1:η^1-N_2 (Fe^2,Fe^6)

The protonation of a µ-S ligand (**12** → **13**) weakens the Fe–S bonds and can facilitate coordination of N_2 (**13** → **14**) and the cleavage of the Fe–S–Fe bridge (**14** → **15**). The formation of a µ-N_2 complex is also to be considered (**14** → **11c**).

$$[Fe]\overset{S}{\frown}[Fe] \xrightarrow{+ H^+} [Fe]\overset{H}{\underset{\oplus}{\overset{S}{\frown}}}[Fe] \xrightarrow{+ N_2} [Fe]\overset{H\;N_2}{\underset{\oplus}{\overset{S}{\frown}}}[Fe] \longrightarrow [Fe]\text{-SH} + N_2\text{-}[Fe]^{\oplus}$$

$$\quad\;\;12 \quad\quad\quad\quad\quad\quad\quad 13 \quad\quad\quad\quad\quad\quad 14 \quad\quad\quad\quad\quad\quad\quad\quad\quad 15$$

Experiments on model iron and ruthenium "Sellmann-type" complexes demonstrate that protonation of their multidentate organosulfur ligands can also be important for H transfers to coordinated dinitrogen [412]. For the nitrogenase, however, it is not clear whether protonation of a µ-S atom is a *necessary* precondition for N_2 coordination.

Although N_2 coordination to the molybdenum atom of the FeMo cofactor cannot be definitively ruled out, it probably plays an important structure-determining role for the complex. It links the cluster with the protein via a histidine and hydrogen bonds involving the homocitrate ligand. The conformation thus achieved seems to be important for the transfer of protons to the cofactor. Furthermore, the hydrophilic homocitrate ligand fosters the water pool in the NH_3 product exit domain. The interstitial N atom seems to be of significance for the conformative stability of the cluster [411].

14.3.3
A Prebiotic Nitrogen-Fixing System?

Ammonia is the most important bioaccessible nitrogen compound. In addition to water vapor, hydrogen, and methane, ammonia was a component of the gas mixture that served S. L. Miller (1953) as a model for the primordial atmosphere. In the presence of highly energetic electric discharges, the formation of small organic molecules, among them amino acids, could be demonstrated. It was not yet possible to answer with certainty the question of how ammonia was emitted into the atmosphere under prebiotic conditions. Volcanic processes and ammonium-containing minerals may have played a role. According to a theory from G. Wächtershäuser, the reducing effect of iron sulfide surfaces is of critical importance. The reaction of FeS with H_2S and N_2, yielding pyrite by oxidation and forming ammonia, is exergonic:

$$N_2 + 3\,FeS_{(s)} + 3\,H_2S_{aq} \longrightarrow 2\,NH_{3\,(g)} + 3\,FeS_{2\,(s)} \qquad \Delta G^\ominus < 0$$

The experimental proof that such a reaction can run has recently been found: Freshly precipitated FeS in aqueous solution in the presence of H_2S reacts with N_2 (1 bar) at 70–80 °C to form ammonia with a yield of 0.1% (based on 3 mol FeS). With this, it was possible to find a model for a prebiotic nitrogen-fixing system. This is an important step for a chemoautotrophic origin[6] of life [413] and further evidence of the significance of iron–sulfur compounds in the development of life on earth [414].

14.4
Homogeneously Catalyzed Nitrogen Fixation

14.4.1
Stoichiometric Reduction of N_2 Complexes

Experiments on the reactivity of dinitrogen transition metal complexes since 1965 have made it clear that the formation of complexes is associated with activation of the N–N triple bond. That has opened new possibilities for reducing N_2 to ammonia under mild conditions. Experiments on stoichiometric reduction represent a necessary first step toward being able to develop a catalytic reaction in a targeted manner. Metal-assisted stoichiometric reduction of N_2 to NH_3 can run, in principle, according to one of the following (simplified) reaction paths:

$$[M]\!-\!N_2^{+n} \begin{array}{c} \xrightarrow{+\,6\,H^+} [M]^{+n+6} + 2\,NH_3 \quad (a) \\ \xrightarrow{+\,6\,H^+ +\,6\,e^-} [M]^{+n} + 2\,NH_3 \quad (b) \\ \xrightarrow{+\,3\,H_2} [M]^{+n} + 2\,NH_3 \quad (c) \end{array}$$

- (a) Protons serve as the source of hydrogen. If no reductant is present, the oxidation number of M increases by 6. If such a reaction is to be the basis of a

---

6) Cells/organisms that use redox reactions as energy sources are called chemotrophic, and those that use CO_2 as their only source of carbon are called autotrophic.

catalytic reaction, in a subsequent step, the oxidized complex $[M^{+n+6}]$ must be reduced to $[M^{+n}]$ and then converted to $[M^{+n}]\text{-}N_2$.
- **(b)** Protons serve as the source of hydrogen and the reactions are performed in the presence of a reductant. Nitrogenases work according to this principle.
- **(c)** H_2 serves as the source of hydrogen and as the reductant. This corresponds to the Haber-Bosch process. In principle, a homogeneously catalytic reaction can also be performed on this basis, but none has yet been implemented in practice.

For the formation of hydrazine, a common side product or even the main product, the following applies: In total, 4 hydrogens (4 H^+ or 2 H_2) are needed, and for **b**, 4 additional reduction equivalents are required.

In 1964, M. E. Vol'pin and V. B. Shur found that in reactions of transition metal halides such as $CrCl_3$, $MoCl_5$, $FeCl_3$, or $TiCl_4$ with strong reducing agents such as $LiAlH_4$, EtMgBr, or $Al(i\text{-}Bu)_3$ in aprotic solvents under pressure in an N_2 atmosphere (100–150 bar), ammonia is formed after hydrolytic workup with yields of up to 25% (relative to the MCl_x). In the system $[TiCl_2Cp_2]$/EtMgBr, at normal pressure, 67% NH_3 was found after hydrolysis. In further experiments, it was possible to obtain a series of dinuclear dinitrogen complexes such as $[(TiCp_2)_2(\mu\text{-}N_2)]$ and $[(TiRCp_2)_2(\mu\text{-}N_2)]$ (R=alkyl, aryl), and also to characterize them structurally. They are considered as possible intermediates for N_2 activation. Upon protolysis, depending on the conditions, ammonia and/or hydrazine are generated in addition to N_2 [415, 416].

An example of a stoichiometric reduction of N_2 to NH_3 is the reaction of the easily accessible dinitrogen complexes of zero-valent molybdenum and tungsten [M-$(N_2)_2L_4$] (**16**, M = Mo, W; L/2L = mono-/bidentate phosphane ligand) with protonic acids HX, in which one ligand is cleaved off as N_2 and the other is reduced to ammonia and hydrazine in variable quantities. An example is shown in the following reaction scheme (L = PMe_2Ph) (J. Chatt, 1975):

16a $[W(N_2)_2L_4]$ $\xrightarrow{H_2SO_4, \text{MeOH}}$ 2 NH_3 (N_2H_4) + N_2 + $\{W^{VI}\}$ + ...
(90%) (2%)   (94%)

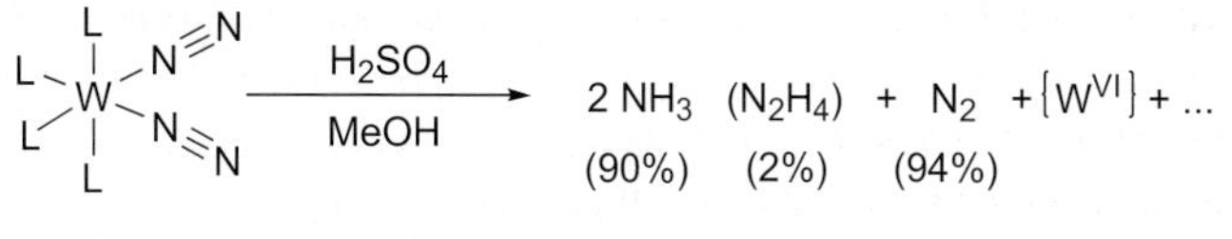

Complexes with monodentate phosphane ligands L and N_2 ligands in mutual *cis* position tend to form more ammonia, while for complexes with bidentate ligands L_2, the quantity of hydrazine formed increases. The reaction proceeds through protonation of the β-N atom. The initial reaction step should thus be understood as electrophilic attack of a proton on an N_2 ligand. Such a reaction is favored by a high negative partial charge on the β-N atom, hence by a low oxidation state of M and a strong back-donation into the π^* orbitals of the N_2 ligand. Diazenido(1−) (**17**), hydrazido(2−)-N,N (**18**) and hydrazidium complexes (**19**) have been found to be intermediates. The monoprotonated species **17** exists in equilibrium with the hydrido complex $[MX(H)(N_2)L_4]$. Further protonation of the β-N atom leads to reduction of the N–N bond order, which is associated with an increase in the M–N bond order. Thus, a metal-to-ligand charge transfer ($d \rightarrow \pi^*$) occurs.

$$N{\equiv}N-[M^0]-N{\equiv}N \xrightarrow[-N_2]{+HX} X-[M^{II}]-N{=}NH \xrightarrow{+H^+} X-[M^{IV}]{=}N-NH_2]^+ \xrightarrow{+H^+} X-[M^{IV}]{=}N-NH_3]^{2+}$$

16   **17**   **18**   **19**

[M] = ML$_4$; M = Mo, W

On the basis of these and additional experiments, a model has been designed to show how, starting from dinitrogen–molybdenum/tungsten(0) complexes **16′** ([M] = ML$_4$, M = Mo, W; 2 L = dppe, …), a catalytic reduction of N$_2$ to NH$_3$ could proceed by successive protonation and reduction (Chatt cycle):

$$N{\equiv}N-[M^0]-N{\equiv}N \xrightarrow[-N_2]{+HX,\ +H^+} X-[M^{IV}]{=}N-NH_2]^+ \xrightarrow[-NH_3]{+H^+,\ +2e^-} X-[M^{IV}]{\equiv}N$$

16′   **18′**   **20′**

$\uparrow$ +2 N$_2$, +2 e$^-$ $\downarrow$ −X$^-$, −NH$_3$ $\downarrow$ +H$^+$

$$X-[M^{II}]-NH_3]^+ \xleftarrow{+H^+} X-[M^{II}]{=}NH_2 \xleftarrow{+H^+,\ +2e^-} X-[M^{IV}]{=}NH]^+$$

23′   **22′**   **21′**

Double protonation of **16′** yields with liberation of N$_2$, via a diazenido complex, a hydrazido complex **18′**. Protonation and reduction of **18′** lead via a hydrazidium complex to the cleavage of the N–N bond, forming a nitrido complex and NH$_3$ (**18′** → **20′**). Successive protonation and reduction of **20′** generate, via imido and amido complexes, an ammine complex (**20′** → … → **23′**), from which, the starting complex **16′** is regenerated by reduction and ligand substitution, with cleavage of ammonia. Experimental evidence of a catalytic reduction of N$_2$ to NH$_3$ with complexes of type **16′** as catalyst, however, has not been found yet [417, 418].

Starting from complexes of type **16**, cyclic but *no* catalytic synthesis of ammonia from nitrogen has been experimentally verified when the protonation of these complexes is coupled with a cathodic reduction of the hydrazido(2–) complexes formed. The *trans* complex **16b** ([W] = W(dppe)$_2$) thus reacts with *p*-toluenesulfonic acid (TsOH) to form the hydrazido(2–) complex **18b**. In a subsequent electrochemical reduction in the presence of N$_2$ (1 bar), ammonia is released and the starting complex **16b** is re-formed. While **16b** is almost fully regenerated, the yield of NH$_3$ amounts to 0.2–0.3 mol per **18b** [419].

$$N_2-[W^0]-N_2 \xrightarrow[-N_2]{+2\ TsOH} TsO-[W^{IV}]{=}N-NH_2](OTs)$$

16b   **18b**

Hg cathode / N$_2$

↓ NH$_3$

In the examples mentioned so far, the formation of ammonia was based on a reaction of N$_2$ with H$^+$/e$^-$. Now, however, cationic dihydrogen ruthenium(II)

14.4 Homogeneously Catalyzed Nitrogen Fixation

complexes *trans*-[RuCl(η^2-H$_2$)(dppp)$_2$]X (**24**, X = weakly coordinating anion such as PF$_6$, BF$_4$, ...; dppp = Ph$_2$P(CH$_2$)$_3$PPh$_2$) have proven themselves acidic enough to protonate the dinitrogen complex **16a**. In an atmosphere of hydrogen, **24** reacts with **16a** to form ammonia, which is obtained in yields of up to 55%. Since complex **24** is formed from [RuCl(dppp)$_2$]X under hydrogen, a stoichiometric reduction of dinitrogen with dihydrogen yielding ammonia has been successfully realized under very mild conditions (55 °C, 1 bar) [420].

$$\mathbf{16a} + 6\,[\text{RuCl}(\eta^2\text{-H}_2)(\text{dppp})_2]\text{X} \; \xrightarrow[\{-\text{N}_2\}]{55\,°\text{C}} \; 2\,\text{NH}_3 + 6\,[\text{RuCl(H)(dppp)}_2] + \{\text{W}^{\text{VI}}\}$$

(with **16a** = L$_3$W(N≡N)(N≡N), L = PMe$_2$Ph; and 6 [RuCl(dppp)$_2$]X + H$_2$ (1 bar) → **24**)

An additional example starts with N$_2$ complexes of electron-deficient transition metals: In dinuclear complexes, a μ-η^2:η^2 coordination of N$_2$ can be associated with a substantial activation of the N–N bond. Thus, in complex **25**, which is created by reduction of [ZrCl$_2$(η^5-C$_5$Me$_4$H)$_2$] with Na/Hg in an atmosphere of nitrogen, the N–N bond is substantially lengthened (1.377(3) Å). At room temperature, **25** reacts with H$_2$, hydrogenating N$_2$ to form a hydrido–diazenido complex **26** (N–N 1.457(3) Å).

25 $\xrightarrow{\text{H}_2,\;22\,°\text{C}}$ **26** $\xrightarrow{-\text{H}_2,\;85\,°\text{C}}$ **27**

26 $\xrightarrow{\text{H}_2,\;85\,°\text{C}}$ 2 **27'** + NH$_3$ (10–15%)

Heating a solution of **26** in heptane causes complete cleavage of the N–N bond, forming the complex **27** with one μ-NH$_2$ and one μ-N ligand, which reacts quantitatively with anhydrous HCl to form [ZrCl$_2$(η^5-C$_5$Me$_4$H)$_2$] and NH$_3$. Ammonia is likewise formed if the thermolysis is performed in an atmosphere of hydrogen (**26** → **27'**). While the yield of NH$_3$ only amounts to 10–15%, it is obtained under very mild conditions from N$_2$ and H$_2$. If, in synthesizing **25**, the η^5-C$_5$Me$_5$ ligand is used instead of the η^5-C$_5$Me$_4$H ligand, a complex forms with an N$_2$ ligand bound end-on (coordination mode: μ-η^1:η^1-N$_2$), which is not as strongly activated as in **25** and shows no corresponding reaction [421, 422].

Hydridotantalum complexes **28a/b** grafted on a mesoporous silica support (see Section 7.5.1 for the structure and reactivity of such complexes) are able to cleave dinitrogen at 250 °C, forming an amido(imido)tantalum(V) complex **29a**. Interestingly, the same complex is formed by the reaction of **28a/b** with ammonia, even at room temperature. This is an impressive demonstration of the unique reactivity of the highly electrophilic tantalum center in **28a/b** (formally 8/10 ve), which is also reflected by a rapid and complete H/D exchange **29a** ⇌ **29b** under mild reaction conditions (60 °C) in an atmosphere of D_2 [423].

Exercise 14.4
(a) How might the H/D exchange **29a** ⇌ **29b** proceed? Propose a mechanism for the formation of **29a** (b) from **28a** and NH_3, as well as (c) from **28a** and N_2/H_2.

14.4.2
Catalytic Reduction of Dinitrogen

A catalytic reduction of dinitrogen to form ammonia under physiological conditions with a defined mononuclear metal complex was first[7] achieved by D. V. Yandulov and R. R. Schrock in 2003. As catalyst, a dinitrogen–molybdenum(III) complex **30** was used. A 2,6-lutidinium salt [LutH][B(Ar)$_4$] served as proton source, and permethylated chromocene [Cr(Cp*)$_2$] served as electron source. The reaction is catalytic with respect to Mo. With an amount of reduction equivalents sufficient to run the cycle six times, ammonia yields of about 64% are achieved. H_2 is obtained as a side product, typically at 33%. Hydrazine is only formed in trace amounts. The catalysis is performed in heptane as solvent. This has the advantage that the lutidinium salt, that serves as proton source, is only slightly soluble in it, and

---

[7] Previously, under physiological conditions, only a catalytic reduction of N_2 to form hydrazine as a main product ($N_2H_4/NH_3 \approx 10/1$) in methanol had been described. Sodium amalgam was used as reductant, and Mo^{III}/Mg(OH)$_2$ was used as catalyst. Details on the mechanism are not known [416].

14.4 Homogeneously Catalyzed Nitrogen Fixation

$$N_2 + 6\,H^+ + 6\,e^- \xrightarrow[\text{heptane (24 °C, 1 bar)}]{[\text{Mo}]-N\equiv N\;(30)} 2\,NH_3$$

6 [LutH][B(Ar)$_4$]   6 Lut
+ 6 [Cr(Cp*)$_2$]   + 6 [Cr(Cp')$_2$][B(Ar)$_4$]

thus a direct reduction of the protons by [Cr(Cp*)$_2$] to form H$_2$ is minimized [424, 425].

The structure of the catalyst [Mo{(hiptNCH$_2$CH$_2$)$_3$N}(η^1-N$_2$)] (**30**) is shown in Figure 14.12. A triamidoamine [(hiptNCH$_2$CH$_2$)$_3$N]$^{3-}$ (hipt = hexaisopropylterphenyl) was used as coligand. Molybdenum is trigonal-bipyramidally coordinated with a dinitrogen ligand bound end-on at an apical position. The trigonal coordination pocket for the N$_2$ ligand is shielded perfectly by the bulky hipt substituents. In the first place, this hinders the formation of relatively stable (and thus probably catalytically inactive) dinuclear complexes with a μ-η^1:η^1-N$_2$ ligand and also contributes to the stabilization of the intermediates that appear in the catalytic cycle. [Mo]–N$_2$ (**30**, [Mo] = Mo{(hiptNCH$_2$CH$_2$)$_3$N}) can be oxidized as well as reduced:

$$[\text{Mo}]-N_2\,]^{\oplus} \xleftarrow{-e^-} [\text{Mo}]-N_2\;(30) \xrightarrow{+e^-} [\text{Mo}]-N_2\,]^{\ominus}$$

ν_{NN} (in cm^{-1}):   2255   1990   1855

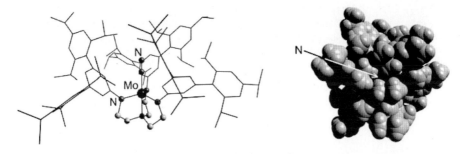

Figure 14.12 Molecular structure of [Mo{(hiptNCH$_2$CH$_2$)$_3$N}(η^1-N$_2$)] (**30**). The hipt substituents (hipt = hexaisopropylterphenyl) are only drawn as a wire model. The catalyst complex in the space-filling depiction (right, viewed from above onto the N$_2$ ligand) shows that the three bulky hipt substituents perfectly enclose the N$_2$ ligand.

14 Nitrogen Fixation

The N≡N stretching frequency shows the expected behavior and demonstrates an increasing π back-donation from left to right. In the cationic complex, comparison with free N_2 (2331 cm^{-1}) reveals a back-donation that is extremely weak and in accord with this, ligand substitution occurs particularly easily. This is an instructive example of how a change in the oxidation state of the central atom in a complex and its total charge affect the electron density distribution and the reactivity [425].

The proposed reaction mechanism of the catalytic reduction of N_2 to form NH_3 on a monomolecular molybdenum complex is shown in Figure 14.13. The following individual reaction steps can be identified [424, 426]:

30 → 31/31' → 32: Reduction of the dinitrogen complex **30**, followed by protonation, would lead via an anionic complex **31**, which can be regarded as a molybdenum(IV) complex with a deprotonated diazenido ligand, to a neutral

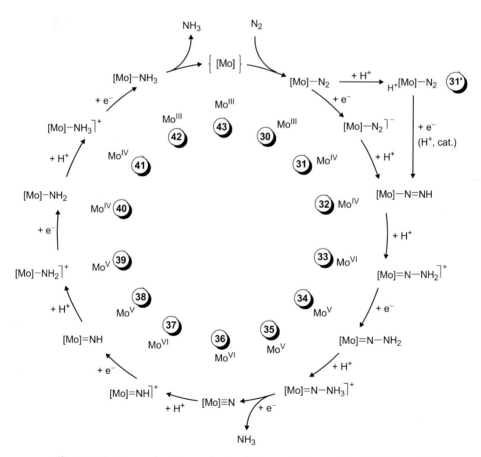

Figure 14.13 Proposed reaction mechanism for the molybdenum-catalyzed reduction of dinitrogen to form ammonia by stepwise addition of protons and electrons ([Mo] = Mo{(hiptNCH$_2$CH$_2$)$_3$N}, hipt = hexaisopropylterphenyl) after Yandulov and Schrock (adapted from [424, 425]).

diazenido complex **32**. However, the reaction probably runs as a proton-catalyzed reductive protonation, where in the first step, one of the hipt ligand's amido nitrogen atoms is protonated (**30** → **31'** → **32**).

32 → **33** → **34**: Protonation of the terminal N atom and reduction lead, via a cationic hydrazido(2−)-N,N complex, to a neutral one.

34 → **35** → **36**: Protonation of the terminal N atom yields a cationic hydrazidium complex, which liberates upon reduction, NH_3 to form a nitridomolybdenum(VI) complex.

36 → **37** → **38**: Protonation of the nitrido ligand and reduction yields an imidomolybdenum(V) complex.

38 → **39** → **40**: Protonation of the imido ligand and reduction yields an amidomolybdenum(V) complex.

40 → **41** → **42**: Protonation of the amido ligand and reduction yields an amminemolybdenum(V) complex.

42 → **43**: Ligand substitution (NH_3 vs. N_2) completes the catalytic cycle. Compare to Exercise 14.5 regarding existence of the intermediate **43** with an "empty" coordination pocket.

Exercise 14.5

- Kinetic experiments demonstrate that the ligand substitution

$$[Mo]-^{15}N_2 + {}^{14}N_2 \longrightarrow [Mo]-^{14}N_2 + {}^{15}N_2$$

(where [Mo] = Mo{(hiptNCH$_2$CH$_2$)$_3$N}) is of first order with respect to [Mo] and, up to about 4 bar, does not depend on the N_2 pressure ($t_{1/2}$ about 30–35 h at 22 °C). Thus, it can be concluded that [Mo] (**43**) is an intermediate. If an even bulkier coligand is used (replacing hexaisopropylterphenyl substituents (hipt) with hexa-*tert*-butylterphenyl substituents (htbt)), the reaction is also pressure-independent, but approximately 20 times slower ($t_{1/2}$ about 750 h), although the strength of the Mo–N bond is comparable (v_{NN} = 1990 cm^{-1} in both complexes). What can be the reason for the very large difference in reaction rate?

- In contrast to the previously cited reaction, there is evidence that the ligand substitution

$$[Mo]-NH_3 + N_2 \rightleftharpoons [Mo]-N_2 + NH_3$$

is a bimolecular reaction in which an intermediate or transition state [Mo](N_2)-(NH_3) appears. Consider what could be a reason for the different mechanism.

It has been possible to isolate and structurally characterize an impressive number of the intermediates in Figure 14.13 (Table 14.2, Figure 14.14). In addition to complex **30**, the complexes **32**, **36** and **41** are particularly frequently used as catalysts. However, the hydrido complex [Mo]–H and alkyl complexes [Mo]–R (R = *n*-hexyl, *n*-octyl), which probably cleave off H_2 or RH to form the catalytically active dinitrogen complex **30** when they react, are also suitable precatalysts. In contrast to the Chatt

Table 14.2 Selected structural parameters (bond lengths in Å, bond angles in °) for complexes [Mo{(hiptNCH$_2$CH$_2$)$_3$N}L] and [Mo{(hiptNCH$_2$CH$_2$)$_3$N}L][B{3,5-(CF$_3$)$_2$C$_6$H$_3$}$_4$] (after [427]).

[Mo]	–N≡N (30)	–N=N⁻ (31)[b]	–N=NH (32)	=N–NH$_2^+$ (33)	≡N (36)	=NH$^+$ (37)	–NH$_3$ (41)	–NH$_3$ (42)
ON(Mo)[a]	+3	+4	+4	+6	+6	+6	+4	+3
	(17 ve)	(18 ve)	(18 ve)	(18 ve)	(18 ve)	(18 ve)	(16 ve)	(17 ve)
Magnetism	param.	diam.	diam.	diam.	diam.	diam.	param.	param.
Mo–N	1.963(5)	1.863(7)	1.780(9)	1.743(4)	1.652(5)	1.631(7)	2.24(1)	2.170(6)
N–N	1.061(7)	1.156(8)	1.30(1)	1.304(6)				
Mo–N–N	179.2(7)	178.7(7)	180(1)	175.4(4)		169[c]		

a) The total number of valence electrons for Mo is given in parentheses. In this context, the (hiptNCH$_2$CH$_2$)$_3$N^{3-} ligand is viewed as a 12-electron donor (8σ + 4π). Two of the six π electrons of the amide nitrogen atoms occupy a nonbonding molecular orbital on the N atoms and are not counted.
b) [MgBr(THF)$_3$]$^+$ acts as a cation with contact to the β-N atom (Mg–N$_\beta$ 2.073(8) Å).
c) Mo–N–H.

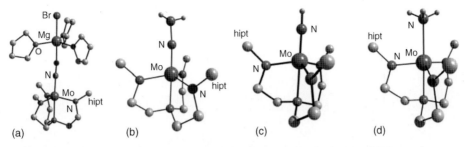

Figure 14.14 Structures of intermediates in the molybdenum-catalyzed reduction of dinitrogen: (a) [Mo]–N=N–MgBr(THF)$_3$ (complex type **31**), (b) [Mo]=N–NH$_2^+$ (**33**), (c) [Mo]=NH$^+$ (**37**), (d) [Mo]–NH$_3^+$ (**41**). Only the (NCH$_2$CH$_2$)$_3$N backbone of the (hiptNCH$_2$CH$_2$)$_3$N^{3-} ligand is shown. The three hipt substituents are each indicated only by the C atom bound to N.

cycle, the molybdenum center in the Schrock cycle passes through physiologically relevant oxidation states.

Exercise 14.6
Give the relevant resonance structures for the intermediates [Mo]–N$_2^-$ (**31**), [Mo]–N=NH$_2^+$ (**33**), and [Mo]=NH$^+$ (**37**) and discuss them.

Quantum-chemical calculations for the course of reaction confirm the essential aspects of the experimental findings. The calculations are based on the following reaction **a** (R = hipt) [428]:

$$N_2 + 6\,[Cr(Cp^*)_2] + 6\,LutH^+ \xrightarrow[\text{a}]{[Mo]-N\equiv N} 2\,NH_3 + 6\,[Cr(Cp^*)_2]^+ + 6\,Lut$$

[Mo]–N≡N ≡ (R = hipt, H)

As Figure 14.15 shows, each of the six consecutive protonation/reduction steps is exothermic, so the reaction leads "downhill" energetically. The detailed calculations show that in every case the protonation occurs first and the reduction follows. According to the calculations, it is most likely that the protonation of the starting compound [Mo]–N$_2$ (**30**) occurs on one of the three amide N atoms in the (hiptNCH$_2$CH$_2$)$_3$N^{3-} ligand (followed by reduction and proton migration N$_{amid}$–H → NNH with formation of **32**), which is in accord with "proton-catalyzed reductive protonation." The energetically least favorable step, the transfer of the second proton and electron (**32** → **34**), is thermoneutral, while the formation of the first ammonia molecule (**34** → **36**) is the most strongly exothermic step. The overall reaction **30** → ... → **30'** is strongly exothermic (−490 kJ/mol).

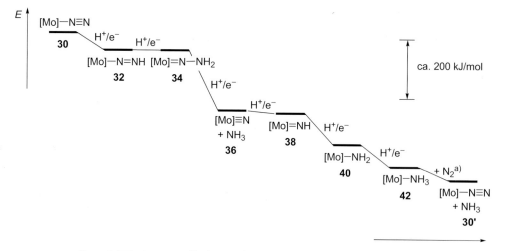

Figure 14.15 Energy profile diagram for the molybdenum-catalyzed reduction of N_2 to NH_3 by means of successive protonation by LutH$^+$ and reduction by [Cr(Cp*)$_2$]. The numbering follows that in Figure 14.13 ([Mo] = Mo{(hiptNCH$_2$CH$_2$)$_3$N}; simplified from Schenk and Reiher [428]).
a) The energies are specified (in particular, for **30–42**, the energy of N_2 is added) such that the reaction **30** → ... → **30'** corresponds to a conversion of N_2 to 2 NH_3 with LutH$^+$/[Cr(Cp*)$_2$] as protonation agent or reductant.

The reaction **30** → ... → **30'** (reaction a, R = H) is exergonic: $\Delta G^\ominus = -831$ kJ/mol. However, in these calculations, the triamidoamine ligand has been simplified by replacing the three hipt substituents with hydrogen (R = H); solvent effects (heptane) are considered in the framework of PCM calculations. Now, we can enter two additional reactions in the diagram:

- *Reaction* **b**. The formation of 2 NH_3 molecules from their elements is exergonic in heptane with $\Delta G^\ominus = -51$ kJ/mol. This reaction is the basis of the Haber-Bosch process.
- *Reaction* **c**. The *formal* formation of 6 [Cr(Cp*)$_2$] (reduced form) and 6 LutH$^+$ (protonated form) from 3 H_2 and 6 [Cr(Cp*)$_2$]$^+$ + 6 Lut is heavily endergonic with $\Delta G^\ominus = 780$ kJ/mol.

Reaction **c** reflects the expenditure of standard Gibbs free energy if NH_3 were to be produced from its elements, catalyzed by molybdenum. A similarly high input of

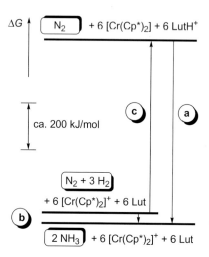

Gibbs free energy for the synthesis of reduced ferredoxin and ATP is required by the nitrogenase [429, 430].

The (triamidoamine)molybdenum system of Schrock enabled the first successful (abiotic) catalytic ammonia synthesis from nitrogen at room temperature and normal pressure. It can be considered a functional model for nitrogenases, although the two mechanisms differ from each other substantially.

14.4.3
Functionalization of Dinitrogen

Within the framework of their experiments on the reduction of N_2 to NH_3 (see Section 14.4.1), M. E. Vol'pin and V. B. Shur found that in the reaction of N_2 with [TiCl$_2$Cp$_2$]/LiPh (after hydrolysis), in addition to ammonia, small quantities of aniline are also formed, so a C–N bond is created. Thus, the activation of N_2 by coordination to a metal can be used not only to protonate N_2, but also to form other E–N bonds (E = C, B, Si, ...). As with protonation, such reactions are based on the attack of an electrophile E^+ on the nucleophilic N atom of an N_2 ligand [431]:

$$[M]-N\equiv N| \curvearrowright E^\oplus \qquad [M]\underset{N}{\overset{N}{\lessgtr}}[M] \quad \overset{E^\oplus}{\underset{E^\oplus}{}}$$

Examples of formation of N–C bonds are reactions of [M(N$_2$)$_2$(dppe)$_2$] (**16'**; M = Mo, W) with acid chlorides to form acyldiazenido complexes **44a**. Alkyl halides react in a radical reaction under the influence of light to form alkyldiazenido complexes **44b**. Protonation of **44a/44b** leads to organylhydrazido complexes **45a/45b**. α,ω-dihaloalkanes react directly to form hydrazido complexes **46** with a heterocyclic terminal N atom [417].

$$N_2\text{-[M]}\text{-}N_2 \;\; \mathbf{16'} \begin{cases} \xrightarrow[-N_2]{+\,R'COCl} Cl\text{-[M]}\text{-}N{=}N\text{-}COR' \xrightarrow{+\,H^+} \left[Cl\text{-[M]}{=}N\text{-}N\substack{COR'\\H}\right]^+ \\ \qquad\qquad\qquad \mathbf{44a} \qquad\qquad\qquad\qquad \mathbf{45a} \\ \xrightarrow[-N_2]{+\,RBr\,(h\nu)} Br\text{-[M]}\text{-}N{=}N\text{-}R \xrightarrow{+\,H^+} \left[Br\text{-[M]}{=}N\text{-}N\substack{R\\H}\right]^+ \\ \qquad\qquad\qquad \mathbf{44b} \qquad\qquad\qquad\qquad \mathbf{45b} \\ \xrightarrow[-N_2]{Br\text{-}(CH_2)_5\text{-}Br\,(h\nu)} Br\text{-[M]}{=}N\text{-}N\!\!\bigcirc\!Br \\ \qquad\qquad\qquad \mathbf{46} \end{cases}$$

[M] = M(dppe)$_2$; M = Mo, W

The formation of trimethylsilylamines from trimethylsilylchloride and nitrogen in the presence of stoichiometric quantities of sodium has proven to be catalytic with regard to cis-[Mo(N$_2$)$_2$(PMe$_2$Ph)$_4$]. Turnover numbers of 24 were reached. The most important side reaction is the coupling to form Me$_3$Si–SiMe$_3$ in a Wurtz-type reaction [432].

$$\text{Me}_3\text{SiCl} + \text{N}_2 + \text{Na} \xrightarrow[\text{THF, 30 °C}]{\textit{cis}\text{-[Mo(N}_2\text{)}_2\text{(PMe}_2\text{Ph)}_4\text{] (0.5 mol\%)}} \underset{(37\%)}{\text{N(SiMe}_3\text{)}_3} + \underset{(1\%)}{\text{NH(SiMe}_3\text{)}_2} + \underset{(39\%)}{\text{Me}_3\text{Si–SiMe}_3}$$

The reaction is probably based on a radical reaction course, where, similarly to the reaction of alkyl halides (16′ → 44b), [Mo]–N=N–SiMe$_3$ is formed as an intermediate. In the sense of a "reductive silylation" of N$_2$, titanium(IV) chloride reacts in tetrahydrofuran in an atmosphere of nitrogen (1 bar) with lithium and trimethylsilylchloride, cleaving the N–N bond to form a mixture of tris(trimethylsilyl)amine, as well as (trimethylsilylimido)- and [bis(trimethylsilyl)amido]titanium complexes (47), which have not been further characterized. Regarding Ti, the reaction is even catalytic; after hydrolysis of the reaction mixture and reaction with benzoyl chloride, 2.5 moles of benzamide were found per mole of Ti (47 → 48).

In place of a pure nitrogen atmosphere, the work can also be done in dry air. Reactions of the reaction mixture 47 with diketones to form nitrogen heterocycles (compare to 47 → 49 as an example) demonstrate that their synthesis starting from N$_2$ under mild reaction conditions is possible in principle [433].

Nucleophiles can attack at the α-N atom. Thus, the consecutive nucleophilic and electrophilic attack on 50 by a carbanion and a carbenium ion, respectively, results in the formation of an azomethane complex 51, which reacts with N$_2$ to regenerate the dinitrogen complex 50, even when extensive repetition of the full cycle is impeded by side reactions.

Moreover, a formation of E–N bonds after protonation of an N$_2$ ligand is possible. Thus, for example, in the presence of catalytic quantities of a protonic acid, hydrazido(2−) complexes 52 ([M] = MX$_2$(PMe$_2$Ph)$_3$; M = Mo, W; X = Cl, Br)

14.4 Homogeneously Catalyzed Nitrogen Fixation | 361

react as nucleophile with ketones and aldehydes in a condensation reaction to form diazoalkane complexes **53**. In the case of the isopropylidene complex (R = R' = Me), the subsequent reductive or protolytic cleavage produces Me_2CHNH_2 and NH_3 or acetonazine ($Me_2C=N-N=CMe_2$) and N_2H_4, respectively [434].

$$[M]=N-NH_2 + O=C\begin{smallmatrix}R\\R'\end{smallmatrix} \xrightarrow[-H_2O]{(H^+)} [M]=N-N=C\begin{smallmatrix}R\\R'\end{smallmatrix}$$

52   **53**

In an analogous manner, in the reaction of **16′**/$H[BF_4]$ with a cyclic acetal (2,5-dihydro-2,5-dimethoxyfuran), being a synthetic equivalent for a dialdehyde, the terminal N atom of a hydrazido complex is incorporated into a heterocycle (**54** → **55**). Reaction of **55** with $LiAlH_4$ yields pyrrole and NH_3 (about 1 : 1) and N-aminopyrrole. The metal complex is obtained as tetrahydrido complex **56**, which with N_2, under photochemical excitation, regenerates the dinitrogen complex **16′**. Although there is no catalytic formation of nitrogen heterocycles from N_2, there is a closed synthetic cycle [434].

$$N_2-[M]-N_2 \xrightarrow[-N_2]{+H[BF_4]} [F-[M]=N-NH_2]^+ \xrightarrow{(H^+)} [F-[M]=N-N\diagup\diagdown]^+$$

16′   **54**   **55**

+ LiAlH$_4$ (MeOH) → pyrrole-N-H / NH_3 ; pyrrole-N-NH$_2$

N_2 ($h\nu$) $[M](H)_4$ **56**

[M] = M(dppe)$_2$; M = Mo, W

μ-η^2:η^2-N_2 complexes of electron-deficient transition metals with highly activated N–N bonds (ligand: N_2^{4-}), and thus with nucleophilic N_2 moieties, can react with suitable substrates via cycloaddition or insertion reactions to form N–C and N–Si bonds. Thus, the hafnocene complex **57a** having η^5-C_5Me_4H ligands and the ansa-zirconocene complex **57b** react with carbon dioxide, forming C–N bonds, to generate the complexes **58a/58b**.[8] These complexes were found to react with Me_3SiI in excess, forming Si–N bonds, to produce $[M]I_2$ (M = Hf, Zr) and the isomeric hydrazine derivatives **59a** and **59b**. Since the dinitrogen complexes **57** are obtained starting from N_2, even if Na/Hg is used as a strong reductant, this is ultimately the synthesis of a

---

[8] MO calculations on the Zr complexes have revealed a π-bond fraction in the M–N bonds, thus establishing their "imido character" (compare with the contribution of the resonance structures **IIa/b** in the framework of VB theory), which facilitates 1,2-additions across the M–N bonds.

$$[M]\diagup\diagdown[M] \longleftrightarrow [M]=N-N=[M] \longleftrightarrow [M]-N=N-[M]$$

I   **IIa**   **IIb**

hydrazine derivative from N_2 and CO_2, two very simple molecules of very low reactivity [422, 435].

In contrast to the dinitrogen complexes **57**, no strong reductant is required to synthesize the dinitrogen–tantalum complex **61**. It is obtained by reaction of the dinuclear tetrahydrido complex [{Ta(NPN)}$_2$(μ-H)$_4$] (**60**, NPN = (PhNSiMe$_2$CH$_2$)$_2$PPh) with N_2 (**60** → **61**). The concomitant reductive elimination of H_2 and cleavage of the Ta–Ta bond in **60** generates four electrons, which cause strong activation of N_2 in **61**. In complex **61** there is a less commonly encountered μ-η^1:η^2 coordination of an N_2 ligand. Complex **61** reacts with a number of hydridoboron, -aluminum, and -silicon compounds E–H (E = BR$_2$, AlR$_2$, SiRH$_2$) with formation of an E–N and a Ta–H bond which can be seen as an E–H addition to the terminal Ta–N bond (**61** → **62**). This is followed by a reductive elimination of H_2, which generates the remaining two electrons required for cleavage of the N–N bond (**62** → **63**). The μ-nitrido–μ-imido complexes **63** assumed to be intermediates react further, mostly with decomposition or rearrangement of the NPN ligands [432].

Solutions to Exercises

Exercise 2.1
The reaction rate of a reaction is proportional to the rate constant k, which is associated with the Gibbs free energy of activation $\Delta G^\ddagger$ in the Eyring equation. For the ratio of rate constants k_{noncat}/k_{cat}, the following applies:

$$\frac{k_{noncat}}{k_{cat}} = \frac{e^{-\Delta G^\ddagger_{noncat}/RT}}{e^{-\Delta G^\ddagger_{cat}/RT}} \text{ and hence } \Delta\Delta G^\ddagger = \Delta G^\ddagger_{cat} - \Delta G^\ddagger_{noncat} = RT \ln \frac{k_{noncat}}{k_{cat}}.$$

Thus, accelerating the reaction by a factor of 10, 100, or 1000 corresponds to a decrease in the Gibbs free energy of activation of 5.7, 11.4, or 17.1 kJ/mol, respectively.

Conclusion: A relatively low decrease in energy for the transition state (compare with the C–C rotational barrier in ethane of about 11–13 kJ/mol) is enough to accelerate the reaction greatly.

Exercise 3.1
The different reactivity can be explained stereoelectronically: The olefin insertion passes through a transition state having a coplanar arrangement of the M–C–C–H fragment. In complex **2**, this can only be accomplished after (energy-intensive) isomerization (after R. Crabtree, *Acc. Chem. Res.* **1979**, *12*, 331).

Exercise 3.2
Cleavage of L from the complex produces a 14-*ve* complex (**1** → **2**), which first undergoes a β-hydrogen elimination and then a reductive elimination (**2** → **3** → **4**). Thus, ethane and ethene are formed in a ratio of exactly 1 : 1. In a radical decomposition, ethene and ethane should also be formed in a 1 : 1 ratio from ethyl radicals by H transfer, but recombination should also produce butane, which was not found experimentally (T. J. McCarthy, R. G. Nuzzo, G. M. Whitesides, *J. Am. Chem. Soc.* **1981**, *103*, 1676).

Exercise 3.3

If a nucleophile NuH with a sufficiently acidic H atom is added to an olefin complex, the nucleophile is very easily deprotonated, so a 2-(N-methylamino)ethyliron complex is formed (1 → 2). Protonation of the nitrogen atom yields a better nucleofuge (NH_2Me vs. $(NHMe)^-$), so **2** reacts with HCl in an electrophilic abstraction reaction, regenerating **1** (after L. Busetto, A. Palazzi, R. Ros, U. Belluco, *J. Organomet. Chem.* **1970**, *25*, 207).

$$Cp(CO)_2Fe-\|\;]^+ \quad \underset{+2\,H^+,\,-[NH_3Me]^+}{\overset{+2\,NH_2Me,\,-[NH_3Me]^+}{\rightleftharpoons}} \quad Cp(CO)_2Fe-CH_2-CH_2-\bar{N}HMe$$

    1   **2**

Exercise 3.4

a) Starting from the ^{13}C-labeled acetyl complex **1** ($^{13}C = {^*C}$), an isotopically labeled methyl–carbonyl complex **2** is obtained, while the acetyl complex formation in the presence of labeled CO does not lead to a labeled acetyl ligand (**3**). In both cases, the *cis* configuration (**2**: *CO/Me or **3**: *CO/COMe) demonstrates that CO insertion and deinsertion proceed uniformly with regard to stereochemistry.

b) Deinsertion of CO from an acetyl group requires previous cleavage of a *cis* CO ligand. Since all four *cis* CO ligands are equivalent, a total of 50% *cis*, 25% *trans*, and 25% unlabeled complex must be formed from **3** by methyl migration. This matches the experimental findings.

Note: Consider that upon CO migration, 75% cis complex and 25% unlabeled complex would be expected (after F. Calderazzo, *Angew. Chem. Int. Ed. Engl.* 1977, 16, 299).

Exercise 4.1
A radical chain reaction would run according to the following reaction scheme:

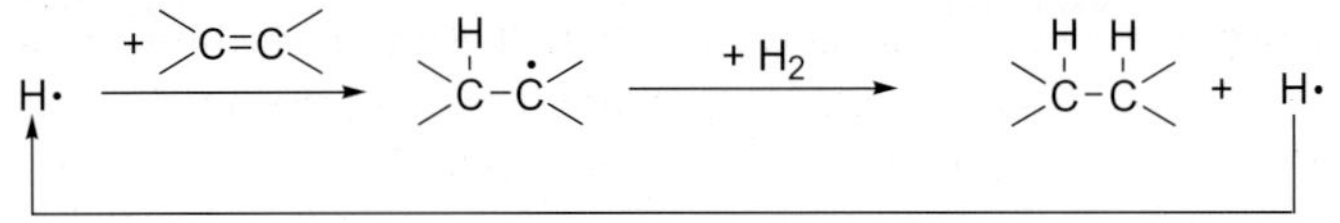

H• + >C=C< ⟶ >C-C<(H)(•) + H₂ ⟶ >C-C<(H)(H) + H•

The H radical is reactive enough to cleave the π bond ($\Delta H^\ominus = -148$ kJ/mol). In contrast, the homolytic cleavage of the H$_2$ bond requires a very high energy (bond enthalpy H–H > C–H), so the second step becomes endothermic ($\Delta H^\ominus = 24$ kJ/mol). Thus, it is evident that olefin hydrogenation requires a catalyst.

Note: A radical chain reaction with sufficiently long chains consists of a series of reaction steps with low activation enthalpy. However, with an endothermic partial reaction, $\Delta H^\ominus$ is now the lower limit for the activation enthalpy of this reaction step. This means that only weakly endothermic partial steps (compensated by exothermic chain steps!) with a radical chain reaction can be tolerated for long chains. Mean bond dissociation enthalpies are useful for *estimating* whether all reaction steps of chain propagation are rapid enough for such a chain reaction (after F. A. Carey, R. J. Sundberg, *Advanced Organic Chemistry*, Part B, 5th ed., Springer, New York 2008, p. 965).

Exercise 4.2
a) Starting from a hydrido–olefin complex **1** double-bond isomerization proceeds in the following sequence: Insertion (**1** → **2**) and β-hydride elimination of H' (**2** → **3**). Thus, a β-hydride elimination is required (in Figure 4.2, cf. Section 4.2.2, which would be the reaction **12** → **11**).

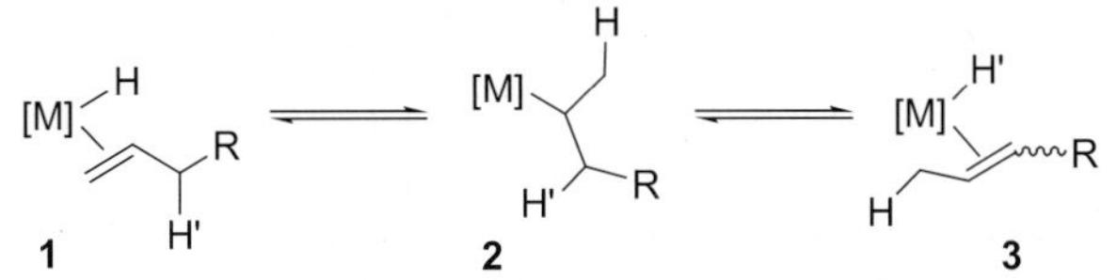

b) This is a migratory insertion (**11** → **12**) with migration of the *cis* hydrido ligand to the olefin ligand. Thus, a *trans* alkyl–hydrido complex **12'** is initially formed. (The numbering of species refers to Figure 4.2.)

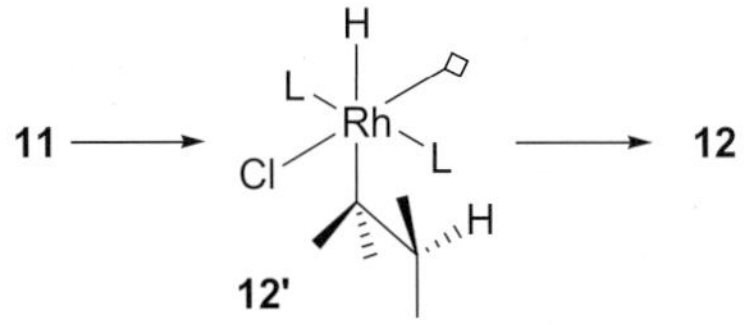

Exercise 4.3

a) The α-C atom of the amino acid determines the designation of the configuration. The reference substance is serine:

Fischer projection: L-serine

$$H_2N-\underset{CH_2OH}{\overset{COOH}{\underset{|}{C}}}-H \qquad H-\underset{CH_2OH}{\overset{COOH}{\underset{|}{C}}}-NH_2$$ D-serine

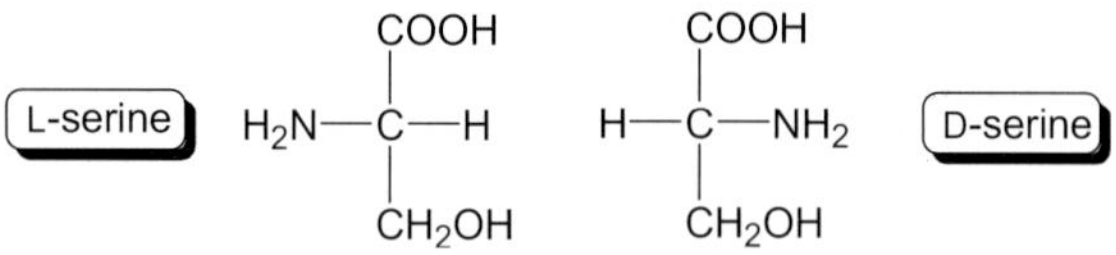

Hint for L-DOPA: Write the stereoformula (wedge projection) for the amino acid as in (a), so the Fischer projection (b) can be derived from it:

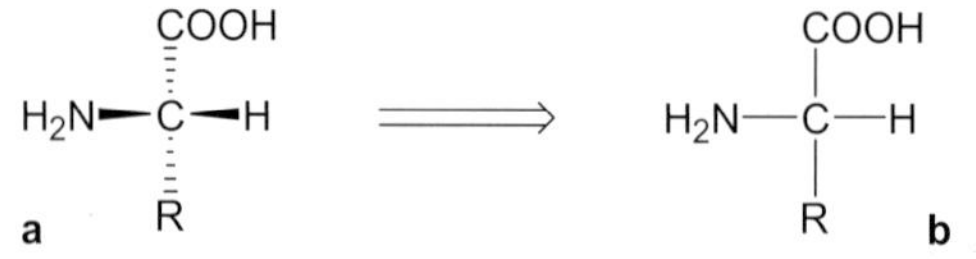

$$H_2N\blacktriangleright\underset{R}{\overset{COOH}{\underset{|}{C}}}-H \qquad \Longrightarrow \qquad H_2N-\underset{R}{\overset{COOH}{\underset{|}{C}}}-H$$

a   b $(R = CH_2C_6H_3(OH)_2)$

The comparison with serine shows that this is L-DOPA.

b) Determine the priorities of the substituents of the chiral C atom according to the CIP system

Example:	general	a		b		c		d
	serine	NH_2	>	COOH	>	CH_2OH	>	H
	DOPA	NH_2	>	COOH	>	$CH_2C_6H_3(OH)_2$	>	H

decreasing priority →

and look at the molecule with the substituent of lowest priority (d) pointing away from you. If the a → b → c ordering then corresponds to clockwise rotation, the chiral center is given the symbol R. For counterclockwise rotation, the symbol S is written. The formula **c**, drawn in perspective, makes clear that L-DOPA has the S-configuration.

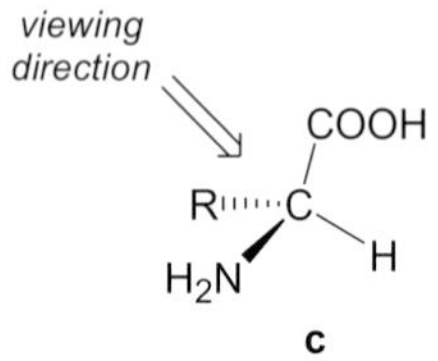

viewing direction

$$R\blacktriangleright\underset{H_2N}{\overset{COOH}{\underset{|}{C}}}\,H$$

c

Note: For α-aminocarboxylic acids, with the exception of cysteine/cystine, the L form corresponds to the S-configuration and the D form corresponds to the R-configuration of the α-C atom.

Exercise 4.4

- This is a highly enantioselective hydrogenation ($ee_{max} = 97.6\%$), using as coligand a *monodentate* phosphite ligand derived from BINOL. Interestingly, the hydrogenation follows the "lock-and-key principle," so there is no kinetically controlled enantioselectivity.

- Since all three catalyst complexes have the same thermodynamic stability, they are statistically distributed. The two homochiral catalyst complexes $[RhL^R_2]^+$ and $[RhL^S_2]^+$ are enantiomeric and consequently display identical catalytic activity and give products with opposite configurations. Since $[RhL^R L^S]^+$ is catalytically inactive, a (+)-NLE is the result. (The plot for the (+)-NLE shown in Figure 4.7 corresponds to the example introduced here.) Regarding general derivation of the relationship, see [65]. To support the statement, we calculate an example (where, for the sake of simplicity, $ee_{max} = 100.0\%$): L^R is present in excess ($ee_{aux} = 50.0\%$), resulting in a mole fraction of $x = 3/4$ for L^R and $x = 1/4$ for L^S. Thus, the following mole fractions can be calculated for the catalyst complexes (compare with Exercise 8.5): $x([RhL^R_2]^+) = 3/4 \cdot 3/4 = 9/16$; $x([RhL^S_2]^+) = 1/4 \cdot 1/4 = 1/16$; $x([RhL^R L^S]^+) = 2 \cdot 3/4 \cdot 1/4 = 6/16$. From the concentration ratio 9 : 1 for the two active complexes, it follows that $ee_{prod} = 80.0\%$, so there is a (+)-NLE (after M. T. Reetz, A. Meiswinkel, G. Mehler, K. Angermund, M. Graf, W. Thiel, R. Mynott, D. G. Blackmond, *J. Am. Chem. Soc.* **2005**, *127*, 10 305).
- There are $n(n-1)/2$ possible hetero-combinations, so for $n = 50$, there are 1225. *Note:* Proceed stepwise: From 3 elements *a, b, c* you can form 6 heteropairs (*ab, ac; ba, bc; ca, cb*). However, since the sequence of the elements in a pair does not play a role, this must be divided by 2. From 4 elements, 12 heteropairs (6 combinations without regard to sequence) are the result, and so on.
- Since the three catalysts can have a different thermodynamic stability, the composition of the catalyst mixture can deviate substantially from the statistical composition (**1a** : **1b** : **2** = 1 : 1 : 2). The enantioselectivity of the hetero-combination **2** is definitely greater than those of the two homo-combinations **1a/1b**. The relative activity of the three catalysts cannot be evaluated from the existing data. To draw conclusions, the activity of the two homo-combinations, the composition of the catalyst mixture, and the corresponding plots of conversion vs. time would all have to be known. More extensive conclusions have been obtained by variation of the ratio L^a/L^b from 5/1 to 1/5 (after [62]).

Exercise 4.5

1st step: Draw the olefin and specify the *Re* and *Si* faces. (In the representation of **1**, look at the *Re* face.)

2nd step: Draw the two H_2 addition products. From the mechanism it is clear that H_2 is added to the face the olefin is coordinated to (see formula drawing **2**, in which **1** is coordinated at the *Re* face).

3rd step: Determine the configuration of the hydrogenated product.

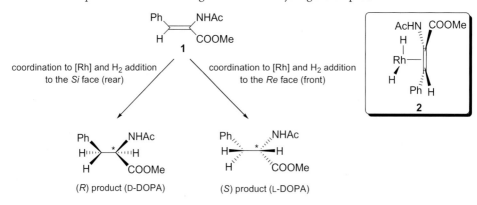

Exercise 4.6

- The reaction rate of the two diastereomers $[M_L]$ and $[M_D]$ is given by $r_L = k_L \cdot c_L \cdot p_{H2}$ and $r_D = k_D \cdot c_D \cdot p_{H2}$. With $K_{L/D} = \dfrac{c_D}{c_L}$, it follows that $\dfrac{r_D}{r_L} = \dfrac{k_D}{k_L} K_{L/D}$. Considering that $k = \dfrac{k_B T}{h} e^{-\Delta G^{\ddagger}/RT}$ and $\Delta\Delta G_{L/D} = \Delta G_D - \Delta G_L = -RT \ln K_{L/D}$, the following can be derived: $\dfrac{r_D}{r_L} = \dfrac{e^{-\Delta G_D^{\ddagger}/RT}}{e^{-\Delta G_L^{\ddagger}/RT}} e^{-\Delta\Delta G_{L/D}/RT} = e^{(-\Delta G_D^{\ddagger} + \Delta G_L^{\ddagger} - \Delta\Delta G_{L/D})/RT} = e^{-\Delta\Delta G_{L\ddagger/D\ddagger}/RT}$. Thus, the ratio of the reaction rates is determined by the difference between the Gibbs free energies of the transition states and is due to both the difference in the Gibbs free energies of activation and the fact that the reactants are present in different concentrations (with ratio $K_{L/D}$). The incorrect statement that the ratio of the reaction rates is determined solely by the difference between the Gibbs free energies of activation does not consider that the thermodynamically less stable reactant (here $[M_L] + H_2$) is present in lower concentration.

- The products P_D and P_L form in the ratio of the reaction rates $r_D : r_L$, which is determined by the difference between the Gibbs free energies of the transition states. We rearrange the equation above to solve for $\Delta\Delta G_{L\ddagger/D\ddagger}$, yielding $\Delta\Delta G_{L\ddagger/D\ddagger} = -RT \ln(r_D/r_L)$, and we use the corresponding values ($T = 298$ K, $R = 8.31441$ J/(mol K)):

% ee	Product ratio P_D/P_L	$|\Delta\Delta G_{L\ddagger/D\ddagger}|$ (in kJ/mol)
90	95/5	7.3
99	99.5/0.5	13.1
99.9	99.95/0.05	18.8

A difference between the Gibbs free energies of the transition states of only 13.1 kJ/mol (compare to C–C rotation barrier in ethane: Exercise 2.1) is enough to yield an enantiomeric excess of 99%.

Exercise 4.7
The Hammond principle states that an interconversion of two states of similar energy content that occur consecutively in a reaction process (e.g., a transition state and an unstable intermediate) will involve only a small reorganization of molecular structure. For the sake of simplicity, we restrict our discussion to the energy.

a. The reaction step is exothermic, and the activation barrier is low. According to the Hammond postulate, the transition states are structurally similar to the reactants, so the energy difference between the transition states corresponds roughly to the energy difference between the reactants. The thermodynamically more stable diastereomer determines the enantioselectivity. The enantiomeric ratio of the products corresponds roughly to that of **1** and **1'**.

b/c. The reaction step is endothermic, and the activation barrier for the reverse reaction is low. According to the Hammond postulate, the transition states are structurally similar to the products **2/2'**. The energy difference between the transition states roughly corresponds to that between the diastereomeric product complexes **2/2'**. In **b**, the thermodynamically more stable reactant complex **1** corresponds to the thermodynamically more stable product complex **2**, while in **c**, the opposite is true. With **b**, the enantiomer formed in excess comes from the thermodynamically more stable reactant complex **1** (albeit without the expectation that the enantiomeric ratio of the products corresponds to the stability difference between the reactants). By contrast, with **c** as a result of kinetically controlled enantioselectivity the enantiomer formed in excess (major enantiomer) is generated from the diastereomer that exists in deficit (see B. Bosnich, *Acc. Chem. Res.* **1998**, *31*, 667).

Exercise 5.1
In $[CoH(CO)_{4-n}(PR_3)_n]$ complexes, the following applies for $PR_3 = P(n\text{-Bu})_3$:

n	0	1	2	3
$T_{dec.}$ (in °C)	−20	20	160	80
ν_{CO} (in cm^{-1})	2043–2121	1933–2050	1902–1978	1883
pK_a	1	7 ($PR_3 = PPh_3$)		

(Data from [M15a], Vol. 5, p. 10 ff and literature cited there.) Substitution of CO, a weakly σ-basic but strongly π-acidic ligand, in $[CoH(CO)_4]$ by PR_3, a strong σ donor but weak π acceptor, leads to a strengthening of the π back-donation to the remaining CO ligands, and thus to a strengthening of the Co–CO bonds. Substitution of CO by PR_3 increases the basicity of the anion, and as a result the acidity of the Co–H compound decreases.

Exercise 5.2
- A possible reaction mechanism is based on oxidative addition of H_2 to the carbene complex **15'** (**15'** → **1**). The alcohol could then be formed by H transfer from the

metal to the carbene ligand followed by reductive C–H elimination (1 → 2 → 3). Another oxidative H_2 addition and deprotonation leads to complex 11, which then reacts as in Figure 5.3 to form 15. Protonation of 15 finally generates 15'.

$$[Rh]=C\begin{matrix}OH\\CH_2CH_2R\end{matrix}\Big]^+ \quad \xrightleftharpoons{+H_2} \quad [Rh]=C\begin{matrix}H\\|\\|\\H\end{matrix}\begin{matrix}OH\\CH_2CH_2R\end{matrix}\Big]^+ \quad \rightleftharpoons \quad [Rh]-C\begin{matrix}H\\|\\H\end{matrix}\begin{matrix}OH\\CH_2CH_2R\end{matrix}\Big]^+ \quad \xrightarrow{-RCH_2CH_2CH_2OH}$$

15'   1   2

$[Rh] = Rh(CO)(PEt_3)_2$

$$[Rh]^+ \quad \xrightleftharpoons{+H_2} \quad [Rh]\begin{matrix}H\\H\end{matrix} \quad \xrightleftharpoons{-H^+} \quad [Rh]-H$$

3   4   11

- The two competing reactions are oxidative addition of H_2 to 15 and protonation of 15 to form 15', which open the reaction channel to formation of the aldehydes and alcohols, respectively. The protonation of the acyl oxygen atom in EtOH is favored by a high electron density at the O atom, and thus by basic phosphanes (PEt_3). With L = PEt_3, the bis(phosphane) complexes $[Rh\{C(O)CH_2CH_2CH_2R\}(CO)L_2]$ are predominant. The bulkiness of $P(i\text{-}Pr)_3$ favors the formation of monophosphane complexes $[Rh\{C(O)CH_2CH_2CH_2R\}(CO)_2L]$ in which the acetyl oxygen atom displays a lower basicity, so the formation of 15' is no longer the favored reaction.

Further explanations and a discussion of the n/iso selectivity can be found in P. Cheliatsidou, D. F. S. White, A. M. Z. Slawin, D. J. Cole-Hamilton, *Dalton Trans.* **2008**, 2389.

Exercise 5.3
Due to the lower thermodynamic stability, the terminal olefins are only present in low concentrations. The catalyst must thus, in addition to possessing a high double-bond isomerization activity, also hydroformylate terminal double bonds significantly faster than inner ones, and produce a good n/iso ratio.

Exercise 5.4
In a toluene–water mixture, 1 is dissolved in the toluene phase. Bubbling CO_2 through the reaction mixture protonates the catalyst, making it soluble in the aqueous phase (1 → 1'). The reaction is reversible: By purging with N_2 at 60 °C, the CO_2 is liberated and the catalyst returns to the organic phase. The catalyst is suitable for reactions in both phases. Oct-1-ene is hydroformylated in the organic phase, and allyl alcohol in the aqueous CO_2-saturated phase. To isolate the product, the pH-dependent solubility of the catalyst is put to advantage by separating the catalyst from the product with CO_2/H_2O (octene → nonanal) or with toluene after purging with N_2 (allyl alcohol → 2-hydroxytrihydrofuran). This might be a possible solution for the problem of mass transport limitations that, in the classical two-phase process (Rhône-Poulenc), hamper the hydroformylation of higher olefins, which display poor solubility in water (after S. L. Desset, D. J. Cole-Hamilton, *Angew. Chem. Int. Ed.* **2009**, *48*, 1472).

Exercise 6.1

The oxidative addition proceeds as an S_N2 reaction. The square-planar complex $[RhI_2(CO)_2]^-$ (Rh^I; d^8) has a doubly occupied d_{z^2} orbital and reacts as a nucleophile. This would lead one to expect that an increase in the electron density at the Rh atom (e.g., by introduction of P- or S-donor ligands) leads to an acceleration of the oxidative addition. However, the higher electron density also results in a stronger Rh–CO bond, so the subsequent insertion step becomes slower. This calls for a special ligand design in which the long-term stability must also be considered in order to achieve a catalyst system suited for industrial applications [106].

Exercise 6.2

First, the catalyst complex $[RhI_2(CO)_2]^-$ forms by reaction of RhI_3 with H_2O/CO. However, it cannot be ruled out that the water-gas shift reaction might also occur, forming H_2, which would then function as a reductant.

Exercise 6.3

The co-carbonylation reaction of methanol and methyl acetate occurs in three steps (after P. Torrence in [M6], p. 104):

- Methanol is the more reactive component, so first, starting from MeOH, according to reaction **b** (cycle I, Figure 6.4), MeI and H_2O are formed, and according to **d** (cycle II, Figure 6.4), MeCOI reacts with methanol to form MeCOOMe.
- MeI is generated from MeCOOMe according to reaction **e** (cycle II, Figure 6.4), and MeCOI is converted by water (formed in the first reaction phase), according to **c**, to form acetic acid (cycle I, Figure 6.4). Both reactions generate acetic acid.
- Carbonylation of methyl acetate occurs, forming acetic anhydride (cycle III/IV, Figure 6.5). This does not change the concentration of acetic acid, so ultimately a mixture of MeCOOH and $(MeCO)_2O$ is formed.

Exercise 6.4

The reaction of phosgene with MeOH produces dimethyl carbonate in a smooth reaction, but the high toxicity of phosgene is a principal disadvantage. The oxidative carbonylation of methanol proceeds according to the following equation:

$$2\ \text{MeOH} + \text{CO} + \tfrac{1}{2}\text{O}_2 \xrightarrow[120–160\ °C,\ 25–35\ \text{bar}]{\text{CuCl}} \text{MeO-C(=O)-OMe} + \text{H}_2\text{O}$$

The most important side product is CO_2, which is formed through hydrolysis of dimethyl carbonate, hence ultimately a copper-catalyzed oxidation of CO to form CO_2 occurs. The mechanism of the reaction is neither known in detail nor proven. One mechanism that seems plausible involves activation of CO by formation of a carbonylcopper(I) complex **2** and a reaction of MeOH with Cu^I in the presence of O_2 to form a methoxocopper(II) complex **3**. Then **2** and **3** could react to form a dinuclear mixed-valence copper complex with a μ-OMe ligand of type **1**. The reaction **1** → **4** can be understood as an oxidative transfer of an OMe radical to the carbonyl carbon atom. This results in a reduction of Cu^{II} to Cu^I as well as the formation of a (methoxycarbonyl)copper(II) complex **4** with oxidation of Cu^I. The reaction of **4** with **3** to form the dimethyl carbonate should be regarded as (bimolecular) reductive elimination, where a dinuclear copper complex could appear as intermediate (after V. Raab, M. Merz, J. Sundermeyer, *J. Mol. Catal. A.* **2001**, *175*, 51; see also P. Tundo, M. Selva, *Acc. Chem. Res.* **2002**, *35*, 706).

$$2\ [Cu^I] \xrightarrow[\ -\text{H}_2\text{O}\ ]{+\ 2\ \text{MeOH},\ +\ \tfrac{1}{2}\text{O}_2} 2\ [Cu^{II}]\text{-OMe}\ (\mathbf{3})$$

$$[Cu^I] \xrightarrow{+\ \text{CO}} [Cu^I]\text{-CO}\ (\mathbf{2})$$

$$\mathbf{2} + \mathbf{3} \longrightarrow [Cu^I]\text{-}\underset{\text{Me}}{\overset{\text{O}}{|}}\text{-}[Cu^{II}]\ (\mathbf{1},\ \text{C=O bridge}) \xrightarrow{-[Cu^I]} [Cu^{II}]\text{-C(OMe)=O}\ (\mathbf{4}) \xrightarrow[-2\ [Cu^I]]{+\ [Cu^{II}]\text{-OMe}} \text{MeO-C(=O)-OMe}$$

Exercise 6.5

a) Solvation of the iodide anion via hydrogen bonds promotes its cleavage from the complex.

b) The strong *trans* influence and *trans* effect of CO in **12′** labilizes the Ir–Me bond and stabilizes the transition state, respectively. Both of which lower the activation energy of the reaction **12′** → **13′**.

Exercise 7.1

The homometathesis of symmetrically substituted olefins is nonproductive. For $R_2C=CR_2$, no change in substance occurs, while in the metathesis of RHC=CHR, a *cis/trans* isomerization is expected. In the cross-metathesis of RHC=CHR′ with R″HC=CHR‴, the four different alkylidene groups can, in principle, be linked in every imaginable way.

	=CHR	=CHR'	=CHR''	=CHR'''
RHC=	x			
R'HC=		x		
R''HC=			x	
R'''HC=				x
RHC=		x		
R'HC=			x	
R''HC=				x
RHC=			x	
R'HC=				x
R''HC=				x

This results in 10 different olefins, each of which can occur as *cis* and *trans* isomer.

Exercise 7.2

The formation of a cyclobutane from two olefins (using ethene here as an example) in a concerted reaction corresponds to a $[2\pi + 2\pi]$ cycloaddition. From the orbital correlation diagram, it follows that a bonding level of the reactants (SA) correlates with an antibonding level of the product (SA) and vice versa (AS $\leftrightarrow$ AS). As a result, in a concerted reaction there is no direct path from the ground state of the two ethene molecules to the ground state of the cyclobutane in which the orbital symmetry is conserved. Thus, in a thermal reaction, symmetry relationships cause a very high barrier that must be overcome, so the reaction is symmetry-forbidden (after R. B. Woodward, R. Hoffmann, *Angew. Chem. Int. Ed. Engl.* **1969**, *8*, 781).

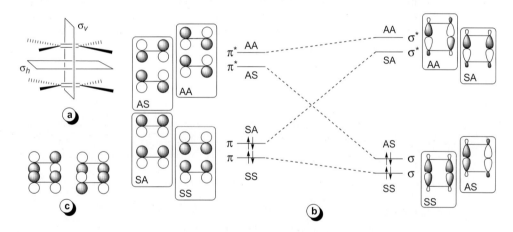

a) Arrangement of two olefin molecules prior to the cyclization. The two molecules approach each other such that σ_v and σ_h are symmetry planes. The symmetry of the orbitals involved is classified with respect to the mirror planes σ_v and σ_h (A = antisymmetric, S = symmetric).
b) Orbital correlation diagram. The two-letter abbreviations specify the symmetry with respect to the mirror plane σ_v (first letter) and σ_h (second letter).
c) The fact that the concerted cyclobutane formation is symmetry-forbidden also follows from an analysis of the two possible HOMO–LUMO interactions, for which, considering symmetry aspects, the overlap integral is $S = 0$. Hence, no bond formation can occur in this way.

Exercise 7.3

At the beginning of the reaction ($t=0$), no C6 olefin has yet been formed ($c_{C6}=0$), so according to the pairwise mechanism, c_{C14} should be equal to 0. As a result, an extrapolation of the product ratios $c_{C14} : c_{C12}$ and $c_{C14} : c_{C16}$ that are determined at different points in time should give a value of zero at $t=0$. Experimentally, values *not equal to zero* have been found, so a pairwise mechanism can be ruled out.

According to the Chauvin mechanism the products C12, C14 and C16 exist in statistical distribution during the whole reaction (for $c_{butene} = c_{octene}$, the following applies: $c_{C12} : c_{C14} : c_{C16} = 1 : 2 : 1$). The very product that, according to the pairwise mechanism, should not be formed at all at the beginning of the reaction (C14) is the main product. Thus, in accord with the experimental findings, the following must apply: $c_{C14}/c_{C12} \cdot c_{C14}/c_{C16} = 4$.

This discussion has been simplified. A detailed kinetic model can be found in T. J. Katz, J. McGinnis, *J. Am. Chem. Soc.* **1977**, *99*, 1903.

Exercise 7.4

- The three metathesis products RCH=CHR' (3), RCH=CHR (4), and R'CH=CHR' (5) are formed in statistical distribution. For an equimolar ratio of 1 and 2, the yield of target product comes to 50 mol%, and the same quantity of homocoupling products is produced. In the second case (10 mol 1/1 mol 2), 5.5 mol of products are formed. Mole fractions of alkylidene fragments (also compare with Exercise 8.5): $x(RCH=) = 10/11$, $x(R'CH=) = 1/11$. Mole fractions of the products: $x(3) = 2 \cdot 10/11 \cdot 1/11 = 20/121$, $x(4) = 10/11 \cdot 10/11 = 100/121$, $x(5) = 1/11 \cdot 1/11 = 1/121$. Thus, $5.5 \cdot 20/121 = 0.91$ mol of **3** and 4.59 mol of homocoupling products (**4/5**) are formed. *Conclusion*: For an acceptable yield of 91 mol % of target product, five times that quantity is generated in side products if the catalyst cannot distinguish between hetero- and homocoupling!

- As the first step, the formation of a Fischer carbene complex **1** by metathesis has been demonstrated (C → 1). Although not proven in detail, a β-silyl elimination could follow with formation of a formyl complex (1 → 2). (The reverse reaction, the formation of a siloxycarbene complex from a silyl–acyl complex, has been described.) Reductive Si–Cl elimination then generates a coordinatively unsaturated formyl complex (2 → 3), which reacts in the sense of a CO extrusion to form the carbonyl–hydrido complex **15**. For the mechanism of the double-bond isomerization, see the cited literature and Section 8.2 (after B. Schmidt, *Eur. J. Org. Chem.* **2004**, 1865).

```
        Cl                         Cl                    Cl                               CO
        |          +    OSiMe₃     |        – PCy₃       |    O                    O      |
[Ru]=CHPh  ───────────────→  [Ru]=CHOSiMe₃ ⇌    [Ru]–C       ─────────→  [Ru]–C   + PCy₃  [Ru]–H
        |          –    Ph         |        + PCy₃       |    H  – Me₃SiCl         H      |
        PCy₃                       PCy₃                  SiMe₃                             PCy₃

         C                          1                     2                        3       15
```

[Ru] = RuCl(Mes₂Imid)

Exercise 7.5

"Back-biting" leads to the formation of cyclic oligomers and a shorter polymer chain bound as a carbene to the metal (**1** → **2**; **O** = oligomer unit, **P** = growing polymer chain). In the corresponding intermolecular reaction (**3** → **4**), polymers are formed. Thus, the products possess bimodal molecular weight distributions (after A. J. Amass in [M1], Vol. 4, p. 109).

Exercise 7.6

The double bond in the cyclobutene ring of **1** is highly strained and selectively undergoes a ROM polymerization to form **1'**, which generates polyacetylene **1''** in a retro Diels-Alder reaction. The product is called "Durham polyacetylene" after the discoverers Feast et al. (1980) from Durham University (U.K.). Benzvalene **2**, under ROM polymerization conditions, generates a highly strained polymer **2'**, which isomerizes with metal salts such as $HgCl_2$ to form polyacetylene **2''** (Grubbs et al., 1988) (after W. J. Feast in [M2], Vol. 4, p. 135; T. M. Swager, R. H. Grubbs, *J. Am. Chem. Soc.* **1989**, *111*, 4413).

Exercise 7.7

The norbornene double bond (8/9) of DCPD is highly strained and thus more reactive than that in the cyclopentene ring (3/4) (**1** → **2**). The double bonds in polymer **2** can be *cis* or *trans* configured. For tacticity, which results from the orientation of the bicyclic system with relation to the main chain, see Exercise 9.9. If the other double bond of DCPD is also included in the ROM polymerization, three structural units can be formed (without considering *cis/trans* isomerism), namely structural units **3** and **4** via the exclusive opening of the norbornene ring or of the cyclopentene ring, respectively, and structural unit **5** via opening of both rings (after [128]).

Exercise 7.8

- If the reaction is carried out with the ratio **S1** : **S2** = 1 : 1, the two expected alkylidyne complexes **56** and **56′** are formed in the ratio 1 : 1. With a 10-fold excess of **S2**, only **56** is generated. In this case, the reaction is accompanied by metathesis; under these conditions the **56′** formed probably reacts very rapidly with **S2** to form **56** and butyne, which is removed in vacuo from the equilibrium (after M. L. Listermann, R. R. Schrock, *Organometallics* **1985**, *4*, 74).

- In both cases, these are W^{VI} complexes with 14 valence electrons. *Ligand sphere of* **A** (Figure 7.1 in Section 7.1.3): $2 \times RO^-$ [2] + CHR^{2-} [4] + RN^{2-} [6]. *Ligand sphere of* **60**: $2 \times RO^-$ [2] + CR^{3-} [6] + $R_2C{=}N^-$ [4]. The numbers in square brackets show the numbers of electrons that each ligand provides to the valence shell of the central atom. Since this is a W^{VI} complex (d^0), their sum yields the total valence electron number. This makes clear that for (formal) substitution (**A** ⇒ **60**) of a doubly negatively charged arylimido ligand by a singly negatively charged iminato ligand with simultaneous conversion of an M–C double bond to an M–C triple bond, the principal structure and electronic properties are preserved.

  The N-bound coligand in **60** is, in principle, a deprotonated imine. Its electronic structure can be described by the resonance structures **60′a–c**. The "ylidic" resonance structures **60′b/c** reveal the capability of the imidazolium ring for effective stabilization of a positive charge that leads to highly basic ligands **60′** with a strong electron-donating capacity. Interestingly, the two complex types **A** and **60** possess both a strong nitrogen donor ligand and two electron-withdrawing alkoxo ligands $(CF_3)_2MeCO^-$, resulting in a "push-pull situation" (after S. Beer, C. G. Hrib, P. G. Jones, K. Brandhorst, J. Grunenberg, M. Tamm, *Angew. Chem. Int. Ed.* **2007**, *46*, 8890).

Exercise 7.9

For the reaction course, both alkyne–nitrile cross-metathesis cycles **a** and **b**, as well as the alkyne cross-metathesis **c**, are relevant. (All reactions are reversible, although not explicitly shown.) Reaction of the two catalyst complexes **1** and **2** in cycle **a** with the reactants Et–C≡C–Et and Ar–C≡N leads to formation of Ar–C≡C–Et and Et–C≡N. The inclusion of the unsymmetrically substituted alkyne Ar–C≡C–Et initially formed leads in an alkyne–nitrile cross-metathesis to the formation of the symmetric alkyne Ar–C≡C–Ar (cycle **b**). Finally, the "normal" alkyne cross-metathesis according to **c** has to be considered. It proceeds – as usual – via metallacyclobutadiene intermediates **4**, while quantum-chemical calculations for the alkyne–nitrile cross-metatheses make azametallacyclobutadiene intermediates **3** likely. (The substituents R, R', R" are each specified under the reaction arrows.)

If alcoholate ligands with greater donor strength were used, the nitrido complex **1** would be thermodynamically more stable than the alkylidyne complexes **2/2'**, so the reversibility of the reactions in the cycles **a** and **b** would not be ensured. The use of alcoholate ligands with weaker donor strength (X = $OCMe(CF_3)_2$, $OCMe_2CF_3$) stabilizes the alkylidyne complexes, so **1** and **2/2'** have approximately the same stability. The favored formation of the symmetric alkyne Ar–C≡C–Ar is due to, among other factors, the fact that the formation of Ar–C≡C–Ar/Et–C≡N is thermodynamically more favored than that of Ar–C≡C–Et/Et–C≡N, and the fact that Et–C≡C–Et is removed from the equilibrium (2 Ar–C≡C–Et ⇌ Ar–C≡C–Ar + Et–C≡C–Et) by alkyne polymerization as a side reaction (after A. M. Geyer, E. S. Wiedner, J. B. Gary, R. L. Gdula, N. C. Kuhlmann, M. J. A. Johnson, B. D. Dunietz, J. W. Kampf, *J. Am. Chem. Soc.* **2008**, *130*, 8984).

Exercise 7.10
In oxidative addition reactions, the metal contributes two d electrons (see arrow) to the formation of the M–X and M–Y bonds, which raises the oxidation number of the metal by two. Since d⁰ complexes have no d electrons available, oxidative addition reactions to these metals are not possible.

$$[M] + \overset{X}{\underset{Y}{|}} \longrightarrow [M]{<}^{X}_{Y}$$

Exercise 7.11
The proposed reaction mechanism starts with a β-alkyl transfer, the reverse of the insertion of an olefin into an M–C bond, which generates an alkyl–olefin complex (**92** → **92′**). Hydrogenolysis, which occurs by σ-bond metathesis (reaction **a**, Section 7.4), yields a hydrido–olefin complex and a saturated hydrocarbon P′–H (**P, P′** = polymer/oligomer chain) (**92′** → **92″**). Via olefin insertion, an alkyl complex is formed (**92″** → **92‴**), whose hydrogenolysis leads, with regeneration of the catalyst complex **88**, to another saturated hydrocarbon P–CH$_2$CH$_3$.

The product distribution for hydrogenolytic degradation of the polymer shows that the C–H activation (**88** → **92**) proceeds nonselectively, so a statistical cleavage of the C–C bonds of the polymer chain occurs. The thermodynamics are determined by the overall result of the reaction, in the course of which hydrogenation of the olefins formed as intermediates creates very stable alkanes, so the overall reaction becomes exothermic. Thus, from a thermodynamic standpoint, a β-alkyl elimination as endothermic partial reaction is possible in principle (after V. Dufaud, J.-M. Basset, *Angew. Chem. Int. Ed.* **1998**, *37*, 806 and D. V. Besedin, L. Y. Ustynyuk, Y. A. Ustynyuk, V. V. Lunin, *Top. Catal.* **2005**, *32*, 47).

Exercise 7.12
The ethyltantalum complex **97** reacts with ethane in a σ-bond metathesis (C–C linkage according to **c**, Section 7.4) via a four-membered cyclic transition state **TS** to form propane and a grafted methyltantalum(III) complex **97′** (**97** → **TS** → **97′**). Through an additional σ-bond metathesis (alkyl exchange according to **b**, Section 7.4) with ethane via a four-membered cyclic transition state **TS′**, methane and the ethyl-tantalum complex **97** are formed (**97′** → **TS′** → **97**).

All these complexes are grafted Ta^{III} complexes, so a reaction path via oxidative C–C addition/reductive C–C elimination is possible, in which case grafted Ta^V complexes **97″** would appear as intermediates (**97** → **97″** → **97′**). However, further experiments have shown that neither of these reaction paths is followed (after [143]).

Exercise 7.13

A plausible reaction is that of **107** with the H-acceptor olefin **109'** to form **109** via insertion and reductive C–H elimination, ultimately generating the 14-ve complex **107'**. Oxidative addition of the substrate **108** to **107'** (which evidently occurs very selectively at the primary C–H bonds of the methyl group) would once again form an alkyl(hydrido)iridium(III) complex, from which, with β-H elimination and regeneration of the catalyst **107**, the product **108'** is cleaved (after F. Liu, E. B. Pak, B. Singh, C. M. Jensen, A. S. Goldman, *J. Am. Chem. Soc.* **1999**, *121*, 4086).

Exercise 8.1

Six isomeric hexenes can be formed:

Exercise 8.2

The ethene dimerization proceeds via insertion of ethene into the W–C single bond and formation of but-1-ene by β-hydride elimination (3 → 3a → 3b). Insertion of ethene into the W–H bond regenerates the catalyst (3b → 3). The double-bond isomerization occurs by an insertion of but-1-ene into the W–H bond of 3b, forming a sec-butyltungsten complex, from which but-2-ene is cleaved by β-hydride elimination (3b → 3c → 3b). The olefin metathesis occurs according to the usual mechanism: Reaction of ethene with the ethylidene complex 3 generates propene and a methylene complex 3d, which reacts with but-2-ene, releasing propene to form the ethylidene complex 3 (after M. Taoufik, E. Le Roux, J. Thivolle-Cazat, J.-M. Basset, *Angew. Chem. Int. Ed.* **2007**, *46*, 7202).

Exercise 8.3

The catalyst complex is a coordinatively unsaturated Ti^{II} compound **1**, formed by reaction of $Ti(OR)_4$ with AlR'_3. Reaction with ethene yields a bis(η^2-ethene) complex **2**, which generates the product (but-1-ene) via oxidative coupling (2 → 3) and β-H elimination/reductive elimination (3 → 4). Competition of ethene and butene for coordination to titanium gives rise to the formation of a butene–ethene complex (2 → 5), which likewise undergoes oxidative coupling (5 → 6). β-Hydrogen elimination/reductive elimination of C–H and of C–H' leads to, respectively, 2-ethylbut-1-ene (6 → 7) and 3-methylpent-1-ene (6 → 7') [158].

Exercise 8.4

Starting from the 7-membered metallacycle **31** (see Figure 8.3 in Section 8.3), the 9-membered metallacycle **32'** is formed via ethene coordination and insertion. β-H elimination and reductive C–H elimination produces **27** and oct-1-ene; refer to Figure 8.3. For alternative mechanisms, see the cited literature. In the tetramerization catalyst, the metallacycloheptane **31** is relatively stable against elimination of hex-1-ene, in comparison to the corresponding metallacycloheptane complex in a trimerization catalyst. However, as a consequence, not only hex-1-ene can be formed via β-H elimination and reductive elimination (**31** → **27**), but β-H elimination could also be followed by a cyclization (**31** → **31'**). (Compare with the polymerization of hexa-1,5-diene, Exercise 9.9) The **31'** formed could in turn react to form **2'** via reductive elimination, or to form **3'** via another β-H elimination. However, that cannot be brought into accord with the observed 1 : 1 ratio of the two products and other experimental findings; likewise, the formation of a free radical and its disproportionation can be ruled out. The mechanism may be bimolecular (see M. J. Overett et al., *J. Am. Chem. Soc.* **2005**, *127*, 10 723) [162].

Exercise 8.5

Metathesis is entropy-driven, so the equilibrium composition can be determined from a statistical redistribution of the alkylidene groups.

a) From the mole fractions (x) of the starting material (x(C10=C10) = 1/2, x(C2=C2) = 1/2), the mole fractions of the alkylidene fragments (x(C10=) = 1/2, x(C2=) = 1/2) are determined and the equilibrium concentrations are calculated:

Olefin	Number of C atoms	Mole fraction	c (in mol%)
C10=C10	20	1/2 · 1/2 = 1/4	25.0
C10=C2	12	2 · 1/2 · 1/2 = 1/2	50.0
C2=C2	4	1/2 · 1/2 = 1/4	25.0

b) Mole fractions of the starting material: x(C15=C5) = 1/2, x(C2=C2) = 1/2. Mole fractions of the alkylidene fragments: x(C15=) = 1/4, x(C5=) = 1/4, x(C2=) = 1/2. Equilibrium concentrations:

Olefin	Number of C atoms	Mole fraction	c (in mol%)
C15=C15	30	$1/4 \cdot 1/4 = 1/16$	6.25
C15=C5	20	$2 \cdot 1/4 \cdot 1/4 = 2/16$	12.5
C15=C2	17	$2 \cdot 1/4 \cdot 1/2 = 2/8$	25.0
C5=C5	10	$1/4 \cdot 1/4 = 1/16$	6.25
C5=C2	7	$2 \cdot 1/4 \cdot 1/2 = 2/8$	25.0
C2=C2	4	$1/2 \cdot 1/2 = 1/4$	25.0

The product distribution for the metathesis of C19=C1 (c) and of C10=C10 (second part of exercise) is calculated analogously:

Olefin	Number of C atoms	c (in mol%)	Olefin	Number of C atoms	c (in mol%)
C19=C19	38	6.25	C10=C10	20	1.0
C19=C2	21	25.0	C10=C2	12	18.0
C19=C1	20	12.5	C2=C2	4	81.0
C2=C2	4	25.0			
C2=C1	3	25.0			
C1=C1	2	6.25			

Note: If you find the probability calculation difficult, remember that the probability of throwing the combination 6 + 6 with two dice comes to $1/6 \cdot 1/6 = 1/36$. The probability of throwing the combination 5 + 1, by contrast, amounts to $2 \cdot 1/6 \cdot 1/6 = 2/36$.

Exercise 9.1
1,4-polymerization of 2,3-dimethylbutadiene and subsequent hydrogenation generate H–H polypropylene.

It can also be made by alternating copolymerization of *cis*-but-2-ene and ethene with, for example, $VCl_4/Al(C_8H_{17})_3$ as catalyst (O. Vogl, M. F. Qin, A. Zilkha, *Prog. Polym. Sci.* **1999**, *24*, 1481).

Exercise 9.2
Treatment of complex **1** with the Lewis acid $B(C_6F_5)_3$ results, with abstraction of a benzyl ligand, in the formation of **1'** ($R = CH_2Ph$). Thus, **1'** has a coordination

site that is either vacant or occupied by a solvent molecule. The oligomerization is terminated by β-hydride elimination, so in the active catalyst complex **1′**, a hydrogen atom or the growing oligomer chain takes the place of R.

Vinylene end groups on oligomers (**2**, M = Ti) are an indication of a secondary insertion (2,1-insertion), and vinylidene end groups (**3**, M = Zr, Hf) indicate a primary insertion (1,2-insertion). Interestingly, the structure of the saturated end of the oligomers **2** makes it clear that, even with the titanium complex **1′** (M = Ti, R = H), the *first* insertion occurs as a primary insertion, which could be the result of a low steric influence from R = H. This is in accord with the structure of the dimers when [1-^{13}C]-labeled hex-1-ene is used: For M = Ti and M = Zr/Hf, the main products obtained were, respectively, **4** (where ^{13}C labeling is marked by a star) or **5** (after B. Lian, K. Beckerle, T. P. Spaniol, J. Okuda, *Angew. Chem. Int. Ed.* **2007**, *46*, 8507 and J. Okuda, personal communication).

Exercise 9.3

Polymers of type **1** are typical for stereochemical chain end control. If, in a growing chain with the configuration (...RRRR)$_{rel.}$, a stereoerror occurs (...RRRRS)$_{rel.}$, the chain will continue to grow with *S*-configuration of the stereocenters (...RRRRSSS...)$_{rel.}$, until the next stereoerror occurs, and so on. Polymers of type **2** are typical for stereocontrol by the catalyst active site (enantiomorphic site control). A stereoerror is immediately corrected on the next insertion step, so polymers are formed with statistically distributed isolated error sites. Thus, stereoerrors propagate in the first case, but not in the second case.

Exercise 9.4

The polymer structures with stereoerrors (indicated by arrows) and the corresponding triads/pentads are listed below (after [195]).

- Stereocontrol by the catalyst active site:
  1 – isotactic: Triads: mr + rr + mr.   Pentads: mmmr + mmrr + mrrm + mmrr + mmmr.
  2 – syndiotactic: Triads: mr + mm + mr.   Pentads: rrrm + mmrr + rmmr + mmrr + rrrm.

- Stereocontrol by the chain end:
  3 – isotactic: Triads: *mr* + *mr*. Pentads: *mmmr* + *mmrm* + *mmrm* + *mmmr*.
  4 – syndiotactic: Triads: *mr* + *mr*. Pentads: *rrrm* + *rrmr* + *rrmr* + *rrrm*.

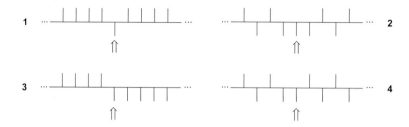

Exercise 9.5

The stoichiometry in the interior of the crystals is given by the linkage pattern of the MCl$_6$ octahedra (M = Ti, Mg). For the listed crystal faces, the following applies:

Crystal	Face	CN	Coordination		Stoichiometry
TiCl$_3$	(110)	5	Ti(μ-Cl)$_4$Cl	TiCl$_{4/2}$Cl$_1$	TiCl$_3$
MgCl$_2$	(100)	5	Mg(μ_3-Cl)$_3$(μ-Cl)$_2$	MgCl$_{3/3}$Cl$_{2/2}$	MgCl$_2$
MgCl$_2$	(110)$^{a)}$	4	Mg(μ_3-Cl)$_2$(μ-Cl)$_2$	MgCl$_{2/3}$Cl$_{2/2}$	Mg$_2$Cl$_4$ = MgCl$_2$
	b)	6	Mg(μ_3-Cl)$_4$(μ-Cl)$_2$	MgCl$_{4/3}$Cl$_{2/2}$	

a) Surface atom.
b) Layer below it, which deserves separate consideration, since the coordination of the Mg ions still differs from that in the interior of the crystal (Mg(μ_3-Cl)$_6$ = MgCl$_{6/3}$).

Note: Prepare virtual 3D models of the structures with a typical crystal structure visualization program such as DIAMOND (data from: Inorganic Crystal Structure Database, ICSD, FIZ Karlsruhe, Karlsruhe 2004):

MgCl$_2$: Crystal system: trigonal/rhombohedral; space group: $P\bar{3}m1$ (no. 164); cell parameters: $a = b = 3.641$ Å, $c = 5.927$ Å; coordinates: Mg (0, 0, 0) [multiplicity/Wyckoff letter: 1a], Cl (1/3, 2/3, 0.23) [2d] (ICSD #17063).

α-TiCl$_3$: Crystal system: trigonal/rhombohedral; space group: $P\bar{3}1m$ (no. 162); cell parameters: $a = b = 6.14$ Å, $c = 5.85$ Å; coordinates: Ti (1/3, 2/3, 0) [2c], Cl (1/3, 0, −0.25) [6k] (ICSD #29035).

Exercise 9.6

Metallocene dichlorides display distorted tetrahedral coordination of two chloro ligands and two η^5-bound cyclopentadienyl ligands. When rotation of the cyclopentadienyl ligands around the M–Cp$_{cg}$ axis is unhindered, the time-averaged symmetry is C_{2v}.

Exercise 9.7

- Reflection at a plane (σ) generates from **53** the other enantiomer **53**$_{ent}$. In **53**, propene is coordinated to the *Si* face and in **53**$_{ent}$, to the *Re* face.

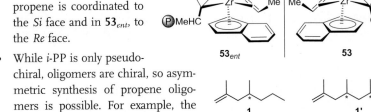

- While *i*-PP is only pseudo-chiral, oligomers are chiral, so asymmetric synthesis of propene oligomers is possible. For example, the trimer **1** and the tetramer **1'** (formed by primary insertion, chain termination via β-H elimination) contain only one or two chiral C atoms, respectively. Isotactic oligostyrenes with measurable optical activity have been obtained up to a polymerization degree of 45 (after [198] and K. Beckerle, R. Manivannan, B. Lian, G.-J. M. Meppelder, G. Raabe, T. P. Spaniol, H. Ebeling, F. Pelascini, R. Mülhaupt, J. Okuda, *Angew. Chem. Int. Ed.* **2007**, *46*, 4790).

Exercise 9.8
Four polymers with well-defined microstructures are obtained:

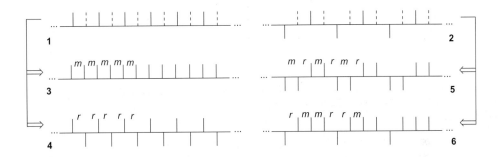

Starting from the hemi-isotactic polymer **1** an isotactic arrangement of the stereocenters 2, 4, 6, ... ("iso–iso" combination) leads either to the isotactic polymer **3** or to the syndiotactic polymer **4**. Starting from the hemi-syndiotactic polymer **2**, a syndiotactic arrangement of the stereocenters 2, 4, 6, ... ("syndio–syndio" combination) leads to the heterotactic polymer **5**, characterized by alternating between *m* and *r* diads. Eventually, the "syndio–iso" combination (**2** → **6**) or "iso–syndio" combination (**1** → **6**), respectively, leads to the bi-heterotactic polymer **6** (after [178], Vol. 1).

Exercise 9.9

- *Inter*molecular primary insertion leads to a polymer strand with a but-3-enyl substituent at C2 (**1** → **2**). Subsequent *intra*molecular primary insertion results in cyclopolymerization, yielding a cyclopentane ring within the main chain of the polymer (2 → 3).

The polymer **4** has two stereocenters per constitutional monomeric unit (cyclopentane-1,2-diylmethylene), and is thus ditactic. The two stereocenters of a cyclopentane ring can form an *r* or an *m* diad. The polymers are called *threo* (**4a/b**) or *erythro* (**4c/d**), respectively (see footnote 3, Section 13.1.2; instead of the terms *threo/erythro*, *trans/cis* are also common). Stereocenters of the same type ($*1, *1'$ or $*2, *2'$) in consecutive structural units are either both isotactic (**4a/d**) or both syndiotactic (**4b/c**). Hence, there is a *threo*-diisotactic (**4a**) and a *threo*-disyndiotactic (**4b**) polymer, as well as an *erythro*-disyndiotactic (**4c**) and an *erythro*-diisotactic polymer (**4d**) (after [194]).

- The reaction of the catalyst complex **5** bearing the growing polymer chain with silanes and boranes should, via a four-membered cyclic transition state, lead to detachment of a polymer with an R_3Si or R_2B end group, without deactivating the catalyst.

With electron-rich chain transfer agents such as $H-PR_2$, the reaction must, due to the opposite polarity of the element-hydrogen bond (H–P vs. H–Si/H–B), proceed with a fundamentally different mechanism. A possible reaction sequence, starting from a phosphidometal catalyst complex **6**, involves olefin insertion into the M–P bond followed by chain growth (**6** → **7**) and the protolysis of the M–C bond by the phosphane via a four-membered cyclic transition state, with regeneration of the catalyst complex (**7** → **8** → **6**).

This is a versatile pathway for efficient and selective functionalization ("end-cappping") of polyolefins during the polymerization process itself, which can lead to polyolefins with promising applicability (after S. B. Amin, T. J. Marks, *Angew. Chem. Int. Ed.* **2008**, *47*, 2006).

Exercise 9.10
Complexes of type [Mo^{III}Cl$_2$CpL$_2$] (**3**, L$_2$ = (PMe$_3$)$_2$, dppe) are 17-*ve* half-sandwich complexes that can react, in principle, with alkyl bromides RBr by bimolecular oxidative addition (one-electron oxidative addition; see Section 3.2) to form 18-*ve* electron complexes **4** and **5**.

The weak Mo–Br bond in **4** enables a so-called atom transfer radical polymerization (ATRP) of styrene, started by the reaction of 1-bromoethylbenzene with **3** (**3** → **4**). Apart from the radical recombination (2 R• → R–R; R = growing polymer chain), the reaction **4** → **3** and the possible reiinitiation **3** → **4** lead to a controlled radical polymerization process.

In the radical polymerization of styrene (conventionally initiated with AIBN), the Mo^{III} complex **3** functions as spin trap for the growing polymer radical (**3** → **5**). As a consequence of the very weak Mo–C bond in **5**, this reaction is reversible under the given conditions, so an organometallic radical polymerization (OMRP) occurs. Another catalytic chain transfer that has been observed is not discussed here (after R. Poli, *Angew. Chem. Int. Ed.* **2006**, *45*, 5058; regarding terminology for living radical polymerization, see G. Moad, E. Rizzardo, S. H. Thang, *Acc. Chem. Res.* **2008**, *41*, 1133).

Exercise 9.11
The reaction rate of the alternating copolymerization is calculated as $r_1 = k_{100 \to 101} \cdot c_{100}$ and that of the double ethene insertion is calculated as $r_2 = k_{105 \to 106} \cdot c_{105}$. Thus, the result is

$$\frac{r_1}{r_2} = \frac{c_{100} \cdot k_{100 \to 101}}{c_{105} \cdot k_{105 \to 106}} = K \frac{c_{CO}}{c_{C2H4}} \frac{k_{100 \to 101}}{k_{105 \to 106}} \approx 10^4 \frac{7.3 \cdot 10^{-3}}{0.11} \cdot 10^2 = 6.6 \cdot 10^4 \approx 10^5.$$

Exercise 9.12

The terminology is derived from the relative configuration of the asymmetric C atoms (see Background, Section 9.3.1). In the isotactic polymer, all these C atoms have either R- or S-configuration. (In **113**, an ...SSS... polymer is shown.) In the syndiotactic polymer **114**, the asymmetric C atoms alternately display R- and S-configuration.

Exercise 10.1

- No! In this case, the relative position of the reference proton and of the substituent R are not changed. Only the two H atoms of the CH_2 group exchange their places, which does not affect the *syn/anti* designation.
- Due to the *anti/syn* isomerization, four topomers occur with η^3-allyl ligands:

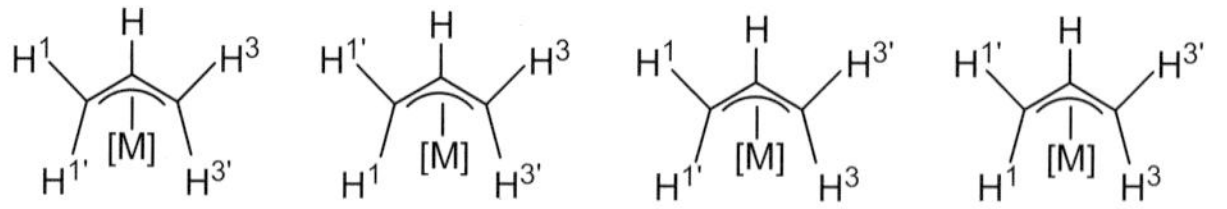

[Structures of four η^3-allyl topomers with [M] center showing H^1, H$^{1'}$, H^3, H$^{3'}$ arrangements]

Exercise 10.2

- Retrosynthetic analysis shows the reaction sequence **35d** ⇒ **34d** ⇒ **33c**. The formation of **33c** requires the coupling of two *s-cis*-butadienes in a complex of type **32** or, starting from **33b**, isomerization of the η^3-*syn* allyl group to form an η^3-*anti* allyl group. Then *s-cis*-butadiene would have to insert into the η^3-*anti* allyl group (**33c** → **34d**). All of these reaction steps are energetically or kinetically disfavored, so this reaction path is not followed.

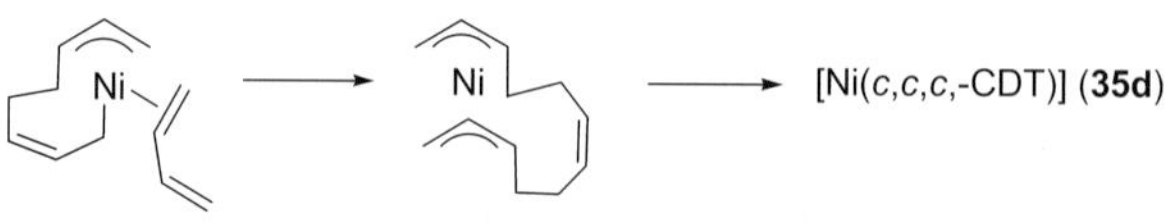

[Scheme: 33c → 34d (bis(η^3-*anti*),Δ-*cis*) → [Ni(c,c,c,-CDT)] (35d)]

- It has been shown experimentally that the reaction

$$[Ni(t,t,t\text{-CDT})] + c,c,c\text{-CDT} \xrightarrow{\text{ether}} [Ni(c,c,c\text{-CDT})] + t,t,t\text{-CDT}$$

35a   **38d**   **35d**   **38a**

runs smoothly in diethyl ether as solvent. For $\Delta G = -RT\ln K$, with $\Delta\Delta G = -20$ kJ/mol, it can be calculated that the complex formation constant for [Ni(c,c,c,-CDT)] (**35d**) is about 3200 times larger than that for **35a** (after K. Jonas, P. Heimbach, G. Wilke, *Angew. Chem. Int. Ed. Engl.* **1968**, *7*, 949; S. Tobisch, *Chem. Eur. J.* **2003**, *9*, 1217).

Exercise 10.3

The structurally characterized complex in Figure 10.11 is the η^3-$syn,\eta^1(C^1),\Delta$-cis isomer with 1,4-linkage of the two isoprene molecules (**2b**). The resulting 2,6-dimethyl substitution suggests a bis(isoprene) complex **1b** with two differently coordinated isoprene molecules (1,2- vs. 3,4-coordination) as precursor. NMR experiments show that the 3,6-dimethyl substituted complex **2a** is formed first. Above 10 °C, **2a** rearranges into a 4 : 1 mixture of **2b** and **2c**. This requires cleavage of the C4–C5 bond, so an equilibrium of the bis(isoprene) complexes **1a**, **1b** and **1c** must be assumed (after R. Benn, B. Büssemeier, S. Holle, P. W. Jolly, R. Mynott, I. Tkatchenko, G. Wilke, *J. Organomet. Chem.* **1985**, *279*, 63)

L = PCy₃

Exercise 10.4

Octa-1,3,7-triene and two functionalized octadienes are formed as products.

Exercise 10.5

The formation of the octadienyl complex **47b** (L = P(OEt)₃; X = CF₃COO⁻) in the usual way is followed by insertion of the terminal double bond into the Ni–C bond and cleavage of product from the complex by β-hydride elimination (after J. Furukawa, J. Kiji, H. Konishi, K. Yamamoto, *Makromol. Chem.* **1973**, *174*, 65).

Exercise 10.6
Assuming head-to-tail linkage, *cis*-1,4- (**1**) and *trans*-1,4-polyisoprene (**2**) as well as *iso*- and *syndio*-1,2- (**3**/**4**) or *iso*- and *syndio*-3,4-polyisoprene (**5**/**6**) can form.

Exercise 10.7
The polymer chain is bound to M via an allyl group. During hydrogenolysis, this is cleaved off as an olefin. Insertion of ethene (or an α-olefin) in place of butadiene in the allyl-bound polymer chain does not lead to regeneration of this allyl group. The alkyl ligand formed readily undergoes a β-H elimination, so that also a polymer with an olefinic end group is formed. In both cases, a hydridometal complex is formed which is capable of inserting butadiene to start a new polymer chain (after L. Porri, A. Giarrusso in [M1], Vol. 4, p. 53).

Exercise 10.8
The polymer chain coordinated to nickel in **83** contains a vinyl substituent resulting from a 1,2-insertion. In accord with the experiment, quantum-chemical calculations that also consider an occasional 1,2-insertion show that **83** is more stable than **82'**, so it is the "true" catalyst resting state. Accordingly, **83** shows a lower rate than **82'** for the subsequent reaction (coordination of butadiene followed by 1,4-insertion into the allyl group). This generates a complex of type **82'** that displays the "normal" high insertion rate until a 1,2-insertion generates a complex of type **83** again. Experimentally, a content of about 1% of 1,2-polybutadiene is found, so one 1,2-insertion occurs per approximately 100 1,4-insertion steps (after A. R. O'Connor, P. S. White, M. Brookhart, *J. Am. Chem. Soc.* **2007**, *129*, 4142; S. Tobisch, R. Taube, *Organometallics* **2008**, *27*, 2159).

Exercise 10.9
The ionization of **1** would lead to a primary carbocation significantly richer in energy than the allylic carbocation **3**, so **1** is a "protected" monomer that cannot be activated by the enzyme that ionizes **2**. Thus, in this process every propagation step is followed by deactivation, accompanied by the release of HX.

The starting reaction is an enzymatically catalyzed double-bond isomerization, of which there are many. Thus, the IPP isomerase catalyzes the isomerization of 1 to form the dimethylallyl derivative **5**. The reaction of **5** with **1** is the chain initiation reaction (compare with the reaction **2** → **3** in Exercise 10.9, Section 10.5.3), so **5** has the function of an initiator, but in biochemical literature it is designated a cosubstrate (after J. E. Puskas, E. Gautriaud, A. Deffieux, J. P. Kennedy, *Prog. Polym. Sci.* **2006**, *31*, 533).

Exercise 11.1
The decrease in reactivity in **2** can be traced to the boron atom, which, due to strong π-B–N bonds, is not a strong Lewis acid. A series of Suzuki couplings (2 mol% [Pd-{P(*t*-Bu)$_3$}$_2$], 2 equiv. CsF; THF, 60 °C), using easily accessible electrophilic coupling partners with masked boron functionality (**2a**/**2b**) and subsequent demasking, generates the target molecule **3** highly selectively, in high yield (79%) (after M. Tobisu, N. Chatani, *Angew. Chem. Int. Ed.* **2009**, *48*, 3565; C. Wang, F. Glorius, *Angew. Chem. Int. Ed.* **2009**, *48*, 5240).

Exercise 11.2
Via oxidative addition and transmetallation, **3** is formed as the central intermediate, undergoing reductive C–C elimination to form the coupling product **1** (isopropylbenzene) and the catalyst complex [Pd(bpy)]. By contrast, a β-H elimination leads to styrene (**2**) and [PdMe(H)(bpy)], which, with reductive C–H elimination, decomposes into methane and [Pd(bpy)]. Addition of fumaronitrile now blocks a coordination site on the Pd, thus hindering the β-H elimination. Furthermore, it is assumed that the electron-deficient character of the double bond decreases the electron density at the metal and favors the reductive C–C elimination (after [263]). The special electronic properties of olefin ligands have led to initial applications for chiral olefins as steering ligands in asymmetric catalysis (C. Defieber, H. Grützmacher, E. M. Carreira, *Angew. Chem. Int. Ed.* **2008**, *47*, 4482).

Exercise 11.3

The critical reaction step is the base-assisted deprotonation of an alkyne activated by coordination. Electron-withdrawing substituents in the alkyne Ar′–C≡C–H facilitate deprotonation, so the (direct) "anionic" path (**25** → **26**) is followed. Electron-donating substituents hamper proton cleavage, so further activation (increasing the acidity!) by forming a cationic complex **25′** via ligand substitution is required. Calculations on a model system show that – in accord with qualitative considerations – the cationic intermediate **25′** with Ar′ = p-Me$_2$NC$_6$H$_4$ is energetically favored over the anionic intermediate **26**, while the opposite is true for Ar′ = p-O$_2$NC$_6$H$_4$. Experimental studies using the Hammett equation show that, depending on the substituents in Ar′, two different mechanisms may occur (after T. Ljungdahl, T. Bennur, A. Dallas, H. Emtenäs, J. Mårtensson, *Organometallics* **2008**, *27*, 2490; see also H. Plenio, *Angew. Chem. Int. Ed.* **2008**, *47*, 6954).

Exercise 11.4

NiCl$_2$ reacts with the Grignard compound to form an alkylnickel(II) compound, which decomposes with formation of Ni^0, which, as we know, reacts with butadiene with oxidative coupling to form a bis(π-allyl)nickel(II) complex **1**. This reacts with the Grignard compound to form an anionic alkylnickelate(II) complex **2**, which reacts with R′X to form the coupling product **3** with regeneration of **1**. This is not common for cross-coupling reactions, since the reaction with RMgX occurs first, followed by the reaction with R′X! The reaction **2** → **3** could run as oxidative addition (more precisely: nucleophilic substitution of X$^-$ by Ni) followed by reductive C–C elimination (**2** → **4** → **3**). However, a direct nucleophilic substitution with participation of the Lewis acid XMg$^+$ could also occur as depicted in **4′**. Although the mechanism has not been demonstrated in detail, it does explain important experimental findings: The high nucleophilicity of Ni in the anionic alkyl complex **2** accelerates the subsequent reaction, and β-H eliminations are suppressed by the substantial coordinative saturation in **2** (after J. Terao, N. Kambe, *Acc. Chem. Res.* **2008**, *41*, 1545).

Exercise 11.5

a) Hydroboration of **1** with 9-BBN (9-borabicyclo[3.3.1]nonane) yields **2′**, which, in a carbonylative Suzuki coupling, reacts to form **2** (after T. Ishiyama, N. Miyaura, A. Suzuki, *Bull. Chem. Jpn.* **1991**, *64*, 1999).

b) The "classical" Stille coupling using benzoyl chlorides XC₆H₄COCl (X = OH, COOH, NH₂, …) cannot be performed, since these are not accessible. By contrast, carbonylative Stille coupling yields **3** (after F. Karimi, J. Barletta, B. Långstöm, *Eur. J. Org. Chem.* **2005**, 2374).

Exercise 11.6

- Oxidative addition of **61** to Pd⁰ leads to **61'** and subsequent olefin insertion forms **61''**. The wedge projection **61'** makes clear that the Pd–C unit attacks the olefinic double bond from behind (**61'** → **61''**), thus forming *cis*-decalin. Elimination of β-H' (**61''** → **62**) is favored by the formation of a stable conjugated double-bond system (after A. Kojima, C. D. J. Boden, M. Shibasaki, *Tetrahedron Lett.* **1997**, *38*, 3459).

- **63** reacts with the catalyst complex to form **A'** via oxidative addition. It can thus be assumed that the precursor complex **A** for the insertion of the double bond into the Pd–C_Ar bond is formed from **A'**.
  A is a cationic square-planar palladium(II) complex with a σ-bound aryl ligand and an olefin ligand. It lies in the plane of the complex ("in-plane"), thus fulfilling a precondition for a facile insertion reaction, namely a coplanar M–C⌒C–C arrangement. C–C bond formation C1–C2' and C1–C5' is ruled out due to excessive strain for in-plane olefin coordination. Formation of the C1–C6' bond leads to product **64**, while formation of a C1–C1' bond would lead to product **B**, but it has not been found experimentally. The geometrical structure of the coordination pocket, and thus the chiral ligand, is responsible for discriminating between these two reaction paths (after K. Kondo, M. Sodeoka, M. Shibasaki, *Tetrahedron: Asymmetry* **1995**, *6*, 2453).

Exercise 11.7
Nucleophilic attack on the C2 atom of the allyl ligand leads to a palladacyclobutane complex **1'**, which decomposes in the presence of CO, with reductive elimination to form the corresponding cyclopropane (after H. M. R. Hoffmann, A. R. Otte, A. Wilde, S. Menzer, D. J. Williams, *Angew. Chem. Int. Ed. Engl.* **1995**, *34*, 100; see also [288]).

Exercise 11.8
- The allyl carbonate **3** (or allyl acetate) synthesized by conventional methods reacts with [Pd$_2$(dba)$_3$]/dppe to form the allylpalladium compound **4**, which reacts in a Tsuji-Trost reaction with 2-amino-6-chloropurine (**5**) to form **1** and **1'**. Fortunately, the reaction to form **1'** is reversible, so eventually the desired isomer **1** is obtained with high regioselectivity (97%) (after R. Freer, G. R. Geen, T. W. Ramsay, A. C. Share, G. R. Slater, N. M. Smith, *Tetrahedron* **2000**, *56*, 4589).

- Reaction of cyclohexanone with ClMg{N(*i*-Pr)$_2$} yields the prochiral ketone enolate **8**, which reacts with the allyl acetate **9** in the presence of the chiral palladium catalyst **10** in a Tsuji-Trost reaction to form **7**. The reaction is diastereoselective (*de* = 98%) and enantioselective (*ee* = 99%) [292].

Exercise 12.1
If the same mechanism as for hydrocyanation of olefins applies, insertion of an alkyne into an Ni–H bond is followed by reductive elimination of a vinyl cyanide. Both reactions proceed consistently from a stereochemical point of view as a *syn* insertion and *syn* elimination (after J. Podlech in *Science of Synthesis*, Vol. 19, Thieme, Stuttgart 2004; p. 325, [301].

Exercise 12.2
The formation of aminonitriles is catalyzed by a Brønsted acid; *ee* values of up to 99% (toluene, −40 °C) are achieved. The postulated catalytic cycle involves protonation of the aldimine by the chiral (enantiomerically pure) BINOL phosphate **1** to form an iminium cation, which forms a chiral contact ion pair **2** with deprotonated **1**. The subsequent addition of HCN then produces the corresponding aminonitrile enantioselectively with regeneration of **1**.

$$\underset{R'H}{\overset{R}{N}\!=\!C} \;\xrightarrow{+\,\mathbf{1}}\; \mathbf{2} \;\xrightarrow[-\,\mathbf{1}]{+\,HCN}\; \underset{R'H}{\overset{H\;\;R}{N-C^{*}\!-\!CN}}$$

One of the two faces (*Re* vs. *Si*) of the prochiral iminium cation in the chiral ion pair **2** is better shielded by **1**, so the nucleophilic cyanide preferentially attacks from the other side. This is a precondition for an enantioselective reaction, which is fulfilled particularly well when there are bulky *ortho* substituents on **1** (Ar = 9-phenanthryl). Since the stereochemical information is transferred through a contact ion pair, nonpolar aprotic solvents are appropriate. In accord, when the reaction occurs in acetonitrile and tetrahydrofuran, only racemic mixtures are obtained (after M. Rueping, E. Sugiono, S. A. Moreth, *Adv. Synth. Catal.* **2007**, *349*, 759). Combining enantioselective Brønsted acids (organocatalysts) as described above with metal catalysts can lead to catalyst systems which comprise two well-differentiated and parallel catalytic cycles ("cooperative catalysis"); see P. de Armas, D. Tejedor, F. García-Tellado, *Angew. Chem. Int. Ed.* **2010**, *49*, 1013.

Exercise 12.3
The reaction enthalpy $\Delta H^{\ominus}$ can be estimated from the difference between the enthalpies of the formed and broken bonds: $\Delta H^{\ominus} = (-311 - 412 + 318 + 612 - 348)$ kJ/mol $= -141$ kJ/mol. A radical chain reaction would run according to the following reaction scheme:

$$R_3Si\cdot \;\xrightarrow{+\,H_2C=CH_2}\; H_2\dot{C}\!-\!CH_2SiR_3 \;\xrightarrow{+\,H-SiR_3}\; H_3C\!-\!CH_2SiR_3 + R_3Si\cdot$$

The R_3Si and 2-silylethyl radicals are reactive enough to cleave the π-C=C bond ($\Delta H^{\ominus} = -47$ kJ/mol) and the Si–H bond ($\Delta H^{\ominus} = -94$ kJ/mol), respectively. Both reactions are exothermic, so hydrosilylation initiated by AIBN or peroxides proceeds according to a radical chain mechanism. For terminal olefins, the chain reaction proceeds via the more stable of the two possible radicals (namely, the secondary C radical), so primary silyl compounds are formed in an anti-Markovnikov reaction (B. Marciniec (ed.): *Comprehensive Handbook of Hydrosilylation*, Pergamon, Oxford 1992).

Exercise 12.4

The silylene complexes **1** and **2** can be illustrated by the two resonance structures **1a/b** or **2a/b** ([Ru] = RuCp*{P(i-Pr)$_3$}). In *cationic* complexes, the silicenium structure (**1b**/**2b**) becomes more important. It can be described as containing an sp^2-hybridized Si atom with a vacant p orbital and is isoelectronic with a borane. Thus, as with hydroboration, direct addition of the Si–H bond across the C=C bond via a four-membered transition state **3** could be possible. Cleavage of the product from the complex by H transfer from the metal to the silicon atom and reductive Si–H elimination (**2** → **4**), as well as oxidative Si–H addition of H$_3$SiPh and an α-H shift (**4** → **5** → **1**), could close the catalytic cycle. The reductive Si–H elimination (**2** → **4**) is probably associated with the oxidative Si–H addition (**4** → **5**), so that **4** does not appear as intermediate (after R. Waterman, P. G. Hayes, T. D. Tilley, *Acc. Chem. Res.* **2007**, *40*, 712). For a silylene–iridium complex that also catalyzes hydrosilylation, see E. Calimano, T. D. Tilley, *J. Am. Chem. Soc.* **2008**, *130*, 9226.

Exercise 12.5

The monomer **1** reacts to form a hyperbranched polymer **2**. In the hydrosilylation, Si–CH=CH–Si groups are formed. It proceeds strictly regioselectively; Si$_2$C=CH$_2$ groups have not been observed. In the polymer, dendritic, linear, and terminal structural units are found (**2a–2c**). **2** is highly soluble and stable against air and moisture, while under the effects of light and heat, further cross-linking occurs via the ethynyl groups. The high reactivity of these groups also allows further functionalization (after M. Häußler, B. Z. Tang, *Adv. Polym. Sci.* **2007**, *209*, 1).

Exercise 12.6

The high stability of α-amino alkyl radicals is the result of (2-orbital–3-electron) π-overlap, which in VB theory is understood as resonance stabilization according to **1a** ↔ **1b**. Aminyl radicals do not experience such a stabilization.

Both the N–H addition (reaction a) and the C–H addition (reaction b) of MeNH$_2$ to ethene are exothermic (−28 and −103, respectively; all values under standard conditions in kJ/mol). The addition of an aminyl radical to ethene (a) is exothermic (−41), but the subsequent H abstraction by the C radical is not (+13). By contrast, in aminomethylation (b), both reactions are exothermic (−84, −19). Thus, a radically initiated aminomethylation of ethene occurs (see D. Steinborn, R. Taube, Z. Chem. 1986, 26, 349).

$$H_2C=CH_2 \xrightarrow{+ \text{MeNH}\cdot} \cdot CH_2-CH_2-NHMe \xrightarrow[- \text{MeNH}\cdot]{+ \text{MeNH}_2} H-CH_2-CH_2-NHMe \quad (a)$$

$$H_2C=CH_2 \xrightarrow{+ \cdot CH_2NH_2} \cdot CH_2-CH_2-CH_2NH_2 \xrightarrow[- \cdot CH_2NH_2]{+ \text{MeNH}_2} H-CH_2-CH_2-CH_2NH_2 \quad (b)$$

Exercise 12.7

With [TiCp$_2$Me$_2$] (1) as precatalyst, the mechanism has been studied in detail. Reaction of 1 with ArNH$_2$ (Ar = 2,6-Me$_2$C$_6$H$_3$, 4-MeC$_6$H$_4$) forms a monocyclopentadienyl–amido–imido complex [Ti(=NAr)(NHAr)Cp] (2, [Ti]=NAr). Notably, the reaction proceeds with protolytic cleavage of a cyclopentadienyl ligand of the complex. If this reaction is carried out in the presence of pyridine, it generates the pyridine adduct [Ti(=NAr)(NHAr)Cp(py)] (3), which under catalytic conditions cleaves off pyridine to become 2. Reaction of 2 with the alkyne leads in a reversible [2 + 2] cycloaddition to form an azatitanacyclobutene complex 4. A protolytic cleavage of the Ti–C bond in 4 by ArNH$_2$ in excess yields an amido complex 5. Hydrogen transfer from the NHAr ligand to the other N atom (compare to carbene complex formation according to R–[M]–CHR′$_2$ → [M]=CR′$_2$ + RH, Section 3.5) cleaves the product from the complex and regenerates the catalyst complex 2 (after F. Pohlki, S. Doye, Chem. Soc. Rev. 2003, 32, 104).

Exercise 13.1

- The ligand substitution is controlled by the *trans* effect H$_2$C=CH$_2$ > Cl$^-$.
- The acidity of water is substantially increased by coordination to an electrophilic Pd^{II} center. [Pd(H$_2$O)$_4$]$^{2+}$ (pK_a < 1) is a very strong acid. In [Pd{H$_2$N(CH$_2$)$_n$NH$_2$}-(H$_2$O)$_2$]$^{2+}$ (n = 2, 3), the electrophilicity of Pd^{II} is reduced by the strongly basic bidentate amine ligand, but nevertheless, these complexes are still weak acids (pK_a about 5.6) (after M. R. Shehata, Trans. Met. Chem. 2001, 26, 198).
- The decisive factor is the charge of the complex. In the neutral complex, the π back-donation is weaker than in the anionic complex, which facilitates the addition of the nucleophile (water).

Solutions to Exercises

Exercise 13.2

Methoxypalladation of **12′** in a *syn* addition yields the complex (*R*,*R*)-**13** and in an *anti* addition yields the complex (*R*,*S*)-**13**. The "normal" Wacker reaction at low Cl⁻ concentrations now leads to formation of (*R*)-**14** or (*S*)-**14**. By contrast, at high chloride ion concentrations, the reverse of hydroxypalladation occurs. It proceeds with *syn* elimination to form (*R*)-(*E*)-**15**, or with *anti* elimination (i.e., it proceeds as a heterolytic fragmentation) to form (*S*)-(*Z*)-**15**. Note that in one case, the (*E*)-isomer is formed and in the other case, the (*Z*)-isomer is formed; analysis of the reaction with a virtual 3D model may make this easier to understand.

a) Complex charges are not considered in the scheme.

(*R*)-**14** has been found experimentally at low chloride ion concentrations, and (*S*)-(*Z*)-**15** at high chloride concentrations. This demonstrates that under the conditions of the Wacker process an intramolecular insertion of ethene into the Pd–OH bond is very likely. The full discussion can be found in [351], see also [346].

Exercise 13.3

After coordination of the olefin to Pd^{II}, aminopalladation (**1** → **2**) occurs, either by addition of the nucleophile to the activated double bond with subsequent deprotonation or by formation of an amidopalladium complex, followed by insertion of the double bond into the Pd–N bond. After this, β-hydride elimination takes place, which can lead to an enamide or an allyl amide. The latter is formed (**2** → **3**). Deprotonation of the palladium hydride species yields Pd⁰, which, notably, is oxidized in DMSO by

O_2 (1 bar) without an additional reoxidant to form Pd^{II} (after A. Minatti, K. Muñiz, *Chem. Soc. Rev.* **2007**, *36*, 1142).

Exercise 13.4

Terminal oxo ligands O^{2-} are both comparatively hard Lewis bases and strong π donors; accordingly, they form particularly strong bonds to high-valent early (hard) transition-metal ions. In corresponding complexes, electrons can be delocalized from oxygen to the empty metal d orbitals. As we move from the early to the late transition metals, the *d* orbitals fill with valence electrons, which repulse the oxo ligands. Thus, in complexes with more than two d electrons (as for oxidation states smaller than + IV), oxo ligands are usually found in bridging positions between two or more metal centers. Consequently, only a few stable (isolable) d^4 M=O complexes exist.

The first structurally characterized Fe^{IV}=O complex – an oxidation state that already is considered highly oxidizing for iron – is *trans*-[Fe(O)(TMC)(MeCN)](OTf)$_2$ (**1**; TMC = 1,4,8,11-tetramethyl-1,4,8,11-tetraazacyclotetradecane). In the meantime, a d^6 Pt^{IV}=O complex has been isolated (**2**) that does not even contain coligands with pronounced acceptor properties, which had previously been considered indispensable for stabilizing such complexes (after C. Limberg, *Angew. Chem. Int. Ed.* **2009**, *48*, 2270, cited in part).

There are even oxoiron(V) complexes. Thus, in the reaction of $[Fe^{II}(bqen)]^{2+}$ (bqen = N,N'-dimethyl-N,N'-bis(8-quinolyl)ethane-1,2-diamine, a neutral N_4 ligand) with peracetic acid under catalytic conditions (C–H activation of alkanes and alcohols), it is very likely that the non-heme oxoiron(V) complex $[Fe^V(O)(bqen)]^{3+}$ appears as intermediate, and reacts by accepting an electron to form the iron(IV) complex $[Fe^{IV}(O)(bqen)]^{2+}$ (J. Yoon, S. A. Wilson, Y. K. Jang, M. S. Seo, K. Nehru, B. Hedman, K. O. Hodgson, E. Bill, E. I. Solomon, W. Nam, *Angew. Chem. Int. Ed.* **2009**, *48*, 1257).

Exercise 13.5

- The reaction is carried out in the presence of a base (K_2HPO_4) and benzoquinone in order to accelerate the transmetallation/reductive elimination. Of the two *ortho*-C–H bonds, the sterically less shielded one is metallated. For the C–H activation, κ^1 coordination of the carboxylate to the Pd^{II} seems to be essential, since it alone can bring the Pd into proximity with the *o*-C–H bond. Co-coordination of K^+ hinders the formation of the more stable κ^2 coordination of the carboxylate to Pd^{II}.

Solutions to Exercises

[Reaction scheme: benzoic acid derivative + K[BF₃Ph] with Pd(OAc)₂ (10 mol%), air/O₂, t-BuOH, 100 °C → biaryl product; 52% (air, 1 bar, 72 h); 65% (O₂, 1 bar, 72 h); 83% (air, 20 bar, 24 h)]

The mechanism corresponds to that discussed previously (80 → ... → 83; Section 13.3.2.1), which is based on a change in oxidation state Pd^0/Pd^{II}. It cannot be strictly ruled out that first the transmetallation between Pd^{II} and K[BF₃Ph] takes place, and then the C–H activation. O_2 is used as oxidant for Pd^0 (after D.-H. Wang, T.-S. Mei, J.-Q. Yu, *J. Am. Chem. Soc.* **2008**, *130*, 17 676; [379]).

- A possible mechanism starts with orthometallation of the acetanilide 1 (1 → 4, R = Me, Et), followed by a *meta*-C–H activation of 2. Further experiments indicate that steric effects seem to be decisive for the markedly high regioselectivity in C–H activation of 2 and that a proton abstraction pathway as indicated in 5 may be applicable (4 → 5 → 6). Reductive C–C elimination (6 → 3) and copper-catalyzed reoxidation of the Pd^0 formed complete the catalytic cycle (after B.-J. Li, S.-L. Tian, Z. Fang, Z.-J. Shi, *Angew. Chem. Int. Ed.* **2008**, *47*, 1115).

[Catalytic cycle scheme showing intermediates 1, 4, 2, 5, 6, 3 with Pd species, + [Pd(O₂CR)₂], − RCOOH, + H, − RCOOH, − Pd⁰]

Exercise 14.1

The enthalpies given in the exercise lead to the accompanying diagram (all values in kJ/mol). The reaction of the formation of 2 TiN from Ti_2N_2 is exothermic (−1112 kJ/mol). From the presented experimental findings, it can be concluded that the activation barrier for the formation of solid titanium nitride from crystalline titanium and N_2 comes from the thermal energy required to form Ti_2 or other small titanium clusters. Thermal cleavage of the N–N bond in N_2 ($\Delta_d H^{\ominus} = 945$ kJ/mol), which requires fairly close to double the energy of atomic Ti from $Ti_{(s)}$

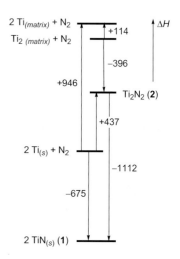

($\Delta_{at}H^{\ominus} = 473$ kJ/mol), is thus not a precondition for the formation of TiN. The high reactivity of the matrix-isolated Ti$_2$N$_2$ (2) is significant in that, preserving the Ti$_2$N$_2$ unit, it reacts with N$_2$ to form [{Ti(η^1-N$_2$)$_4$}$_2$(μ-N)$_2$(μ-η^2:η^2-N$_2$)] (T about 9 K). The vibrational frequency $v_{NN} = 1407$ cm^{-1} shows that the μ-η^2:η^2-N$_2$ ligand should be seen as an N$_2^{2-}$ ligand, so formally a Ti^{IV} complex is present (after H.-J. Himmel, M. Reiher, *Angew. Chem. Int. Ed.* **2006**, *45*, 6264; L. Manceron, O. Hübner, H.-J. Himmel, *Eur. J. Inorg. Chem.* **2009**, 595).

Exercise 14.2

If the M $\cdots$ M axis is considered the z-axis, the d_{z^2} and $d_{x^2-y^2}$ orbitals are involved in the σ-M–N and σ-M–L bonds. Thus, the d_{xy}, d_{xz}, and d_{yz} orbitals still remain to be considered.

By linear combination of the two d_{yz} orbitals of M with the two p_y orbitals of nitrogen (or with the π_{yz}/π^*_{yz} orbitals of N$_2$), the four-center molecular orbitals indicated in the diagram are obtained. Since, in addition to the orbitals shown, there exists an equivalent set of π molecular orbitals, formed from the d_{xz} orbitals of M and the p_x orbitals of nitrogen (or the π_{xz}/π^*_{xz} orbitals of N$_2$), the energy levels are degenerate (1e–4e). Furthermore, the d_{xy} orbitals of M form δ bonds (1b/2b), although these scarcely contribute to the bonding. For energetic reasons, the 1e and 4e levels correspond largely to the bonding and antibonding π-MOs, respectively, of the N$_2$ molecule. The 2e and 3e levels possess mainly metal character, as do the 1b and 2b levels.

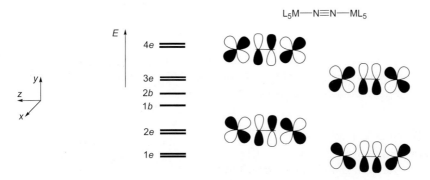

The strength of the M–N and N–N bonds now chiefly depends on the occupancy of the e levels (see the schematic MO diagram). The four π electrons of N$_2$ occupy the 1e level, which is bonding for the N–N bond as well as the M–N bonds. Due to its partial metal character, however, a weaker N–N bond results than in the free N$_2$ molecule.

a) Bond analysis for [{Ru(NH$_3$)$_5$}$_2$(μ-N$_2$)]$^{4+}$: This is a Ru^{II} complex, so 12 d electrons need to be entered in the diagram. Thus, the antibonding 2e and bonding 3e levels (with regard to the N–N bond) are occupied. The filling of the 1b/2b levels has almost no effect on the N–N bond strength and is not considered here. Altogether there are thus five bonding filled orbitals (the bonding character of the filled M–N–N–M σ orbital must of course also be taken into account here) as against two antibonding filled orbitals with respect to the N–N bond. In comparison with the mononuclear complex $[Ru(NH_3)_5(N_2)]^{2+}$, the strength of the N–N bond scarcely changes (v_{NN} 2100 vs. 2130 cm^{-1}).

b) Bond analysis for $[\{ReCl(PMe_2Ph)_4\}\{CrCl_3(THF)_2\}(\mu\text{-}N_2)]$ (v_{NN} = 1875 cm^{-1}) and $[\{ReCl(PMe_2Ph)_4\}(TaCl_5)(\mu\text{-}N_2)]$ (v_{NN} = 1695 cm^{-1}): Only 9 or 6 d electrons of the metals are available, respectively, so the 3e level is occupied by only one electron or remains fully unoccupied. In comparison with the mononuclear complex $[\{ReCl(PMe_2Ph)_4\}(N_2)]$ (v_{NN} = 1925 cm^{-1}), this gives a reason for a pronounced weakening of the N–N bond (abbreviated from D. Sellmann, *Angew. Chem. Int. Ed. Engl.* **1974**, *13*, 639; cited in part).

Exercise 14.3

The symbol (*hkl*) indicates the Miller indices of a crystal face or a lattice plane. The indices are based on the reciprocal of the plane intercepts of the axes. The (111) plane intercepts the axes at 1 *a*, 1 *b*, 1 *c*. For the (211) plane, the values are 1/2 *a*, 1 *b*, 1 *c*; for (100) the values are 1 *a*, ∞ *b*, ∞ *c*, and so on. The latter plane intercepts the *b*- and *c*-axes at infinity. Since Miller indices do not distinguish between parallel planes, but only specify the direction of the normals to the planes, the planes drawn below should be considered as representatives of the infinite set of parallel planes. In α-Fe, the most closely packed face is (110), which has two atoms on an area of $\sqrt{2}a^2$ (*a* = 2.8662 Å; *a* – lattice constant, edge length of the cube). This corresponds to surface coverage of about 83%. Simple geometric considerations show that, in contrast, the top layer of the (111) surface is only 34% covered by Fe atoms.

Note: Prepare a virtual 3D model of the structure of α-iron using a common crystal structure visualization program such as DIAMOND. Crystal system: cubic, space group: $Im\bar{3}m$ (No. 229), *a* = 2.8662 Å, atomic parameters: (0, 0, 0).

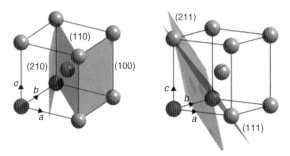

Exercise 14.4

a) According to quantum-chemical calculations on tantalum complexes with a bis(siloxy) ligand, which models the bidentate coordination to the SiO$_2$ surface, a synchronous addition of D$_2$ proceeds via four-membered transition states **29ts** or **29ts'** with formation of a [Ta]$_s$D(NH)(NH$_2$D) or [Ta]$_s$D(NHD)(NH$_2$) complex, respectively. Thus, a heterolytic cleavage of the D–D bond occurs such that D$^+$ adds to the nitrogen atom and D$^-$ to the tantalum center.

b) After precoordination of NH$_3$ to the tantalum(III) complex **28a**, oxidative N–H addition could occur (**28a → 1**). The amidotantalum(V) complex formed undergoes a 1,2-H$_2$ elimination (**1 → 2**; see Section 3.4 for the nomenclature of eliminations), which causes it to pass through a transition state that resembles the one in **29ts'**. Formally (!) this reaction is to be regarded as α-H elimination with coupled reductive H–H elimination; compare this to an analogous reaction, which proceeds with α-C–H elimination (Section 3.5). In molecular chemistry, there is a model reaction for this: [Ta(H)(NH$_2$)(t-Bu$_3$SiO)$_3$] → [Ta(NH)(t-Bu$_3$SiO)$_3$] + H$_2$. NH$_3$ coordination and another 1,2-H$_2$ elimination (compare with the transition state **29ts**) generate **29a**. With excess ammonia, the coordinatively and electronically (10 ve) unsaturated complex **29a** reacts to form an NH$_3$ adduct **3**.

c) According to quantum-chemical calculations, a side-on coordination of N$_2$ occurs, followed by the formation of a diazenido complex via H transfer (**28a → 4 → 5**). Via H$_2$ complex formation, oxidative addition of H$_2$ and another H transfer occur (**5 → 6 → 7 → 8**). For the strongly exergonic N–N bond cleavage **8 → 29a** ($\Delta G = -351$ kJ/mol), only a relatively low activation barrier ($\Delta G^{\ddagger} = 68$ kJ/mol) has been calculated.

The relatively rare (side-on) η^2 coordination of N$_2$ to a *single* metal center in **4** is noteworthy. This η^2 coordination is essential for the subsequent hydrogen transfer, since it is associated with a substantial activation of the N–N bond by an effective $d \to \pi^*$ back-donation (N–N 1.216 Å; compare to 1.166 Å in [Ta]$_s$(η^1-N$_2$)(H)). The η^2 coordination also leads to the fact that, as opposed to the Chatt and Schrock cycles (Sections 14.4.1 and 14.4.2), an NH–NH$_2$ intermediate **8** appears (P. Avenier, A. Lesage, M. Taoufik, A. Baudouin, A. De Mallmann, S. Fiddy, M. Vautier, L. Veyre, J.-M. Basset, L. Emsley, E. A. Quadrelli, *J. Am. Chem. Soc.* **2007**, *129*, 176; J. Li, S. Li, *Angew. Chem. Int. Ed.* **2008**, *47*, 8040; [423]).

Exercise 14.5

- It seems plausible that, due to greater steric demand of the htbt-substituted ligand, the $^{15}N_2$ to be released cannot leave the coordination pocket as easily as in the other complex, and that the coordination pocket is not large enough for the released $^{15}N_2$ *and* the entering $^{14}N_2$.
- In a bimolecular ligand substitution ($^{15}N_2$ vs. $^{14}N_2$ or NH_3 vs. N_2), an intermediate or a transition state [Mo](N$_2$)L (L = N$_2$ or NH$_3$) appears. In the trigonal coordination pocket of Mo, there are three orbitals available for binding ligands, namely d_{z^2} and d_{xz}/d_{yz} (A$_1$ or E, respectively, in C_{3v}; z-axis corresponds to the threefold symmetry axis). Thus, 3σ, $2\sigma + 1\pi$, or $1\sigma + 2\pi$ bonds can be formed. However, an intermediate or transition state with two π-acidic ligands such as [Mo](N$_2$)$_2$ requires four bonds ($2\sigma + 2\pi$). However, if one σ-bonding ligand is involved, as in [Mo](N$_2$)(NH$_3$), these three orbitals are sufficient for binding the two ligands.

The discussions are simplified; other explanations are also possible ([425, 426], R. R. Schrock, *Acc. Chem. Res.* **1997**, *30*, 9).

Exercise 14.6

In all complexes introduced, there is a markedly delocalized π-electron system, so using only a single Lewis structure to describe them may not sufficiently reflect the electron distribution.

Complex **31**: There is a linear M–N–N arrangement, so the α-N atom should be seen as *sp*-hybridized. Resonance formula **31a** reflects none of the substantial π back-donation, but it is overemphasized in **31b/31c**. In **31c** (often shortened to [Mo]–N≡N$^-$), it must be clarified that the nonbonding electron pair on the α-N atom displays π-symmetry, and thus (according to VB theory) is localized in a *p* orbital. This must not be confused with an electron pair of σ-symmetry; see the discussion on the linear and bent imido complexes (see below).

$$\left[[Mo]-N\equiv N| \;\longleftrightarrow\; [Mo]=N=N \;\longleftrightarrow\; [Mo]-\overline{N}=N \right]^-$$
$$\quad\text{31a} \qquad\qquad\qquad \text{31b} \qquad\qquad\qquad \text{31c}$$

If the resonance formula **31a** were to dominate, an anionic N$_2$ complex of Mo^{II} would be present, while **31b** and **31c** represent an Mo^{IV} complex with a deprotonated diazenido ligand. A useful discussion on the oxidation state of metals in complexes with N$_2{}^{n-}$ ligands (especially $n = 3$) can be found in W. Kaim, B. Sarkar, *Angew. Chem. Int. Ed.* **2009**, *48*, 9409.

Complex **33**: As with **31**, a linear M–N–N arrangement is present and thus the nonbonding electron pair on the α-N atom in **33c** (often shortened to [Mo]=N–NH$_2{}^+$) has π-symmetry. The resonance structure **33b** implies that the β-N atom is to be considered (approximately) as sp^2-hybridized.

$$\left[[Mo]\equiv N-\overline{N}H_2 \;\longleftrightarrow\; [\overline{Mo}]=N=NH_2 \;\longleftrightarrow\; [Mo]=\overline{N}-NH_2 \right]^+ \qquad [Mo]=N{\diagdown}_{NH_2}$$
$$\quad\text{33a} \qquad\qquad\qquad \text{33b} \qquad\qquad\qquad \text{33c} \qquad\qquad\qquad \text{33}_{ang}$$

By contrast, the corresponding neutral complexes are bent (Mo–N–N about 144°, calculated) and can be described by a Lewis structure **33**$_{ang}$. The α-N atom, which can be assigned an sp^2-hybridization, thus displays a nonbonding electron pair of σ-symmetry and hence can also be protonated. A useful discussion, which also helps to understand the structural difference between the highly electrophilic cationic Mo^{VI} complex (**33a–c**) and the neutral Mo^V complex (**33**$_{ang}$), can be found in K. Mersmann, K. H. Horn, N. Böres, N. Lehnert, F. Studt, F. Paulat, G. Peters, I. Ivanovic-Burmazovic, R. van Eldik, F. Tuczek, *Inorg. Chem.* **2005**, *44*, 3031.

Complex **37**: A nearly linear Mo–N–H arrangement is present, so an *sp*-hybridized N atom with a nonbonding π-electron pair in **37b** can be assumed. Although the Mo–N bond in **37** is even shorter than in **36** (see the data in Table 14.2, Section 14.4.2), it is often shortened to [Mo]=NH$^+$, following the general convention that in imido complexes a metal–nitrogen double bond is drawn.

$$\left[[\text{Mo}]\equiv\overset{\oplus}{\text{N}}-\text{H} \longleftrightarrow [\text{Mo}]=\overset{\oplus}{\overline{\text{N}}}-\text{H} \right]^+ \qquad \text{M}=\text{N}\diagdown_R$$

37a　　　　**37b**　　　　**37**$_{ang}$

Most imido complexes M=NR display a linear or only weakly bent M–N–C unit (160–180°), and thus correspond to complex **37**. For the less commonly encountered bent imido complexes (M–N–C about 120–140°), Lewis structures according to **37**$_{ang}$ should be given, in which the nitrogen atoms are described as sp^2-hybridized with a nonbonding σ-electron pair. The resonance structure **37b** must never be confused with the Lewis structure **37**$_{ang}$ (see R. A. Eikey, M. M. Abu-Omar, *Coord. Chem. Rev.* **2003**, *243*, 83; Y. Li, W.-T. Wong, *Coord. Chem. Rev.* **2003**, *243*, 191).

Bibliography and Sources

This bibliography includes relevant textbooks and monographs as well as additional literature, preferentially review articles, and recent papers. The titles of the papers are given so that readers can find the aspects of most interest to them.

Textbooks on Organometallic Catalysis and Organometallic Chemistry

[T1] Astruc, D. (2007) *Organometallic Chemistry and Catalysis*, Springer, Berlin.
[T2] Behr, A. (2008) *Angewandte homogene Katalyse*, Wiley-VCH, Weinheim.
[T3] Bhaduri, S. and Mukesh, D. (2000) *Homogeneous Catalysis*, Wiley-Interscience, New York.
[T4] Crabtree, R.H. (2009) *The Organometallic Chemistry of the Transition Metals*, 5th edn, John Wiley & Sons, Hoboken NJ.
[T5] Elschenbroich, C. (2006) *Organometallics*, 3rd edn, Wiley-VCH, Weinheim.
[T6] Hagen, J. (2006) *Industrial Catalysis*, 2nd edn, Wiley-VCH, Weinheim.
[T7] Hartwig, J. (2009) *Organotransition Metal Chemistry: From Bonding to Catalysis*, University Science Books, Sausalito (Calif., USA).
[T8] Hegedus, L.S. and Söderberg, B.C.G. (2009) *Transition Metals in the Synthesis of Complex Organic Molecules*, 3rd edn, University Science Books, Sausalito (Calif., USA).
[T9] van Leeuwen, P.W.N.M. (2004) *Homogeneous Catalysis*, Kluwer, Dordrecht.
[T10] Moulijn, J.A., van Leeuwen, P.W.N.M., and van Santen, R.A. (eds) (1993) *Catalysis, an Integrated Approach to Homogeneous, Heterogeneous and Industrial Catalysis (Studies in Surface Science and Catalysis, Vol. 79)*, Elsevier, Amsterdam.
[T11] Noyori, R. (1994) *Asymmetric Catalysis in Organic Synthesis*, Wiley-Interscience, New York.
[T12] Parshall, G.W. and Ittel, S.D. (1992) *Homogeneous Catalysis*, 2nd edn, Wiley-Interscience, New York.
[T13] Rothenberg, G. (2008) *Catalysis, Concepts and Green Applications*, Wiley-VCH, Weinheim.
[T14] Sheldon, R.A., Arends, I., and Hanefeld, U. (2007) *Green Chemistry and Catalysis*, Wiley-VCH, Weinheim.
[T15] Taube, R. (1988) *Homogene Katalyse*, Akademie-Verlag, Berlin.

Monographs on Organometallic Catalysis and Organometallic Chemistry

[M1] Allen, G. and Bevington, J.C. (eds) (1989) *Comprehensive Polymer Science*, vol. 3–4 (Chain Polymerization), Pergamon, Oxford.
[M2] Beller, M. and Bolm, C. (eds) (2004) *Transition Metals for Organic Synthesis*,

vol. 1–2, 2nd edn, Wiley-VCH, Weinheim.

[M3] Blaser, H.U. and Schmidt, E. (eds) (2003) *Asymmetric Catalysis on Industrial Scale*, Wiley-VCH, Weinheim.

[M4] Cornils, B. and Herrmann, W.A. (eds) (1998) *Aqueous-Phase Organometallic Catalysis*, Wiley-VCH, Weinheim.

[M5] Cornils, B., Herrmann, W.A., Schlögl, R., and Wong, C.-H. (eds) (2003) *Catalysis from A to Z*, 2nd edn, Wiley-VCH, Weinheim.

[M6] Cornils, B. and Herrmann, W.A. (eds) (2002) *Applied Homogeneous Catalysis with Organometallic Compounds*, 2nd edn, vol. 1–3, Wiley-VCH, Weinheim.

[M7] Cornils, B., Herrmann, W.A., Horváth, I.T., Leitner, W., Mecking, S., Olvier-Bourbigou, H., and Vogt, D. (eds) (2005) *Multiphase Homogeneous Catalysis*, vol. 1–2, Wiley-VCH, Weinheim.

[M8] Heaton, B. (ed.) (2005) *Mechanisms in Homogeneous Catalysis*, Wiley-VCH, Weinheim.

[M9] Horvath, I.T. (ed.) (2003) *Encyclopedia of Catalysis*, vol. 1–6, Wiley-Interscience, Hoboken NJ.

[M10] Maseras, F. and Lledós, A. (eds) (2002) *Computational Modeling of Homogeneous Catalysis*, Kluwer, Dordrecht.

[M11] McCleverty, J.A. and Meyer, T.J. (eds) (2004) *Comprehensive Coordination Chemistry II*, vol. 9 (*Applications of Coordination Chemistry*), Elsevier, Oxford.

[M12] de Meijere, A. and Diederich, F. (eds) (2004) *Metal-Catalyzed Cross-Coupling Reactions*, 2nd edn, vol. 1–2, Wiley-VCH, Weinheim.

[M13] Morokuma, K. and Musaev, D.G. (eds) (2008) *Computational Modeling for Homogeneous and Enzymatic Catalysis*, Wiley-VCH, Weinheim.

[M14] (2006) *Ullmann's Encyclopedia of Industrial Chemistry*, 7th edn, Electronic Release, Wiley-VCH, Weinheim.

[M15] (a) Wilkinson, G., Stone, F.G.A., and Abel, E.W. (eds) (1982) *Comprehensive Organometallic Chemistry*, vol. 1–9, Pergamon, Oxford; (b) Abel, E.W., Stone, F.G.A., and Wilkinson, G. (eds) (1995) *Comprehensive Organometallic Chemistry II*, vol. 1–14, Pergamon/Elsevier, Oxford; (c) Crabtree, R.H. and Mingos, D.M.P. (eds) (2007) *Comprehensive Organometallic Chemistry III*, vol. 1–13, Elsevier, Oxford.

References

1 Mittasch, A. (1936) *Über Katalyse und Katalysatoren in Chemie und Biologie*, Springer, Berlin.

2 Mittasch, A. (1939) *Kurze Geschichte der Katalyse in Praxis und Theorie*, Springer, Berlin.

3 Mittasch, A. (1951) *Döbereiner, Goethe und die Katalyse*, Hippokrates-Verlag, Stuttgart.

4 Ostwald, W. (2003) *Lebenslinien – Eine Selbstbiographie* (ed K. Hansel), Hirzel, Stuttgart/Leipzig.

5 Denmark, S.E. and Beutner, G.L. (2008) Lewis base catalysis in organic synthesis. *Angewandte Chemie – International Edition*, **47** (9), 1560.

6 Berkessel, A. and Gröger, H. (2005) *Asymmetric Organocatalysis*, Wiley-VCH, Weinheim.

7 Dondoni, A. and Massi, A. (2008) Asymmetric organocatalysis: From infancy to adolescence. *Angewandte Chemie – International Edition*, **47**, 4638.

8 Afagh, N.A. and Yudin, A.K. (2010) Chemoselectivity and the curious reactivity preferences of functional groups. *Angewandte Chemie – International Edition*, **49**, 262.

9 Christmann, M. and S. Bräse (eds) (2008) *Asymmetric Synthesis – The Essentials*, 2nd edn, Wiley-VCH, Weinheim.

10 Carey, F.A. and Sundberg, R.J. (2008) *Advanced Organic Chemistry*, 5th edn, Springer, New York.

11 Frenking, G. (ed.) (2005) *Theoretical Aspects of Transition Metal Catalysis* (Topics in Organometallic Chemistry 2005, **12**)

12 Koch, W. and Holthausen, M.C. (2000) *A Chemist's Guide to Density Functional Theory*, Wiley-VCH, Weinheim.

13 Senn, H.M. and Thiel, W. (2009) QM/MM methods for biomolecular systems. *Angewandte Chemie – International Edition*, **48**, 1198.

14 Brown, J.M. and Deeth, R.J. (2009) Is enantioselectivity predictable in asymmetric catalysis? *Angewandte Chemie – International Edition*, **48**, 4476.

15 Maseras, F. and Lledós, A. (2002) Computational methods for homogeneous catalysis, in [M10], p. 1.

16 Bell, A.T. (2004) Challenges for the application of quantum chemical calculations to problems in catalysis. *Molecular Physics*, **102**, 319.

17 Laidler, K.J. (1996) A glossary of terms used in chemical kinetics, including reaction dynamics. *Pure and Applied Chemistry*, **68**, 149.

18 Mikami, K. and Yamanaka, M. (2003) Symmetry breaking in asymmetric catalysis: Racemic catalysis to autocatalysis. *Chemical Reviews*, **103**, 3369.

19 Cruickshank, F.R., Hyde, A.J., and Pugh, D. (1977) Free energy surfaces and transition state theory. *Journal of Chemical Education*, **54**, 288.

20 Espenson, J.H. (1995) *Chemical Kinetics and Reaction Mechanisms*, 2nd edn, McGraw-Hill, New York.

21 Murdoch, J.R. (1981) What is the rate-limiting step of a multistep reaction? *Journal of Chemical Education*, **58**, 32.

22 Soai, K., Shibata, T., and Sato, I. (2000) Enantioselective automultiplication of chiral molecules by asymmetric autocatalysis. *Accounts of Chemical Research*, **33**, 382.

23 Koelle, U. (1992) Transition metal catalyzed proton reduction. *New Journal of Chemistry*, **16**, 157.

24 Steinborn, D. (2004) The concept of oxidation states in metal complexes. *Journal of Chemical Education*, **81**, 1148.

25 Huheey, J.E., Keiter, E.A., and Keiter, R.L. (1993) *Inorganic Chemistry: Principles of Structure and Reactivity*, 4th edn, Harper Collins, New York.

26 Brookhart, M., Green, M.L.H., and Wong, L.-L. (1988) Carbon–hydrogen–transition metal bonds. *Progress in Inorganic Chemistry*, **36**, 1.

27 Scherer, W. and McGrady, G.S. (2004) Agostic interactions in d^0 metal alkyl complexes. *Angewandte Chemie – International Edition*, **43**, 1782.

28 Brookhart, M., Green, M.L.H., and Parkin, G. (2007) Agostic interactions in transition metal compounds. *Proceedings of the National Academy of Sciences of the United States of America*, **104**, 6908.

29 Atwood, J.D. (1997) *Inorganic and Organometallic Reaction Mechanisms*, 2nd edn, VCH, New York.

30 Bartlett, K.L., Goldberg, K.I., and Borden, W.T. (2000) A computational study of reductive elimination reactions to form C–H bonds from Pt(II) and Pt(IV) centers. Why does ligand loss precede reductive elimination from six-coordinate but not four-coordinate platinum? *Journal of the American Chemical Society*, **122**, 1456.

31 Procelewska, J., Zahl, A., Liehr, G., van Eldik, R., Smythe, N.A., Williams, B.S., and Goldberg, K.I. (2005) Mechanistic information on the reductive elimination from cationic trimethylplatinum(IV) complexes to form carbon–carbon bonds. *Inorganic Chemistry*, **44**, 7732 (and references cited therein).

32 Steinborn, D. (2005) The unique chemistry of platina-β-diketones. *Dalton Transactions*, 2664.

33 Grubbs, R.H., Miyashita, A., Liu, M., and Burk, P. (1978) Preparation and reactions of phosphine nickelocyclopentanes. *Journal of the American Chemical Society*, **100**, 2418.

34 O'Connor, J.M., Closson, A., and Gantzel, P. (2002) Hydrotris(pyrazolyl)-borate metallacycles: Conversion of a late-metal metallacyclopentene to a stable metallacyclopentadiene–alkene complex. *Journal of the American Chemical Society*, **124**, 2434.

35 Paquette, L.A. (ed.) (1995) *Encyclopedia of Reagents for Organic Synthesis*, vol. 1–8, John Wiley & Sons, Chichester.

36 Grubbs, R.H. (2004) Olefin metathesis. *Tetrahedron*, **60**, 7117.

37 Romeo, R., Alibrandi, G., and Scolaro, L.M. (1993) Kinetic study of β-hydride elimination from monoalkyl solvento

complexes of platinum(II). *Inorganic Chemistry*, **32**, 4688.
38 Wipf, P. and Jahn, H. (1996) Synthetic applications of organochlorozirconocene complexes. *Tetrahedron*, **52**, 12853.
39 Roesky, H.W. (2004) Hydroalumination reactions in organic chemistry. *Aldrichimica Acta*, **37**, 103.
40 Negishi, E.-i. and Tan, Z. (2004) Diastereoselective, enantioselective, and regioselective carboalumination reactions catalyzed by zirconocene derivatives. *Topics in Organometallic Chemistry*, **8**, 139.
41 Green, J.C. and Jardine, C.N. (2001) Hydrogen shifts in $[W(\eta\text{-}C_5H_5)_2(CH_3)]^+$; a density functional study. *Journal of The Chemical Society-Dalton Transactions*, 274.
42 Fellmann, J.D., Schrock, R.R., and Traficante, D.D. (1982) α-Hydride vs. β-hydride elimination. An example of an equilibrium between two tautomers. *Organometallics*, **1**, 481.
43 Schrock, R.R. (1986) High-oxidation-state molybdenum and tungsten alkylidyne complexes. *Accounts of Chemical Research*, **19**, 342.
44 Grob, C.A. and Schiess, P.W. (1967) Heterolytic fragmentation. A class of organic reactions. *Angewandte Chemie – International Edition in English*, **6**, 1.
45 Grob, C.A. (1969) Mechanisms and stereochemistry of heterolytic fragmentation. *Angewandte Chemie – International Edition in English*, **8**, 535.
46 Steinborn, D. (1992) On the influence of heteroatoms in α- and β-functionalized alkyl transition-metal compounds. *Angewandte Chemie – International Edition in English*, **31**, 401.
47 Eckert, H. and Ugi, I. (1979) Spaltung β-halogenierter Urethane mit Kobalt(I)-phthalocyanin: Eine neue Schutzgruppentechnik für Peptid-Synthesen. *Liebigs Annalen der Chemie*, 278.
48 Taube, R. (1974) New aspects of the chemistry of transition metal phthalocyanines. *Pure and Applied Chemistry*, **38**, 427.
49 Halpern, J. (1981) Mechanistic aspects of homogeneous catalytic hydrogenation and related processes. *Inorganica Chimica Acta*, **50**, 11.
50 Koga, N. and Morokuma, K. (1989) Ab initio molecular orbital studies of intermediates and transition states of organometallic elementary reactions and homogeneous catalytic cycles. *Topics in Physical Organometallic Chemistry*, **3**, 1.
51 Eliel, E.L. (1980) Stereochemical non-equivalence of ligands and faces (heterotopicity). *Journal of Chemical Education*, **57**, 52.
52 Blaser, H.-U., Malan, C., Pugin, B., Spindler, F., Steiner, H., and Studer, M. (2003) Selective hydrogenation for fine chemicals: Recent trends and new developments. *Advanced Synthesis & Catalysis*, **345**, 103.
53 McCulloch, B., Halpern, J., Thompson, M.R., and Landis, C.R. (1990) Catalyst–substrate adducts in asymmetric catalytic hydrogenation. Crystal and molecular structure of [((R,R)-1,2-bis{phenyl-o-anisoylphosphino}ethane)(methyl (Z)-β-propyl-α-acetamidoacrylate)] rhodium tetrafluoroborate, [Rh(DIPAMP)-(MPAA)]BF$_4$. *Organometallics*, **9**, 1392.
54 Drexler, H.-J., Baumann, W., Schmidt, T., Zhang, S., Sun, A., Spannenberg, A., Fischer, C., Buschmann, H., and Heller, D. (2005) Are β-acylaminoacrylates hydrogenated in the same way as α-acylaminoacrylates? *Angewandte Chemie – International Edition*, **44**, 1184.
55 Ohta, T., Takaya, H., Kitamura, M., Nagai, K., and Noyori, R. (1987) Asymmetric hydrogenation of unsaturated carboxylic acids catalyzed by BINAP-ruthenium(II) complexes. *The Journal of Organic Chemistry*, **52**, 3176.
56 Noyori, R. (2002) Asymmetric catalysis: Science and opportunities (Nobel Lecture). *Angewandte Chemie – International Edition*, **41**, 2008.
57 Hofer, R. (2005) In Kaisten, Syngenta operates the world's largest plant in which an enantioselective catalytic hydrogenation is performed. How did this come about? *Chimia*, **59**, 10.
58 Jäkel, C. and Paciello, R. (2006) High-throughput and parallel screening

methods in asymmetric hydrogenation. *Chemical Reviews*, **106**, 2912.

59 Teichert, J.F. and Feringa, B.L. (2010) Phosphoramidites: Privileged ligands in asymmetric catalysis. *Angewandte Chemie – International Edition*, **49**, 2486.

60 Eberhardt, L., Armspach, D., Harrowfield, J., and Matt, D. (2008) BINOL-derived phosphoramidites in asymmetric hydrogenation: Can the presence of a functionality in the amino group influence the catalytic outcome? *Chemical Society Reviews*, **37**, 839.

61 Terrett, N.K. (1998) *Combinatorial Chemistry*, Oxford Univ. Press, New York.

62 Reetz, M.T. (2008) Combinatorial transition-metal catalysis: Mixing monodentate ligands to control enantio-, diastereo-, and regioselectivity. *Angewandte Chemie – International Edition*, **47**, 2556.

63 Maier, W.F. (1999) Combinatorial chemistry – challenge and chance for the development of new catalysts and materials. *Angewandte Chemie – International Edition*, **38**, 1216.

64 Reetz, M.T. (2001) Combinatorial and evolution-based methods in the creation of enantioselective catalysts. *Angewandte Chemie – International Edition*, **40**, 284.

65 Satyanarayana, T., Abraham, S., and Kagan, H.B. (2009) Nonlinear effects in asymmetric catalysis. *Angewandte Chemie – International Edition*, **48**, 456.

66 Seeman, J.I. (1983) Effect of conformational change on reactivity in organic chemistry. Evaluations, applications, and extensions of Curtin–Hammett/Winstein–Holness kinetics. *Chemical Reviews*, **83**, 83.

67 Caddick, S. and Jenkins, K. (1996) Dynamic resolutions in asymmetric synthesis. *Chemical Society Reviews*, **25**, 447.

68 Feldgus, S. and Landis, C.R. (2000) Large-scale computational modeling of [Rh(DuPHOS)]$^+$-catalyzed hydrogenation of prochiral enamides: Reaction pathways and the origin of enantioselection. *Journal of the American Chemical Society*, **122**, 12714.

69 de Vries, J.G. and C.J. Elsevier (eds) (2007) *The Handbook of Homogeneous Hydrogenation*, vol. 1–3, Wiley-VCH, Weinheim.

70 Krische, M.J. and Sun, Y. (eds) (2007) Special issue on hydrogenation and transfer hydrogenation. *Accounts of Chemical Research*, **40** (12), 1237.

71 Peruzzini, M. and R. Poli (eds) (2001) *Recent Advances in Hydride Chemistry*, Elsevier, Amsterdam.

72 Schlaf, M., Lough, A.J., Maltby, P.A., and Morris, R.H. (1996) Synthesis, structure, and properties of the stable and highly acidic dihydrogen complex trans-[Os(η^2-H$_2$)(CH$_3$CN)(dppe)$_2$](BF$_4$)$_2$. Perspectives on the influence of the *trans* ligand on the chemistry of the dihydrogen ligand. *Organometallics*, **15**, 2270.

73 Adams, R.D. and Captain, B. (2008) Hydrogen activation by unsaturated mixed-metal cluster complexes: New directions. *Angewandte Chemie – International Edition*, **47**, 252.

74 Stephan, D.W. and Erker, G. (2010) Frustrated Lewis pairs: Metal-free hydrogen activation and more. *Angewandte Chemie – International Edition*, **49**, 46.

75 Rauchfuss, T.B. (2004) Research on soluble metal sulfides: From polysulfido complexes to functional models for hydrogenases. *Inorganic Chemistry*, **43**, 14.

76 Noyori, R., Yamakawa, M., and Hashiguchi, S. (2001) Metal–ligand bifunctional catalysis: A nonclassical mechanism for asymmetric hydrogen transfer between alcohols and carbonyl compounds. *The Journal of Organic Chemistry*, **66**, 7931.

77 Roelen, O. (1977) Die Entdeckung der Synthese von Aldehyden aus Olefinen, Kohlenoxid und Wasserstoff – ein Beitrag zur Psychologie der naturwissenschaftlichen Forschung. *Chemie, Experiment und Didaktik*, **3**, 119.

78 Solà, M. and Ziegler, T. (1996) Theoretical study on acetaldehyde and ethanol elimination from the hydrogenation of CH$_3$(O)CCo(CO)$_3$. *Organometallics*, **15**, 2611.

79 Orchin, M. (1981) HCo(CO)$_4$, the quintessential catalyst. *Accounts of Chemical Research*, **14**, 259.

80 Borovikov, M.S., Kovács, I., Ungváry, F., Sisak, A., and Markó, L. (1992) Kinetics and equilibrium of the olefin-promoted interconversion of *n*-butyryl- and isobutyrylcobalt tetracarbonyl. The aldehyde isomer ratio in the cobalt-catalyzed olefin hydroformylation. *Organometallics*, **11**, 1576.

81 Bohnen, H.-W. and Cornils, B. (2002) Hydroformylation of alkenes: An industrial view of the status and importance. *Advances in Catalysis*, **47**, 1.

82 Matsubara, T., Koga, N., Ding, Y., Musaev, D.G., and Morokuma, K. (1997) Ab initio MO study of the full cycle of olefin hydroformylation catalyzed by a rhodium complex, $RhH(CO)_2(PH_3)_2$. *Organometallics*, **16**, 1065.

83 Nozaki, K., Sakai, N., Nanno, T., Higashijima, T., Mano, S., Horiuchi, T., and Takaya, H. (1997) Highly enantioselective hydroformylation of olefins catalyzed by rhodium(I) complexes of new chiral phosphine–phosphite ligands. *Journal of the American Chemical Society*, **119**, 4413.

84 Gleich, D. and Herrmann, W.A. (1999) Why do many C_2-symmetric bisphosphine ligands fail in asymmetric hydroformylation? Theory in front of experiment. *Organometallics*, **18**, 4354.

85 Agbossou, F., Carpentier, J.-F., and Mortreux, A. (1995) Asymmetric hydroformylation. *Chemical Reviews*, **95**, 2485.

86 Diéguez, M., Pàmies, O., and Claver, C. (2004) Recent advances in Rh-catalyzed asymmetric hydroformylation using phosphite ligands. *Tetrahedron: Asymmetry*, **15**, 2113.

87 Freixa, Z. and van Leeuwen, P.W.N.M. (2003) Bite angle effects in diphosphine metal catalysts: Steric or electronic? *Dalton Transactions*, 1890.

88 Carbó, J.J., Maseras, F., Bo, C., and van Leeuwen, P.W.N.M. (2001) Unraveling the origin of regioselectivity in rhodium diphosphine catalyzed hydroformylation. A DFT QM/MM study. *Journal of the American Chemical Society*, **123**, 7630.

89 van Leeuwen, P.W.N.M., Kamer, P.C.J., Reek, J.N.H., and Dierkes, P. (2000) Ligand bite angle effects in metal-catalyzed C–C bond formation. *Chemical Reviews*, **100**, 2741.

90 Kamer, P.C.J., van Leeuwen, P.W.N.M., and Reek, J.N.H. (2001) Wide bite angle diphosphines: Xantphos ligands in transition metal complexes and catalysis. *Accounts of Chemical Research*, **34**, 895.

91 Mehnert, C.P. (2005) Supported ionic liquid catalysis. *Chemistry – A European Journal*, **11**, 50.

92 Haumann, M. and Riisager, A. (2008) Hydroformylation in room temperature ionic liquids (RTILs): Catalyst and process developments. *Chemical Reviews*, **108**, 1474.

93 Wilke, J.S. (2004) Properties of ionic liquid solvents for catalysis. *Journal of Molecular Catalysis A*, **214**, 11.

94 Rogers, R.D. and Voth, G.A. (eds) (2007) Special issue on ionic liquids. *Accounts of Chemical Research*, **40** (11), 1077.

95 Weingärtner, H. (2008) Understanding ionic liquids at the molecular level: Facts, problems, and controversies. *Angewandte Chemie – International Edition*, **47**, 654.

96 Giernoth, R. (2010) Task-specific ionic liquids. *Angewandte Chemie – International Edition*, **49**, 2834.

97 Waloch, C., Wieland, J., Keller, M., and Breit, B. (2007) Self-assembly of bidentate ligands for combinatorial homogeneous catalysis: Methanol-stable platforms analogous to the adenine–thymine base pair. *Angewandte Chemie – International Edition*, **46**, 3037.

98 Šmejkal, T. and Breit, B. (2008) A supramolecular catalyst for regioselective hydroformylation of unsaturated carboxylic acids. *Angewandte Chemie – International Edition*, **47**, 311.

99 Hejl, A., Trnka, T.M., Day, M.W., and Grubbs, R.H. (2002) Terminal ruthenium carbido complexes as σ-donor ligands. *Chemical Communications*, 2524.

100 Overett, M.J., Hill, R.O., and Moss, J.R. (2000) Organometallic chemistry and surface science: Mechanistic models for the Fischer–Tropsch synthesis. *Coordination Chemistry Reviews*, **206–207**, 581.

101 Schulz, H. (2003) Major and minor reactions in Fischer–Tropsch synthesis on cobalt catalysts. *Topics in Catalysis*, **26**, 73.

102 Maitlis, P.M. (2004) Fischer–Tropsch, organometallics, and other friends. *Journal of Organometallic Chemistry*, **689**, 4366.

103 Whyman, R., Wright, A.P., Iggo, J.A., and Heaton, B.T. (2002) Carbon monoxide activation in homogeneously catalysed reactions: The nature and roles of catalytic promoters. *Journal of The Chemical Society-Dalton Transactions*, 771.

104 Jones, J.H. (2000) The Cativa™ process for the manufacture of acetic acid. *Platinum Metals Review*, **44**, 94.

105 Maitlis, P.M., Haynes, A., Sunley, G.J., and Howard, M.J. (1996) Methanol carbonylation revisited: Thirty years on. *Journal of The Chemical Society-Dalton Transactions*, 2187.

106 Thomas, C.M. and Süss-Fink, G. (2003) Ligand effects in the rhodium-catalyzed carbonylation of methanol. *Coordination Chemistry Reviews*, **243**, 125.

107 Cheong, M. and Ziegler, T. (2005) Density functional study of the oxidative addition step in the carbonylation of methanol catalyzed by $[M(CO)_2I_2]^-$ (M=Rh, Ir). *Organometallics*, **24**, 3053.

108 Ellis, P.R., Pearson, J.M., Haynes, A., Adams, H., Bailey, N.A., and Maitlis, P.M. (1994) Oxidative addition of alkyl halides to rhodium(I) and iridium(I) dicarbonyl diiodides: Key reactions in the catalytic carbonylation of alcohols. *Organometallics*, **13**, 3215.

109 Haynes, A., Maitlis, P.M., Morris, G.E., Sunley, G.J., Adams, H., Badger, P.W., Bowers, C.M., Cook, D.B., Elliott, P.I.P., Ghaffar, T., Green, H., Griffin, T.R., Payne, M., Pearson, J.M., Taylor, M.J., Vickers, P.W., and Watt, R.J. (2004) Promotion of iridium-catalyzed methanol carbonylation: Mechanistic studies of the Cativa process. *Journal of the American Chemical Society*, **126**, 2847.

110 Ford, P.C. and Rokicki, A. (1988) Nucleophilic activation of carbon monoxide: Applications to homogeneous catalysis by metal carbonyls of the water gas shift and related reactions. *Advances in Organometallic Chemistry*, **28**, 139.

111 Evans, D.J. (2005) Chemistry relating to the nickel enzymes CODH and ACS. *Coordination Chemistry Reviews*, **249**, 1582.

112 Lindahl, P.A. (2008) Implications of a carboxylate-bound C-cluster structure of carbon monoxide dehydrogenase. *Angewandte Chemie – International Edition*, **47**, 4054.

113 Siebeneicher, H. and Doye, S. (2000) Dimethyltitanocene Cp_2TiMe_2: A useful reagent for C−C and C−N bond formation. *Journal für Praktische Chemie*, **342**, 102.

114 Grubbs, R.H. (2006) Olefin-metathesis catalysts for the preparation of molecules and materials (Nobel Lecture). *Angewandte Chemie – International Edition*, **45**, 3760.

115 Schrock, R.R. (2006) Multiple metal-carbon bonds for catalytic metathesis reactions (Nobel Lecture). *Angewandte Chemie – International Edition*, **45**, 3748.

116 Burtscher, D. and Grela, K. (2009) Aqueous olefin metathesis. *Angewandte Chemie – International Edition*, **48**, 442.

117 Dinger, M.B. and Mol, J.C. (2002) High turnover numbers with ruthenium-based metathesis catalysts. *Advanced Synthesis & Catalysis*, **344**, 671.

118 Donohoe, T.J., O'Riordan, T.J.C., and Rosa, C.P. (2009) Ruthenium-catalyzed isomerization of terminal olefins: Applications to synthesis. *Angewandte Chemie – International Edition*, **48**, 1014.

119 Herrmann, W.A., Weskamp, T., and Böhm, V.P.W. (2001) Metal complexes of stable carbenes. *Advances in Organometallic Chemistry*, **48**, 1.

120 Canac, Y., Soleilhavoup, M., Conejero, S., and Bertrand, G. (2004) Stable non-N-heterocyclic carbenes (non-NHC): Recent progress. *Journal of Organometallic Chemistry*, **689**, 3857.

121 Hahn, F.E. and Jahnke, M.C. (2008) Heterocyclic carbenes: Synthesis and coordination chemistry. *Angewandte Chemie – International Edition*, **47**, 3122.

122 Díez-González, S., Marion, N., and Nolan, S.P. (2009) N-Heterocyclic carbenes in late transition metal catalysis. *Chemical Reviews*, **109**, 3612.

123 Dröge, T. and Glorius, F. (2010) The measure of all rings – N-Heterocyclic carbenes. *Angewandte Chemie – International Edition*, **49**, 6940.

124 Arduengo, A.J. and Bertrand, G. (2009) (eds) Special issue on carbenes. *Chemical Reviews*, **109** (8), 3209.

125 Sanford, M.S., Love, J.A., and Grubbs, R.H. (2001) Mechanism and activity of ruthenium olefin metathesis catalysts. *Journal of the American Chemical Society*, **123**, 6543.

126 Vyboishchikov, S.F., Bühl, M., and Thiel, W. (2002) Mechanism of olefin metathesis with catalysis by ruthenium carbene complexes: Density functional studies on model systems. *Chemistry – A European Journal*, **8**, 3962.

127 Straub, B.F. (2005) Origin of the high activity of second-generation Grubbs catalysts. *Angewandte Chemie – International Edition*, **44**, 5974.

128 Mol, J.C. (2004) Industrial applications of olefin metathesis. *Journal of Molecular Catalysis A*, **213**, 39.

129 Wallace, D.J. (2005) Relay ring-closing metathesis – A strategy for achieving reactivity and selectivity in metathesis chemistry. *Angewandte Chemie – International Edition*, **44**, 1912.

130 Hoveyda, A.H. and Schrock, R.R. (2001) Catalytic asymmetric olefin metathesis. *Chemistry – A European Journal*, **7**, 945.

131 Goumans, T.P.M., Ehlers, A.W., and Lammertsma, K. (2005) The asymmetric Schrock olefin metathesis catalyst. A computational study. *Organometallics*, **24**, 3200.

132 Klare, H.F.T. and Oestreich, M. (2009) Asymmetric ring-closing metathesis with a twist. *Angewandte Chemie – International Edition*, **48**, 2085.

133 Hoveyda, A.H., Malcolmson, S.J., Meek, S.J., and Zhugralin, A.R. (2010) Catalytic enantioselective olefin metathesis in natural product synthesis. Chiral metal-based complexes that deliver high enantioselectivity and more. *Angewandte Chemie – International Edition*, **49**, 34.

134 Zhang, W. and Moore, J.S. (2007) Alkyne metathesis: Catalysts and synthetic applications. *Advanced Synthesis & Catalysis*, **349**, 93.

135 Schrock, R.R. and Czekelius, C. (2007) Recent advances in the syntheses and applications of molybdenum und tungsten alkylidene and alkylidyne catalysts for the metathesis of alkenes and alkynes. *Advanced Synthesis & Catalysis*, **349**, 55.

136 Fürstner, A. and Davies, P.W. (2005) Alkyne metathesis. *Chemical Communications*, 2307.

137 Bunz, U.H.F. and Kloppenburg, L. (1999) Alkyne metathesis as a new synthetic tool: Ring-closing, ring-opening, and acyclic. *Angewandte Chemie – International Edition*, **38**, 478.

138 Bunz, U.H.F. (2001) Poly(*p*-phenyleneethynylene)s by alkyne metathesis. *Accounts of Chemical Research*, **34**, 998.

139 Diver, S.T. (2007) Ruthenium vinyl carbene intermediates in enyne metathesis. *Coordination Chemistry Reviews*, **251**, 671.

140 Lippstreu, J.J. and Straub, B.F. (2005) Mechanism of enyne metathesis catalyzed by grubbs ruthenium–carbene complexes: A DFT study. *Journal of the American Chemical Society*, **127**, 7444.

141 Basset, J.M., Copéret, C., Lefort, L., Maunders, B.M., Maury, O., Le Roux, E., Saggio, G., Soignier, S., Soulivong, D., Sunley, G.J., Taoufik, M., and Thivolle-Cazat, J. (2005) Primary products and mechanistic considerations in alkane metathesis. *Journal of the American Chemical Society*, **127**, 8604.

142 Perutz, R.N. and Sabo-Etienne, S. (2007) The σ-CAM mechanism: σ-Complexes as the basis of σ-bond metathesis at late-transition-metal centers. *Angewandte Chemie – International Edition*, **46**, 2578.

143 Copéret, C., Chabanas, M., Saint-Arroman, R.P., and Basset, J.-M. (2003) Homogeneous and heterogeneous catalysis: Bridging the gap through surface organometallic chemistry.

144 Le Roux, E., Taoufik, M., Copéret, C., de Mallmann, A., Thivolle-Cazat, J., Basset, J.-M., Maunders, B.M., and Sunley, G.J. (2005) Development of tungsten-based heterogeneous alkane metathesis catalysts through a structure–activity relationship. *Angewandte Chemie – International Edition*, **44**, 6755.

145 Vidal, V., Théolier, A., Thivolle-Cazat, J., and Basset, J.-M. (1997) Metathesis of alkanes catalyzed by silica-supported transition metal hydrides. *Science*, **276**, 99.

146 Basset, J.-M., Copéret, C., Soulivong, D., Taoufik, M., and Thivolle-Cazat, J. (2006) From olefin to alkane metathesis: A historical point of view. *Angewandte Chemie – International Edition*, **45**, 6082.

147 Schinzel, S., Chermette, H., Copéret, C., and Basset, J.-M. (2008) Evaluation of the carbene hydride mechanism in the carbon–carbon bond formation process of alkane metathesis through a DFT study. *Journal of the American Chemical Society*, **130**, 7984.

148 Goldman, A.S., Roy, A.H., Huang, Z., Ahuja, R., Schinski, W., and Brookhart, M. (2006) Catalytic alkane metathesis by tandem alkane dehydrogenation – Olefin metathesis. *Science*, **312**, 257.

149 Blanc, F., Copéret, C., Thivolle-Cazat, J., and Basset, J.-M. (2006) Alkane metathesis catalyzed by a well-defined silica-supported Mo imido alkylidene complex: [($\equiv$SiO)Mo(=NAr)(=CHtBu)-(CH$_2t$Bu)]. *Angewandte Chemie – International Edition*, **45**, 6201.

150 Albrecht, M. and van Koten, G. (2001) Platinum group organometallics based on "Pincer" complexes: Sensors, switches, and catalysts. *Angewandte Chemie – International Edition*, **40**, 3750.

151 van der Boom, M.E. and Milstein, D. (2003) Cyclometalated phosphine-based pincer complexes: Mechanistic insight in catalysis, coordination, and bond activation. *Chemical Reviews*, **103**, 1759.

152 Martin, H. and Bretinger, H. (1992) High-molecular-weight polyethylene: Growth reactions at Bis(dichloroaluminium)ethane and trialkylaluminium. *Die Makromolekulare Chemie*, **193**, 1283.

153 Fischer, K., Jonas, K., Misbach, P., Stabba, R., and Wilke, G. (1973) The "Nickel-Effect". *Angewandte Chemie – International Edition in English*, **12**, 943.

154 Wilke, G. (2003) Fifty years of Ziegler catalysts: Consequences and development of an invention. *Angewandte Chemie – International Edition*, **42**, 5000.

155 Janiak, C. (2006) Metallocene and related catalysts for olefin, alkyne and silane dimerization and oligomerization. *Coordination Chemistry Reviews*, **250**, 66.

156 Bogdanović, B. (1979) Selectivity control in nickel-catalyzed olefin oligomerization. *Advances in Organometallic Chemistry*, **17**, 105.

157 Bogdanović, B., Spliethoff, B., and Wilke, G. (1980) Dimerization of propylene with catalysts exhibiting activities like highly-active enzymes. *Angewandte Chemie – International Edition in English*, **19**, 622.

158 Commereuc, D., Forestière, A., Gaillard, J.F., and Hugues, F. (1997) GMK Tagungsbericht 9705 C$_4$ Chemistry – Manufacture and Use of C$_4$ Hydrocarbons, Aachen 1997, S. 141. Highly selective 1-butene production from ethylene: The IFP-Sabic Alphabutol™ process, in *DGMK Tagungsbericht 9705, C$_4$ Chemistry – Manufacture and Use of C$_4$ Hydrocarbons*, Aachen, p. 141.

159 Dixon, J.T., Green, M.J., Hess, F.M., and Morgan, D.H. (2004) Advances in selective ethylene trimerisation – A critical overview. *Journal of Organometallic Chemistry*, **689**, 3641.

160 Tobisch, S. and Ziegler, T. (2003) Catalytic linear oligomerization of ethylene to higher α-olefins: Insight into the origin of the selective generation of 1-hexene promoted by a cationic cyclopentadienyl-arene titanium active catalyst. *Organometallics*, **22**, 5392.

161 McGuinness, D.S., Brown, D.B., Tooze, R.P., Hess, F.M., Dixon, J.T., and Slawin, A.M.Z. (2006) Ethylene trimerization with Cr−PNP and Cr−SNS complexes: Effect of ligand structure, metal oxidation

state, and role of activator on catalysis. *Organometallics*, **25**, 3605.
162 Wass, D.F. (2007) Chromium-catalysed ethene trimerisation and tetramerisation – Breaking the rules in olefin oligomerisation. *Dalton Transactions*, 816.
163 Slone, C.S., Weinberger, D.A., and Mirkin, C.A. (1999) The transition metal coordination chemistry of hemilabile ligands. *Progress in Inorganic Chemistry*, **48**, 233.
164 Bassetti, M. (2006) Kinetic evaluation of ligand hemilability in transition metal complexes. *European Journal of Inorganic Chemistry*, 4473.
165 Grützmacher, H. (2008) Cooperating ligands in catalysis. *Angewandte Chemie – International Edition*, **47**, 1814.
166 Kuhn, P., Sémeril, D., Matt, D., Chetcuti, M.J., and Lutz, P. (2007) Structure–reactivity relationships in SHOP-type complexes: Tunable catalysts for the oligomerisation and polymerisation of ethylene. *Dalton Transactions*, 515.
167 Bölt, H.V. and Fritz, P.M. (2004) Base stock for plastics and detergents, in *Linde Technology (Reports on Science and Technology)*, Dec. 2004, 38.
168 Fritz, P.M. and Bölt, H.V. (2005) A new process is born. *Process – Worldwide* (1), 26.
169 Ziegler, K., Holzkamp, E., Breil, H., and Martin, H. (1955) Das Mülheimer Normaldruck-Polyäthylen-Verfahren. *Angewandte Chemie*, **67**, 541.
170 Grubbs, R.H. and Coates, G.W. (1996) α-Agostic interactions and olefin insertion in metallocene polymerization catalysts. *Accounts of Chemical Research*, **29**, 85.
171 Szabo, M.J., Berke, H., Weiss, T., and Ziegler, T. (2003) Is the polymerization of linear α-olefins by transition-metal carbene complexes a viable process? A theoretical study based on density functional theory. *Organometallics*, **22**, 3671.
172 Guerra, G., Cavallo, L., Moscardi, G., Vacatello, M., and Corradini, P. (1996) Back-skip of the growing chain at model complexes for the metallocene polymerization catalysis. *Macromolecules*, **29**, 4834.
173 Weckhuysen, B.M. and Schoonheydt, R.A. (1999) Olefin polymerization over supported chromium oxide catalysts. *Catalysis Today*, **51**, 215.
174 Amor Nait Ajjou, J. and Scott, S.L. (2000) A kinetic study of ethylene and 1-hexene homo- and copolymerization catalyzed by a silica-supported Cr(IV) complex: Evidence for propagation by a migratory insertion mechanism. *Journal of the American Chemical Society*, **122**, 8968.
175 Schmid, R. and Ziegler, T. (2000) Ethylene-polymerization by surface supported Cr(IV) species: Possible reaction mechanisms revisited by theoretical calculations. *Canadian Journal of Chemistry*, **78**, 265.
176 Theopold, K.H. (1997) Understanding chromium-based olefin polymerization catalysts. *Chemtech*, **27**, 26.
177 Theopold, K.H. (1998) Homogeneous chromium catalysts for olefin polymerization. *European Journal of Inorganic Chemistry*, 15.
178 Elias, H.-G. (2005–2009) *Macromolecules*, vol. 1–4, Wiley-VCH, Weinheim.
179 Jenkins, A.D. (1981) Stereochemical definitions and notations relating to polymers. *Pure and Applied Chemistry*, **53**, 733.
180 Mattice, W.L. and Helfer, C.A. (2003) Conformation and configuration, in *Encyclopedia of Polymer Science and Technology*, vol. 2 (ed. J.I. Kroschwitz), Wiley-Interscience, Hoboken NJ, p.97.
181 Bochmann, M. (2005) in [M8] p. 311.
182 Randall, J.C. (1987) Microstructure, in *Encyclopedia of Polymer Science and Engineering*, vol. 9 (ed. J.I. Kroschwitz), Wiley-Interscience, New York, p. 795.
183 Corradini, P., Guerra, G., and Cavallo, L. (2004) Do new century catalysts unravel the mechanism of stereocontrol of old Ziegler–Natta catalysts? *Accounts of Chemical Research*, **37**, 231.
184 Fink, G., Mülhaupt, R., and H.H. Brintzinger (eds) (1995) *Ziegler Catalysts*, Springer, Berlin.
185 Brintzinger, H.-H., Fischer, D., Mülhaupt, R., Rieger, B., and Waymouth,

R. (1995) Stereospecific olefin polymerization with chiral metallocene catalysts. *Angewandte Chemie – International Edition in English*, **34**, 1143.

186 Chen, E.Y.-X. and Marks, T.J. (2000) Cocatalysts for metal-catalyzed olefin polymerization: Activators, activation processes, and structure–activity relationships. *Chemical Reviews*, **100**, 1391.

187 Xu, Z., Vanka, K., Firman, T., Michalak, A., Zurek, E., Zhu, C., and Ziegler, T. (2002) Theoretical study of the interactions between cations and anions in group IV transititon-metal catalysts for single-site homogeneous olefin polymerization. *Organometallics*, **21**, 2444.

188 Chan, M.S.W., Vanka, K., Pye, C.C., and Ziegler, T. (1999) Density functional study on activation and ion-pair formation in group IV metallocene and related olefin polymerization catalysts. *Organometallics*, **18**, 4624.

189 Margl, P., Deng, L., and Ziegler, T. (1998) A unified view of ethylene polymerization by d^0 and d^0f^n transition metals. Part 2: Chain propagation. *Journal of the American Chemical Society*, **120**, 5517.

190 Chan, M.S.W. and Ziegler, T. (2000) A combined density functional and molecular dynamics study on ethylene insertion into the $Cp_2ZrEt-MeB(C_6F_5)_3$ ion-pair. *Organometallics*, **19**, 5182.

191 Kaloustian, S.A. and Kaloustian, M.K. (1975) Determining homotopic, enantiotopic, and diastereotopic faces in organic molecules. *Journal of Chemical Education*, **52**, 56.

192 Steinborn, D. (1993) *Symmetrie und Struktur in der Chemie*, VCH, Weinheim.

193 Kaminsky, W. (1998) Highly active metallocene catalysts for olefin polymerization. *Journal of The Chemical Society-Dalton Transactions*, 1413.

194 Coates, G.W. (2000) Precise control of polyolefin stereochemistry using single-site metal catalysts. *Chemical Reviews*, **100**, 1223.

195 Resconi, L., Cavallo, L., Fait, A., and Piemontesi, F. (2000) Selectivity in propene polymerization with metallocene catalysts. *Chemical Reviews*, **100**, 1253.

196 Lin, S. and Waymouth, R.M. (2002) 2-Arylindene metallocenes: Conformationally dynamic catalysts to control the structure and properties of polypropylenes. *Accounts of Chemical Research*, **35**, 765.

197 Busico, V., Van Axel Castelli, V., Aprea, P., Cipullo, R., Segre, A., Talarico, G., and Vacatello, M. (2003) "Oscillating" metallocene catalysts: What stops the oscillation? *Journal of the American Chemical Society*, **125**, 5451.

198 Kaminsky, W. (2001) Olefin polymerization catalyzed by metallocenes. *Advances in Catalysis*, **46**, 89.

199 Britovsek, G.J.P., Gibson, V.C., and Wass, D.F. (1999) The search for new-generation olefin polymerization catalysts: Life beyond metallocenes. *Angewandte Chemie – International Edition*, **38**, 428.

200 Gromada, J., Carpentier, J.-F., and Mortreux, A. (2004) Group 3 metal catalysts for ethylene and α-olefin polymerization. *Coordination Chemistry Reviews*, **248**, 397.

201 Li, H. and Marks, T.J. (2006) Nuclearity and cooperativity effects in binuclear catalysts and cocatalysts for olefin polymerization. *Proceedings of the National Academy of Sciences of the United States of America*, **103**, 15295.

202 Gibson, V.C. and Spitzmesser, S.K. (2003) Advances in non-metallocene olefin polymerization catalysis. *Chemical Reviews*, **103**, 283.

203 Bianchini, C., Giambastiani, G., Guerrero Rios, I., Mantovani, G., Meli, A., and Segarra, A.M. (2006) Ethylene oligomerization, homopolymerization and copolymerization by iron and cobalt catalysts with 2,6-(bis-organylimino)-pyridyl ligands. *Coordination Chemistry Reviews*, **250**, 1391.

204 Gibson, V.C., Redshaw, C., and Solan, G.A. (2007) Bis(imino)pyridines: Surprisingly reactive ligands and a gateway to new families of catalysts. *Chemical Reviews*, **107**, 1745.

205 Mecking, S. (2001) Olefin polymerization by late transition metal complexes – A

root of Ziegler catalysts gains new ground. *Angewandte Chemie – International Edition*, **40**, 534.

206 Michalak, A. and Ziegler, T. (2002) The key steps in olefin polymerization catalyzed by late transition metals, in [M10] p. 57.

207 Rieger, B., Baugh, L.S., Kacker, S. and Striegler, S. (eds) (2003) *Late Transition Metal Polymerization Catalysis*, Wiley-VCH, Weinheim.

208 Coates, G.W., Hustad, P.D., and Reinartz, S. (2002) Catalysts for the living insertion polymerization of alkenes: Access to new polyolefin architectures using Ziegler-Natta chemistry. *Angewandte Chemie – International Edition*, **41**, 2236.

209 Kempe, R. (2007) How to polymerize ethylene in a highly controlled fashion? *Chemistry – A European Journal*, **13**, 2764.

210 Sita, L.R. (2009) Ex uno plures ("out of one, many"): New paradigms for expanding the range of polyolefins through reversible group transfers. *Angewandte Chemie – International Edition*, **48**, 2464.

211 Zintl, M. and Rieger, B. (2007) Novel olefin block copolymers through chain-shuttling polymerization. *Angewandte Chemie – International Edition*, **46**, 333.

212 Rix, F.C., Brookhart, M., and White, P.S. (1996) Mechanistic studies of the palladium(II)-catalyzed copolymerization of ethylene with carbon monoxide. *Journal of the American Chemical Society*, **118**, 4746.

213 Drent, E. and Budzelaar, P.H.M. (1996) Palladium-catalyzed alternating copolymerization of alkenes and carbon monoxide. *Chemical Reviews*, **96**, 663.

214 Bianchini, C. and Meli, A. (2002) Alternating copolymerization of carbon monoxide and olefins by single-site metal catalysis. *Coordination Chemistry Reviews*, **225**, 35.

215 Sen, A. (1993) Mechanistic aspects of metal-catalyzed alternating copolymerization of olefins with carbon monoxide. *Accounts of Chemical Research*, **26**, 303.

216 Shultz, C.S., Ledford, J., DeSimone, J.M., and Brookhart, M. (2000) Kinetic studies of migratory insertion reactions at the (1,3-bis(diphenylphosphino)propane)-Pd(II) center and their relationship to the alternating copolymerization of ethylene and carbon monoxide. *Journal of the American Chemical Society*, **122**, 6351.

217 Hearly, A.K., Nowack, R.J., and Rieger, B. (2005) New single-site palladium catalysts for the nonalternating copolymerization of ethylene and carbon monoxide. *Organometallics*, **24**, 2755.

218 Haras, A., Michalak, A., Rieger, B., and Ziegler, T. (2005) Theoretical analysis of factors controlling the nonalternating CO/C_2H_4 copolymerization. *Journal of the American Chemical Society*, **127**, 8765.

219 Bettucci, L., Bianchini, C., Claver, C., Garcia Suarez, E.J., Ruiz, A., Meli, A., and Oberhauser, W. (2007) Ligand effects in the non-alternating CO–ethylene copolymerization by palladium(II) catalysis. *Dalton Transactions*, 5590.

220 Vrieze, K. and van Leeuwen, P.W.N.M. (1971) Studies of dynamic organometallic compounds of the transition metals by means of nuclear magnetic resonance. *Progress in Inorganic Chemistry*, **14**, 1.

221 Erker, G., Kehr, G., and Fröhlich, R. (2004) Some selected chapters from the (butadiene)zirconocene story. *Journal of Organometallic Chemistry*, **689**, 4305.

222 Erker, G., Kehr, G., and Fröhlich, R. (2005) The (butadiene)zirconocene route to active homogeneous olefin polymerization catalysts. *Journal of Organometallic Chemistry*, **690**, 6254.

223 Yasuda, H. and Nakamura, A. (1987) Diene, alkyne, alkene, and alkyl complexes of early transition metals: Structures and synthetic applications in organic and polymer chemistry. *Angewandte Chemie – International Edition in English*, **26**, 723.

224 Tobisch, S. and Ziegler, T. (2002) [Ni^{0}L]-catalyzed cyclodimerization of 1,3-butadiene: A comprehensive density functional investigation based on the generic [$(C_4H_6)_2Ni^0PH_3$] catalyst. *Journal of the American Chemical Society*, **124**, 4881.

225 Méndez, M., Cuerva, J., Gómez-Bengoa, E., Cárdenas, D.J., and Echavarren, A.M.

(2002) Intramolecular coupling of allyl carboxylates with allyl stannanes and allyl silanes: A new type of reductive elimination reaction? *Chemistry – A European Journal*, **8**, 3620.

226 Wilke, G. (1963) Cyclooligomerization of butadiene and transition metal π-complexes. *Angewandte Chemie – International Edition in English*, **2**, 105.

227 Wilke, G., Bogdanović, B., Hardt, P., Heimbach, P., Keim, W., Kröner, M., Oberkirch, W., Tanaka, K., Steinrücke, E., Walter, D., and Zimmermann, H. (1966) Allyl-transition metal systems. *Angewandte Chemie – International Edition in English*, **5**, 151.

228 Tobisch, S. (2003) Structure–reactivity relationships in the cyclo-oligomerization of 1,3-butadiene catalyzed by zerovalent nickel complexes. *Advances in Organometallic Chemistry*, **49**, 167.

229 Ring, W. and Gaube, J. (1966) Zur technischen Synthese von Cyclododecatrien-(1,5,9). *Chemie Ingenieur Technik*, **10**, 1041.

230 Heimbach, P., Kluth, J., Schenkluhn, H., and Weimann, B. (1980) Ligand-property control in the nickel(0)/butadiene/P-ligand catalytic system: Dominance of "steric" factors in the control of oligomer distribution. *Angewandte Chemie – International Edition in English*, **19**, 569.

231 Heimbach, P., Kluth, J., Schenkluhn, H., and Weimann, B. (1980) Ligand-property control in the nickel(0)/butadiene/P-ligand catalytic system: "Electronic" factors in the control of cyclodimer distribution. *Angewandte Chemie – International Edition in English*, **19**, 570.

232 Tolman, C.A. (1977) Steric effects of phosphorus ligands in organometallic chemistry and homogeneous catalysis. *Chemical Reviews*, **77**, 313.

233 Gusev, D.G. (2009) Donor properties of a series of two-electron ligands. *Organometallics*, **28**, 763.

234 Fernandez, A.L., Reyes, C., Prock, A., and Giering, W.P. (2000) The stereoelectronic parameters of phosphites. The quantitative analysis of ligand effects (QALE). *Journal of The Chemical Society-Perkin Transactions 2*, 1033. See also "www.bu.edu/qale."

235 Bunten, K.A., Chen, L., Fernandez, A.L., and Poë, A.J. (2002) Cone angles: Tolman's and Plato's. *Coordination Chemistry Reviews*, **233–234**, 41.

236 Gordon III, B. and Loftus, J.E. (1989) Telomerization, in *Encyclopedia of Polymer Science and Engineering*, vol. 16 (ed. J.I. Kroschwitz), Wiley-Interscience, New York, p. 533.

237 Tsuji, J. (1979) Palladium-catalyzed reactions of butadiene and isoprene. *Advances in Organometallic Chemistry*, **17**, 141.

238 Jolly, P.W., Mynott, R., Raspel, B., and Schick, K.-P. (1986) Intermediates in the palladium-catalyzed reactions of 1,3-dienes. 3. The reaction of (η^1,η^3-octadienediyl)palladium complexes with acidic substrates. *Organometallics*, **5**, 473.

239 Behr, A. (1984) Telomerization of dienes by homogeneous transition metal catalysts. *Aspects of Homogeneous Catalysis*, **5**, 3.

240 Behr, A., Becker, M., Beckmann, T., Johnen, L., Leschinski, J., and Reyer, S. (2009) Telomerization: Advances and applications of a versatile reaction. *Angewandte Chemie – International Edition*, **48**, 3598.

241 Zapf, A. and Beller, M. (2002) Fine chemical synthesis with homogeneous palladium catalysts: Examples, status and trends. *Topics in Catalysis*, **19**, 101.

242 Tobisch, S. (2002) Theoretical investigation of the mechanism of cis-trans regulation for the allylnickel(II)-catalyzed 1,4 polymerization of butadiene. *Accounts of Chemical Research*, **35**, 96.

243 Fischbach, A. and Anwander, R. (2006) Rare-earth metals and aluminium getting close in Ziegler-type organometallics. *Advances in Polymer Science*, **204**, 155.

244 Tamao, K., Hiyama, T., and E.-i. Negishi (eds) (2002) Special issue: 30 years of the cross-coupling reaction. *Journal of the Organometallic Chemistry*, **653**, 1; see especially the "Historical Notes" by K. Tamao (p. 23), S.-I. Murahashi (p. 27), E.-i. Negishi (p. 34), K. Sonogashira (p. 46), N. Miyaura (p. 54), and T. Hiyama (p. 58).

245 Negishi, E.-i. (2005) A quarter of a century of explorations in organozirconium chemistry. *Dalton Transactions*, 827.

246 Buchwald, S.L. (ed.) (2008) Special issue on cross coupling. *Accounts of Chemical Research*, **41** (11), 1439.

247 Stambuli, J.P., Incarvito, C.D., Bühl, M., and Hartwig, J.F. (2004) Synthesis, structure, theoretical studies, and ligand exchange reactions of monomeric, T-shaped arylpalladium(II) halide complexes with an additional, weak agostic interaction. *Journal of the American Chemical Society*, **126**, 1184.

248 Espinet, P. and Echavarren, A.M. (2004) The mechanisms of the Stille reaction. *Angewandte Chemie – International Edition*, **43**, 4704.

249 Kozuch, S., Shaik, S., Jutand, A., and Amatore, C. (2004) Active anionic zero-valent palladium catalysts: Characterization by density functional calculations. *Chemistry – A European Journal*, **10**, 3072.

250 Kozuch, S., Amatore, C., Jutand, A., and Shaik, S. (2005) What makes for a good catalytic cycle? A theoretical study of the role of an anionic palladium(0) complex in the cross-coupling of an aryl halide with an anionic nucleophile. *Organometallics*, **24**, 2319.

251 Negishi, E.-i. (ed.) (2002) *Handbook of Organopalladium Chemistry for Organic Synthesis*, vol. 1–2, Wiley-Interscience, New York.

252 Bolm, C., Legros, J., Le Paih, J., and Zani, L. (2004) Iron-catalyzed reactions in organic synthesis. *Chemical Reviews*, **104**, 6217.

253 Sherry, B.D. and Fürstner, A. (2008) The promise and challenge of iron-catalyzed cross coupling. *Accounts of Chemical Research*, **41**, 1500.

254 Kleimark, J., Hedström, A., Larsson, P.-F., Johansson, C., and Norrby, P.-O. (2009) Mechanistic investigation of iron-catalyzed coupling reactions. *ChemCatChem*, **1**, 152.

255 Fürstner, A. (2009) From oblivion into the limelight: Iron (domino) catalysis. *Angewandte Chemie – International Edition*, **48**, 1364.

256 Molander, G.A. and Canturk, B. (2009) Organotrifluoroborates and monocoordinated palladium complexes as catalysts – A perfect combination for Suzuki–Miyaura coupling. *Angewandte Chemie – International Edition*, **48**, 9240.

257 Kotha, S., Lahiri, K., and Kashinath, D. (2002) Recent applications of the Suzuki–Miyaura cross-coupling reaction in organic synthesis. *Tetrahedron*, **58**, 9633.

258 Suzuki, A. (2005) Carbon–carbon bonding made easy. *Chemical Communications*, 4759.

259 Alonso, F., Beletskaya, I.P., and Yus, M. (2008) Non-conventional methodologies for transition-metal catalysed carbon–carbon coupling: A critical overview. Part 2: The Suzuki reaction. *Tetrahedron*, **64**, 3047.

260 Denmark, S.E. and Regens, C.S. (2008) Palladium-catalyzed cross-coupling reactions of organosilanols and their salts: Practical alternatives to boron- and tin-based methods. *Accounts of Chemical Research*, **41**, 1486.

261 Stille, J.K. (1986) The palladium-catalyzed cross-coupling reactions of organotin reagents with organic electrophiles. *Angewandte Chemie – International Edition in English*, **25**, 508.

262 Álvarez, R., Faza, O.N., de Lera, A.R., and Cárdenas, D.J. (2007) A density functional theory study of the Stille cross-coupling via associative transmetalation. The role of ligands and coordinating solvents. *Advanced Synthesis & Catalysis*, **349**, 887.

263 Johnson, J.B. and Rovis, T. (2008) More than bystanders: The effect of olefins on transition-metal-catalyzed cross-coupling reactions. *Angewandte Chemie – International Edition*, **47**, 840.

264 Chinchilla, R. and Nájera, C. (2007) The Sonogashira reaction: A booming methodology in synthetic organic chemistry. *Chemical Reviews*, **107**, 874.

265 Doucet, H. and Hierso, J.-C. (2007) Palladium-based catalytic systems for the synthesis of conjugated enynes by Sonogashira reactions and related alkynylations. *Angewandte Chemie – International Edition*, **46**, 834.

266 Farina, V. (2004) High-turnover palladium catalysts in cross-coupling and Heck chemistry: A critical overview. *Advanced Synthesis & Catalysis*, **346**, 1553.

267 Fu, G.C. (2008) The development of versatile methods for palladium-catalyzed coupling reactions of aryl electrophiles through the use of P(*t*-Bu)$_3$ and PCy$_3$ as ligands. *Accounts of Chemical Research*, **41**, 1555.

268 Christmann, U. and Vilar, R. (2005) Monoligated palladium species as catalysts in cross-coupling reactions. *Angewandte Chemie – International Edition*, **44**, 366.

269 Kantchev, E.A.B., O'Brien, C.J., and Organ, M.G. (2007) Palladium complexes of N-heterocyclic carbenes as catalysts for cross-coupling reactions – A synthetic chemist's perspective. *Angewandte Chemie – International Edition*, **46**, 2768.

270 Cárdenas, D.J. (1999) Towards efficient and wide-scope metal-catalyzed alkyl–alkyl cross-coupling reactions. *Angewandte Chemie – International Edition*, **38**, 3018.

271 Cárdenas, D.J. (2003) Advances in functional-group-tolerant metal-catalyzed alkyl–alkyl cross-coupling reactions. *Angewandte Chemie – International Edition*, **42**, 384.

272 Frisch, A.C. and Beller, M. (2005) Catalysts for cross-coupling reactions with non-activated alkyl halides. *Angewandte Chemie – International Edition*, **44**, 674.

273 Rudolph, A. and Lautens, M. (2009) Secondary alkyl halides in transition-metal-catalyzed cross-coupling reactions. *Angewandte Chemie – International Edition*, **48**, 2656.

274 Jensen, A.E. and Knochel, P. (2002) Nickel-catalyzed cross-coupling between functionalized primary or secondary alkylzinc halides and primary alkyl halides. *The Journal of Organic Chemistry*, **67**, 79.

275 Netherton, M.R. and Fu, G.C. (2004) Nickel-catalyzed cross-couplings of unactivated alkyl halides and pseudohalides with organometallic compounds. *Advanced Synthesis & Catalysis*, **346**, 1525 (and references cited therein).

276 Tietze, L.F., Ila, H., and Bell, H.P. (2004) Enantioselective palladium-catalyzed transformations. *Chemical Reviews*, **104**, 3453.

277 Glorius, F. (2008) Asymmetric cross-coupling of non-activated secondary alkyl halides. *Angewandte Chemie – International Edition*, **47**, 8347.

278 Brennführer, A., Neumann, H., and Beller, M. (2009) Palladium-catalyzed carbonylation reactions of aryl halides and related compounds. *Angewandte Chemie – International Edition*, **48**, 4114.

279 Beletskaya, I.P. and Cheprakov, A.V. (2000) The Heck reaction as a sharpening stone of palladium catalysis. *Chemical Reviews*, **100**, 3009.

280 Crisp, G.T. (1998) Variations on a theme – Recent developments on the mechanism of the Heck reaction and their implications for synthesis. *Chemical Society Reviews*, **27**, 427.

281 Cabri, W. and Candiani, I. (1995) Recent developments and new perspectives in the Heck reaction. *Accounts of Chemical Research*, **28**, 2.

282 Amatore, C. and Jutand, A. (2000) Anionic Pd(0) and Pd(II) intermediates in palladium-catalyzed Heck and cross-coupling reactions. *Accounts of Chemical Research*, **33**, 314.

283 Schenck, H.v., Åkermark, B., and Svensson, M. (2003) Electronic control of the regiochemistry in the Heck reaction. *Journal of the American Chemical Society*, **125**, 3503.

284 Beletskaya, I.P. and Cheprakov, A.V. (2004) Palladacycles in catalysis – A critical survey. *Journal of Organometallic Chemistry*, **689**, 4055.

285 Dupont, J., Consorti, C.S., and Spencer, J. (2005) The potential of palladacycles: More than just precatalysts. *Chemical Reviews*, **105**, 2527.

286 de Vries, J.G. (2006) A unifying mechanism for all high-temperature Heck reactions. The role of palladium colloids and anionic species. *Dalton Transactions*, 421.

287 Trzeciak, A.M. and Ziólkowski, J.J. (2007) Monomolecular, nanosized and heterogenized palladium catalysts for the Heck reaction. *Coordination Chemistry Reviews*, **251**, 1281.

288 Cárdenas, D.J. and Echavarren, A.M. (2004) Mechanistic aspects of C–C bond formation involving allylpalladium complexes: The role of computational studies. *New Journal of Chemistry*, **28**, 338.

289 Consiglio, G. and Waymouth, R.M. (1989) Enantioselective homogeneous catalysis involving transition-metal–allyl intermediates. *Chemical Reviews*, **89**, 257.

290 Trost, B.M. (1996) Designing a receptor for molecular recognition in a catalytic synthetic reaction: Allylic alkylation. *Accounts of Chemical Research*, **29**, 355.

291 Lu, Z. and Ma, S. (2008) Metal-catalyzed enantioselective allylation in asymmetric synthesis. *Angewandte Chemie – International Edition*, **47**, 258.

292 Braun, M. and Meier, T. (2006) Tsuji-Trost allylic alkylation with ketone enolates. *Angewandte Chemie – International Edition*, **45**, 6952.

293 Bini, L., Müller, C., and Vogt, D. (2010) Mechanistic studies on hydrocyanation reactions. *ChemCatChem*, **2**, 590.

294 McKinney, R.J. and Roe, D.C. (1986) The mechanism of nickel-catalyzed ethylene hydrocyanation. Reductive elimination by an associative process. *Journal of the American Chemical Society*, **108**, 5167.

295 Tolman, C.A., McKinney, R.J., Seidel, W.C., Druliner, J.D., and Stevens, W.R. (1985) Homogeneous nickel-catalyzed olefin hydrocyanation. *Advances in Catalysis*, **33**, 1.

296 Göthlich, A.P.V., Tensfeldt, M., Rothfuss, H., Tauchert, M.E., Haap, D., Rominger, F., and Hofmann, P. (2008) Novel chelating phosphonite ligands: Syntheses, structures, and nickel-catalyzed hydrocyanation of olefins. *Organometallics*, **27**, 2189.

297 McKinney, R.J. (1985) Kinetic control in catalytic olefin isomerization. An explanation for the apparent contrathermodynamic isomerization of 3-pentenenitrile. *Organometallics*, **4**, 1142.

298 RajanBabu, T.V., and Casalnuovo, A.L. (1999) Hydrocyanation of carbon-carbon double bonds, in *Comprehensive Asymmetric Catalysis I–III*, vol. 1 (eds E.N. Jacobsen, A. Pfaltz, and H. Yamamoto), Springer, Berlin, p. 367.

299 RajanBabu, T.V., Casalnuovo, A.L., Ayers, T.A., Nomura, N., Jin, J., Park, H., and Nandi, M. (2003) Ligand tuning as a tool for the discovery of new catalytic asymmetric processes. *Current Organic Chemistry*, **7**, 301.

300 Flanagan, S.P. and Guiry, P.J. (2006) Substituent electronic effects in chiral ligands for asymmetric catalysis. *Journal of Organometallic Chemistry*, **691**, 2125.

301 Jackson, W.R. and Perlmutter, P. (1986) The hydrocyanation of alkynes. *Chemistry in Britain*, 338.

302 Limberg, C. (2007) The SOHIO process as an inspiration for molecular organometallic chemistry. *Topics in Organometallic Chemistry*, **22**, 79.

303 Brunel, J.-M. and Holmes, I.P. (2004) Chemically catalyzed asymmetric cyanohydrin syntheses. *Angewandte Chemie – International Edition*, **43**, 2752.

304 Paull, D.H., Abraham, C.J., Scerba, M.T., Alden-Danforth, E., and Lectka, T. (2008) Bifunctional asymmetric catalysis: Cooperative Lewis acid/base systems. *Accounts of Chemical Research*, **41**, 655.

305 Gröger, H. (2003) Catalytic enantioselective Strecker reactions and analogous syntheses. *Chemical Reviews*, **103**, 2795.

306 Connon, S.J. (2008) The catalytic asymmetric Strecker reaction: Ketimines continue to join the fold. *Angewandte Chemie – International Edition*, **47**, 1176.

307 Marciniec, B. (2002) Catalysis of hydrosilylation of carbon-carbon multiple bonds: Recent progress. *Silicon Chemistry*, **1**, 155.

308 Roy, A.K. (2008) A review of recent progress in catalyzed homogeneous hydrosilation (hydrosilylation). *Advances in Organometallic Chemistry*, **55**, 1.

309 Sakaki, S., Mizoe, N., Sugimoto, M., and Musashi, Y. (1999) Pt-catalyzed hydrosilylation of ethylene. A theoretical study of the reaction mechanism. *Coordination Chemistry Reviews*, **190–192**, 933.

310 Sakaki, S., Sumimoto, M., Fukuhara, M., Sugimoto, M., Fujimoto, H., and Matsuzaki, S. (2002) Why does the rhodium-catalyzed hydrosilylation of

alkenes take place through a modified Chalk–Harrod mechanism? A theoretical study. *Organometallics*, **21**, 3788.

311 Marciniec, B. (2000) Silicometallics and catalysis. *Applied Organometallic Chemistry*, **14**, 527.

312 Marciniec, B. (1997) Dehydrogenative coupling of olefins with silicon compounds catalyzed by transition-metal complexes. *New Journal of Chemistry*, **21**, 815.

313 Grate, J.W. and Kaganove, S.N. (1999) Hydrosilylation: A versatile reaction for polymer synthesis. *Polymer News*, **24**, 149.

314 Hayashi, T. (2000) Axially chiral monophosphine ligands (MOPs) and their use for palladium-catalyzed asymmetric hydrosilylation of olefins. *Catalysis Today*, **62**, 3.

315 Hayashi, T. (2000) Chiral monodentate phosphine ligand MOP for transition-metal-catalyzed asymmetric reactions. *Accounts of Chemical Research*, **33**, 354.

316 Gibson, S.E. and Rudd, M. (2007) The role of secondary interactions in the asymmetric palladium-catalysed hydrosilylation of olefins with monophosphane ligands. *Advanced Synthesis & Catalysis*, **349**, 781.

317 Nishiyama, H. (1999) Hydrosilylation of carbonyl and imino groups, in *Comprehensive Asymmetric Catalysis I–III*, vol. 1 (eds E.N. Jacobsen, A. Pfaltz, and H. Yamamoto), Springer, Berlin, p 267.

318 Riant, O., Mostefaï, N., and Courmarcel, J. (2004) Recent advances in the asymmetric hydrosilylation of ketones, imines and electrophilic double bonds. *Synthesis*, 2943.

319 Díez-González, S. and Nolan, S.P. (2008) Copper, silver, and gold complexes in hydrosilylation reactions. *Accounts of Chemical Research*, **41**, 349.

320 Denmark, S.E. and Ober, M.H. (2003) Organosilicon reagents: Synthesis and application to palladium-catalyzed cross-coupling reactions. *Aldrichimica Acta*, **36**, 75.

321 Vincent, J.L., Luo, S., Scott, B.L., Butcher, R., Unkefer, C.J., Burns, C.J., Kubas, G.J., Lledós, A., Maseras, F., and Tomàs, J. (2003) Experimental and theoretical studies of bonding and oxidative addition of germanes and silanes, $EH_{4-n}Ph_n$ (E = Si, Ge; $n = 0$–3), to Mo(CO)-(diphosphine)$_2$. The first structurally characterized germane σ complex. *Organometallics*, **22**, 5307.

322 Kubas, G.J. (2005) Catalytic processes involving dihydrogen complexes and other sigma-bond complexes. *Catalysis Letters*, **104**, 79.

323 Lin, Z. (2002) Structural and bonding characteristics in transition metal–silane complexes. *Chemical Society Reviews*, **31**, 239.

324 Nikonov, G.I. (2001) Going beyond σ complexation: Nonclassical interligand interactions of silyl groups with two and more hydrides. *Angewandte Chemie – International Edition*, **40**, 3353.

325 Nikonov, G.I. (2001) New types of non-classical interligand interactions involving silicon based ligands. *Journal of Organometallic Chemistry*, **635**, 24.

326 Haak, E. and Doye, S. (1999) Katalytische Hydroaminierung von Alkenen und Alkinen. *Chemie in unserer Zeit*, **33**, 297.

327 Nobis, M. and Drießen-Hölscher, B. (2001) Recent developments in transition metal catalyzed intermolecular hydroamination reactions – A breakthrough? *Angewandte Chemie – International Edition*, **113**, 4105.

328 Müller, T.E., Hultzsch, K.C., Yus, M., Foubelo, F., and Tada, M. (2008) Hydroamination: Direct addition of amines to alkenes and alkynes. *Chemical Reviews*, **108**, 3795.

329 Seayad, J., Tillak, A., Hartung, C.G., and Beller, M. (2002) Base-catalyzed hydroamination of olefins: An environmentally friendly route to amines. *Advanced Synthesis & Catalysis*, **344**, 795.

330 Brunet, J.-J., Chu, N.-C., and Rodriguez-Zubiri, M. (2007) Platinum-catalyzed intermolecular hydroamination of alkenes: Halide-anion-promoted catalysis. *European Journal of Inorganic Chemistry*, 4711.

331 Cochran, B.M. and Michael, F.E. (2008) Mechanistic studies of a palladium–catalyzed intramolecular hydroamination of unactivated alkenes: Protonolysis of a stable palladium alkyl complex is the turnover-limiting step.

Journal of the American Chemical Society, **130**, 2786.

332 Johns, A.M., Utsunomiya, M., Incarvito, C.D., and Hartwig, J.F. (2006) A highly active palladium catalyst for intermolecular hydroamination. Factors that control reactivity and additions of functionalized anilines to dienes and vinylarenes. *Journal of the American Chemical Society*, **128**, 1828.

333 Shekhar, S., Ryberg, P., Hartwig, J.F., Mathew, J.S., Blackmond, D.G., Strieter, E.R., and Buchwald, S.L. (2006) Reevaluation of the mechanism of the amination of aryl halides catalyzed by BINAP-ligated palladium complexes. *Journal of the American Chemical Society*, **128**, 3584.

334 Widenhoefer, R.A. and Han, X. (2006) Gold-catalyzed hydroamination of C—C-multiple bonds. *European Journal of Organic Chemistry*, 4555.

335 Marion, N. and Nolan, S.P. (2008) N-Heterocyclic carbenes in gold catalysis. *Chemical Society Reviews*, **37**, 1776.

336 Hong, S. and Marks, T.J. (2004) Organolanthanide-catalyzed hydroamination. *Accounts of Chemical Research*, **37**, 673.

337 Hunt, P.A. (2007) Organolanthanide mediated catalytic cycles: A computational perspective. *Dalton Transactions*, 1743.

338 Hultzsch, K.C. (2005) Catalytic asymmetric hydroamination of non-activated olefins. *Organic and Biomolecular Chemistry*, **3**, 1819.

339 Hultzsch, K.C. (2005) Transition metal-catalyzed asymmetric hydroamination of alkenes (AHA). *Advanced Synthesis & Catalysis*, **347**, 367.

340 Aillaud, I., Collin, J., Hannedouche, J., and Schulz, E. (2007) Asymmetric hydroamination of non-activated carbon—carbon multiple bonds. *Dalton Transactions*, 5105.

341 Hazari, N. and Mountford, P. (2005) Reactions and applications of titanium imido complexes. *Accounts of Chemical Research*, **38**, 839.

342 Severin, R. and Doye, S. (2007) The catalytic hydroamination of alkynes. *Chemical Society Reviews*, **36**, 1407.

343 Smidt, J., Hafner, W., Jira, R., Sieber, R., Sedlmeier, J., and Sabel, A. (1962) The oxidation of olefins with palladium chloride catalysts. *Angewandte Chemie – International Edition in English*, **1**, 80.

344 Jira, R. (2009) Acetaldehyde from ethylene – A retrospective on the discovery of the Wacker process. *Angewandte Chemie – International Edition*, **48**, 9034.

345 Piera, J. and Bäckvall, J-E. (2008) Catalytic oxidation of organic substrates by molecular oxygen and hydrogen peroxide by multistep electron transfer – A biomimetic approach. *Angewandte Chemie – International Edition*, **47**, 3506.

346 Keith, J.A. and Henry, P.M. (2009) The mechanism of the Wacker reaction: A tale of two hydroxypalladations. *Angewandte Chemie – International Edition*, **48**, 9038.

347 DeKock, R.L., Hristov, I.H., Anderson, G.D.W., Göttker-Schnetmann, I., Mecking, S., and Ziegler, T. (2005) Possible side reactions due to water in emulsion polymerization by late transition metal complexes II: Deactivation of the catalyst by a Wacker-type reaction. *Organometallics*, **24**, 2679.

348 Nelson, D.J., Li, R., and Brammer, C. (2001) Correlation of relative rates of $PdCl_2$ oxidation of functionalized acyclic alkenes versus alkene ionization potentials, HOMOs, and LUMOs. *Journal of the American Chemical Society*, **123**, 1564.

349 Bäckvall, J.E., Åkermark, B., and Ljunggren, S.O. (1979) Stereochemistry and mechanism for the palladium(II)-catalyzed oxidation of ethene in water (the Wacker process). *Journal of the American Chemical Society*, **101**, 2411.

350 Keith, J.A., Nielsen, R.J., Oxgaard, J., and GoddardIII, W.A. (2007) Unraveling the Wacker oxidation mechanisms. *Journal of the American Chemical Society*, **129**, 12342.

351 Hamed, O., Henry, P.M., and Thompson, C. (1999) Palladium(II)-catalyzed exchange and isomerization reactions. 17. Exchange of chiral allyl alcohols with hydroxide, methoxide, and phenyl at high

[Cl⁻]. Stereochemistry of the Wacker reaction. *The Journal of Organic Chemistry*, **64**, 7745.

352 Hosokawa, T. and Murahashi, S.-I. (1990) New aspects of oxypalladation of alkenes. *Accounts of Chemical Research*, **23**, 49.

353 Takacs, J.M. and Jiang, X.-t. (2003) The Wacker reaction and related alkene oxidation reactions. *Current Organic Chemistry*, **7**, 369.

354 Mitsudome, T., Mizumoto, K., Mizugaki, T., Jitsukawa, K., and Kaneda, K. (2010) Wacker-type oxidation of internal olefins using a $PdCl_2/N,N$-dimethylacetamide catalyst systeme under copper-free reaction conditions. *Angewandte Chemie – International Edition*, **49**, 1238.

355 El-Qisairi, A.K., Qaseer, H.A., and Henry, P.M. (2002) Oxidation of olefins by palladium(II). 18. Effect of reaction conditions, substrate structure and chiral ligand on the bimetallic palladium(II) catalyzed asymmetric chlorohydrin synthesis. *Journal of Organometallic Chemistry*, **656**, 168.

356 Stahl, S.S. (2004) Palladium oxidase catalysis: Selective oxidation of organic chemicals by direct dioxygen-coupled turnover. *Angewandte Chemie – International Edition*, **43**, 3400.

357 Gligorich, K.M. and Sigman, M.S. (2006) Mechanistic questions about the reaction of molecular oxygen with palladium in oxidase catalysis. *Angewandte Chemie – International Edition*, **45**, 6612.

358 Deubel, D.V. (2004) From evolution to green chemistry: Rationalization of biomimetic oxygen-transfer cascades. *Journal of the American Chemical Society*, **126**, 996.

359 Mimoun, H. (1982) Oxygen transfer from inorganic and organic peroxides to organic substrates: A common mechanism? *Angewandte Chemie – International Edition in English*, **21**, 734.

360 Deubel, D.V., Sundermeyer, J., and Frenking, G. (2000) Mechanism of the olefin epoxidation catalyzed by molybdenum diperoxo complexes: Quantum-chemical calculations give an answer to a long-standing question. *Journal of the American Chemical Society*, **122**, 10101.

361 Deubel, D.V. and Frenking, G. (2003) [3 + 2] versus [2 + 2] addition of metal oxides across C=C bonds. Reconciliation of experiment and theory. *Accounts of Chemical Research*, **36**, 645.

362 Deubel, D.V., Frenking, G., Gisdakis, P., Herrmann, W.A., Rösch, N., and Sundermeyer, J. (2004) Olefin epoxidation with inorganic peroxides. Solutions to four long-standing controversies on the mechanism of oxygen transfer. *Accounts of Chemical Research*, **37**, 645.

363 Noyori, R., Aoki, M., and Sato, K. (2003) Green oxidation with aqueous hydrogen peroxide. *Chemical Communications*, 1977.

364 Herrmann, W.A. and Kühn, F.E. (1997) Organorhenium oxides. *Accounts of Chemical Research*, **30**, 169.

365 Kühn, F.E., Scherbaum, A., and Herrmann, W.A. (2004) Methyltrioxorhenium and its applications in olefin oxidation, metathesis and aldehyde olefination. *Journal of Organometallic Chemistry*, **689**, 4149.

366 Zuwei, X., Ning, Z., Yu, S., and Kunlan, L. (2001) Reaction-controlled phase-transfer catalysis for propylene epoxidation to propylene oxide. *Science*, **292**, 1139.

367 Gao, S., Li, M., Lv, Y., Zhou, N., and Xi, Z. (2004) Epoxidation of propylene with aqueous hydrogen peroxide on a reaction-controlled phase-transfer catalyst. *Organic Process Research & Development*, **8**, 131.

368 Sharpless, K.B. (2002) Searching for new reactivity (Nobel Lecture). *Angewandte Chemie – International Edition*, **41**, 2024.

369 Katsuki, T. (2002) Chiral metallosalen complexes: Structures and catalyst tuning for asymmetric epoxidation and cyclopropanation. *Advanced Synthesis & Catalysis*, **344**, 131.

370 Larrow, J.F. and Jacobsen, E.N. (2004) Asymmetric processes catalyzed by chiral (salen)metal complexes. *Topics in Organometallic Chemistry*, **6**, 123.

371 Rose, E., Andrioletti, B., Zrig, S., and Quelquejeu-Ethève, M. (2005) Enantioselective epoxidation of olefins with chiral metalloporphyrin catalysts. *Chemical Society Reviews*, **34**, 573.

372 Chatterjee, D. (2008) Asymmetric epoxidation of unsaturated hydrocarbons catalyzed by ruthenium complexes. *Coordination Chemistry Reviews*, **252**, 176.

373 Lane, B.S. and Burgess, K. (2003) Metal-catalyzed epoxidations of alkenes with hydrogen peroxide. *Chemical Reviews*, **103**, 2457.

374 Arends, I.W.C.E. (2006) Metal-catalyzed asymmetric epoxidations of terminal olefins using hydrogen peroxide as the oxidant. *Angewandte Chemie – International Edition*, **38**, 6250.

375 Nam, W. (ed.) (2007) Special issue on dioxygen activation by metalloenzymes and models. *Accounts of Chemical Research*, **40** (7), 465.

376 Meunier, B., de Visser, S.P., and Shaik, S. (2004) Mechanism of oxidation reactions catalyzed by cytochrome P450 enzymes. *Chemical Reviews*, **104**, 3947.

377 Shaik, S., Kumar, D., de Visser, S.P., Altun, A., and Thiel, W. (2005) Theoretical perspective on the structure and mechanism of cytochrome P450 enzymes. *Chemical Reviews*, **105**, 2279.

378 Nam, W. (2007) High-valent iron(IV)-oxo complexes of heme and non-heme ligands in oxygenation reactions. *Accounts of Chemical Research*, **40**, 522.

379 Chen, X., Engle, K.M., Wang, D.-H., and Yu, J.-Q. (2009) Palladium(II)-catalyzed C−H activation/C−C cross-coupling reactions: Versatility and practicality. *Angewandte Chemie – International Edition*, **48**, 5094.

380 Ackermann, L., Vicente, R., and Kapdi, A.R. (2009) Transition-metal-catalyzed direct arylation of (hetero)arenes by C−H bond cleavage. *Angewandte Chemie – International Edition*, **48**, 9792.

381 Labinger, J.A. and Bercaw, J.E. (2002) Understanding and exploiting C−H bond activation. *Nature*, **417**, 507.

382 Goldman, A.S. and Goldberg, K.I. (2004) Organometallic C−H bond activation: An introduction. *ACS Symposium Series*, **885**, 1.

383 Crabtree, R.H. (2004) Organometallic alkane CH activation. *Journal of Organometallic Chemistry*, **689**, 4083.

384 Fekl, U. and Goldberg, K.I. (2003) Homogeneous hydrocarbon C−H bond activation and functionalization with platinum. *Advances in Organometallic Chemistry*, **54**, 259.

385 Chen, G.S., Labinger, J.A., and Bercaw, J.E. (2007) The role of alkane coordination in C−H bond cleavage at a Pt(II) center. *Proceedings of the National Academy of Sciences of the United States of America*, **104**, 6915.

386 Davies, H.M.L. and Beckwith, R.E.J. (2003) Catalytic enantioselective C−H activation by means of metal–carbenoid-induced C−H insertion. *Chemical Reviews*, **103**, 2861.

387 Davies, H.M.L. (2006) Recent advances in catalytic enantioselective intermolecular C−H functionalization. *Angewandte Chemie – International Edition*, **45**, 6422.

388 Crabtree, R.H. (2001) Alkane C−H activation and functionalization with homogeneous transition metal catalysts: A century of progress – A new millennium in prospect. *Journal of The Chemical Society-Dalton Transactions*, 2437.

389 Stahl, S.S., Labinger, J.A., and Bercaw, J.E. (1998) Homogeneous oxidation of alkanes by electrophilic late transition metals. *Angewandte Chemie – International Edition*, **37**, 2180.

390 Lersch, M. and Tilset, M. (2005) Mechanistic aspects of C−H activation by Pt complexes. *Chemical Reviews*, **105**, 2471.

391 Periana, R.A., Taube, D.J., Evitt, E.R., Löffler, D.G., Wentrcek, P.R., Voss, G., and Masuda, T. (1993) A mercury-catalyzed, high-yield system for the oxidation of methane to methanol. *Science*, **259**, 340.

392 Periana, R.A., Taube, D.J., Gamble, S., Taube, H., Satoh, T., and Fujii, H. (1998) Platinum catalysts for the high-yield oxidation of methane to a methanol derivative. *Science*, **280**, 560.

393 Zhan, C.-G., Nichols, J.A., and Dixon, D.A. (2003) Ionization potential, electron affinity, electronegativity, hardness, and electron excitation energy: Molecular properties from density functional theory orbital energies. *The Journal of Physical Chemistry A*, **107**, 4184.

394 Ertl, G. (2003) Ammonia Synthesis – Heterogeneous, in [M9], vol. 1, p. 329.

395 Schlögl, R. (2003) Catalytic synthesis of ammonia – A "Never-Ending Story"?

Angewandte Chemie – International Edition, **42**, 2004.

396 Schlögl, R. (2008) Ammonia synthesis, in *Handbook of Heterogeneous Catalysis*, 2nd edn (eds G. Ertl, H. Knözinger, F. Schüth, and J. Weitkamp), Wiley-VCH, Weinheim, p. 2501.

397 Zeise, H. (1954) *Thermodynamik*, Vol 3/1, Hirzel Verlag, Leipzig.

398 Ertl, G. (2001) Heterogeneous catalysis: From "Black art" to atomic understanding, in *Chemistry for the 21st Century* (eds E. Keinan and I. Schechter), Wiley-VCH, Weinheim, S. 54.

399 Honkala, K., Hellman, A., Remediakis, I.N., Logadottir, A., Carlsson, A., Dahl., S., Christensen, C.H., and Nørskov, J.K. (2005) Ammonia synthesis from first-principles calculations. *Science*, **307**, 555.

400 Jacobsen, C.J.H., Dahl, S., Boisen, A., Clausen, B.S., Topsøe, H., Logadottir, A., and Nørskov, J.K. (2002) Optimal catalyst curves: Connecting density functional theory calculations with industrial reactor design and catalyst selection. *Journal of Catalysis*, **205**, 382 (and references cited therein).

401 Bielawa, H., Hinrichsen, O., Birkner, A., and Muhler, M. (2001) The ammonia-synthesis catalyst of the next generation: Barium-promoted oxide-supported ruthenium. *Angewandte Chemie – International Edition*, **40**, 1061.

402 van Santen, R.A. and Neurock, M. (2006) *Molecular Heterogeneous Catalysis*, Wiley-VCH, Weinheim.

403 Chorkendorff, I. and Niemantsverdriet, J.W. (2003) see Further Reading, Chap. 2.

404 Nørskov, J.K., Bligaard, T., Hvolbæk, B., Abild-Pedersen, F., Chorkendorff, I., and Christensen, C.H. (2008) The nature of the active site in heterogeneous metal catalysis. *Chemical Society Reviews*, **37**, 2163.

405 Lee, S.C. and Holm, R.H. (2004) The clusters of nitrogenase: Synthetic methodology in the construction of weak-field clusters. *Chemical Reviews*, **104**, 1135.

406 Hinnemann, B. and Nørskov, J.K. (2006) Catalysis by enzymes: The biological ammonia synthesis. *Topics in Catalysis*, **37**, 55.

407 Einsle, O., Tezcan, F.A., Andrade, S.L.A., Schmid, B., Yoshida, M., Howard, J.B., and Rees, D.C. (2002) Nitrogenase MoFe-protein at 1.16 Å resolution: A central ligand in the FeMo-cofactor. *Science*, **297**, 1696.

408 Lukoyanov, D., Pelmenschikov, V., Maeser, N., Laryukhin, M., Chin Yang, T., Noodleman, L., Dean, D.R., Case., D.A., Seefeldt, L.C., and Hoffman, B.M. (2007) Testing if the interstitial atom, X, of the nitrogenase molybdenum–iron cofactor is N or C: ENDOR, ESEEM, and DFT studies of the $S=3/2$ resting state in multiple environments. *Inorganic Chemistry*, **46**, 11437.

409 Hoffman, B.M., Dean, D.R., and Seefeldt, L.C. (2009) Climbing nitrogenase: Toward a mechanism of enzymatic nitrogen fixation. *Accounts of Chemical Research*, **42**, 609.

410 Barney, B.M., Lee, H.-I., Dos Santos, P.C., Hoffman, B.M., Dean, D.R., and Seefeldt, L.C. (2006) Breaking the N_2 triple bond: Insights into the nitrogenase mechanism. *Dalton Transactions*, 2277.

411 Dance, I. (2007) The mechanistically significant coordination chemistry of dinitrogen at FeMo-co, the catalytic site of nitrogenase. *Journal of the American Chemical Society*, **129**, 1076.

412 Reiher, M. and Hess, B.A. (2004) Quantum chemical investigations into the problem of biological nitrogen fixation: Sellmann-type metal–sulfur model complexes. *Advances in Inorganic Chemistry*, **56**, 55.

413 Dörr, M., Käßbohrer, J., Grunert, R., Kreisel, G., Brand, W.A., Werner, R.A., Geilmann, H., Apfel, C., Robl, C., and Weigand, W. (2003) A possible prebiotic formation of ammonia from dinitrogen on iron sulfide surfaces. *Angewandte Chemie – International Edition*, **42**, 1540.

414 Rees, D.C. and Howard, J.B. (2003) The interface between the biological and inorganic world: Iron-sulfur metalloclusters. *Science*, **300**, 929.

415 Vol'pin, M.E., Shur, V.B., and Berkovich, E.G. (1998) Transformations of molecular nitrogen into aromatic amines under the

action of titanium compounds. *Inorganica Chimica Acta*, **280**, 264.
416 Shilov, A.E. (2003) Catalytic reduction of molecular nitrogen in solutions. *Russian Chemical Bulletin, International Edition*, **52**, 2555.
417 Chatt, J., Dilworth, J.R., and Richards, R.L. (1978) Recent advances in the chemistry of nitrogen fixation. *Chemical Reviews*, **78**, 589.
418 Leigh, G.J. (1992) Protonation of coordinated dinitrogen. *Accounts of Chemical Research*, **25**, 177.
419 Pickett, C.J., Ryder, K.S., and Talarmin, J. (1986) Electron-transfer reactions in nitrogen fixation. Part 2. The electrosynthesis of ammonia: Identification and estimation of products. *Journal of The Chemical Society-Dalton Transactions*, 1453.
420 Nishibayashi, Y., Iwai, S., and Hidai, M. (1998) Bimetallic system for nitrogen fixation: Ruthenium-assisted protonation of coordinated N_2 on tungsten with H_2. *Science*, **279**, 540.
421 Pool, J.A., Lobkovsky, E., and Chirik, P.J. (2004) Hydrogenation and cleavage of dinitrogen to ammonia with a zirconium complex. *Nature*, **427**, 527.
422 Ohki, Y. and Fryzuk, M.D. (2007) Dinitrogen activation by group 4 metal complexes. *Angewandte Chemie – International Edition*, **46**, 3180.
423 Avenier, P., Solans-Monfort, X., Veyre, L., Renili, F., Basset, J.-M., Eisenstein, O., Taoufik, M., and Quadrelli, E.A. (2009) H/D exchange on silica-grafted tantalum(V) imido amido $[(\equiv SiO)_2Ta^{(V)}(NH)(NH_2)]$ synthesized from either ammonia or dinitrogen: IR and DFT evidence for heterolytic splitting of D_2. *Topics in Catalysis*, **52**, 1482.
424 Schrock, R.R. (2005) Catalytic reduction of dinitrogen to ammonia at a single molybdenum center. *Accounts of Chemical Research*, **38**, 955.
425 Weare, W.W., Dai, X., Byrnes, M.J., Min Chin, J., Schrock, R.R., and Müller, P. (2006) Catalytic reduction of dinitrogen to ammonia at a single molybdenum center. *Proceedings of the National Academy of Sciences of the United States of America*, **103**, 17099.

426 Schrock, R.R. (2008) Catalytic reduction of dinitrogen to ammonia by molybdenum: Theory versus experiment. *Angewandte Chemie – International Edition*, **47**, 5512.
427 Yandulov, D.V. and Schrock, R.R. (2005) Studies relevant to catalytic reduction of dinitrogen to ammonia by molybdenum triamidoamine complexes. *Inorganic Chemistry*, **44**, 1103.
428 Schenk, S., Le Guennic, B., Kirchner, B., and Reiher, M. (2008) First-principles investigation of the Schrock mechanism of dinitrogen reduction employing the full $HIPTN_3N$ ligand. *Inorganic Chemistry*, **47**, 3634.
429 Neese, F. (2006) The Yandulov/Schrock cycle and the nitrogenase reaction: Pathways of nitrogen fixation studied by density functional theory. *Angewandte Chemie – International Edition*, **45**, 196.
430 Studt, F. and Tuczek, F. (2005) Energetics and mechanism of a room-temperature catalytic process for ammonia synthesis (Schrock cycle): Comparison with biological nitrogen fixation. *Angewandte Chemie – International Edition*, **44**, 5639.
431 Fryzuk, M.D. and Johnson, S.A. (2000) The continuing story of dinitrogen activation. *Coordination Chemistry Reviews*, **200–202**, 379.
432 Shaver, M.P. and Fryzuk, M.D. (2003) Activation of molecular nitrogen: Coordination, cleavage and functionalization of N_2 mediated by metal complexes. *Advanced Synthesis & Catalysis*, **345**, 1061.
433 Mori, M. (2004) Activation of nitrogen for organic synthesis. *Journal of Organometallic Chemistry*, **689**, 4210.
434 Hidai, M. (1999) Chemical nitrogen fixation by molybdenum and tungsten complexes. *Coordination Chemistry Reviews*, **185–186**, 99.
435 Knobloch, D.J., Toomey, H.E., and Chirik, P.J. (2008) Carboxylation of an *ansa*-zirconocene dinitrogen complex: Regiospecific hydrazine synthesis from N_2 and CO_2. *Journal of the American Chemical Society*, **130**, 4248.

Further Reading

Chapter 1

Berzelius, J. (1836) *Jber Berz*, **15**, 242.
Ertl, G. (2009) Wilhelm Ostwald: Founder of physical chemistry and Nobel laureate 1909. *Angewandte Chemie – International Edition*, **48**, 6600.
Ertl, G. and Gloyna, T. (2003) Katalyse: Vom Stein der Weisen zu Wilhelm Ostwald. *Zeitschrift Fur Physikalische Chemie*, **217**, 1207.
Hunt, L.B. (1958) The ammonia oxidation process for nitric acid manufacture. *Platinum Metals Review*, **2**, 129.
Mittasch, A. (1935) *Berzelius und die Katalyse*, Akademische Verlagsgesellschaft, Leipzig.
Ostwald, W. (1910) Über Katalyse, in *Les Prix Nobel en 1909*, Stockholm, p. 1.
Ostwald, W. (1923) *Über Katalyse (Ostwald's Klassiker der exakten Wissenschaften)* (ed. G. Bredig), Akademische Verlagsgesellschaft, Leipzig.
Ostwald, W., Über Katalyse (Lecture held at the 73rd Naturforscherversammlung in Hamburg on 26 September 1901), in Ostwald, W. (1923), p. 23.
Robertson, A.J.B. (1975) The early history of catalysis. *Platinum Metals Review*, **19**, 64.
Taube, R. (2004) Wilhelm Ostwald und die Katalyse, in *Jahrbuch 2003 der Deutschen Akademie der Naturforscher Leopoldina (Halle/Saale)*, Leopoldina (R. 3), **49**, 369.
Walden, P. (1930) Berzelius und wir. *Zeitschrift für Angewandte Chemie*, **43**, 325; **43**, 351; **43**, 366.

Chapter 2

Chorkendorff, I. and Niemantsverdriet, J.W. (2003) *Concepts of Modern Catalysis and Kinetics*, Wiley-VCH, Weinheim.
Espenson, J.H. (2003) Kinetics of catalyzed reactions – homogeneous, in [M9], vol. 4, p. 490.
Haim, A. (1989) Catalysis: New reaction pathways, not just a lowering of the activation energy. *Journal of Chemical Education*, **66**, 935.
Herrmann, W.A. and Cornils, B. (1997) Organometallic homogeneous catalysis – Quo vadis? *Angewandte Chemie – International Edition in English*, **36**, 1049.
Laidler, K.J. (1988) Rate-controlling step: A necessary or useful concept? *Journal of Chemical Education*, **65**, 250.
Trost, B.M. (1995) Atom economy – A challenge for organic synthesis: Homogeneous catalysis leads the way. *Angewandte Chemie – International Edition in English*, **34**, 259.

Chapter 3

Cotton, F.A., Wilkinson, G., Murillo, C.A., and Bochmann, M. (1999) *Advanced Inorganic Chemistry*, 6th edn, John Wiley & Sons, New York.
Jordan, R.B. (2007) *Reaction Mechanisms of Inorganic and Organometallic Systems*, 3rd edn, University Press, Oxford.

Chapter 4

Bakhmutov, V.I. (2005) Proton transfer to hydride ligands with formation of dihydrogen complexes: A physicochemical view. *European Journal of Inorganic Chemistry*, 245.
Börner, A. (2001) The effect of internal hydroxy groups in chiral diphosphane rhodium(I) catalysts on the asymmetric hydrogenation of functionalized olefins. *European Journal of Inorganic Chemistry*, 327.
Brown, J.M. and Giernoth, R. (2000) New mechanistic aspects of the asymmetric homogeneous hydrogenation of alkenes. *Current Opinion in Drug Discovery & Development*, **3**, 825.
Crabtree, R.H. and Hamilton, D.G. (1988) H−H, C−H, and related sigma-bonded groups as ligands. *Advances in Organometallic Chemistry*, **28**, 299.
Cui, X. and Burgess, K. (2005) Catalytic homogeneous asymmetric hydrogenations of largely unfunctionalized alkenes. *Chemical Reviews*, **105**, 3272.
Daniel, C., Koga, N., Han, J., Fu, X.Y., and Morokuma, K. (1988) Ab initio MO study of the full catalytic cycle of olefin hydrogenation by the Wilkinson catalyst $RhCl(PR_3)_3$. *Journal of the American Chemical Society*, **110**, 3773.

Heinekey, D.M. and OldhamJr., W.J. (1993) Coordination chemistry of dihydrogen. *Chemical Reviews*, **93**, 913.

James, B.R. (1979) Hydrogenation reactions catalyzed by transition metal complexes. *Advances in Organometallic Chemistry*, **17**, 319.

Knowles, W.S. (2002) Asymmetric hydrogenations (Nobel Lecture). *Angewandte Chemie – International Edition*, **41**, 1998.

Komarov, I.V. and Börner, A. (2001) Highly enantioselective or not? – Chiral monodentate monophosphorus ligands in the asymmetric hydrogenation. *Angewandte Chemie – International Edition*, **40**, 1197.

Kubas, G.J. (2004) Heterolytic splitting of H–H, Si–H, and other σ bonds on electrophilic metal centers. *Advances in Inorganic Chemistry*, **56**, 127.

Noyori, R. and Hashiguchi, S. (1997) Asymmetric transfer hydrogenation catalyzed by chiral ruthenium complexes. *Accounts of Chemical Research*, **30**, 97.

Noyori, R. and Ohkuma, T. (2001) Asymmetric catalysis by architectural and functional molecular engineering: Practical chemo- and stereoselective hydrogenation of ketones. *Angewandte Chemie – International Edition*, **40**, 40.

Pfaltz, A., Blankenstein, J., Hilgraf, R., Hörmann, E., McIntyre, S., Menges, F., Schönleber, M., Smidt, S.P., Wüstenberg, B., and Zimmermann, N. (2003) Iridium-catalyzed enantioselective hydrogenation of olefins. *Advanced Synthesis & Catalysis*, **345**, 33.

Sánchez-Delgado, R.A. and Rosales, M. (2000) Kinetic studies as a tool for the elucidation of the mechanisms of metal complex-catalyzed homogeneous hydrogenation reactions. *Coordination Chemistry Reviews*, **196**, 249.

Zassinovic, G., Mestroni, G., and Gladiali, S. (1992) Asymmetric hydrogen transfer reactions promoted by homogeneous transition metal catalysts. *Chemical Reviews*, **92**, 1051.

Chapter 5

Clarke, M.L. (2005) Branched selective hydroformylation: A useful tool for organic synthesis. *Current Organic Chemistry*, **9**, 701.

Dry, M.E. (2002) The Fischer–Tropsch process: 1950–2000. *Catalysis Today*, **71**, 227.

Geerlings, J.J.C., Wilson, J.H., Kramer, G.J., Kuipers, H.P.C.E., Hoek, A., and Huisman, H.M. (1999) Fischer–Tropsch technology – From active site to commercial process. *Applied Catalysis A*, **186**, 27.

Herrmann, W.A. and Kohlpaintner, C.W. (1993) Water-soluble ligands, metal complexes, and catalysts: Synergism of homogeneous and heterogeneous catalysis. *Angewandte Chemie – International Edition in English*, **32**, 1524.

Jang, H.-Y. and Krische, M.J. (2004) Catalytic C–C bond formation via capture of hydrogenation intermediates. *Accounts of Chemical Research*, **37**, 653.

Kalck, P., Peres, Y., and Jenck, J. (1991) Hydroformylation catalyzed by ruthenium complexes. *Advances in Organometallic Chemistry*, **32**, 121.

Koga, N. and Morokuma, K. (1989) Potential energy surface of olefin hydrogenation by Wilkinson catalyst. *ACS Symposium Series*, **394**, 77.

Kohlpaintner, C.W., Fischer, R.W., and Cornils, B. (2001) Aqueous biphasic catalysis: Ruhrchemie/Rhône-Poulenc oxo process. *Applied Catalysis A*, **221**, 219.

Laurent, P., Le Bris, N., and des Abbayes, H. (2002) Rhodium complexes of bidentate and potentially hemilabile phosphorus ligands for hydroformylation of styrene at low temperature. *Trends in Organometallic Chemistry*, **4**, 131.

Lunsford, J.H. (2000) Catalytic conversion of methane to more useful chemicals and fuels: A challenge for the 21st century. *Catalysis Today*, **63**, 165.

Maitlis, P.M., Quyoum, R., Long, H.C., and Turner, M.L. (1999) Towards a chemical understanding of the Fischer–Tropsch reaction: Alkene formation. *Applied Catalysis A*, **186**, 363.

Masters, C. (1979) The Fischer-Tropsch reaction. *Advances in Organometallic Chemistry*, **17**, 61.

Musaev, D.G., Matsubara, T., Mebel, A.M., Koga, N., and Morokuma, K. (1995) Ab initio molecular orbital studies of elementary reactions and homogeneous catalytic cycles with organometallic compounds. *Pure and Applied Chemistry*, **67**, 257.

Pidun, U. and Frenking, G. (1998) [HRh(CO)₄]-catalyzed hydrogenation of CO: A systematic ab initio quantum-chemical investigation of the reaction mechanism. *Chemistry – A European Journal*, **4**, 522.

Pruett, R.L. (1979) Hydroformylation. *Advances in Organometallic Chemistry*, **17**, 1.

Schulz, H. (1999) Short history and present trends of Fischer–Tropsch synthesis. *Applied Catalysis A*, **186**, 3.

Trzeciak, A.M. and Ziolkowski, J.J. (1999) Perspectives of rhodium organometallic catalysis. Fundamental and applied aspects of hydroformylation. *Coordination Chemistry Reviews*, **190–192**, 883.

Ujaque, G. and Maseras, F. (2004) Applications of hybrid DFT/molecular mechanics to homogeneous catalysis. *Structure & Bonding*, **112**, 117.

Chapter 6

Drennan, C.L., Doukov, T.I., and Ragsdale, S.W. (2004) The metalloclusters of carbon monoxide dehydrogenase/acetyl-CoA synthase: A story in pictures. *Journal of Biological Inorganic Chemistry: JBIC*, **9**, 511.

Ford, P.C. (1981) The water gas shift reaction: Homogeneous catalysis by ruthenium and other metal carbonyls. *Accounts of Chemical Research*, **14**, 31.

Forster, D. (1979) Mechanistic pathways in the catalytic carbonylation of methanol by rhodium and iridium complexes. *Advances in Organometallic Chemistry*, **17**, 255.

Hallinan, N. and Hinnenkamp, J. (2001) Rhodium catalyzed methanol carbonylation: New low water technology. *Chemistry & Industry (Dekker)*, **82**, 545.

Hill, A.F. (2000) "Simple" carbonyls of ruthenium: New avenues from the Hieber base reaction. *Angewandte Chemie – International Edition*, **39**, 130.

Kinnunen, T. and Laasonen, K. (2001) Reaction mechanism of the reductive elimination in the catalytic carbonylation of methanol. A density functional study. *Journal of Organometallic Chemistry*, **628**, 222.

Sunley, G.J. and Watson, D.J. (2000) High productivity methanol carbonylation catalysis using iridium. The Cativa™ process for the manufacture of acetic acid. *Catalysis Today*, **58**, 293.

Torrent, M., Solà, M., and Frenking, G. (2000) Theoretical studies of some transition-metal-mediated reactions of industrial and synthetic importance. *Chemical Reviews*, **100**, 439.

Volbeda, A. and Fontecilla-Camps, J.C. (2005) Structural bases for the catalytic mechanism of Ni-containing carbon monoxide dehydrogenases. *Dalton Transactions*, 3443.

Chapter 7

Basset, J.-M., Lefebvre, F., and Santini, C. (1998) Surface organometallic chemistry: Some fundamental features including the coordination effects of the support. *Coordination Chemistry Reviews*, **178–180**, 1703.

Buchmeiser, M.R. (2000) Homogeneous metathesis polymerization by well-defined group VI and group VIII transition-metal alkylidenes: Fundamentals and applications in the preparation of advanced materials. *Chemical Reviews*, **100**, 1565.

Chauvin, Y. (2006) Olefin metathesis: The early days (Nobel Lecture). *Angewandte Chemie – International Edition*, **45**, 3740.

Fomine, S., Vargas, S.M., and Tlenkopatchev, M.A. (2003) Molecular modeling of ruthenium alkylidene mediated olefin metathesis reactions. DFT study of reaction pathways. *Organometallics*, **22**, 93.

Grubbs, R.H. (ed.) (2003) *Handbook of Metathesis*, vol. 1–3, Wiley-VCH, Weinheim.

Lam, W.H., Jia, G., Lin, Z., Lau, C.P., and Eisenstein, O. (2003) Theoretical studies on the metathesis processes, [Tp(PH₃)MR-(η^2-H–CH₃)] → [Tp(PH₃)M(CH₃)(η^2-H–R)] (M=Fe, Ru, and Os; R=H and CH₃). *Chemistry – A European Journal*, **9**, 2775.

Le Roux, E., Chabanas, M., Baudouin, A., de Mallmann, A., Copéret, C., Quadrelli, E.A., Thivolle-Cazat, J., Basset, J.-M., Lukens, W., Lesage, A., Emsley, L., and Sunley, G.J. (2004) Detailed structural investigation of the grafting of [Ta(=CH*t*Bu)(CH₂*t*Bu)₃] and [Cp*TaMe₄] on silica partially dehydroxylated at 700 °C and the activity of the grafted complexes toward alkane metathesis. *Journal of the American Chemical Society*, **126**, 13391.

Sattely, E.S., Cortez, G.A., Moebius, D.C., Schrock, R.R., and Hoveyda, A.H. (2005)

Enantioselective synthesis of cyclic amides and amines through Mo-catalyzed asymmetric ring-closing metathesis. *Journal of the American Chemical Society*, **127**, 8526.

Schrock, R.R. (1990) Living ring-opening metathesis polymerization catalyzed by well-characterized transition metal alkylidene complexes. *Accounts of Chemical Research*, **23**, 158.

Schrock, R.R. (1998) Olefin metathesis by well-defined complexes of molybdenum and tungsten. *Topics in Organometallic Chemistry*, **1**, 1.

Schrock, R.R. (2005) High oxidation state alkylidene and alkylidyne complexes. *Chemical Communications*, 2773.

Solans-Monfort, X., Clot, E., Copéret, C., and Eisenstein, O. (2005) Understanding structural and dynamic properties of well-defined rhenium-based olefin metathesis catalysts, Re($\equiv$CR)(=CHR)(X)(Y), from DFT and QM/MM calculations. *Organometallics*, **24**, 1586.

Thieuleux, C., Copéret, C., Dufaud, V., Marangelli, C., Kuntz, E., and Basset, J.-M. (2004) Heterogeneous well-defined catalysts for metathesis of inert and not so inert bonds. *Journal of Molecular Catalysis A*, **213**, 47.

Trnka, T.M. and Grubbs, R.H. (2001) The development of L_2X_2Ru=CHR olefin metathesis catalysts: An organometallic success story. *Accounts of Chemical Research*, **34**, 18.

Van Veldhuizen, J.J., Gillingham, D.G., Garber, S.B., Kataoka, O., and Hoveyda, A.H. (2003) Chiral Ru-based complexes for asymmetric olefin metathesis: Enhancement of catalyst activity through steric and electronic modifications. *Journal of the American Chemical Society*, **125**, 12502.

Vyboishchikov, S.F. and Thiel, W. (2005) Ring-closing olefin metathesis on ruthenium carbene complexes: Model DFT study of stereochemistry. *Chemistry – A European Journal*, **11**, 3921.

Zhu, J., Jia, G., and Lin, Z. (2006) Theoretical investigation of alkyne metathesis catalyzed by W/Mo alkylidyne complexes. *Organometallics*, **25**, 1812.

Chapter 8

Deckers, P.J.W., Hessen, B., and Teuben, J.H. (2002) Catalytic trimerization of ethene with highly active cyclopentadienyl–arene titanium catalysts. *Organometallics*, **21**, 5122.

Hessen, B. (2004) Monocyclopentadienyl titanium catalysts: Ethene polymerisation versus ethene trimerisation. *Journal of Molecular Catalysis A*, **213**, 129.

Speiser, F., Braunstein, P., and Saussine, L. (2005) Catalytic ethylene dimerization and oligomerization: Recent developments with nickel complexes containing P,N-chelating ligands. *Accounts of Chemical Research*, **38**, 784.

Wilke, G. (1988) Contributions to organo-nickel chemistry. *Angewandte Chemie – International Edition in English*, **27**, 185.

Chapter 9

Alt, H.G. (ed.) (2006) Metallocene complexes as catalysts for olefin polymerization. *Coordination Chemistry Reviews*, **250**, 1–272.

Alt, H.G. and Köppl, A. (2000) Effect of the nature of metallocene complexes of group IV metals on their performance in catalytic ethylene and propylene polymerization. *Chemical Reviews*, **100**, 1205.

Angermund, K., Fink, G., Jensen, V.R., and Kleinschmidt, R. (2000) Toward quantitative prediction of stereospecificity of metallocene-based catalysts for α-olefin polymerization. *Chemical Reviews*, **100**, 1457.

Butenschön, H. (2000) Cyclopentadienylmetal complexes bearing pendant phosphorus, arsenic, and sulfur ligands. *Chemical Reviews*, **100**, 1527.

Chan, M.S.W., Deng, L., and Ziegler, T. (2000) Density functional study of neutral salicylaldiminato nickel(II) complexes as olefin polymerization catalysts. *Organometallics*, **19**, 2741.

Coates, G.W. (2002) Polymerization catalysis at the millennium: Frontiers in stereoselective, metal-catalyzed polymerization. *Journal of The Chemical Society-Dalton Transactions*, 467.

Deng, L., Margl, P., and Ziegler, T. (1997) A density functional study of nickel(II) diimide catalyzed polymerization of ethylene. *Journal of the American Chemical Society*, **119**, 1094.

Deng, L., Woo, T.K., Cavallo, L., Margl, P.M., and Ziegler, T. (1997) The role of bulky substituents in Brookhart-type Ni(II) diimine catalyzed olefin polymerization: A combined density functional theory and molecular mechanics study. *Journal of the American Chemical Society*, **119**, 6177.

Guan, Z. (ed.) *Metal Catalysts in Olefin Polymerization* (Topics in Organometallic Chemistry 2009, **26**).

Guerra, G., Longo, P., Cavallo, L., Corradini, P., and Resconi, L. (1997) Relationship between regiospecificity and type of stereospecificity in propene polymerization with zirconocene-based catalysts. *Journal of the American Chemical Society*, **119**, 4394.

Hlatky, G.G. (2000) Heterogeneous single-site catalysts for olefin polymerization. *Chemical Reviews*, **100**, 1347.

Leek, Y.v.d., Angermund, K., Reffke, M., Kleinschmidt, R., Goretzki, R., and Fink, G. (1997) On the mechanism of stereospecific polymerization – Development of a universal model to demonstrate the relationship between metallocene structure and polymer microstructure. *Chemistry – A European Journal*, **3**, 585.

Margl, P., Deng, L., and Ziegler, T. (1998) A unified view of ethylene polymerization by d^0 and d^0f^n transition metals. 1. Precursor compounds and olefin uptake energetics. *Organometallics*, **17**, 933.

Margl, P., Deng, L., and Ziegler, T. (1999) A unified view of ethylene polymerization by d^0 and d^0f^n transition metals. 3. Termination of the growing polymer chain. *Journal of the American Chemical Society*, **121**, 154.

McKnight, A.L. and Waymouth, R.M. (1998) Group 4 *ansa*-cyclopentadienyl-amido catalysts for olefin polymerization. *Chemical Reviews*, **98**, 2587.

Mecking, S., Held, A., and Bauers, F.M. (2002) Aqueous catalytic polymerization of olefins. *Angewandte Chemie – International Edition*, **41**, 544.

Natta, G. (1964) Von der stereospezifischen Polymerisation zur asymmetrischen autokatalytischen Synthese von Makromolekülen. *Angew Chem*, **76**, 553 (Nobel-Vortrag). The English version "From the stereospecific polymerization to the asymmetric autocatalytic synthesis of macromolecules" see "http://nobelprize.org."

Rappe, A.K., Skiff, W.M., and Casewit, C.J. (2000) Modeling metal-catalyzed olefin polymerization. *Chemical Reviews*, **100**, 1435.

Scheirs, J. and Kaminsky, W. (eds) (1999) *Metallocene-based Polyolefins*, vol. 1–2, John Wiley & Sons, Chichester.

Siemeling, U. (2000) Chelate complexes of cyclopentadienyl ligands bearing pendant *O*-donors. *Chemical Reviews*, **100**, 1495.

Stephan, D.W. (2005) The road to early-transition-metal phosphinimide olefin polymerization catalysts. *Organometallics*, **24**, 2548.

Yoshida, T., Koga, N., and Morokuma, K. (1996) A combined ab initio MO–MM study on isotacticity control in propylene polymerization with silylene-bridged group 4 metallocenes. C_2 symmetrical and asymmetrical catalysts. *Organometallics*, **15**, 766.

Yoshida, Y., Matsui, S., and Fujita, T. (2005) Bis(pyrrolide-imine) Ti complexes with MAO: A new family of high performance catalysts for olefin polymerization. *Journal of Organometallic Chemistry*, **690**, 4382.

Ziegler, K. (1964) Folgen und Werdegang einer Erfindung (Nobel-Vortrag). *Angewandte Chemie*, **76**, 545. The English version "Consequences and development of an invention" see "http://nobelprize.org."

Chapter 10

Bönnemann, H. (1973) The allylcobalt system. *Angewandte Chemie – International Edition in English*, **12**, 964.

Erker, G. (2003) The (butadiene)metal complex/B(C_6F_5)$_3$ pathway to homogeneous single component Ziegler–Natta catalyst systems. *Chemical Communications*, 1469.

Heuck, C. (1970) Ein Beitrag zur Geschichte der Kautschuk-Synthese: Buna-Kautschuk

IG (1926–1945). *Chemie Zeitung*, **94**, 147.

Thiele, S.K.-H. and Wilson, D.R. (2003) Alternate transition metal complex based diene polymerization. *J Macromol Sci*, **C43**, 581.

Tobisch, S. (2003) Ni⁰-catalyzed cyclotrimerization of 1,3-butadiene: A comprehensive density functional investigation on the origin of the selectivity. *Chemistry – A European Journal*, **9**, 1217.

Tobisch, S. and Ziegler, T. (2002) [Ni⁰L]-catalyzed cyclodimerization of 1,3-butadiene: A density functional investigation on the influence of electronic and steric factors on the regulation of the selectivity. *Journal of the American Chemical Society*, **124**, 13290.

Tsuji, J. (1973) Addition reactions of butadiene catalyzed by palladium complexes. *Accounts of Chemical Research*, **6**, 8.

Chapter 11

Beletskaya, I.P. and Cheprakov, A.V. (2004) Copper in cross-coupling reactions. The post-Ullmann chemistry. *Coordination Chemistry Reviews*, **248**, 2337.

Bellina, F., Carpita, A., and Rossi, R. (2004) Palladium catalysts for the Suzuki cross-coupling reaction: An overview of recent advances. *Synthesis*, 2419.

Dedieu, A. (2000) Theoretical studies in palladium and platinum molecular chemistry. *Chemical Reviews*, **100**, 543.

King, A.O. and Yasuda, N. (2004) Palladium-catalyzed cross-coupling reactions in the synthesis of pharmaceuticals. *Topics in Organometallic Chemistry*, **6**, 205.

Lin, B.-L., Liu, L., Fu, Y., Luo, S.-W., Chen, Q., and Guo, Q.-X. (2004) Comparing nickel- and palladium-catalyzed Heck reactions. *Organometallics*, **23**, 2114.

Littke, A.F. and Fu, G.C. (2002) Palladium-catalyzed coupling reactions of aryl chlorides. *Angewandte Chemie – International Edition*, **41**, 4176.

Molander, G.A. and Canturk, B. (2009) Organotrifluoroborates and monocoordinated palladium complexes as catalysts – A perfect combination for Suzuki–Miyaura coupling. *Angewandte Chemie – International Edition*, **48**, 9240.

Shinokubo, H. and Oshima, K. (2004) Transition metal-catalyzed carbon–carbon bond formation with Grignard reagents – Novel reactions with a classic reagent. *European Journal of Organic Chemistry*, 2081.

Tykwinski, R.R. (2003) Evolution in the palladium-catalyzed cross-coupling of sp- and sp²-hybridized carbon atoms. *Angewandte Chemie – International Edition*, **42**, 1566.

Zapf, A. and Beller, M. (2002) see Ref. [241].

Chapter 12

Beller, M., Seayad, J., Tillak, A., and Jiao, H. (2004) Catalytic Markovnikov and anti-Markovnikov functionalization of alkenes and alkynes: Recent developments and trends. *Angewandte Chemie – International Edition*, **43**, 3368.

Brunner, H. (2004) A new hydrosilylation mechanism – New preparative opportunities. *Angewandte Chemie – International Edition*, **43**, 2749.

Burling, S., Field, L.D., Messerle, B.A., and Turner, P. (2004) Intramolecular hydroamination catalyzed by cationic rhodium and iridium complexes with bidentate nitrogen-donor ligands. *Organometallics*, **23**, 1714.

Bytschkov, I. and Doye, S. (2003) Group-IV metal complexes as hydroamination catalysts. *European Journal of Organic Chemistry*, 935.

Doye, S. (2004) Development of the Ti-catalyzed intermolecular hydroamination of alkynes. *Synlett*, 1653.

Hayashi, T. (1999) Hydrosilylation of carbon-carbon double bonds, in *Comprehensive Asymmetric Catalysis I–III*, vol. 1 (eds E.N. Jacobsen, A. Pfaltz, and H. Yamamoto), Springer, Berlin, p. 319.

Kim, B.-H., Cho, M.-S., and Woo, H.-G. (2004) Si–Si/Si–C/Si–O/Si–N coupling of hydrosilanes to useful silicon-containing materials. *Synlett*, 761.

Marciniec, B. and Pietraszuk, C. (2004) Synthesis of silicon derivatives with ruthenium catalysts. *Topics in Organometallic Chemistry*, **11**, 197.

Molander, G.A. and Romero, J.A.C. (2002) Lanthanocene catalysts in selective organic synthesis. *Chemical Reviews*, **102**, 2161.

Ozawa, F. (2000) The chemistry of organo-(silyl)platinum(II) complexes relevant to catalysis. *Journal of Organometallic Chemistry*, **611**, 332.

Pohlki, F. and Doye, S. (2003) The catalytic hydroamination of alkynes. *Chemical Society Reviews*, **32**, 104.

Roesky, P.W. and Müller, T.E. (2003) Enantioselective catalytic hydroamination of alkenes. *Angewandte Chemie – International Edition*, **42**, 2708.

Senn, H.M., Blöchl, P.E., and Togni, A. (2000) Toward an alkene hydroamination catalyst: Static and dynamic ab initio DFT studies. *Journal of the American Chemical Society*, **122**, 4098.

Thiel, W.R. (2003) On the way to a new class of catalysts – High-valent transition-metal complexes that catalyze reductions. *Angewandte Chemie – International Edition*, **42**, 5390.

Chapter 13

Adam, W. and Wirth, T. (1999) Hydroxy group directivity in the epoxidation of chiral allylic alcohols: Control of diastereoselectivity through allylic strain and hydrogen bonding. *Accounts of Chemical Research*, **32**, 703.

Bodkin, J.A. and McLeod, M.D. (2002) The sharpless asymmetric aminohydroxylation. *Journal of The Chemical Society-Perkin Transactions 1*, **1**, 2733.

Jia, C., Kitamura, T., and Fujiwara, Y. (2001) Catalytic functionalization of arenes and alkanes via C–H bond activation. *Accounts of Chemical Research*, **34**, 633.

Sen, A. (1998) Catalytic functionalization of carbon–hydrogen and carbon–carbon bonds in protic media. *Accounts of Chemical Research*, **31**, 550.

Senn, H.M. and Thiel, W. (2009), see Ref. [13].

Tsuji, J. (1999) Recollections of organopalladium chemistry. *Pure and Applied Chemistry*, **71**, 1539.

Chapter 14

Dance, I. (2007) Elucidating the coordination chemistry and mechanism of biological nitrogen fixation. *Chemistry – An Asian Journal*, **2**, 936.

Dos Santos, P.C., Dean, D.R., Hu, Y., and Ribbe, M.W. (2004) Formation and insertion of the nitrogenase iron–molybdenum cofactor. *Chemical Reviews*, **104**, 1159.

Ertl, G. (2008) Reactions at surfaces: From atoms to complexity (Nobel Lecture). *Angewandte Chemie – International Edition*, **47**, 3524.

Holland, P.L. (2004) Nitrogen fixation, in [M11], vol. 8, p. 569.

Howard, J.B. and Rees, D.C. (2006) How many metals does it take to fix N_2? A mechanistic overview of biological nitrogen fixation. *Proceedings of the National Academy of Sciences of the United States of America*, **103**, 17088.

Jacobsen, C.J.H., Dahl, S., Clausen, B.S., Bahn, S., Logadottir, A., and Nørskov, J.K. (2001) Catalyst design by interpolation in the periodic table: Bimetallic ammonia synthesis catalysts. *Journal of the American Chemical Society*, **123**, 8404.

Kreisel, G., Wolf, C., Weigand, W., and Dörr, M. (2003) Wie entstand das Leben auf der Erde? *Chemie in unserer Zeit*, **37**, 306.

Leigh, G.J. (2004) A personal account of some dinitrogen and organometallic chemistry research at the University of Sussex. *Journal of Organometallic Chemistry*, **689**, 3999.

MacKay, B.A. and Fryzuk, M.D. (2004) Dinitrogen coordination chemistry: On the biomimetic borderlands. *Chemical Reviews*, **104**, 385.

Sellmann, D., Utz, J., Blum, N., and Heinemann, F.W. (1999) On the function of nitrogenase FeMo cofactors and competitive catalysts: Chemical principles, structural blue-prints, and the relevance of iron sulfur complexes for N_2 fixation. *Coordination Chemistry Reviews*, **190–192**, 607.

Spencer, L.P., MacKay, B.A., Patrick, B.O., and Fryzuk, M.D. (2006) Inner-sphere two-electron reduction leads to cleavage and functionalization of coordinated dinitrogen. *Proceedings of the National Academy of Sciences of the United States of America*, **103**, 17094.

Tuczek, F. (2009) Electronic structure calculations: Dinitrogen reduction in nitrogenase and synthetic model systems, in *Encyclopedia of Inorganic Chemistry* (ed. R.H. Crabtree), Wiley-VCH, Weinheim.

Sources for Structures and Production Quantities of Key Chemicals

The CSD reference codes (Cambridge Crystallographic Data Centre: www.ccdc.cam.ac.uk) are given along with the sources of illustrations of the molecular structures.

Figure 4.11: Wasserman, H.J., Kubas, G.J., and Ryan, R.R. (1986) *Journal of the American Chemical Society*, **108**, 2294 (*CSD*: CEJDEA). Brammer, L., Howard, J.A.K., Johnson, O., Koetzle, T.F., Spencer, J.L., and Stringer, A.M. (1991) *Chemical Communications*, 241 (*CSD*: KILPEA).

Figure 5.6: Herrmann, W.A., Bauer, C., Huggins, J.M., Pfisterer, H., and Ziegler, M.L. (1983) *Journal of Organometallic Chemistry*, **258**, 81 (*CSD*: BELKOS10). Leung, P., Coppens, P., McMullan, R.K., and Koetzle, T.F. (1981) *Acta Crystallographica*, **37B**, 1347 (*CSD*: MEDYCO01). Boehme, R.F. and Coppens, P. (1981) *Acta Crystallographica*, **37B**, 1914 (*CSD*: BAHDIX).

Figure 6.9: Gloux, J., Gloux, P., and Laugier, J. (1996) *Journal of the American Chemical Society*, **118**, 11644 (*CSD*: RAPSEG).

Figure 7.1: Schrock, R.R., DePue, R.T., Feldman, J., Yap, K.B., Yang, D.C., Davis, W.M., Park, L., DiMare, M., Schofield, M., Anhaus, J., Walborsky, E., Evitt, E., Krüger, C., and Betz, P. (1990) *Organometallics*, **9**, 2262 (*CSD*: TACGAF). Schwab, P., Grubbs, R.H., and Ziller, J.W. (1996) *Journal of the American Chemical Society*, **118**, 100 (*CSD*: ZETLOZ10). Love, J.A., Sanford, M.S., Day, M.W., and Grubbs, R.H. (2003) *Journal of the American Chemical Society*, **125**, 10103 (*CSD*: VOMQUJ).

Figure 8.1: Kaschube, W., Pörschke, K.-R., Angermund, K., Krüger, C., and Wilke, G. (1988) *Chemische Berichte*, **121**, 1921 (*CSD*: GAYTOP01).

Figure 9.4: Yang, X., Stern, C.L., and Marks, T.J. (1994) *Journal of the American Chemical Society*, **116**, 10015 (*CSD*: YEKKKII).

Figure 9.6: Bajgur, C.S., Tikkanen, W.R., and Petersen, J.L. (1985) *Inorganic Chemistry*, **24**, 2539 (*CSD*: DEBZUF). Herrmann, W.A., Rohrmann, J., Herdtweck, E., Spaleck, W., and Winter, A. (1989) *Angewandte Chemie – International Edition in English*, **28**, 1511 (*CSD*: KEDMEL). Razavi, A. and Ferrara, J. (1992) *Journal of Organometallic Chemistry*, **435**, 299 (*CSD*: JUDFUJ).

Figure 9.7: Kaminsky, W., Rabe, O., Schauwienold, A.-M., Schupfner, G.U., Hanss, J., and Kopf, J. (1995) *Journal of Organometallic Chemistry*, **497**, 181 (*CSD*: ZEHKIT).

Figure 9.9: Schleis, T., Spaniol, T.P., Okuda, J., Heinemann, J., and Mülhaupt, R. (1998) *Journal of Organometallic Chemistry*, **569**, 159 (*CSD*: HIWCOF).

Figure 10.2: Huffman, J.C., Laurent, M.P., and Kochi, J.K. (1977) *Inorganic Chemistry*, **16**, 2639 (*CSD*: ALBPPT). Henc, B., Jolly, P.W., Salz, R., Stobbe, S., Wilke, G., Benn, R., Mynott, R., Seevogel, K., Goddard, R., and Krüger, C. (1980) *Journal of Organometallic Chemistry*, **191**, 449 (*CSD*: ALPHNI).

Figure 10.4: Mayer, W., Wilke, G., Benn, R., Goddard, R., and Krüger, C. (1985) *Monatshefte für Chemie*, **116**, 879 (*CSD*: DODYOQ). Huttner, G., Neugebauer, D., and Razavi, A. (1975) *Angewandte Chemie – International Edition in English*, **14**, 352 (*CSD*: BUTMNC). King Jr., J.A. and Vollhardt, K.P.C. (1983) *Organometallics*, **2**, 684 (*CSD*: CAGHOH).

Figure 10.11: Barnett, B., Büssemeier, B., Heimbach, P., Jolly, P.W., Krüger, C., Tkatchenko, I., and Wilke, G. (1972) *Tetrahedron Letters*, 1457 (*CSD*: IPRNIP).

Figure 10.15: Taube, R., Langlotz, J., Sieler, J., Gelbrich, T., and Tittes, K. (2000) *Journal of Organometallic Chemistry*, **597**, 92 (*CSD*: KODZOS). O'Connor, A.R., White, P.S., and Brookhart, M. (2007) *Journal of the American Chemical Society*, **129**, 4142.

Figure 12.4: Hitchcock, P.B., Lappert, M.F., and Warhurst, N.J.W. (1991) *Angewandte Chemie – International Edition*, **30**, 438 (*CSD*: TALDOZ).

Figure 12.6: [321] (*CSD*: ESOTEL).

Figure 14.1: Ghosh, R., Kanzelberger, M., Emge, T.J., Hall, G.S., and Goldman, A.S. (2006) *Organometallics*, **25**,

5668 (*CSD*: KEVJEB). Scheer, M., Müller, J., Schiffer, M., Baum, G., and Winter, R. (2000) *Chemistry – A European Journal*, **6**, 1252 (*CSD*: MAZRIO). Fryzuk, M.D., Haddad, T.S., Mylvaganam, M., McConville, D.H., and Rettig, S.J. (1993) *Journal of the American Chemical Society*, **115**, 2782 (*CSD*: SIMMEG10).

Figure 14.8/Figure 14.10: [407] (*PDB*: 1MIN).

Figure 14.12/Figure 14.14: Yandulov, D.V., Schrock, R.R., Rheingold, A.L., Ceccarelli, C., and Davies, W.M. (2003) *Inorganic Chemistry*, **42**, 796 (*CSD*: HUTQOC, HUTQIW, HUTRAB). [427] (*CSD*: FIVDIY, FIWLED).

Production quantities and capacities of key chemicals were kindly provided by:
SRI Consulting, a unit of IHS, Inc., Chemical Economics Handbook (Menlo Park, CA, USA): http://www.sriconsulting.com.
PlasticsEurope Deutschland e. V. (Frankfurt, Germany): http://www.plasticseurope.de.
International Rubber Study Group (Singapore): http://www.rubberstudy.com.

Index

a

absolute configuration 172
π-acceptor ligands/strength 29, 30, 123, 227, 228, 229, 256, 261, 369
acetaldehyde 21, 22, 295–304, 310
– annual production 297
– coal-chemistry based synthesis 21, 296
– synthesis by Wacker process 22, 295–302 see also Wacker process
acetamidocinnamic acid esters 54, 63
acetic acid 23, 101, 102, 297, 319, 326 see also carbonylation of methanol
– annual production 101
– biotechnological production 101
– Cativa process 23, 102, 108–111
– high-pressure BASF process 101, 102
– Hoechst-Celanese process 105, 106
– Monsanto process 23, 102–107, 112
acetic anhydride 106–108, 371
acid/base catalysis 9, 10, 395
acid cycle 107
acrylonitrile 241, 278
activator 18
activity of catalysts 12
activity of enzymes 24, 151, 187
acyclic diene metathesis (ADMET) polymerization/depolymerization 128, 129
acyclic diyne metathesis (ADIMET) polymerization 133
addition of nucleophiles 42–44, 45, 213, 288, 289, 290, 298, 300, 364
adenosine triphosphate (ATP) 329, 343, 344, 359
ADIMET polymerization 133
adiponitrile 274, 276
ADMET polymerization/depolymerization 128, 129

ae coordination 85, 87, 89
agostic C−H···M interactions 31, 38, 39, 40, 70, 79, 153, 155, 165, 166, 185, 196, 254, 299, 319
aldol reaction 80, 88, 297
ALFOL process 146
alkanes
– activation see C−H activation of alkanes
– dehydrogenation 141, 142
– functionalization see C−H functionalization of alkanes
alkyl–alkyl coupling reactions 255–257
alkylating function/agent 149, 164, 180, 193
alkyl–carbene complexes 40
β-alkyl elimination/transfer 38, 137, 378
alkyl–hydrido complexes 52, 53, 153, 282, 290, 321, 365, 379
alkylidene complexes 24, 40, 41, 97, 120, 121, 124, 129, 134, 140–142, 168, 323, 380
alkylidene groups, transposition/redistribution 117–119, 158, 372, 374, 381
alkylidyne complexes 41, 97, 131–133, 376, 377
alkylidyne groups, exchange 131
alkyne–nitrile cross-metathesis 133, 377
alkynes
– addition of nucleophiles 43
– complexes 32, 33, 131, 134
– hydroamination 293, 294, 397
– hydrocyanation see hydrocyanation of alkynes
– hydrogenation 51, 132, 304
– hydrosilylation see hydrosilylation of alkynes
– hydrozirconation 39
– insertion reactions 38
– metathesis see metathesis of alkynes
– oxidative coupling reactions 35, 36
– polymerization 131, 133, 377

Fundamentals of Organometallic Catalysis. First Edition. Dirk Steinborn
Copyright © 2012 WILEY-VCH Verlag GmbH & Co. KGaA, Weinheim

– Sonogashira coupling 253
allyl alcohol 24, 91, 301, 302, 370
– epoxidation of 313, 314
allylation reactions, asymmetric 267–269
   see also allylic alkylation/substitution
allyl complexes 207–211, 215–217, 222, 230, 267–269
– *anti/syn* isomerization 210, 211, 221, 222, 235, 237, 239, 240, 388
– NMR spectra 210
– rearrangements (π–σ/η^3–η^1; σ–π/η^1–η^3) 42, 208, 209, 267
– *syn/anti* isomerization 209, 221, 230, 235, 236
– *syn/anti* notation 209
allylic alkylation 264–269
– enantioselective 268
– mechanism 265, 266
allylic substitution 264
– chirality transfer in 267–269
allyl insertion 215–216, 221, 234, 235, 240
– mechanism 215, 216
allyl isomerization 221, 224, 230 see also allyl complexes, isomerization
Alphabutol process 152, 153
alternating copolymerization of olefins and CO 200–205, 387
– double CO insertion 201, 202, 204
– double ethene insertion 201, 202, 204, 387
– stereoselectivity 204
aluminum alkyls 145, 146, 147, 149, 158, 162, 164, 169, 175, 178, 193, 236, 242
aluminum hydride 136, 145
aluminum triethyl 145, 147
aluminum trimethyl 120, 179
amido complexes 131, 194, 294, 350, 352, 355, 358, 360, 397, 398, 403
amino acids 55, 56, 60, 88, 316, 329, 345, 348, 366
α-amino alkyl radicals 396
aminomethylation 397
aminopalladation 398
ammonia see also nitrogen fixation
– decomposition 2, 4
– synthesis 7, 23, 329, 330, 334, 342, 352
ammoxidation of propene 278
anionic polymerization 162, 207, 242
ansa-metallocenes 24, 182, 183, 187
anti addition 42, 43, 300–303, 398
anti/cis correlation 218, 221, 235
anti-lock-and-key relationship 57
anti-Markovnikov addition 273, 276, 280, 283, 285, 395

anti/syn isomerization see allyl complexes, isomerization
ARCM 130
arene–ruthenium complexes 75
AROM 130
arylpropionic acids 88, 277
associative hydride mechanism 53
associative mechanism/reaction 28, 33, 53, 124, 166, 343
asymmetric see also enantioselective reactions
– allylic alkylation 268, 269
– autocatalysis 18, 20
– cross coupling 256–257
– epoxidation 24, 313–315
– Heck reaction 263–264
– hydrocyanation 276–277
– hydroformylation 84–88
– hydrogenation 23, 54–68, 366
– hydrosilylation 284–285
– metathesis 130
– organocatalysis 9, 279, 395
– oxypalladation 305
– ring-closing metathesis (ARCM) 130
– ring-opening metathesis (AROM) 130
– transfer hydrogenation 75
atom transfer radical polymerization (ATRP) 387
atropisomerism 56, 256, 284
ATRP 387
Aufbau reaction see Ziegler growth reaction
autocatalysis 18, 20
auto-oxidation 20
azametallacyclobutadiene complexes 377
aza-Wacker cyclization 304, 398

b
back biting 126, 375
π back-donation 10, 29, 33, 68, 124, 286, 291, 338, 349, 354, 369, 397, 403
back skip 167, 188
balata 242
Beckmann rearrangement 224
Berry pseudorotation 210
Berzelius, annual reports 5
bifunctional catalysts 75, 278
bifunctional monomers 192, 385
BINAP 56, 59, 90, 263, 264, 284, 291, 305
BINAPHOS 85, 87
BINOL derivatives 279, 366, 395
biocatalysis 9
biomass 96, 329
biphasic catalysis 11, 24, 91, 92, 152, 157
(2,2'-bipyrimidine)platinum(II) complexes 326

BISBI 90
bis(2-ethylhexyl) phthalate 88
bis(imino)pyridine ligands 194
bite angle of chelating ligands 87, 89, 90
block copolymers 127, 198–200
bond dissociation enthalpies 49, 108, 200, 279, 288, 319, 330, 331, 337, 365
π-bond metathesis 140–143 *see also* metathesis
σ-bond metathesis 78, 135–137, 140, 293, 322
– of alkanes 135, 139, 323, 378
– of H_2 71, 135
– in hydrogenolysis of polyethylene 136
– of silanes 136
boronic acid derivatives 250, 251
Brønsted acid/base catalysis 9, 10, 395
Brønsted-Evans-Polanyi relation 340–342
Buna 207, 241
butadiene
– C–C linkage 22, 207, 208, 392
– cocyclization 232
– complexes 211–214, 219
– conformations 211, 212
– cyclo-cooligomerization 218, 219
– cyclodimerization 224–230
– cyclooligomerization 226–230
–– ligand control 227
–– quantum-chemical calculations 229
– cyclotrimerization 218–223
–– CDT, industrial synthesis 224
–– mechanism 220, 221
–– quantum-chemical calculations 223
– hydrocyanation 274–275
– hydrodimerization 231, 233
– linear oligomerization 230–233
– oligomerization 207, 208, 219
– polymerization *see* polymerization of butadiene
– telomerization 230–233
– trimerization 232
butanediol (butane-1,4-diol) 156, 157, 304

C

Cahn-Ingold-Prelog (CIP) rules 55, 366
calcium cyanamide 330
σ-CAM 135, 136
cane sugar, inversion of 6
carbene conformations, active/inactive 124, 125
carbene–hydrido complexes 40, 139–142, 166
carbene insertion reactions 40, 41, 140, 323
carbene ligands 36, 40, 84, 119–125, 140, 370
carbene mechanism 118
carbene transfer 119

carbido ligands 96, 97
carboalumination 39
carbometallation 38, 39
carbon dioxide 93, 94, 112, 114, 306, 361
– reduction of 93
carbon monoxide 45, 80, 93, 96, 105, 112–114, 200–202, 258, 331
carbon monoxide conversion 112, 114
carbon monoxide dehydrogenases (CODH) 114, 115
carbonylation of methanol 19, 23, 101–106, 108–111, 297, 372
– acid and salt cycle 107
– cobalt catalyzed 101, 102
– iodide cycle 103, 105–107, 111
– iridium catalyzed 23, 102, 108–111
– mechanism 103, 104, 110, 111
– process parameters 102
– production capacities 101
– quantum-chemical calculations 104, 108, 109
– rhodium catalyzed 23, 102–107, 112
– rhodium cycle 103
– selectivity 102, 105
carbonylation of methyl acetate 106–108, 371
carbonylative cross-coupling 258
carbonyl–hydrido complexes 29, 113, 369, 374
catalysis
– acceleration of a reaction by 6, 363
– kinetic definition 6
catalysis constant 12
catalysis with
– Ag 4, 307
– Al 145, 146
– Au 291
– Co 21, 77, 80, 101, 102, 135, 242, 243, 273, 311, 319
– Cr 101, 152, 154, 167, 168, 242
– Cu 23, 108, 254, 278, 295, 303, 305, 372, 400
– Fe 4, 7, 95, 113, 114, 195, 199, 249, 316, 326, 335, 338, 341, 344
– Hg 21, 325
– Ir 23, 59, 75, 102, 108, 142, 280, 290, 379, 396
– Ln 51, 72, 182, 194, 199, 207, 243, 292, 294
– Mn 101, 311, 314, 319
– Mo 24, 25, 121, 130, 131, 143, 157, 305, 309, 347, 352, 358
– Ni 22, 23, 24, 115, 147–151, 156–158, 162, 164, 196, 198, 200, 207, 218–232, 237–241, 243, 245, 249, 255, 272–278, 335, 392
– Os 338

- Pd 22, 55, 201, 204, 232, 245–269, 284, 290, 295–306, 320, 326, 391–394, 398, 400
- Pt 2, 3, 8, 87, 279–286, 290, 324, 326
- Re 119, 157, 312
- Rh 22, 23, 50–68, 75, 80–94, 102–109, 280, 285, 290
- Ru 24, 59, 72, 75, 94, 102, 113, 120, 121, 127, 134, 283, 340
- Ta 25, 138, 139, 378
- Ti 21, 22, 120, 152, 153, 162–165, 169, 175–178, 182, 224, 242, 243, 294, 309, 311, 313, 315, 380, 383,
- U 338
- V 194, 198, 242
- W 24, 119, 121, 131, 132, 133, 138, 152, 312, 380
- Zn 73
- Zr 136, 158, 165, 182–191, 294, 383

catalyst
- deactivation 11, 13, 163, 166
- generation/formation 11, 18, 28, 149
- optimization 10, 16, 25, 92, 122

catalyst complex 18
- chiral 175

catalyst library 59, 60, 61, 63, 94
catalyst–substrate complex 57, 58, 66, 85
Catalytica system 325, 326
catalytic cycle 11, 12, 16, 17
catalytic force 5
catalytic research, history 1–8
- Berzelius, catalysis concept 5, 6
- historical overview 2
- Ostwald's definition of catalysis 6–8
cation–anion interactions 181, 191, 237, 268
cationic polymerization 161, 244
Cativa process 23, 102, 108–111
- mechanism 110, 111
C–C activation 30, 32, 135
C–C coupling 21, 97, 99
C–C cross-coupling see cross-coupling
cellulose acetate 101
chain branching 167, 196, 198
chain carrier 161
chain end control, stereochemical 174, 175, 187, 188, 383
chain growth/propagation 98, 99, 127, 148, 163, 165, 184, 196, 198, 199, 240, 242, 386
chain initiation 98, 99, 244, 391
chain running 167, 195
chain shuttling polymerization 199
chain termination 98, 99, 148, 149, 163, 385
see also polymerization, chain termination
chain transfer 126, 127, 163 see also polymerization, chain transfer

Chalk-Harrod mechanism 281, 282
Chatt cycle 350, 355, 403
C–H activation 30, 32, 75, 135, 137, 143, 378, 400
- of alkanes 135, 319–325
- of aromatics 320, 400
Chauvin mechanism 22, 119, 374
chemical affinity 5, 6
chemisorption, dissociative 96, 335–337, 339, 341
chemoselective reactions 14, 15, 87, 131, 303
C–H functionalization of alkanes 143, 319, 321, 323–326
chiral oligomer 385
chlorohydrin process 308, 309
CIP rules 55, 366
classification of homogeneously catalyzed reactions 10
classification of ligands 29–30
C–N bond formation 288, 289
coal chemistry 22, 112, 296
coal liquefaction 95
co-carbonylation of MeOH/MeCOOMe 108, 371
cocatalyst 10, 18, 19, 103, 120, 147, 149, 151, 162, 164, 169, 171, 179, 198, 231, 289
CODH see carbon monoxide dehydrogenases
combinatorial catalysis 59–61, 94, 367
comonomer 152, 159, 169, 192, 195, 198, 233
σ-complex assisted metathesis (σ-CAM) 135, 136
π complexes 10, 28–30, 32, 35, 82, 211, 272
σ complexes 10, 28–31, 68–71, 135, 136, 286, 287, 322
constrained geometry catalysts (CGC) 194
contact catalyst 3, 9
contact reaction 2, 3
control/spectator ligand 10, 144, 227
conversion-time plots 13
coordination pocket 56, 57, 63, 67, 85, 144, 186–191, 196, 269, 353, 355, 393, 404
coordination polymerization 162, 165, 169, 170, 173, 198
coordinative chain transfer polymerization 199
copolymerization 155, 169, 178, 192, 199, 241, 382
- of olefins and CO see alternating copolymerization of olefins and CO
Cossee-Arlman mechanism 163
cross-coupling reaction
- after Hiyama 246, 251, 252, 257
- after Kumada 23, 245, 246, 249, 257
- after Murahashi 246, 249

- after Negishi 23, 25, 246, 249, 257
- after Sonogashira 246, 253, 254, 392
- after Stille 23, 246, 249, 252, 253, 258, 393
- after Suzuki 23, 25, 246, 249–252, 254, 256–258, 391, 392
- alkyl–alkyl 255, 256
- carbonylative 258, 392, 393
- dehydrogenative 321, 400
- enantioselective 256, 257
- iron catalyzed 249
- ligand effects 254, 255
- mechanism 246–248
- nickel catalyzed 23, 245, 246, 255, 256, 392
- overview 245, 246
- palladium catalyzed 245–258
- synthesis potential 248
- with Grignard reagents 23, 245, 246, 249
- with organolithium reagents 246, 249
- with organozinc reagents 23, 246, 249
cross-metathesis 117, 119, 122, 126, 133, 372, 377
CSD reference code 436
Curtin-Hammett principle 65, 67, 68, 202, 222, 228
cyanohydrins 278
2-cyano-4′-methylbiphenyl 250
cyclic transition state 49, 72, 74, 135, 253, 293, 378, 386, 396, 403
cycloaddition 35, 36, 118, 124, 135, 141, 277, 310, 361, 373, 397
cyclododecatriene (CDT) 218–223, 226–230, 232, 388
- industrial synthesis 224
cyclometallation 320, 321
cyclooctadiene (COD) 219, 224–230, 232
cyclooligomerization of butadiene see butadiene, cyclooligomerization
cyclopolymerization 192, 386
cyclopropanation 23, 265
cycloreversion 35, 118, 124, 135, 141, 310
cytochrome P-450 316–318, 326, 327

d

DACH 305
DBD-DIOP 87
DBFphos 90
dehydrocoupling of silanes 136
dehydrogenative silylation 282
deinsertion/extrusion of CO 45–46, 78, 82, 104, 205, 364, 374
dendrimers 284
density functional theory (DFT) 17, 25
depolymerization 129, 137
detergents 159

Dewar-Chatt-Duncanson model 33
diacetoxybutenes 93, 304
diad 172, 191, 385, 386
dialkyl tartrates (DAT) 24, 313
diastereomeric excess 14
diastereoselective reactions 14, 132, 269, 394
diastereotopic coordination pockets/sites 187–191
diastereotopic fragments/ligands 85, 183
dibenzylideneacetone 232, 247
dicyclopentadiene (DCPD) 128, 375
Difasol process 152
dihydrogen
- activation 68, 71–73
- activation in hydrogenases 73
- σ-bond metathesis 71, 72, 135, 136
- complexes 30, 34, 51, 68–71, 79
-- synthesis 69
- coproduction 329, 345, 352
- heterolytic cleavage 72, 73
- homolytic cleavage 72, 365
- oxidative addition 30, 34, 68
diimine ligands 194, 196
(diimine)nickel complexes 196–198
Dimersol process 151
2,3-dimethylbutadiene 241, 382
dimethyl carbonate 108, 372
dinitrogen see also nitrogen fixation
- complexes 331–333, 355–357, 403, 404
-- coordination modes of N_2 ligands 332, 333, 403, 404
-- functionalization 359–362
-- reduction 348–352
- electronic and bond characteristics 331
- functionalization 359–362
- reduction 344, 346, 352
dinitrogenase 342
dinitrogenase-reductase 342
dioctyl phthalate 88
DIOP 23, 56, 57, 87, 90, 305
DIPAMP 56–58
dissociative chemisorption see chemisorption, dissociative
dissociative hydride mechanism 51–52
dissociative mechanism/reaction 27, 33, 52, 81, 96, 110, 124, 166, 335–337, 339–341, 343, 344
divinylcyclobutane (DVCB) 224, 225
Döbereiner's lamp 4
domino reaction 250, 286 see also tandem reaction
π-donor ligands/strength 29, 33, 122, 155, 399
σ-donor ligands/strength 30, 120, 123, 124, 132, 141, 155, 227, 228, 261, 265, 286

DOPA 23, 55, 56, 59, 63, 64, 366, 368
double-bond isomerization 52, 88, 122, 143, 149, 152, 156–158, 263, 275, 276, 365, 370, 380, 391
double-bond shift, nonoxidative 301
double CO insertion 201, 202, 204
double displacement reaction 246
double ethene insertion 201, 202, 204, 387
DPEphos 90
dppb 90
dppe 90
DPPF 90, 291
dppm 90
dppp 90
DuPHOS 56, 90
DuPont adiponitrile process 274–276
Durham polyacetylene 375

e

edge-face arrangement 57
ee coordination 85, 87, 89
elastomers 127, 191, 192
electrofuge, electrofugal group 43
electronegativity 32, 252, 318, 323
electronic effects of phosphorous ligands 227, 254
electronic ligand parameter 16, 227, 228
electron transfer 10, 34, 287, 311, 317, 319, 324, 326, 343, 344
electron transfer mediators 296
electron-variable complexes 46, 114, 318
electrophilic abstraction 42, 364
electrophilic catalysis 9, 10
electrophilicity 123, 203, 204, 265, 291, 306, 311, 326, 397
electrophilic substitution 253, 322, 325, 326
elementary steps in organometallic catalysis *see* organometallic catalysis, elementary steps
elimination of CO 45
emulsion (co)polymerization 241
enantiofacial differentiation/selection 266, 285, 315
enantiomeric excess 14, 61, 66, 368
enantiomorphic site control 174, 175, 383
enantioselective *see also* asymmetric
enantioselective coordination sites 187–189
enantioselective reaction 14–15, 61–63
enantioselectivity, kinetically controlled *see* kinetically controlled enantioselectivity
enantiotopic coordination pockets/sites 186–188, 190
enantiotopic fragments/ligands 183, 184, 257

end groups of polymers/oligomers 127, 163, 165, 166, 172, 174, 203, 235, 242, 244, 383, 386, 390
enyne metathesis 133–135
enzyme catalysis 5, 9, 14, 57, 73, 95, 114, 244, 296, 305, 316, 318, 326, 329, 342–347, 390
EPDM elastomers 192
epoxidation
– of allyl alcohols, asymmetric 24, 313–314
– enantioselective 24, 313–315
– of ethene and propene 307–309, 311–313
– of olefins 306–315, 317
– – concerted mechanism 310
– – enantioselective 314–315
– – mechanism 309–311
– – quantum-chemical calculations 310
– – stepwise mechanism 310
– – with dioxygen 307
– – with hydrogen peroxide 311–313
– – with hydroperoxides 308, 309
– with oxometal complexes 306, 307
– with peroxometal complexes 306, 307
EPR 179, 198, 199
erythro, definition 301
erythro polymer 386
ester hydrolysis, acid catalyzed 6
ethene
– copolymerization 169, 178, 192, 198
– copolymerization with CO 200–205
– dimerization 147–152, 380
– direct transformation to propene 152, 380
– epoxidation *see* epoxidation of ethene
– hydroamination 288
– hydrocyanation 273, 274
– hydroformylation 77, 83
– hydrogenation 49, 51, 53
– metathesis 125, 126, 129
– oligomerization 24, 145–147, 156–158
– oxidation *see* Wacker process
– polymerization *see* polymerization of ethene
– tetramerization 155, 381
– trimerization 152–155
ethene–propene copolymer/rubber (EPR) 179, 198, 199
ethylene oxide 242, 300, 301, 307
ethyl process 146
external donor 178
extrusion of CO *see* deinsertion of CO
Eyring equation 12, 363

f

Famciclovir 269
fatty alcohols/acids 145, 146, 243
FeMo cofactor 342–347

Fe protein (cycle) 342–344
Fe–S clusters 73, 114, 115, 342–345, 347
Fischer carbene complexes 45, 122, 374
Fischer projection 173, 301, 366
Fischer-Tropsch synthesis 77, 95–100, 112
– alkenyl mechanism 98
– alkylidene mechanism 99, 100
– alkyl mechanism 97
fluidized-bed process 170, 171
fluxional molecules 208, 210, 212
forward donation 29
four-center transition state 135, 163, 323
Frank-Caro process 330
frustrated Lewis pairs 73

g

gasoline 96, 151, 162, 164
Gibbs free energy of activation 12, 21, 65, 363
Gibbs free energy of transition states 65, 66, 68, 83, 228, 368
glass (transition) temperature 127, 178, 192, 243
GLUP 56, 277
government rubber 241
Green-Rooney mechanism 166
Grignard formation reaction 44, 249
Grob fragmentation 43
Grubbs catalysts 24, 120–125, 127, 129, 133, 134
– quantum-chemical analysis 124, 125
Gulf process 146
Gulftene 146
gutta-percha 242

h

Haber-Bosch process 7, 329, 334–341, 349, 358
– model catalyst 337, 341
– technical catalyst 338–340
Halcon-ARCO process 308
half-sandwich complexes 194, 387
Hammett equation 392
Hammond postulate 68, 369
hard nucleophiles 264, 265, 267, 269
Hastelloy 102
HDPE see polyethylene, high-density
head-to-head linkage 151, 171, 172, 382
head-to-tail linkage 151, 171, 390
Heck reactions 23, 25, 246, 253, 258–264
– anion influence 260
– enantioselective 263–264, 393
– ligand effects 261, 262
– mechanism 259–261
– nonpolar route 260, 261

– polar route 260, 261
helical chirality 56
helical structure 185
heme group/ligand 316, 326, 342, 399
hemilabile ligands 150, 154, 155
hemitactic polymers 190, 191, 385
heterocubane structure 114, 312, 342, 344
N-heterocyclic carbene (NHC) see NHC ligand
heterogeneous catalysis 3, 9, 25
heterolytic cleavage of H_2 72, 73, 403
heterolytic fragmentation 42–44, 298, 302, 398
heteroolefins 35, 38, 232
Hieber base reaction 45, 112, 115
high-density polyethylene see polyethylene, high-density (HDPE)
high-impact polystyrene 241
high-temperature shift 112
high-throughput screening 60
Hiyama coupling 246, 251, 252, 257
Hoechst-Celanese process 105, 106
homogeneous catalysis 1, 9, 11, 21
HOMO–LUMO interactions 49, 50, 271, 311, 373
homolytic cleavage of H_2 72, 365
homometathesis 117, 124, 125, 372
homotopic coordination sites 183, 186, 187
homotopic fragments/ligands 183, 184
Hoveyda-Schrock catalysts 130
HPPO process 312
hybrid polymer 133
hydrazido complexes 333, 349, 350, 355, 359–361, 404
α-hydride elimination see α-hydrogen elimination
β-hydride elimination see β-hydrogen elimination
hydrido–alkylidene complexes 40, 139–142, 166
hydrido–olefin complexes 37, 38, 51, 53, 167, 197, 198, 298, 299, 322, 363, 365, 378
hydroalumination 39
hydroamination 287–294
– alkali metal amide catalysts 289
– of alkynes 294, 397
– of aminoalkynes 293, 294
– of aminoolefins 292, 293
– asymmetric 291, 293
– gold catalysts 291, 292
– lanthanoid catalysts 292–294
– mechanistic principles 288, 289
– of olefins 271, 287–289, 397
– platinum group metal catalysts 289–291
– of vinyl aromatics 291, 292
hydroboration 250, 283, 392, 396

hydrocarbonylation 83, 84, 102, 369
hydrocarboxylation of olefins 112
hydrocyanation
– of acetylene 278
– of alkynes 277, 278, 394
– of butadiene 274, 275
– DuPont adiponitrile process 274–276
– enantioselective 276, 277, 395
– mechanism 272–274
– of olefins 272–274, 394
– of pentenenitriles 275, 276
– of polar C=X bonds 278, 279, 395
– reaction profile 274
hydrodimerization of butadiene 231, 233
hydroformylation 21, 77–95, 156, 158, 233
– biphasic catalysis 91, 92, 370
– of branched olefins 88
– with carbon dioxide 93, 94
– cobalt catalysts 77–80
– C_4 selectivity 80, 91
– dissociative mechanism 81
– and double-bond isomerization 88, 89
– enantioselective 84–88
– of higher olefins 80, 88, 91, 94, 370
– mechanism 78, 79, 81, 82
– n/iso ratio 79, 80, 82, 89
– phosphane-modified cobalt catalysts 79, 80
– phosphane-modified rhodium catalysts 80–84
– platinum catalysts 87
– process parameters 80
– production quantities 88
– of propene 88, 89, 91
– quantum-chemical calculations 79, 83
– side reactions 80, 81
– unmodified rhodium catalysts 93
hydroformylation catalysts, hydrogenation activity of 80, 94
hydrogen see dihydrogen
hydrogen acceptor 74
hydrogenases 70, 73, 345
hydrogenation
– of aldehydes/ketones 72, 74, 75, 80, 88, 94
– of alkynes 51, 132, 304
– of arenes 51
– of dienes 51, 233
– of enamides 54, 63, 66, 67
– of imines 59, 73
– of olefins 22, 49–54, 233, 378, 382
– – enantioselective 23, 25, 54–58, 63–68, 87, 366
– – heterogeneous 49
– – hydride mechanism 54
– – – associative 53

– – – dissociative 51, 52
– – mechanism 51–54
– – monohydride mechanism 59
– – olefin mechanism 54
– – quantum-chemical calculations 53, 66
hydrogenation catalysts 50, 51, 55, 59, 60, 71, 72, 85, 87, 143
hydrogen bond 31, 94, 95, 347, 372
hydrogen donor 74
α-hydrogen elimination 40–41, 119, 139, 403
– coupled with reductive C–H elimination 40–41, 120, 152
β-hydrogen elimination 37–39, 41, 52, 74, 78, 82, 97, 113, 139, 145, 148, 155, 157, 158, 165, 195, 215, 231, 248, 256, 260, 263, 285, 298, 303, 322, 380, 392
– coupled with reductive C–H elimination 39, 41, 150, 153, 282, 363
hydrogenolysis of CO 96
hydrogenolysis of polyethylene 136, 137, 378
hydrogen peroxide, catalytic decomposition of 2, 4
hydrogen peroxide to propylene oxide (HPPO) process 312
hydrogen storage systems 71
hydrogen transfer 74, 75, 163, 235, 290, 311, 347, 369, 396, 397, 403
hydrometallation 38, 39
hydroperoxo complexes 309, 311, 316
hydrosilanes, σ complexes 286
hydrosilylation 271, 279–286
– of alkynes 285, 286
– enantioselective 284, 285
– industrial significance 283, 284
– mechanism 280–282
– of olefins 132, 279, 280, 282–284
– quantum-chemical calculations 281
– radical 279, 395
– transition metal catalyzed 279, 280
hydroxycarbonyl complexes 112, 113
hydrozirconation 39
hyperbranched polymer 286, 396

i

Ibuprofen 88
ICSD reference code 384
imido complexes 120, 130, 143, 194, 294, 350, 352, 355, 360, 362, 376, 397, 404
induction period 13, 18, 107
inhibitor 18, 20, 111, 303, 326
initiator 18, 20, 127, 161, 162, 170, 198, 200, 207, 242, 387, 391, 395
inorganic Grignard reagents 249

1,2-insertion (primary insertion)   173, 174, 192, 383, 385, 390
2,1-insertion (secondary insertion)   173, 174, 383
cis insertion   37, 184, 301, 302
insertion
– of alkynes   38, 39, 134, 394
– of allyl alcohols   302, 398
– of butadiene   215–218, 221, 222, 232, 235, 242, 245
– of carbon dioxide   113
– of CO   45–46, 78, 82, 104, 110, 201, 202, 204, 258, 364, 371
– of dioxygen   306
– of ethene   145, 148, 152, 157, 196, 201, 202, 205, 298, 398
– of heteroolefins   38
– notation   37
– of olefins   37–39, 52, 72, 78, 82, 85, 137, 140, 149, 163, 165, 167, 259, 261, 272, 277, 280, 288, 310, 363, 378, 383, 390, 393
– of propene   173, 174, 184–186
– of vinyl alcohol   300
insertionless migration   167, 188, 189, 190
insertion scheme   188, 189
in situ functionalization of polyolefins   192, 386
in situ polymerization   128
interchange mechanism   28
intermediate complex   9, 11, 16, 17, 19, 25, 28, 34, 40, 44–47
internal donor   177, 178
interstitial atom   97, 344, 347
iodide cycle   103, 105–107, 111
ionic liquid   91, 92, 94, 151, 152, 290
ion pair, chiral   395
iron, modifications   337
iron, surface structure   338, 339, 402
isomerization, kinetically controlled   275, 276
isoprene   207, 208, 225, 226, 234, 241, 242, 256, 389
isotactic index   175

j
Jeffrey-Larock conditions   262

k
Karstedt catalyst   280, 281, 396
Kellogg advanced ammonia process (KAAP)   341
kinetically controlled enantioselectivity   57, 58, 63–68, 366, 369
kinetically controlled isomerization   275, 276
kinetic inhibition   8, 282, 307

Kumada coupling   23, 245, 246, 249, 257
Kutscheroff process   21, 22

l
LAO   159
law of constant and multiple proportions   1
L-DOPA   23, 55, 56, 63, 64, 366, 368
LDPE *see* polyethylene, low-density
lead chamber process   2
Le Chatelier's principle   334
letter Buna   241
Lewis acid-base interaction/reaction   10, 29, 72, 149
Lewis acidity   149, 164, 180, 311
ligand attachment   27–30, 51, 78, 81, 82, 104, 113, 152, 220, 246, 298
ligand-centered molecular orbital   47
ligand cleavage   27–30, 31, 33, 51, 78, 81, 82, 104, 113, 152, 228, 238, 246, 265, 298
ligand dissociation   27
ligand effects   90, 227, 228, 254, 261
ligand-free catalyst systems   232, 262
ligand library   59–61, 63, 94
ligand-property control   227, 228 *see also* ligand tuning
ligands, classification of   29–30
ligand substitution   27–30, 36, 47, 221, 240, 265, 297, 298, 324, 331, 355, 397, 404
ligand tuning   10, 227 *see also* ligand-property control
Lindlar catalyst   132
linear α-olefins (LAOs)   91, 147, 156, 159
linear low-density polyethylene *see* polyethylene, linear low-density (LLDPE)
linear oligomerization of butadiene *see* butadiene, linear oligomerization
living polymerization   127, 198–200, 242, 244, 387
LLDPE *see* polyethylene, linear low-density
lock-and-key relationship   57, 58, 366
low-density polyethylene *see* polyethylene, low-density (LDPE)
low-temperature shift   112
lubricants   159

m
magnesium chloride, crystal and surface structure   176, 177, 384
magnetite (Fe_3O_4)   339
manganese, salen complexes   314
MAO *see* methylaluminoxanes
Markovnikov addition   84, 273, 276, 278, 284, 285, 291
Meerwein-Pondorf-Verley reduction   74

memory effect 268, 269
metal-carbonyl complexes 47, 113
metal-catalyzed polymerization 148, 162
metal-centered molecular orbital 46
metal–intermediate complex 11
metallacycles 35, 143, 185, 320, 380
metallacyclo-
– butadiene complexes 36, 131, 377
– butane/butene complexes 36, 118–120, 124, 134, 135, 141, 166, 394, 397
– heptane complexes 152, 154, 381
– pentane/pentene complexes 36, 150, 152–154, 212, 310
– propane/propene complexes 32, 33
metallocene catalysts 24, 149, 165, 179–192 see also polymerization
– activation and cocatalysts 179–182
– activity 186, 187
– C_2-symmetric 182–187
– C_s-symmetric 182–187
– with diastereotopic coordination pockets 187–191
– and polymer structure 189, 192
– productivity 187
– significance 165, 191, 192
– stereoregulation 184–186
– symmetry relationships in 187
metalloradical 323
metal–substrate complex 11
metathesis 22, 24, 25, 117–144
– of acyclic dienes 128–130
– of alkanes 25, 137–144, 378
– – mechanism 138–141, 378
– – quantum-chemical calculations 141, 142
– – via tandem reactions 141–143, 379
– of alkynes 131–133, 376, 377
– – mechanism 131
– of alkynes and nitriles 133, 377
– of cycloalkenes 125–128
– of cylcoalkynes 133
– enantioselective 130
– entropy driven 118, 381
– of enynes 133–135
– equilibrium composition 118, 158, 374, 381, 382
– of ethene oligomers 156–158
– non-pairwise (carbene) mechanism 118, 119, 374
– of olefins 22, 24, 25, 117–130, 152, 156–158, 372, 380
– – mechanism 118, 119, 123–125
– – quantum-chemical calculations 124, 125
– pairwise mechanism 119, 374
metathesis catalysts

– first generation 119, 121
– heterogeneous 117, 121, 157
– single-component 120, 121, 132
metathesis-like reaction (double displacement reaction) 246
methane activation and functionalization 319, 326
methane oxidation 325, 326
methanol 73, 203, 233, 309, 325 see also carbonylation of methanol
methoxycarbonyl complexes 203, 372
methyl acetate 105, 106, 108, 371
methylaluminoxanes (MAO) 24, 149, 153, 154, 165, 168, 179, 180, 186, 190, 193, 195, 198, 199
methylene ligand/complexes 40, 96, 97, 120, 124, 142
methyl rubber 241
methyltrioxorhenium(VII) 312
Metolachlor 59
microstructure of polymers 128, 168, 173, 175, 178, 186, 187, 189, 190, 191, 195, 199, 200, 385
migratory insertion 45, 78, 82, 84, 104, 110, 163, 167, 186, 189, 364, 365
Miller indices 176, 402
modified Chalk-Harrod mechanism 282
MoFe protein (cycle) 342–347
molar mass control see polymerization of olefins, molar mass control
molar mass distribution 127, 192, 243, 375
monohydride mechanism 59
monoligand palladium(0) intermediates 254, 255, 261, 262
monooxygenases 305, 315–318
monophosphane ligands, chiral (MOPs) 284, 285
Monsanto process 23, 102–107, 112
– mechanism 103, 104
MOPs 284, 285
Mortreux catalysts 131
Mülheim catalysts 163
Murahashi coupling 246, 249

n
nanoparticles 262, 339
Naproxen 59, 88, 277
n complex/donor 29, 30, 155
Negishi coupling 23, 25, 246, 249, 257
Newman projection 174, 301
NHC ligand 120, 122–124, 228, 255, 262, 280, 291
N-heterocyclic carbene (NHC) see NHC ligand
nickel effect 147, 162

nicotinamide adenine dinucleotide (NADH)   315
nicotinamide adenine dinucleotide phosphate (NADPH)   315
n/iso ratio   79, 80, 87, 89, 91, 94, 370
nitrido complexes   133, 345, 350, 355, 362, 377
nitrogen *see* dinitrogen
nitrogenases   329, 330, 334, 342–347, 349, 359
nitrogen fixation   25, 329
– enzyme-catalyzed   342–348
– – alternating reaction path   345, 346
– – distal reaction path   345, 346
– – Fe protein cycle   342–344
– – mechanism   345–346
– – MoFe protein cycle   344–347
– heterogeneously catalyzed   334–342 *see also* Haber-Bosch process
– – industrial catalyst   338–340
– – mechanism   335–338
– – ruthenium catalysts   340–342
– homogeneously catalyzed   348, 352–359
– – mechanism   354–355
– – quantum-chemical calculations   358
– prebiotic   347–348
NLE   61–63, 367
Nobel Prize laureates
– Bosch, C.   334
– Chauvin, Y.   22, 25, 118
– Ertl, G.   25, 335
– Fischer, E. O.   24, 45
– Grignard, V.   22, 245,
– Grubbs, R. H.   24, 25, 118
– Haber, F.   338
– Heck, R. F.   23, 25, 246
– Knowles, W. S.   23, 25, 56
– Kohn, W.   17, 25
– Langmuir, I.   7
– Merrifield, R. B.   60
– Natta, G.   21, 22, 163
– Negishi, E.-i.   23, 25, 246
– Noyori, R.   23, 25, 56
– Ostwald, W.   6, 7
– Pople, J. A.   17, 25
– Sabatier, P.   22, 49, 341
– Schrock, R. R.   24, 25, 118, 352
– Sharpless, K. B.   24, 25, 313
– Staudinger, H.   161
– Suzuki, A.   23, 25, 246
– Wilkinson, G.   22, 24, 49
– Ziegler, K.   21, 22, 163
nodule bacteria   342
nonalternating copolymerization   204, 205
noncyclic carbene ligands   123

nonlinear effects (NLE)   61–63, 367
nonmetallocene catalysts   193–200 *see also* polymerization
– catalyst formation   193
– of early transition metals   194
– of late transition metals   194–198
nonproductive metathesis   117, 124, 125, 142, 372
nonselective coordination site   186–191
NORPHOS   90
Norsorex   127
nucleofuge, nucleofugal group   43, 364
nucleophiles, addition of *see* addition of nucleophiles
nucleophilic addition/attack   109, 112, 115, 242, 265, 310, 311, 313, 324, 360, 394
nucleophilic catalysis   9, 10
nucleophilicity   92, 123, 392
number Buna   241
nylon   224, 274

O

octadienediylnickel complexes   214, 221, 222, 224, 225, 226, 228, 230
octadienylnickel complexes   231, 389
octane number   96, 151
octanol   233
octatrienes   230, 231, 232, 389
Octinoxate   262
olefin mechanism   54, 57
olefins
– activation of   78, 82, 272, 303
– addition of nucleophiles   42–44, 288, 289, 290, 298, 300, 364
– epoxidation *see* epoxidation of olefins
– hydroamination *see* hydroamination of olefins
– hydrocyanation *see* hydrocyanation of olefins
– hydroformylation *see* hydroformylation
– hydrogenation *see* hydrogenation of olefins
– hydrosilylation *see* hydrosilylation of olefins
– insertion of *see* insertion of olefins
– metathesis *see* metathesis of olefins
– oligomerization *see* oligomerization of olefins
– polymerization *see* polymerization of olefins
– substitution by nucleophiles   42
oligomerization
– of butadiene *see* butadiene, oligomerization
– of ethene *see* ethene, oligomerization
– of olefins   145–159, 174, 383
– of propene *see* propene, oligomerization
OMRP   200, 387

ON see oxidation number/state
one-electron oxidative addition 34, 387
one-electron reduction and oxidation 46, 47
orbital correlation diagram 373
organoboronic acids 250
organocatalysis/-catalysts 9, 279, 395
organometallic catalysis 9
– catalysts
– – activity and productivity 12, 13
– – conversion-time plots 13
– – selectivity and specificity 14, 15
– catalytic cycles 11, 12
– catalytic mechanisms, determination of 15–17
– development 21–25
– elementary steps
– – addition of nucleophiles 42–44
– – allyl insertion 215, 216, 218
– – butadiene insertion 215
– – carbene insertion reactions 40, 41
– – cleavage and coordination of ligands 27–30
– – heterolytic fragmentation 42–44
– – α-hydrogen elimination 40, 41
– – β-hydrogen elimination 37–39
– – insertion and extrusion of CO 45, 46
– – olefin insertion 37–39
– – one-electron reduction and oxidation 46, 47
– – oxidative addition 30–35, 216, 217
– – oxidative coupling 35–37, 214, 215
– – reductive cleavage 35–37, 214, 215
– – reductive elimination 30–35, 216, 217
organometallic inner complexes 143, 320
organometallic mixed catalysts 21, 163, 171, 207, 234
organometallic pincer complexes 142–144
organometallic radical polymerization (OMRP) 200, 387
origin of life 348
orthogonal functionalization 251
orthometallation 319, 320, 321, 400
orthopalladation 320
oscillating catalysts 191
Ostwald process 8
overlap integral 49, 271, 373
oxene 306, 307, 318
oxenoids 307
oxidases 305, 306
oxidation number/state
– of Cr in oligo-/polymerization catalysts 154, 168
– of Fe in cross-coupling catalysts 249
– of Fe in cytochrome P-450 enzymes 316, 317
– of metals in catalysts 72, 132, 307, 322, 354, 400, 405
– of metals in dinitrogen and related complexes 333, 354, 404, 405
– of metals in electron-variable complexes 46, 47, 318
– of metals in olefin/alkyne complexes 32, 33
– of metals in oxo/dioxygen complexes 318, 399
– of Pt in hydrosilylation catalysts 282
– of Ta in grafted complexes 138, 141, 378
oxidative addition 30–35
– of alkyl halides 255, 256, 392
– of allyl derivatives 216, 265, 266, 267
– bimolecular 34, 200, 387
– of C–C bonds 225, 275, 378
– of C–H bonds 52, 319, 321, 322, 379
– of C–X bonds 246, 248, 252, 254, 259, 261, 320, 391, 393
– of H_2 51, 53, 57, 68, 70, 71, 79, 82, 113, 136, 369, 403
– of HCN 273, 277
– of H–X bonds 272
– of MeI 103, 104, 109, 110, 371
– of N–H bonds 288, 290, 403
– of Si–H bonds 136, 280, 281, 286, 287
oxidative carbonylation 108, 372
oxidative coupling 35–37, 150, 152, 154, 214, 221, 222, 224, 226, 230, 380, 389, 392
oxidative cyclization 303
oxo(porphyrin)iron(IV) complexes 317
oxo synthesis 21, 77, 82
oxygenases 305, 315, 316
oxygenates 95
oxygen transfer 307–313, 326
oxyhydrogen gas 1, 3, 4
oxypalladation of olefins 303–305, 398
– with AcOH as nucleophile 304
– enantioselective 305
– with H_2O as nucleophile 303
– with ROH as nucleophile 303, 398

p

PAEs 133
palladium colloids 262
palladium oxidase catalysis 305, 306
parallel synthesis 59
Parkinson's disease 55
P cluster 342–344
pentad 173, 175, 383, 384
pentenenitriles 275, 276

peroxometal complexes 306, 307, 309, 312, 318
petrochemistry 22, 96
phase-transfer catalysis 262, 312
Phillips catalysts 167–171
Phillips triolefin process 118
phosgene 108, 372
phosphane-modified catalysts 79–84
2-(phosphinophenyl)oxazolines 264, 268
phosphorus ligands, steric and electronic effects 90, 227
PHOX ligands 264, 268
phthalocyanine–metal complexes 44, 46
pincer ligands/complexes 142–144
plasticizer 88, 127, 159, 231, 233
platina-β-diketon 34
platinum colloids 280
platinum sponge, catalytic effect 3, 4
polarized continuum model (PCM) 17
polyacetylene 127, 375
polyalkenamers 125
polyalkynamers 133
poly(arylene-ethynylene)s (PAEs) 133
polybutadiene 22, 129, 207, 234, 238 see also rubber, synthetic
– microstructure 15, 234
– properties 241–243
– synthesis of 1,2-polybutadienes 241–243
– synthesis of 1,4-polybutadienes 22, 241–243
polybutadiene rubber 241–243
poly-2,3-dimethylbutadiene 241, 382
polyethylene
– annual production 169
– high-density (HDPE) 162, 169, 170, 178
– hydrogenolysis 136
– linear low-density (LLDPE) 152, 159, 169, 170, 192, 233
– low-densitiy (LDPE) 169, 170
– polymer/structure types 169–171
polyisoprene 22, 207, 234, 241, 242, 390
polyketones 200, 201, 203
polymerization
– of butadiene 22, 207, 234–244
– – catalyzed by allylnickel(II) complexes 237–241
– – chain termination 235
– – mechanism 234–236, 238–240
– – quantum-chemical calculations 239
– – regio- and stereoselectivity 234, 235, 238, 243, 390
– of cycloolefins 192
– of α,ω-dienes 192, 381, 385
– of 2,3-dimethylbutadiene 241, 382

– of ethene 21, 24, 146, 162–171, 224, 386
– – high-pressure process 163, 169, 170
– – low-pressure process 21, 162, 169, 170
– – mechanism 163, 165–167
– – with metallocene catalysts 179, 192
– – with nonmetallocene catalysts 194, 196, 197
– – with Phillips catalysts 167–169
– – process specifications 169–171
– – quantum-chemical calculations 181, 197
– – radically initiated 163, 169, 170
– – with Ziegler catalysts 162–165
– of isoprene 22, 207, 241, 242, 244, 390
– of olefins 21, 148, 149, 161, 162
– – chain termination 148, 149, 163, 165, 166, 196, 197, 198, 385
– – chain transfer 163, 165, 166, 192, 194, 198, 199, 386, 387
– – insertion scheme 188, 189
– – mechanism 161, 162
– – molar mass control 165, 195
– of propene 21, 24, 171–192
– – mechanism 173–175, 184–189
– – process specifications 178, 179
– – regioselectivity 171
– – stereoselectivity 171–175, 187, 189, 383–385
– – with metallocene catalysts 184–192
– – with Ziegler-Natta catalysts 175–178
– of styrene 161, 192, 200, 241, 387
polymerization catalysts, tailor-made 179, 192
polyolefins, in situ functionalization 192, 386
poly(1-oxotrimethylene) 200
poly(p-phenylene-ethynylene)s (PPEs) 133
poly(p-phenylene-vinylene) (PPV) 133
polypropylene
– amorphous 175
– annual production 178
– atactic 172
– bi-heterotactic 385
– configuration 172
– head-to-head 172, 382
– hemi-isotactic 190
– hemi-syndiotactic 190
– hemitactic 190
– isotactic 171, 174, 175, 178, 183, 190
– microstructure, analysis of 173
– polymer types 178, 179
– properties 178, 179
– regioerror 187
– stereoerror 175
– syndiotactic 24, 171, 174, 175, 183, 186, 204
polystyrene 20, 192, 241, 242

polyvinyl acetate   101
pool/split procedure   59, 60
porphyrinatorhodium(II) complexes, dinuclear   323
PPEs   133
precatalyst   11, 16, 18, 19, 21, 28, 29
preferred asymmetric induction   86
primary insertion (1,2-insertion)   173, 174, 192, 383, 385, 390
primordial atmosphere   347
principle of least structural variation   218
prochiral
– allyl alcohol   302, 313
– allyl ligand   213
– butadiene   214, 215
– imine   75, 395
– ketone   75
– metal enolate   269, 394
– olefin   14, 23, 54, 56, 57, 63, 84, 171, 174, 183, 187
productive metathesis   117, 131
productivity of catalysts   12, 13
promoter   18, 20, 106, 107, 111, 261, 273, 275, 276, 336, 339, 340, 341
prone orientation   213
propene
– dimerization, non-regioselective   151, 379
– epoxidation *see* epoxidation of propene
– hydroformylation *see* hydroformylation of propene
– metathesis   117
– oligomerization   185, 385
– polymerization *see* polymerization of propene
propylene oxide   308, 309, 311, 312
prostereogenic faces   55, 174, 266
prostereogenicity   55
protein–protein complex   343, 344
pseudochiral polymer   185, 385

q

quantum-chemical calculations on
– alkane metathesis   141, 142
– alkyne–nitrile cross metathesis   377
– butadiene polymerization   239
– cyclooligomerization of butadiene   229
– cyclotrimerization of butadiene   223
– epoxidation   310
– ethene polymerization   181, 197
– Haber-Bosch process   336
– hydroformylation   79, 83
– hydrosilylation   281
– methanol carbonylation   104, 108, 109
– nitrogen fixation   358
– nonalternating copolymerization   204
– olefin hydrogenation   53, 63
– olefin metathesis   124, 125
– oxidative coupling of butadiene   214
– reductive elimination of diallyl   217
– Wacker process   299

r

radical chain reaction   49, 279, 365, 395
radical oxidation reaction   101, 306
radical polymerization   161, 162, 163, 169, 170, 171, 200, 241, 387
rate-determining step   16, 19
RCAM   132
RCEYM   134
RCM   126, 128, 129, 134
reaction-controlled phase-transfer catalysis   312
reaction injection molding (RIM) technology   128
reaction mechanism, determination of   15–17
reaction profile diagram   17, 19
π–σ/σ–π (η^3–η^1/η^1–η^3) rearrangements *see* allyl complexes, rearrangements
redox catalysis   9, 10
reductive
– C–C elimination   32, 136, 150, 217, 221, 223, 225, 247, 248, 253, 258, 265, 320, 378, 391, 400
– C–H elimination   32, 52, 79, 82, 98, 153, 272, 280, 290, 321, 322, 363, 381, 391
– C–I elimination   104, 110
– C–N elimination   288
– C–X elimination   230, 232, 246, 272
– Si–C elimination   280, 285
– Si–H elimination   396
reductive cleavage/decoupling   35–37, 152, 214–215, 221
reductive elimination   30–35
– of allyl derivatives   216–218
– bimolecular   97, 246, 372
– of H_2 51, 71, 82, 113, 282, 362, 403
– of HX   260, 272
– quantum-chemical calculations   217
– of RCN   273, 275, 277, 394
Re face   55, 174, 183, 213, 313, 367
regioselective reaction   14
regioselectivity   14, 15, 79, 87, 171, 173, 234, 261, 269, 273, 278, 286, 319, 396, 400
reinsertion   149, 167, 195, 196, 198, 298
relative configuration   171–173, 185, 191, 388
relay ring-closing metathesis (relay RCM)   129

reoxidation 2, 295, 296, 303, 305, 306, 325, 400
Reppe chemistry 21
Reppe synthesis of acrylonitrile 278
resonance structures 32, 33, 123, 208, 265, 286, 316–318, 332, 357, 361, 376, 396, 404, 405
resting state 19, 20, 82, 110, 196, 241, 261, 290, 316, 344, 390
retro Diels-Alder reaction 375
reverse water-gas shift reaction 93
rhodium cycle 103
RIM technology 128
ring-closing alkyne metathesis (RCAM) 132
ring-closing enyne metathesis (RCEYM) 134
ring-closing metathesis (RCM) 126, 128, 129, 134
– of acyclic diynes 132
ring-opening metathesis (ROM) 126, 128
ring-opening metathesis polymerization (ROMP) 125–128, 133, 375
root nodule bacteria 342
rubber 128, 241
– annual production 241
– natural 207, 208, 234, 242
– – monomer for biosynthesis 244, 390
– synthetic 161, 207, 241
rubber-modified polystyrene 241, 242
Ruhrchemie/Rhône-Poulenc process 91

s

Sabatier's principle 340, 341
α-Sablin process 158, 159
safety lamps for miners 3
salen complexes 314, 315
salt cycle 107
sandwich complexes, metallorganic 24
Schrock carbene complexes 97, 120, 122
Schrock's catalyst 24, 120, 121, 127, 130, 132, 143
Schwartz reagent 39
secondary insertion (2,1-insertion) 173, 174, 383
cis/trans selectivity 221–223, 236, 240
selectivity of catalysts 14, 15
Sharpless epoxidation 24, 313, 314
Shell Higher Olefin Process (SHOP) 24, 88, 156–158
Shilov catalyst system 324
SHOP see Shell Higher Olefin Process
SHOP catalyst 157, 195
Si face 55, 174, 183, 213, 313, 367
silane complexes 136, 286, 287
silicalite 311

silicon carbide fibers 136
silicons, cross-linking 283
SILP catalysis 91
silylcyanation 278, 279
silylene complexes 284, 396
silylenes 283
single-component catalysts 24, 120, 121, 132, 237, 238
single crystal approach 341
single electron transfer (SET) 34, 311
single-site catalyst 24, 152, 179, 192
soft nucleophiles 264, 265, 267
SOHIO process 278
solid-phase synthesis 60
Sonogashira coupling 246, 253, 254, 392
– copper-free 254, 392
specificity of catalysts 14, 15
spectator/control ligand 10, 144, 227
Speier catalyst 280, 281
spin conservation rule 306
spin trap 200, 387
stable carbenes 122, 123
steering ligand 391 see also spectator/control ligand
stereoblock polymers 175, 189, 191, 192
stereochemical chain end control 174, 175, 187, 188, 383
stereoscrambling 266–268
stereoselective reaction 14, 15
stereoselectivity 54, 86, 90, 163, 171–177, 182, 188, 204, 234, 235, 238, 260, 267, 286, 315
stereospecific reaction 14, 15, 248, 267
steric effects of phosphorous ligands 227, 254 see also bite angle of chelating ligands
steric ligand parameter 16, 227, 228
Stille coupling 23, 246, 249, 252, 253, 258, 393
Strecker synthesis 88, 279, 395
styrene–butadiene rubber (SBR) 241
styrene and styrene derivatives
– in asymmetric catalysis 23, 56, 84, 85, 130, 276, 277, 284, 285
– coproduction in epoxidation reactions 309, 311
– in cross-coupling reactions 253, 256, 391
– hydroamination 291, 292
– oligomerization 385
– polymerization 161, 192, 200, 241, 387
– synthesis in Heck reactions 259
substrate activation 9, 10, 11, 28, 96, 272, 321, 341
sunscreen 262
supercritical state 170

supine orientation   213
supported catalysts/catalysis   11, 91, 169, 170, 175, 177 *see also* surface-bound/grafted complexes
supported ionic liquid-phase (SILP) catalysis   91
supramolecular catalysis   94, 95
surface-bound/grafted
– carbido ligands/C atoms   96, 97
– Cr complexes   168
– H atoms   96–99, 335
– metal amide/imide complexes   143, 352, 403
– M–H/M–R complexes   138, 143
– N atoms/N_2 molecules   335–337
– organo ligands   96–100
– Ta–H/Ta–R complexes   138, 139, 141, 143, 351, 352, 378
– W–H/W–R complexes   138, 152
– Zr–H/Zr–R complexes   136, 137
surface complex   138, 337 *see also* surface-bound/grafted complexes
Suzuki coupling   23, 25, 246, 249–252, 254, 256–258, 391, 392
symmetry equivalence   183, 184, 186, 187, 268
symmetry-forbidden reaction   49, 373
symmetry of orbitals   69, 286, 373, 401, 404, 405
symmetry relationships   182, 187, 373
syn addition   37–39, 42, 54, 163, 174, 260, 261, 278, 280, 300–302, 398
syn/anti isomerization *see* allyl complexes, isomerization
syn/anti notation of allyl complexes   209
synchronous asymmetric induction   86
syn elimination   37, 260, 263, 394, 398
synthesis gas   21, 77, 89, 95, 96, 102
syn/trans correlation   218, 221, 235

t

tailor-made catalysts/polymers   10, 179, 192
tail-to-tail linkage   151, 171
tandem reaction   134, 141, 250 *see also* domino reaction
Tebbe reagent   36, 120
telogen   231
telomerization   22, 207, 208, 219, 230–233
terminal oxo ligands   312, 318, 399
terpenes   207, 208, 263
tetratungstate structure   312
thermal polymerization of dienes   241
thermodynamic oxygen-transfer potential (TOP)   307, 308

thermoplastic elastomers (TPEs)   191
Thorneley-Lowe cycle   346
three-center/four-electron bond (3c–4e)   31
three-center/two-electron bond (3c–2e)   31
threo, definition   301
threo polymer   386
α-titanium(III) chloride, crystal and surface structure   175–177, 384
TOF see turnover frequency
Tolman parameter   16, 227, 228, 254
TON see turnover number
topic relationships between molecular fragments   183, 184
topomers, topomerization   210, 211, 388
trans addition   267, 301
transalkylation   147, 166
transalkylation reactor   146
transfer dehydrogenation   142, 143
transfer hydrogenation   73–75, 142
transition state   17, 19, 25, 38, 49, 65, 85, 95, 119, 135, 163, 165, 185, 212, 240, 253, 271, 279, 298, 311, 323, 335, 363
transition state theory   7
transmetallation   246–248, 265, 391, 399
TRANSPHOS   90
triad   173, 175, 383, 384
Tsuji-Trost reaction   264, 269, 291, 394
turnover frequency (*TOF*)   12, 151, 186, 340
turnover number (*TON*)   12, 13, 151
turnstile mechanism   141

u

Umpolung of the reactivity   245

v

Vaska complex   34
Vestamid   224
vinyl acetate   161, 304
vinylcyclohexene (VCH)   224, 225, 227, 228, 229
vitamin A   93, 208
volcano plot   340, 341

w

Wacker process   22, 295, 297, 303, 398
– mechanism   297–302
– quantum-chemical calculations   299
Wacker reaction   301, 302, 303, 305, 398
water-gas (shift) equilibrium   112
water-gas shift reaction   102, 105, 106, 112, 335, 371
– mechanism   105, 112, 113
– reverse reaction   93
Watson-Crick base pairing   94

weakly coordinating anions 92, 180, 201, 237, 261, 314, 325
Wilke catalyst 150, 151
Wilkinson complex/catalyst 19, 23, 28, 49–54, 71, 75, 150, 151, 280
Woodward-Hoffmann rules 373
Wurtz reaction 245, 360

x
Xantphos 90, 91, 291

z
Zeise's salt 33, 297
Zeonex 127
Ziegler catalysts 10, 21, 127, 158, 162–165, 169, 170, 175, 179, 192, 193, 224, 236, 242
Ziegler growth (Aufbau) reaction 39, 145–147, 199
Ziegler-Natta catalysts 163, 169, 170, 173, 175, 179, 192, 207, 234, 242

Index of Backgrounds

Classification of Ligands   29
Agostic C–H⋯M Interactions   31
Oxidation States of Metals in Olefin and Alkyne Complexes   32
Heterolytic Fragmentation (Grob Fragmentation)   43
Prostereogenicity, Prostereogenic Faces   55
Combinatorial Catalysis and High-Throughput Screening   60
The Curtin-Hammett Principle   65
The "Bite" of P,P-Chelating Ligands   90
Ionic Liquids   92
Stable Carbenes as Ligands   122
Organometallic Pincer Complexes   143
Hemilabile Ligands   155
Configuration of Polypropylene   172
Analysis of the Microstructure of Polypropylene   173
Topic Relationships Between Molecular Fragments   183
Fluxional Molecules   210
Steric and Electronic Effects of Phosphorus Ligands   227
Telomerization   231
Oxidation States of Metals in Complexes   318
Sabatier's Principle and the Brønsted–Evans–Polanyi Relation   341